Methodology for Social Sciences Research in Agriculture

The Authors

Dr. K. Nirmal Ravi Kumar is presently working as a Professor (Agricultural Economics) in Acharya N.G. Ranga Agricultural University. He had a brilliant academic career and he specialized in 'Agricultural Marketing' both in his post-graduate and doctoral programmes. He is actively involved both in agricultural research and teaching activities during the past 13 years in the University. He published 50 articles in popular national and international agricultural journals. He also contributed two technical bulletins on economic aspects of irrigation water management highlighting the research priorities in major irrigation commands of Andhra Pradesh and need based technological interventions to address the same during his active stint in "Andhra Pradesh Water Management Project", an international project funded by The Royal Netherlands Embassy. His interested areas include International trade of Indian agriculture, Farming systems approach, Irrigation water management etc. The following are the other major contributions (books) by the same author:

- Indian agriculture in the 21st century – Challenges and Opportunities
- World Trade Agreement and Indian Agriculture – Implementation Experience
- Microeconomic analysis in Agriculture
- Objective Agricultural Economics (Two Editions)
- Farmers Indebtedness in India – An Economic Analysis
- Agricultural Marketing
- Methodology for Social Sciences Research in Agriculture
- Agricultural Production Economics
- Farm Managerial Economics
- Encyclopedia of Rural Management

Dr. M.Jagan Mohan Reddy is presently working as Professor in the Dept. of Agril. Extension, College of Agriculture, Rajendranagar, Hyderabad. He had 15 years of experience in teaching, research and extension disciplines of Acharya N.G Ranga Agricultural University, Rajendranagar, Hyderabad, Andhra Pradesh. He has published five bulletins on crops and livestock production technologies, vermi compost, fodder crops, green manure crops and seed production. Conducted method and result demonstrations, trials and group discussions for the benefit of farmers. He has published 32 research papers in reputed local and national level journals, written 52 popular articles for various daily news papers. Authored a book on Information Technology - An effective tool in agricultural communication. He has guided five M.Sc. (Ag.) and two Ph.D. students majoring in the area of Agricultural Extension.

Methodology for Social Sciences Research in Agriculture

— *Authors* —

K. Nirmal Ravi Kumar
Associate Professor and Head (Agricultural Economics)
Agricultural College,
Mahanandi

M. Jagan Mohan Reddy
Professor, Department of Agricultural Extension,
College of Agriculture,
Rajendranagar, Hyderabad

2015
Daya Publishing House®
A Division of
Astral International Pvt. Ltd.
New Delhi – 110 002

ISBN 9789351305606

Published by : **Daya Publishing House®**
A Division of
Astral International Pvt. Ltd.
– ISO 9001:2008 Certified Company –
4760-61/23, Ansari Road, Darya Ganj
New Delhi-110 002
Ph. 011-43549197, 23278134
E-mail: info@astralint.com
Website: www.astralint.com

Laser Typesetting : **Classic Computer Services**, Delhi - 110 035

– Dedicated to –

Late K. RAMA CHANDRA RAO

ACHARYA N.G. RANGA AGRICULTURAL UNIVERSITY
ADMINISTRATIVE OFFICE, RAJENDRANAGAR, HYDERABAD-500030

Dr. A. Padma Raju
Vice-Chancellor
Post-Graduate Studies

Phone : +91-40-24015035(O)
+91-40-24015031(F)
Email: angrau_vc@yahoo.co.in
GRAMS: "AGRIVARSITY"

Foreword

Research leads to understanding about reality and hence, it has been a passion to mankind of all times. With continuous changes in political, economical, social and technological factors, the complexity of social life increases and this led to increased significance for conducting social sciences research. Many researchers and even students are in continuous pursuit of conducting social sciences research with the prime objective of understanding social realities. In this context, they should be familiar about the methodology *i.e.,* designing and conducting of the research studies. In fact, several State Agricultural Universities in India have already made an attempt in this direction. In this context, different designs for conducting social research and their underlying methodologies are well conceptualized in this text book. This enhances the seriousness of the researchers and students in making fruitful attempts in identifying, conducting and addressing the social researchable issues. The sound methodology conceptualized in this book is more than a theoretical discussion about research and thus, help them in the context of 'how-to-get-it-done' towards addressing social issues. I complement the sincere efforts of Dr. K. Nirmal Ravi Kumar, Associate Professor (Agricultural Economics) and M. Jagan Mohan Reddy, Professor (Agricultural Extension) in bringing out this text book on *'Methodology for Social Sciences Research in Agriculture'* duly considering the latest syllabi, as prescribed by the IV Deans' Committee recommendations appointed by the Indian Council of Agricultural Research. The authors presented the topics in this book in a cogent, concise and lucid manner, so as to benefit the researchers and students especially, who are untrained and under trained in the methodological aspects of doing research in general and social sciences research in particular. I hope this text book will be very useful and instructive for the post-graduates and doctoral students of social sciences, who engage in research activities during their respective course programmes. It will also be of immense use for the Academicians, as it harbours standard and updated material pertaining to the concepts of Research Methodology.

A. Padma Raju

(A.PADMA RAJU) 3/9/2013

ACHARYA N.G. RANGA AGRICULTURAL UNIVERSITY
ADMINISTRATIVE OFFICE, RAJENDRANAGAR, HYDERABAD-500030

Dr. P. Chandrasekhar Rao
M.Sc. (Ag.), Ph.D, FAPAS, FHAS
Dean of Agriculture

Phone : +91-40-24015197(O)
9866892000
Email: deanagri@hotmail.com
GRAMS: "AGRIVARSITY"

Preface

It In the context of multiplicity of social issues with rapidly changing socio-economic factors, conducting research with sound methodology is essential. However, for conducting such research, the researchers and students must be exposed with various research designs and their underlying methodologies. This orientation enables them to conduct social sciences research in a planned manner, so as to conceptualize and analyze the social realities. For understanding the social realities, it is not the quantity of research, but the quality of research that counts much. Keeping these aspects in view, this text book entitled, *'Methodology for Social Sciences Research in Agriculture'* has been designed to render clarity both in terms of theoretical orientation and practical execution of the concepts. It is essential that, the knowledge gained through this book will not only benefit the scientific community in identifying and addressing the prioritized researchable issues, but also help the students in having a thorough understanding about the importance of this subject in their future scientific investigations. Further, it contains an up-dated and standard material on the various aspects of Research Methodology duly covering the latest syllabi as prescribed by the several State Agricultural Universities in India. Considering these efforts made in this text book, I complement Dr. K. Nirmal Ravi Kumar, Associate Professor (Agricultural Economics) and M. Jagan Mohan Reddy, Professor (Agricultural Extension) in bringing out this important and timely publication.

(P. CHANDRASEKHAR RAO)

ACHARYA N.G. RANGA AGRICULTURAL UNIVERSITY

Email: drknrk@gmail.com
jaganmaligi@yahoo.co.in

Authors' Note

We consider this as a bold attempt to present a comprehensive text material on the fundamental aspects of research methodology for conducting social sciences research. With a limited experience in the field of teaching in the Acharya N.G. Ranga Agricultural University, we will not dare enough to make a tall claim regarding the originality of the material incorporated in this text book entitled, 'Methodology for Social Sciences Research in Agriculture', but we certainly admit that, the concepts are dealt in-detail in a lucid form, so as to enable the students and academicians to learn, understand and conceptualize the same. The present volume can serve as a standardized text book for the post-graduates and doctoral students, who wish to pursue their career in the field of Social Sciences (Agricultural Economics and Agricultural Extension) in different State Agricultural Universities. Utmost care has been taken to present and discuss all the concepts to suit the scientific procedures of the contemporary research and fulfill the needs of the teachers engaged in various fields of academic and programmatic research across the country. We are especially thankful for the friendship and unfailing support of Dr. K.V. Hari Prasad, Assistant Professor (Entomology), who practically demonstrated the true meaning of friendship by standing with us during the pursuit of this work. Acknowledgements are due to Meena, Sri Ram and Janaki for their love and support.

(K. Nirmal Ravi Kumar) ***(M. Jagan Mohan Reddy)***

Contents

Foreword *vii*

Preface *ix*

Author's Note *xi*

Acronyms *xix*

1. **Research Methodology** **1**

Definitions – Importance of research methodology – marketing – Types of marketing

2. **Research and Types of Research** **5**

What is research? – Principles or Characteristics of Quality Research Work – Objectives of research – Qualities of a good researcher – Motivations in research – Ethical considerations of research – Importance of Research – Problems/Limitations of Research in India – Issues and Trends in Research – Typology of research – Basic (Pure or Fundamental) research – Applied research – Practical research – Exploratory research – Descriptive research – Explanatory research or Analytical research – Predictive research – Quantitative research – Qualitative research – Structured approach – Unstructured approach – Deductive research – Inductive research – Cross section research – Longitudinal research – Experimental Research – Survey research – Causal-Comparative Research – Case Study – Correlational Research – Historical Research – Ex-post facto research – Academic levels of research – Research Method vs Research Design – Research Method vs Research Methodology – Research design vs Research methodology – Importance of Research Methodology in Research Studies

3. Research Process **84**

Conception/Identification of Research Problem – Formulation of Research Problem – Review of Literature – Formulation of Testable Hypothesis – Preparation of Research Design (Methodology) – Execution of Research Project – Data Collection – Data Processing and Analysis – Testing of Hypothesis and Interpretation of the Findings – Writing Research Report

4. Conception/Identification of Research Problem **93**

What is research problem – The Five 'W's of research problem – What makes a good research problem? – Sub-problems vs Pseudo-Sub-problems – Factors to be considered, while identifying a research problem – Situations to avoid when identifying a research problem – Applying for funding – Delimiting the research – Practical problem vs Research problem – Research problem vs Topic

5. Formulation of Research Problem **110**

Components involved in formulating the research problem – Statement about the problem (Problem statement) – Functions of a Problem Statement – Problem statements across different research methods – How to write a problem statement – Common Deficiencies of Problem Statement – Justification of study of the selected research problem – How to write a Justification of the selected problem or Significance of the study – Scope of the selected research problem – How to write 'scope of the research study' – General (Main) objective – Specific (Sub) objectives – Features of research objectives – Objectives across different research methods – Importance of framing the research objectives – How to write 'objectives for the research study' – Framing the objectives for the selected problem – Identification of key terms and concepts of the selected problem – Identification of variables relevant to the selected problem – Criteria for selection of variables – Assumptions, Limitations and Delimitations of the research study

6. Review of Literature **131**

Why to collect review of literature? – Checklist for Selection of Literature to be included in Literature Review – Characteristics of a good literature review – Importance of collecting review of literature – Four Ws before writing literature review – Sources of literature review – Primary sources – Secondary sources – Tertiary sources – Search for existing literature – Books – Journals – Internet – Review the literature collected – Steps in reviewing the literature – Characteristics of a Good Research Question for a Literature Review – What the researcher should do before writing the literature review? – Common mistakes the researchers do in reviewing the literature/Problems faced when writing a Literature Review – Advantages of Literature Review – Disadvantages of literature review – What do examiners look for in a literature review?

7. Formulation of Testable Hypothesis **162**

Defintions – Characteristics of 'good' hypothesis – Examples of 'good' hypotheses – Importance of hypothesis – Sources of Hypothesis – Classification of hypothesis – Main hypothesis – Sub-hypothesis – Descriptive hypothesis – Relational hypothesis – Causal hypothesis – Tentative hypothesis – Working hypothesis – Research hypothesis – Directional hypothesis – Non-directional hypothesis – Statistical hypothesis – Null hypothesis – Alternative hypothesis – Directional Alternative hypothesis – Non-directional Alternative hypothesis – Null hypothesis vs Alternative Hypothesis – Declarative hypothesis – Simple hypothesis – Complex Hypothesis – Parametric hypothesis – Non-Parametric Hypothesis – Limitations or difficulties in the formulation of testable hypothesis – Parametric tests and Non-parametric tests

8. Preparation of Research Design (Methodology) 179

Definitions of research design – Steps in planning the research design – Components dealt in research design – Preparation of sampling design – Universe or Population – Sampling – Sample – Features of good sample – Sample unit – Sample design – Statistic and Parameter – Sampling frame – Sampling fraction – Elevation factor – Sampling error – Random sampling error – Non-sampling error – Methods to reduce sampling error – Relation between Random sampling error and Variance – Census study – Census study vis-à-vis Sample study – Precision and Accuracy of research findings – Bias – Steps in sampling process – Determining sample design – Characteristics of good sample design – Types of sampling techniques – Probability (or Random) sampling – Differences between Probability and Non-probability sampling methods – Simple Random Sampling (SRS) – SRS without Replacement (SRSWOR) and SRS with Replacement (SRSWR) – Possible number of samples can be drawn by SRSWOR technique – Possible number of samples can be drawn by SRSWR technique – Methods of selecting simple random sample – Merits of SRS – Demerits of SRS – Stratified Random Sampling – Examples of stratified random sample – Selecting a Stratified Random Sample – Types of Stratified Random Sampling – Proportionate Stratified Random Sampling – Disproportionate Stratified Random Sampling – Disproportionate allocation for within strata – Disproportionate allocation for between-strata – Optimum allocation – Stratified random sampling vis-à-vis SRS – Merits of Stratified random sampling – Demerits of Stratified random sampling – Systematic Sampling – Steps in Selecting a Systematic Sample – Sub-types of systematic sampling – Examples for systematic sampling – Merits of Systematic sampling – Demerits of Systematic sampling – Cluster Sampling – Sub-types of Cluster Sampling – Sub-types Based on Number of Stages of drawing sample – Single-stage cluster sampling – Two-stage cluster sampling – Multi-stage cluster sampling – Sub-types Based on the Proportional Representation of Clusters in Sample – Equi-probability sampling – Probability Proportion to Size (PPS) – Numerical example to derive sample from different clusters based on PPS – Differences between Cluster sampling and Stratified random sampling – Merits of Cluster Sampling – Demerits of Cluster Sampling – When to Use Cluster Sampling – Choice between SRS, stratified random sampling and cluster sampling – Sampling with Probability Proportional to Size – Merits of PPS sampling – Demerits of PPS sampling – Multistage Sampling – Merits of multi-stage sampling – Demerits of multi-stage sampling – Multi-phase sampling – Differences between multi-stage sampling and multi-phase sampling – Merits of multi-phase sampling – Demerits of multi-phase sampling – Non-probability (Non-random or Judgment) sampling – Quota sampling – Merits of Quota sampling – Demerits of Quota sampling – Convenience sampling – Merits of Convenience sampling – Demerits of Convenience sampling – Judgement Sampling – Merits of Judgement sampling – Demerits of Judgement sampling – Snowball Sampling – Merits of Snowball sampling – Demerits of Snowball sampling – Volunteer Sampling – Merits of Volunteer sampling – Demerits of Volunteer sampling – Panel Sampling – Expert Sampling – Opportunity Sampling – Guidelines to use the sampling techniques – Determination of representative sample size for conducting research study – Formulae used to determine sample size – Common misconception in selecting the sample size – Advantages of sampling – Disadvantages of sampling – Identification of variables relevant to the objectives framed for the study – Characteristics of the variables – Types of variables – Independent variable – Dependent variable – Differences between Independent variable and Dependent variable – Qualitative (Categorical) Variable – Nominal variable – Dichotomous variable – Ordinal variable – Quantitative Variable – Continuous variable – Discontinuous or Discrete variable – Extraneous Variable (EV) – Confounded Variable (CV) – Moderating Variable (MV) – Controlled Variable – Mediating Variable – Endogenous Variable – Exogenous Variable – Types of data to be collected –

Primary data and Secondary data – Cross-sectional data and Time series data – Qualitative data and Quantitative data – Univariate data, Bivariate data and Multivariate data – Subjective data and Objective data – Identification of the sources of data – Primary sources – Secondary sources – Tertiary sources – Census survey – Sample survey

9. Execution of Research Project **313**

Schedule or Interview schedule – Open-ended questions and Closed-ended questions – Direct questions and Indirect questions – Primary questions and Secondary questions – Nominal, Ordinal and Interval questions – Contingency questions – Characteristics of good schedule – Design of Schedule – Types of Schedules – Structured schedules – Unstructured schedules – Observation Schedule – Rating Schedules – Document Schedule – Interview Schedule – Village or Community Schdedule – Family or Household Schedule – Opinion or Attitude Schedule – Steps in construction of schedule – Advantages of using schedule in data collection – Disadvantages of using schedule in data collection – Questionnaire – Types of questionnaires – Structured questionnaire – Unstructured questionnaire – Mixed Questionnaire – Disguised typed questionnaire – Closed and Open Questionnaires – Fact and Opinion questionnaires – Pictorial questionnaire – Self-administrated questionnaire – Interviewer-administrated questionnaire – Factual-type questionnaire – Opinion-type questionnaire – Attitude questionnaire – Exploratory questionnaire – Formal standardized questionnaire – Questionnaire Layout or Design of questionnaire or Characteristics of a good questionnaire – Similarities between schedule and questionnaire – Schedule vis-à-vis Questionnaire – Advantages of Questionnaire – Disadvantages of Questionnaire – Physical appearance of the questionnaire – Response Error – Sources of Response Error – Techniques to Reduce Response Errors or Increasing Response Rate.

10. Data Collection **364**

Survey method – Techniques of Survey Method – Telephone survey – Personal survey – Mail Survey – Electronic survey – Types of Surveys – Cross-Sectional Surveys – Longitudinal Surveys – Trend studies – Cohort studies – Panel studies – Merits of Survey Method – Demerits of survey (schedule) method – Different measures to promote response rate of the respondents – Case study – Basic assumptions of a case study – Steps involved in case study method – Sources of evidence in case studies – Types of Case Study Research – Intrinsic case study – Instrumental case study – Collective case study – Depending on the type of research question – Exploratory case study – Explanatory case study – Descriptive case study – Depending on the disciplinary orientation – Ethnographic case study – Historical case study – Psychological case study – Sociological case study – Depending on the overall intent of the study – Descriptive case study – Interpretative case study – Evaluative case study – Pilot case study – Depending on the number of case studies handled by the researcher – Differences between Survey method and Case study – Merits of case study – Demerits of case study – Other data collection methods employed in Social Sciences Research studies – Observation method – Characteristics of observation method – Types of Observation – Unstructured observation – Structured observation – Participant Observation – Non-participant observation – Steps for Effective Observation – Merits of Observation Method – Limitations of Observation Method – Participatory Rural Appraisal Technique – Evolution of PRA – Principles or Key features of PRA – Tools employed in PRA – Successful conductance of PRA – Layout of PRA – Areas of application of PRA – Dangers and drawbacks of PRA

11. Data Processing and Analysis **396**

Processing of data – Checking – Editing – Coding – Classification – Tabulation – Data analysis – Phases of data analysis – Data cleaning (Data cleansing or data scrubbing) – Initial data analysis – Main data analysis phase – Statistical analysis to be employed – Graphical representation of the analytical findings – Concept of Scaling – Types of Scales – Scaling based on Response Form – Categorical scales – Comparative scales – Scaling based on Subjectivity – Scaling based on properties – Nominal Scale – Ordinal Scale – Interval Scale – Ratio scale – Scaling based on judging the attitude – Rating scale – Attitude scale – Thurstone's Scale (Differential scale) – Likert's scale – Guttman's scaling – Bogardus Social Distance Scale – Semantic Differential Scale (Factor Scale) – Choosing the scaling technique

12. Testing of Hypothesis and Interpretation of the Findings **435**

Steps in testing of hypothesis – Specify the statistical hypothesis viz., H_O *and* H_A *– Select the Significant Level – Find the critical value at the required level of significance and degrees of freedom – One Sided or Two Sided Hypothesis – Calculate the Test Statistic and Corresponding Probability (P) Value – Drawing a Conclusion – Errors in drawing conclusions in research – Type I Error – Type II Error – Size of test vs Power of a test – Benefits of Hypothesis testing – Interpretation of analytical findings*

13. Writing Research Report **458**

Layout of the Thesis – Preliminary Pages or Preliminaries – Title page – Declaration – Certificates – Acknowledgements – List of contents – List of tables – List of figures – List of symbols and abbreviations – List of appendices – Abstract – Main Text – Introduction – Review of literature – Material and Methods – Agro-Economic Features – Results and Discussion – Summary and Conclusions – End Matter – Literature cited – Appendices – Other important guidelines the researcher should follow in presenting the style of thesis.

References **501**

Acronyms

AICRP: All India Coordinated Research Project
ANCOVA: Analysis of Covariance
ANOVA: Analysis of Variance
APEDA: Agricultural and Processed Food Products Export Development Authority
APMCs: Agricultural Produce Market Committees
ARIMA: Auto Regressive Integrated Moving Average Model
CV: Coefficient of Variation
DEA: Data Envelopment Analysis
DV: Dependent Variable
EV: Extraneous Variable
FPC: Finite Population Correction
H_A: Alternate hypothesis
H_0: Null Hypothesis
ICAR: Indian Council of Agriculture Research

ICRISAT : International Crops Research Institute for Semi-Arid Tropics
INM: Integrated Nutrient Management
IPM: Integrated Pest Management
IV: Independent Variable
KVK: Krishi Vigyan Kendra
KWD: Krishna Western Delta
LDMU: Law of Diminishing Marginal Utility
LDR: Law of Diminishing Returns
LEMU: Law of Equi-Marginal Utility
MLR: Multiple Linear Regression
M.Sc.: Master of Science
MSPs: Minimum Support Prices
MU: Marginal Utility
MV: Moderating Variable
NPS: New Population Size
NSS: New Sample Size
OPS: Old Population Size
Ph.D: Doctor of Philosophy
PPS: Probability Proportion to Size
PRA: Participatory Rural Appraisal
RTE: Ready To Eat Food
SAUs: State Agricultural Universities
SD: Standard Deviation
SRI: System of Rice Intensification
SRS: Simple Random Sampling
SRSWR: Simple Random Sampling with Replacement
SRSWOR: Simple Random Sampling without Replacement
SS: Sample Size
UGC: University Grants Commission
UNICEF: The United Nations Children's Fund
WTO: World Trade Organization

1

Research Methodology

Research is a process of enquiry and investigation. It is systematic, methodical and ethical. Its conductance can help solve practical problems and increase knowledge. In the well-known nursery rhyme,

Twinkle Twinkle Little Star

How I Wonder What You Are

the use of the words 'how' and 'what' essentially summarizes what research is. So, research is an investigation of finding solutions to scientific and social problems through objective and systematic analysis. It is a search for knowledge ie., a discovery of hidden truths. According to Collis and Hussey, 2003, the purpose of research is to:

- ☆ Review or synthesize existing knowledge
- ☆ Investigate existing situations or problems
- ☆ Provide solutions to problems
- ☆ Explore and analyze more general issues
- ☆ Construct or create new procedures or systems
- ☆ Explain new phenomenon

- ☆ Generate new knowledge
- ☆ or a combination of any of the above!

Note that, research is not confined to science and technology only. There are vast areas of research in other disciplines such as languages, literature, history and sociology. Whatever might be the subject, research has to be an active, diligent and systematic process of inquiry in order to discover, interpret or revise facts, events, behaviours and theories. Applying the outcome of research for the refinement of knowledge in other subjects, or in enhancing the quality of human life also becomes a kind of research and development.

Research methodology is defined as the scientific or systematic way of conducting research by the researcher or investigator towards accomplishing the desired goals and objectives. In simple terms, research methodology gives the clear cut idea for the researcher right from initial conceptualizing of the research problem and up to final preparation of report. For example, in conducting an agro-economic research on *'Marketing efficiency is transacting major agricultural commodities in scarce rainfall zone of Andhra Pradesh'*, the methodology guides the researcher in selecting the research problem; selection of sample with reference to commodities, sample farmers, market intermediaries; data to be collected; techniques to be used; interpreting the analytical findings and final submission of report. Thus, 'methodology' provides the right platform to the researcher in planning to pursuit of his/her research work and thereby, paves the systematic way in solving the research problem. The research methodology constitutes of research methods, selection criterion of research methods used in the context of research study and explanation of using of a particular method or technique in the research study and why other techniques are not used so that, research results are capable of being evaluated either by researcher himself or by others. Further, the research methodology also provides answers to several questions concerning a research problem or study like: *'why a research study has been undertaken, how the research problem has been formulated, what data have been collected, what particular sampling method has been adopted, why a particular technique of analyzing data has been used etc'*. It is essential to note that, in most of the situations, the aim of the research and the research topic won't be same and it varies in terms of formulating objectives and flow of the research. Here, the importance of 'methodology' lies, as it guides and set the things in the right direction towards fitting the objectives of the research in the broad framework of research aims or goals. Thus, a scientific methodology drives the researcher in the right track and forms the basis or right platform for the entire research plan. This makes the researcher or investigator to involve and to be active in his or her particular field of enquiry. Further, a scientific methodology guides the researcher in economizing the use of time, money and other resources in accomplishing the desired goals and objectives. The following are the popular definitions of 'Research Methodology':

Research methodology is defined by Leedy and Ormrod (2001) as 'the general approach the researcher takes in carrying out the research project'.

'Research Methodology is a way to find out the result of a given problem on a specific matter or problem that is also referred as research problem. In Methodology, researcher uses different criteria for solving/searching the given research problem. Different sources use different type of methods for solving the problem. If we think about the word 'Methodology', it is the way of searching or solving the research problem. (Industrial Research Institute, 2010).

'A research methodology defines what the activity of research is, how to proceed, how to measure progress, and what constitutes successes.

'Research methodology refers to the ways or methods that are used by a researcher to conduct or organize a research to attain the desired objectives. The research methodology part is quite important due to identifying the ways in that research that is conducted to solve any research issue. Additionally, to explore the research problem in effective and efficient manner, research methodology plays a fundamental role that also indicates the importance of this part for a research'.

'Research Methodology' comprises of two words *viz., 'Research'* and *'Methodology'*. The former *i.e.,* 'Research' refers to the systematic inquiry aimed at providing information to solve the problems and the latter *i.e., "Methodology'* refers to the scientific procedure that is followed in the pursuit of conceptualized researchable issue. The methodology usually encompasses the procedures followed in the selection of sample, selection of tools and techniques to analyze and interpret the data gathered and to identify the statistical significance in the results. Thus, 'Research Methodology' is a collective term for the structured process of conducting research. These two aspects *viz.,* 'Research' and 'Methodology' are discussed in-detail in the ensuing pages.

Importance of Research Methodology

The following points highlight the importance of research methodology:

- ☆ In research methodology, researcher always tries to explore the formulated research problem in a systematic way and find out all the answers to arrive at till valid conclusions. If researcher does not work systematically on a problem, there would be less possibility to find out the final result. For finding or exploring a research problem, a researcher faces lot of problems that can be effectively resolved with using correct research methodology.
- ☆ Research methodology helps the researcher to identify the relevant strategies and methods to solve the research problem under consideration.
- ☆ Research methodology guide the researcher in identifying and selecting the appropriate research approach, sampling methods, data collection approach, selection of relevant techniques for analyzing the data etc. Thus, research methodology aims towards positive attainment of the formulated objectives of the research investigation.
- ☆ Research methodology enables the researcher in collecting relevant and accurate information or data for the selected parameters under consideration.

Thus, research methodology enables the researcher to develop in-depth understanding towards the research methods that can be used to attain the formulated objectives of the research. To understand the meaning of 'Research methodology', let us go through the following example, where a Ph.D student has followed the following methodology in planning the research problem on 'Performance of Agricultural Produce Market Committees (APMCs) in Andhra Pradesh':

- ☆ The researcher will select two major APMCs purposively in each region of Andhra Pradesh *i.e.,* Coastal Andhra, Rayalaseema and Telangana of Andhra Pradesh, considering the total arrivals and valuation of all agricultural commodities in these APMCs. Thus, six APMCs will be selected for this in-depth study.
- ☆ One hundred farmers from each region of Andhra Pradesh (50 farmers transacting through each APMC) will be selected at random, who are transacting their produce in these APMCs on regular basis. Thus, 300 farmers form the total sample size. Adequate number of market intermediaries will also be selected (approximately one-third of their respective total size) to ascertain both operational and pricing efficiency in the selected APMCs.

- The study will be based on both primary and secondary data. The major part of the data will be primary in nature. The primary data will be collected thorough well-structured pre-tested schedule designed for the purpose.
- The researcher will visit the sample respondents personally to collect the primary information. In designing schedule, every attempt will be made to make it precise so that, the filling up may be completed within shortest possible time.
- The secondary data pertaining to market arrivals and prices of major agricultural commodities will be collected from the Annual Administration Reports of the selected APMCs, journals, books, published articles and the internet. Apart from this, the information will also taken from the various news articles, which will appear in the leading newspapers and published brochures that have been brought out from time to time.
- The collected data will be subjected to statistical analysis by using relevant tools and techniques like correlation, regression analysis, Markov-chain analysis, Auto-Regressive Integrated Moving Average (ARIMA) model etc., based upon the composition of the data as well as the requirements of the research work. The various values that will result by the statistical methods in the report indicated with double star (**) imply that the data is significant at one per cent level and values obtained with single star (*) indicate the data is significant at five per cent level respectively. It will also be mentioned that, values obtained in the report without any star imply that, there is no significant association between the data. The stars that will originate after the calculated values of various methods will be compared with the in-built table values of advanced software packages.
- Suitable policy implications will be drawn from the analytical findings of the study, so as to improve the performance of selected APMCs keeping in view of safeguarding the interests of the farming community.

2

Research and Types of Research

2.1. What is Research?

Research is composed of two syllables, a prefix *'re'* and a verb *'search'*. *'Re'* means again, a new, over again. *'Search'* means to examine closely and carefully, to test and try to probe. These two words form a noun to describe a careful and systematic study in some field of knowledge undertaken to establish facts or principles. Thus, research is an organized and systematic way of finding answers to unknown questions. The word *research* is derived from the Middle French *'recherche'*, which means 'to go about seeking', the term itself being derived from the Old French term *'recerchier'* a compound word from *'re' + 'cerchier', or 'sercher'*, meaning 'search'. The earliest recorded use of the term was in 1577. In fact, each letter of word 'RESEARCH' had a definitive meaning as given below:

- ☆ R – *Rational way of thinking*
- ☆ E – *Expert and Exhaustive treatment*
- ☆ S – *Search for solution*
- ☆ E – *Exactness*
- ☆ A – *Analysis of adequate data*
- ☆ R – *Relationship of facts*

- ☆ C – *Careful recoding and critical observation*
- ☆ H – *Honesty and hard work*

The following are the important definitions of research:

'Research is the process of collecting, analyzing, and interpreting data in order to understand a phenomenon'.

(Leedy and Ormrod)

'Research is a systematic way of asking questions'.

(Drew)

'Research is a way of going about finding answers to questions about the social world'.

(Neuman)

'In the broadest sense of the word, the definition of research includes any gathering of data, information and facts for the advancement of knowledge'.

(Martin Shuttleworth)

'Research is a process of steps used to collect and analyze information to increase our understanding of a topic or issue. It consists of three steps: pose a question, collect data to answer the question, and present an answer to the question'.

(Creswell)

'A studious inquiry or examination; especially : investigation or experimentation aimed at the discovery and interpretation of facts, revision of accepted theories or laws in the light of new facts, or practical application of such new or revised theories or laws'.

(Merriam-Webster Online Dictionary)

'A careful inquiry specially through search for new facts in any branch of knowledge'.

(Oxford Dictionary)

'Research comprises defining and redefining problems, formulating hypothesis or suggested solution; collecting, organizing and evaluating data; making deductions and reaching conclusion; and at last carefully testing the conclusions to determine whether they fit the formulating hypothesis'.

(Clifford Woody)

'Research is a systematized effort to gain new knowledge'.

(Redman and Mory)

'Research is a careful investigation or inquiry specially through search for new facts in any branch of knowledge".

(Advanced Learner's Dictionary of Current English)

'Any honest attempt to study a problem systematically or to add to man's knowledge of a problem may be regarded as research'.

(Theodorson and Theodorson)

'The aim, as far as I can see, is the same in all sciences. Put simply and cursorily, the aim is to make known something previously unknown to human beings. It is to advance human knowledge, to make it more certain or better fitting. the aim is, as I have said, discovery'.

(Elias)

'Research is a point of view, an attitude of inquiry or a frame of mind. It asks questions which have hitherto not been asked, and it seeks to answer them by following a fairly definite procedure. It is not a mere theorising, but rather an attempt to elicit facts and to face them once they have been assembled. Research is likewise not an attempt to bolster up pre-conceived opinions, and it implies a readiness to accept the conclusions to which an inquiry leads, no matter how unwelcome they may prove. When successful, research adds to the scientific knowledge of the subject.'

(Rusk)

'The systematic and scholarly application of the scientific method interpreted in its broader sense, to the solution of social studiesal problems; conversely, any systematic study designed to promote the development of social studies as a science can be considered research.'

(George J. Mouly)

'To be sure the best research is that which is reliable verifiable and exhaustive, so that it provides information in which we have confidence. The main point here is that research is, literally speaking, a kind of human behaviour, an activity in which people engage. By this definition all intelligent human behaviour involves some research.'

(Francis G. Cornell)

The above definitions simply infer research as a movement, a movement from the unknown to the known. It is an organized or systematic, data-based, critical, objective, scientific investigation into a specific problem, undertaken with the purpose of finding answers or solutions to it. The needed information provided by the research guides the researcher (say, business manager) to make informed decisions to successfully deal with problems. The needed information generated from the research could be the result of a careful analysis of data gathered first-hand and of data that are already available. So, research is a logical and systematic search for generating new and useful information on a particular topic and it is an investigation of finding solutions to scientific and social problems. It is, otherwise, defined as *'a search for knowledge, that is, a discovery of hidden truths'*. Thus, research enriches the body of knowledge through leading to new contributions from the existing knowledge. Only through research it is possible to make progress in a any field. Research is done with the help of study, experiment, observation, analysis, comparison and reasoning. Research is in fact ubiquitous. For example, we know that, cigarette smoking is injurious to health; cow dung is a useful source of biogas; chemical fertilizers boost the yields of the crops, rice blast disease is due to a fungus etc., and we all aware of these information only through research.

Social sciences research revolves around four Ps *viz.*, People, Problems, Programmes and Phenomena. Here,

- ☆ ***People:*** refer to a group of individuals involved in the research.
- ☆ ***Problems:*** refer to the identification, prioritization and selection of research problem for the study.
- ☆ ***Programmes:*** refer to the execution of technological interventions in solving the problem and their assessment.
- ☆ ***Phenomena***: refer to causal relationship the researcher analyzes in addressing the research problem.

In practice, most research studies are based upon at least a combination of two *Ps*. These four *Ps* can be conveniently categorized into two aspects *viz.*, population and study area. The former deals with the first P *i.e.*, participants (say, farmers, consumers etc) involved in the research and the latter deals with the subject area, which includes remaining three *Ps*.

It is to be noted that, the research process is systematic in that defining the objectives, managing the data, and communicating the findings occur within established frameworks and in accordance with existing guidelines. The framework and guidelines provide researchers with an indication of what to include in the research, how to perform the research, and what types of inferences are probable based on the data collected and analyzed. Research originates with at least one question about one phenomenon of interest. For example, at what competence the APMCs are functioning in state of Andhra Pradesh, Economic analysis of mango cultivation in Chittoor district of Andhra Pradesh, Export competitiveness of Indian agricultural commodities in the international market etc., are some of the examples of research studies in the field of Agricultural Economics. These research topics help researchers to focus thoughts, manage efforts, and choose the appropriate analytical approach or perspective from which to make reliable conclusions and formulate policy guidelines.

In the above discussion, it is mentioned that, research is a systematic enquiry about a problem or phenomenon. The concept of *'systematic enquiry'* in conducting research can be better understood, if we analyze *'what research is not'*. The following are the interesting examples:

- ✰ A researcher prepared a paper on 'grades and standards followed for different agricultural commodities in Adoni regulated market in Andhra Pradesh' after reviewing literature on the subject available in the Annual administration Reports of the APMC, Adoni and the researcher claimed it as a piece of research.
- ✰ A researcher has researched and completed a document, which gives information about the different age-groups of farmers, their farming experience, their managerial ability, number of years of transacting their agricultural produce in the nearby regulated market etc.
- ✰ A researcher participated in a workshop on curriculum development for B.Sc. (Agriculture) programme and prepared a report through surveying literature and conducting discussion with the participants of the workshop.

It is to be noted that, none of the above examples can be classified under the name 'research' and let us try to know 'what research is'. The answer for this can be better understood by studying the following example:

- ✰ A researcher conducted preliminary survey among the farmers and prioritized the following problems with reference to marketing of paddy:
  - ❖ Presence of more number of middlemen between farmers and ultimate consumers in transacting the produce
  - ❖ Poor market information network
  - ❖ Insufficient storage facilities for the produce
  - ❖ Lack of scientific grading equipment for grading the produce
  - ❖ High transportation costs of produce to the market
- ✰ The researcher reviewed the earlier literature to analyze the top most prioritized problem.
- ✰ The researcher gathered information about different types of intermediaries that intervene between the farmer and the ultimate consumer.
- ✰ The researcher then formulated the problem and generated guesses (hypotheses).

- ☆ The researcher selected a sample of farmers to obtain requisite information pertaining to different types of marketing channels for paddy, marketing costs, marketing margins of different agencies etc.
- ☆ The researcher prepared a pre-tested schedule to collect the requisite data from the selected sample
- ☆ The researcher collect the data from the sample farmers and market intermediaries, compile and analyze the data and interpreted the findings in the light of the hypotheses formulated earlier and reach to conclusions.

In the above explanation, we can easily notice that, the researcher went through a sequence of steps, which were in order and thus, systematic. Further, the researcher did not just jump at the conclusions, but used a scientific method of inquiry in reaching at conclusions (Figure 2.1). Thus, the two important characteristics of research include: it is systematic and secondly it follows a scientific method of enquiry. This approach also infers that, the earlier three examples does not fall into the category of research. We consider research is 'systematic' because, it follows certain steps that are logical in order. These steps include (Figure 2.1):

A. Conception/Identification of Research Problem
B. Formulation of Research Problem
C. Review of Literature
D. Formulation of Testable Hypothesis
E. Preparation of Research Design (Methodology)
F. Execution of Research Project
G. Data Collection
H. Data Processing and Analysis
I. Testing of Hypothesis and Interpretation of the Findings
J. Writing Research Report

Characteristics of Research

From the above discussion, it is easier now to describe what is 'quality' research and the following are the key features:

- ☆ It involves a systematic approach for critical investigation into a phenomenon.
- ☆ It is based on the collection of facts about the (prioritized) problem under consideration.
- ☆ It can be replicated (duplicated).
- ☆ It adopts scientific method.
- ☆ It is based on observable experience or empirical evidence.
- ☆ It is generalizable to other settings.
- ☆ It is based on some logical rationale and tied to theory.
- ☆ It is doable!
- ☆ It generates new questions or is cyclical in nature.
- ☆ It is incremental.

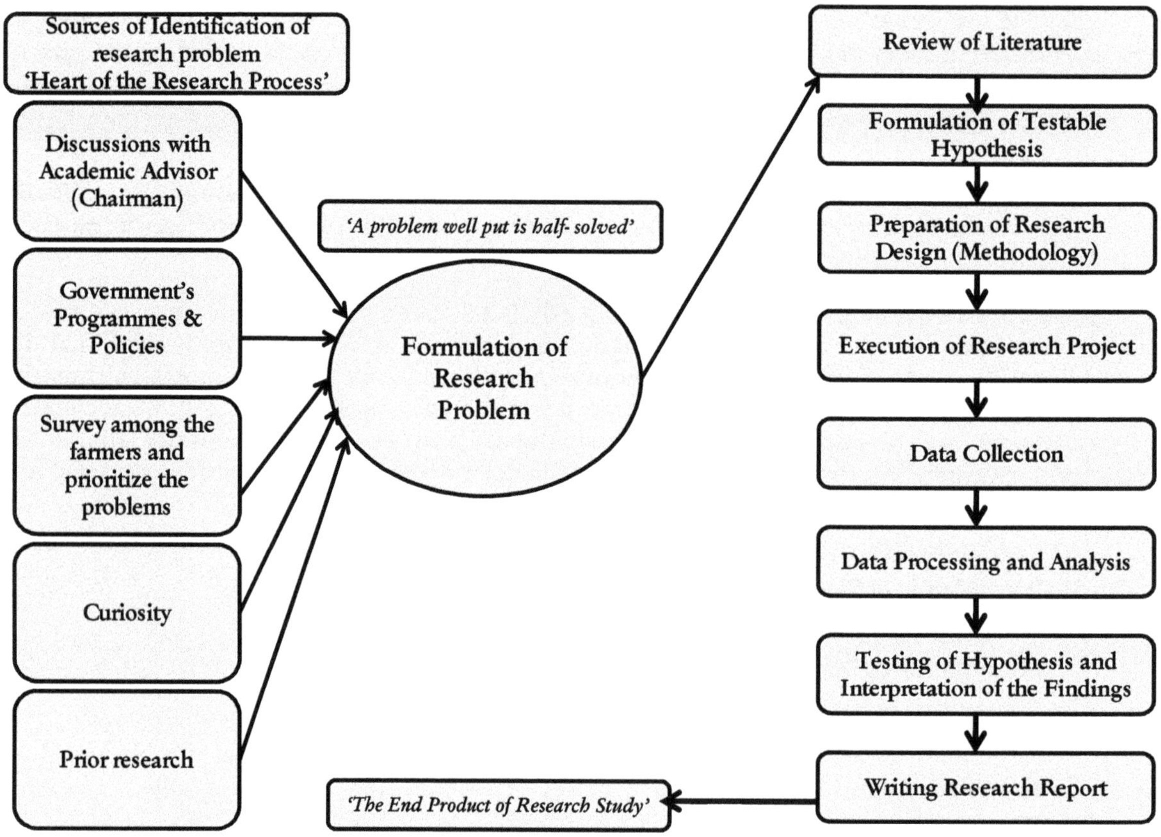

Figure 2.1: Agricultural Research : *'Where Do Researcher Begin'.*

- It should be undertaken for the betterment of society.
- It is directed towards finding answers to pertaining questions and solutions to problems. Note that, the purpose of research is not to arrive at an answer, but the generated findings must be stand up to the test of criticism.

From the above points, we can infer what research is and what are the key features of research. These can be better understood from the following interesting points, where few situations are to be avoided in considering a problem for research purposes:

- ***Research is not mere information gathering:*** Mere collecting of information about the research problem under consideration is not 'research'. For example, if a researcher goes to the library or an internet center and collects information about market prices of major agricultural commodities, it represents mere gathering of data or information or 'information discovery' and it can also be attributed as 'reference skills' of the researcher, but not 'research'.
- ***Research is not mere transportation of facts from one location to another:*** Mere compilation of facts from several sources and placing them at a common place is not 'research'. For example, a researcher collects facts about the defects in present agricultural marketing system in different States in India from several sources and list out the same to write an article on 'Constraints in the marketing of agricultural produce in India'. This does not mean 'true

research', as the researcher simply collected the information and organized it in a certain way for presentation to others. But, he missed the essence of research: *i.e.*, analysis and interpretation of data. Nowhere in the paper does the student studies cause and effect relationship(s), interpret and draw conclusions. The researcher simply gathered information from different States (more realistically be called fact transcription, fact organization, or fact summarization) and try to explain *'this'* about the agricultural marketing system in India.

- ***Research is not merely rummaging for information:*** If the researcher reviews the past information to ascertain the current fact about the problem under consideration, it will not fall under 'research'. For example, a farmer decides to sell away his five acres of land to start a business on his own, he may call upon the realtor to ascertain the current market value of land in that area. Then the realtor do 'some research' in reviewing the past information about the recent sales of land in that area that are comparable to farmer's land. This will help the realtor in fixing a reasonable selling price to the farmer's land. This process of rummaging *i.e.*, reviewing the past information and analyzing the present, is not research and is more accurately called an *'exercise in self-enlightenment'*.
- ***Research is not a catchword used to get attention:*** Simply by using catchwords in the advertising the products, it does not mean 'research'. For example, a caption in the news paper about a fairness cream states that, *'Years of Research Have Produced a fairness cream! Apply to your face and enjoy the glow!.* In this slogan, the phrase *'years of research'* catches the attention of readers. The readers presume that, the product 'must be good', as many years of research have been spent on developing it. But, after purchasing the product, it is not on par with the expectations of the customers. So, the phrase *'years of research'* has cheated the customers through drawing their attention towards the product. That is, *'years of research'* is an attention-getting phrase, but it is misleading!
- ***Research is not meant for achieving self-enlightenment:*** We know, all of us have great gaps in education, and diminishing them is the joy of learning. But *it is not the purpose of research to educate you.* The lack of information may be a *'personal'* problem, and if the researcher want to know more about a certain area of knowledge, then it is laudable.
- ***Problems whose sole purpose is merely to compare two sets of data are not suitable research problems.*** *Similarly, finding a coefficient of correlation between two sets of data merely to show a relationship between those data sets is not acceptable as a problem for research.* This is because, here, the basic requirement for research is ignored *i.e.*, there is no involvement of human mind in drawing the cause and effect relation (whether the relationship is genetical or social or environmental etc) and this statistical operation can be performed by a computer in an infinitely faster and more accurately than a person can.

2.2. Principles or Characteristics of Quality Research Work

Research is a process of collecting, analyzing and interpreting information to answer questions. Although research projects vary in complexity and duration, they typically enjoy distinct characteristics and they are mentioned below:

- The research problem should be 'interesting' and should attract the attention of the researcher and other people to conduct it.
- The research study should be innovative *i.e.*, it should be new, original, and unique towards addressing the prioritized problem. This contributes to the economic development of the country.

- The purpose of the research should be clearly defined.
- Research study taken up should aim at solving the needs and problems of society, but not at gaining personal benefits on the part of the researcher undertaking it.
- The research problem identified must be relevant to the Government thrusts in promoting the welfare of the society.
- The research design should be carefully planned, so as to yield meaningful results for the objectives framed under the study.
- Originates with a question or problem *i.e.*, research demands a clear statement of the problem. For example, if farmers get a low price for their produce in the market, then it puzzles the researcher to pose several questions like 'Why?' 'What's the cause of that?', 'what is the competitive price', 'when the farmers should transact their produce to fetch higher prices' etc. These questions enable the researcher to discover the answers in the light of the available facts. So, the researcher must know clearly what it is to be attempted in the research. We shall discuss more about the research problem in a later chapter.
- Requires clear articulation of a goal and clearly defined objectives. That is, the planned design should be as objective as possible.
- Accepts certain critical assumptions.
- It aims to tackle a particular research problem.
- The procedure adopted to undertake the research investigation must follow a certain logical sequence *i.e.*, a systematic procedure (Figure 2.1). The different steps cannot be taken in a haphazard way.
- The process of research investigation must be foolproof and free from drawbacks. The process adopted and the procedures used must be able to withstand critical scrutiny.
- Highly ethical standards be applied.
- From a practical standpoint, researcher often divides main problem into sub-problems. For example, if the researcher wants to analyze *'Marketing efficiency in transacting paddy in Kurnool district of Andhra Pradesh'*, he can divide this main (research) problem into sub-problems like:
    - Marketing channels for paddy.
    - Marketing costs and margins of different market intermediaries
    - Spatial price integration of paddy markets
    - Marketing problems involved in transacting paddy

  So, what seems like a single question can be divided into smaller questions that must be addressed before the principal question can be resolved. By addressing each of the sub-problems, the researcher can more easily address the main problem.
- It requires a systematic plan *i.e.*, plan the work and work the plan. That is, it is not aimless and undirected activity. Rather, it entails a definite plan, direction and design.
- The research procedure should be in a systematic manner so that, it permit another researcher to repeat and extend the same. So, the research should be carried out in such a manner that, the principle of continuity is ensured.

- ☆ The research design should be planned in a most scientific way and all aspects of resources, time frame, constraints and procedural aspects should be taken into consideration.
- ☆ Research is guided by the formulation of hypotheses. After stating the main problem and its sub-problems, the researcher usually forms one or more hypotheses about what he or she may discover. A research hypothesis provides a tentative explanation for a phenomenon under investigation. It will guide the researcher to think in all possible directions to solve the problem (sub-problems and in the process, the main research problem) under consideration. These hypotheses are certainly not unique to research and they are recurring features of everyday life. They represent the natural working of the human mind. This can be easily explained with the help of the following example: Suppose a farmer went to his field and switched on the electric motor to irrigate paddy crop. But, the irrigation motor is not pumping the water. At this point, the farmer begin to construct a series of reasonable guesses or hypotheses to explain the motor's failure:
    - ❖ The motor is not working.
    - ❖ The electric plug is not plugged well into the plug hole.
    - ❖ Yesterday night's thunderstorm might interrupted the electrical service in that locality.
    - ❖ The farmer has not paid the electricity bill in the last month.
    - ❖ The wire from the power source to the motor is defective.

  Each of these hypotheses hints at a direction to the farmer to act upon to resolve the problem. Now, the farmer tries to collect the information for each and every hypothesis formulated to determine which is correct. If one hypothesis is found correct, the other hypotheses can be rejected. The following data* are collected by the farmer:
    - ❖ The farmer call upon a person, who repair the electric motor. But, it was found that, there is no problem with the electric motor. (So, hypothesis 1 is rejected).
    - ❖ The farmer verified and found that, the plug is plugged well into the plug hole (So, hypothesis 2 is rejected).
    - ❖ The neighbouring farmers are enjoying the use of motors for irrigation purpose. So, other farmers have electrical power. (So, hypothesis 3 is rejected).
    - ❖ The farmer verified the paid electricity bill of the last month (So, hypothesis 4 is rejected)
    - ❖ The farmer verified the connecting wire from the plug to the electric motor and it was found defective. (So, hypothesis 5 is accepted). So, by repairing or replacing the connecting wire, the problem can be resolved.

      Thus, hypotheses are tentative statements formulated in a research to have a guessing about the solution for the researchable problem under consideration. This concept about framing of hypothesis and the testing of hypothesis for its validity are discussed in-detail under Chapters 7 and 12 respectively.
- ☆ The researcher should report the results with complete frankness, interpret the findings with reasons and draw policy implications.
- ☆ Guessing and intuition in arriving at conclusions are rejected.

---

* We have to use the word '*data*' as a plural noun (for instance, we ask 'Where *are* the data?' rather than 'Where *is* the data?'). Contrary to popular usage of the term as a singular noun, *data*, which was originally a Latin word, refers to more than one piece of information. A single piece of information is known as a *datum*, or sometimes as a *data point*.

- The research results should be empirical and should be presented unambiguously. They should be replicable so that, it helps to build a sound basis for decision. The conclusions drawn are based upon information collected and they should be justified by sufficient evidence.
- The selected sample should be adequate and representative of the population. Similarly, the collected data should be sufficiently adequate and the techniques employed should be relevant to draw realistic conclusions. The validity and reliability of data should be checked carefully. Result should be built on existing data and thereby, enriches the body of knowledge.
- Research is cyclical or, more exactly, helical in nature (Figure 2.1). As discussed above, the research cycle begins with identifying a problem, sub-dividing the same into sub-problems each of which fits into the broad framework of overall objective of the research and all of which collectively comprise the principal research problem. What we have been calling the environment out of which the researchable problem arises is more appropriately called the research universe*. The researcher collects the relevant facts or data or information from the universe, which seem to be pertinent to the solution of the problem and its attendant sub-problems. The search for the facts is facilitated by the construction of tentative hypotheses and they guide the researcher in the collection of relevant facts and proceed in the right direction. The collected data or facts are then compiled, analyzed and interpreted towards solving the problem under consideration. This completes one cycle and again the researcher searches for another problem from the universe and this proceed in a cyclical (more as a helix) manner. Thus, in the helical process of solving problems, still new problems are identified and hence, the research continues progressively onward.
- The scope and limitations of the work are to be clearly defined.
- The process to be clearly explained so that, it can be reproduced and verified by other researchers.
- Research should be completed within the allocated financial resources.
- Research should be completed in the stipulated time frame.
- It finds solution to the research problem under consideration. It helps to draw generalizations about the problem and gives further scope to extend the research.
- In exploring the causality among the variables under consideration, the research must be set in such a way that, it minimizes the effects of other variables affecting the relationship. However, this can be achieved to a large extent in the case of laboratory experiments of plant pathology, entomology, soil science etc. However, in the social sciences like Agricultural Economics and Agricultural Extension, it is extremely difficult, as research is carried out on issues related to human beings living in society, where such controls are not possible. Hence, it is difficult to gain control over other external factors.
- The procedures followed in the research to find answers to questions must be relevant, appropriate and justified. Again, the degree of rigor varies markedly between the laboratory experiments and social sciences and also within the social sciences.

---

* The term *universe* is simply an 'area' surrounding the problem, which may contain facts relevant to the problem. Literally, the word signifies the factual area that revolves around the central inquiry, or main problem of the research. The word derives from '*unus*', 'one', and *'vertere'*, 'to turn' that which turns or revolves around one central inquiry.

- ☆ The research should be cost-effective in solving the problem under consideration.
- ☆ Conclusions and policy suggestions drawn from the research must be confined in relevance to the objectives framed and data analyzed under the study.
- ☆ Greater confidence in research is warranted if the researcher is experienced, has a good reputation in research and is a person of integrity.

Thus, for a process to be called as 'good research', it is imperative that, it has to obey the above characteristics.

2.3. Objectives of Research

The following are the objectives of research:

- ☆ One of the goals of research work is to develop a procedure, as general as possible.
- ☆ The prime objectives of research are to gain familiarity with a phenomenon, discover new facts, to verify and test important facts.
- ☆ ***Academic objective:*** We know, several post-graduate and doctoral students conduct research on various topics out of academic interest with the objective of acquisition of knowledge coupled with curiosity that has been the guiding force behind conducting the research work independent of any material incentive.
- ☆ ***Utilitarian objective:*** Research also aims at studying and understanding the organizational culture, social life, social environment, decision making processes etc., and thereby, gain a greater measure of control over human behaviour in the organization and social context.
- ☆ Research helps in portraying accurately the characteristics of a particular individual or a group in the organization and leads to organization redesign, and design of strategies of development.
- ☆ Research may be used to determine the frequency with which a certain thing occurs or with which it is associated with something else. Say, for example, research can be employed to quantify the risk in conducting crop production programme and thereby, helps to design strategies to overcome the same.
- ☆ Research helps in testing a hypothesis or a casual relationship between variables to determine the cause and effect relationships.
- ☆ Research provides solutions to complex problems.
- ☆ Research leads to new discoveries.
- ☆ Research develop new products.
- ☆ Research aims at cost minimization and profit maximization.
- ☆ Research purifies human life and improves its quality.
- ☆ Research enhance human desires.
- ☆ To develop new scientific tools, concepts and theories to solve and understand scientic and non-scientic problems.
- ☆ To find solutions to scientic, non-scientic and social problems, and
- ☆ To overcome or solve the problems occurring in everyday life.

2.4. Qualities of a Good Researcher

They include:

- The researcher should follow a systematic procedure right from identifying a problem and then for working on it, to find a solution for that problem.
- The researcher should aim at the prioritized research problem with in the available resources.
- The researcher should be well aware and should have complete knowledge and information about the field of investigation.
- The researcher should have a good background of the study so that, he can better analyze the problem under consideration
- The researcher must have a vision of his own and should be creative in his approach in solving the problem.
- The researcher should be open minded in approaching towards solutions of the problem.
- The researcher should be stable, having consistent thinking.
- The researcher should be honest, sincere, dedicated and ambitious.
- The researcher should be prompt in working the plan of the research project.

2.5. Motivations in Research

No person would like to do research unless there are some motivating factors. Some of the motivations are given here under.

- Curiosity to know what is unknown.
- To study the cause-effect-relation.
- To discover the facts of real nature.
- To identify and analyze the unusual happenings in the life of man, society and organization, which are sudden and unexpected and do not fit in any known causation pattern.
- To develop new techniques and refinement of existing techniques for problem-solving.
- To get a research degree (M.Sc. (Ag.), Ph.D.) along with its benets like better employment, promotion, increment in salary etc.
- To get a teaching position in a college or university or become a scientist in a research institution in home country or in abroad like USA, Australia, UK, Canada, Germany etc.
- Desire to face the challenge in solving the unsolved and challenging problems.
- Desire to get intellectual joy of doing some creative work.
- To acquire respectability.
- To get recognition in the society.
- Curiosity to find new things.
- To serve the society by solving social problems.

Some students undertake research without any aim possibly because of not being able to think of anything else to do. Such students can also become good researchers by motivating themselves toward a respectable goal. In the words of Prof. P. Balaram [Current Science, 87(2004)1319], *'Ph.D. degree is a passport to a research career. The Ph.D. period often influence a research scholar to make or to break in a scientific career'.*

2.6. Ethical Considerations of Research

Social science research exerts a significant influence over social systems. Hence, a researcher needs to adhere to an ethical code of conduct. These ethical considerations are as follows :

- Though researcher have some obligations towards the funding (sponsoring) agency of the research project, he should not reveal the project data that can lead to unfavourable conclusions and interpretations for the sponsoring agency.
- The researcher should maintain strict confidentiality about the personal information of the respondents and should not mention in any of the records or reports. Only the data pertaining to research information should be revealed and analyzed.
- While collecting primary data and secondary data in conducting research, the researcher should not make use of hidden cameras, microphones, tape-recorders or observers without the respondents' permission.
- In collecting information from the respondents, the researcher should not show bias based on caste, religion etc. All the respondents should be treated same with the sole intention of obtaining requisite data from them.
- If the researcher engages volunteers or data collectors for conducting survey among the farmers, they should be prior trained and well-informed about the objectives of the investigation, structure of schedule, duration of completion of data collection etc.
- The researcher should accept the fact that, there is no compulsion on the part of the respondents to provide all the requisite data for the research investigation. So, while selecting the sample, the researcher has to go for pre-testing of schedule and identify the sampling frame that shows interest in the investigation.
- The researcher should not promise or never try to make undue efforts giving favourable treatment to some respondents (say, giving money, offering farm inputs at free of cost etc.) during the research investigation.
- The periodical surveys conducted by the researcher should not pose any mental harm, stress, disturbing personal life of the respondents etc.
- The researcher should make the collected data available to peers for scrutiny or verification.
- The researcher must be in a position to provide the findings of the analysis from the data collected to the respondents, if they so demand.
- The researcher should give due credit to all those who have helped him/her in the research procedure, data collection, data analysis or preparation of the research report.
- If at all the researcher has made some promises to the respondents say, like installation of sub-surface drainage system, reclamation of saline soils etc., they must be honoured and fulfilled.
- The researcher should accept the fact that, the subjects (participants) have the freedom to decline to participate or to withdraw from the research.

2.7. Importance of Research

With increasing complexity of agri-business, fast changing technology, size of the business, changing paradigm shifts in the fields of production and trade etc., the importance of research especially in solving the operational problems has gained immense significance. So, research serves as an aid to economic policies framed by the country and thus, guide the Government and businesses in the right direction. The following aspects emphasize the importance of research:

- Research is crucial in formulating various Governmental programmes and policies. For example, to match the Government revenues with its expenditures, research is most needed. Research also guides the Government in formulating various welfare measures and investment activities.
- Research is crucial in various sectors of the economy like agriculture, industry, service, defense etc. The allocation of nation's resources among these sectors counts much in the economic prosperity of the country and this too requires sound research.
- Research is essential to analyze the dynamic changes in the economy and in predicting the future. This guides the policy makers to know what is happening in the economy, what changes are needed in the future, formulation of new policy guidelines in the context of changing environment etc.
- Research plays a significant role in various fields like product development (diversification, launching a new product, product improvement, process technologies etc.), personnel management (job redesign, organization restructuring, development of motivational strategies etc) and marketing management (deciding the size of target market, the consumer behaviour in terms of attitudes, life style, pricing policy of the firm, selection of channel of distribution, product mix, promotional strategies etc), finance management (distribution of dividend, capital raising, hedging, taking care of fluctuations in foreign exchange and product cycles etc.) and materials management (selecting the supplier, buying decisions, deciding negotiation strategies etc.).
- Research helps in developing the standards, objectives, long-term goals and growth strategies of the business firms.
- Research is helpful in the welfare and progress of humanity and the society. It aims at solving communal troubles, religious riots, the misnomer of social and racial superiority. Thus, through research, it is possible to do away with all these wrong notions.
- Research guides in providing solutions to various planning, operational and marketing aspects of business. It also helps the business to understand about fluctuations in business cycles, upgradation of technologies, relevance of business policies in relation to human welfare etc.
- Research also helps to analyze the social phenomena, events and the factors that govern and guide them. This helps to understand social values, beliefs, traditions, events etc., and thus guide to control social life and draw predictions of social behaviour.
- Research is important for tackling problems, studying cause and effect relations between the variables, offering explanations and applications. Practically implementable solutions and suggestions are to be suggested for tackling the present problems and new problems that may arise.
- Research provides basis for many Governmental policies. For example, research on cost analysis in the production of agricultural crops helps the Government to recommend Minimum Support Prices (MSPs) for agricultural commodities. This further guides the Government to prepare the budgetary outlay.
- Research leads to a new style of life and makes it delightful and glorious.

2.8. Problems/Limitations of Research in India

Sometimes, the outcomes planned and estimated from research will not be realized as expected and this is due to the following limitations:

- Lack of scientific training to the researchers in the field of methodology of conducting research is a major constraint in India. Many researchers undertake research work without having actual knowledge of the research methods. They just look for similar studies and copy the methodologies listed in it. Even some of the research personnel who guide the students also do not have a thorough knowledge of the various methodologies.
- There are no special incentives for the researchers in conducting research in the country and this affects competency among researchers.
- The sample framed by the researcher will not provide complete and reliable information to the researcher, because of fear of misuse of information. This affects the quality of research outcome.
- Researchers in India have to cope up with the non-availability of adequate and timely secretarial assistance, which affects the schedule of their research study.
- Due to lack of confidence and competence among the researchers, generally they undertake repetitive studies and no new areas are attempted for research.
- Lack of frequent and healthy interactions between researchers, scientific communities in universities, line departments of agriculture, business firms etc., will hamper the availability of full and quality information to the researchers and this, in turn, affect the quality output from research.
- With the intention of completing the assigned researchable issue in time, the researchers does not follow the particular code of conduct. This problem is further aggravated by inter-university and inter-departmental rivalries.
- To complete the research with desired quality output and in right time, it demands huge resources from the funding agency. But, in India, there is paucity of funds.
- There exists lack of coordination among various departments in conducting a multi-disciplinary research and this leads to outcome of poor results.
- Without prioritizing the researchable issues at the beneficiary level (say at farmers level), the researchers often plan the research of their own interest *i.e.*, supply driven research. This leads to wastage of resources, as the research outcome will not address the top most problem faced by the farmers.
- The research agencies spend a huge proportion of total funds in training the researchers (even in abroad) rather than strengthening infrastructure at the farmers level. This will result in short term benefits to the farmers and not in the long run.
- The libraries are not managed systematically. So, much of the precious time of the researchers is spent in searching for books, reports, newspapers, etc., rather than searching relevant information from them.
- Once the researcher completes the research, he has to publish it. But, for publishing the research findings, there is less demand from the publishing houses and hence, the researcher has to sell away the reports even at low price. If the researcher wishes to publish research findings in international journals, it is highly expensive. This discourages most of the researchers from taking up research work.

- The lack of a scientific training in the methodology of research.
- Problems of collection of data, especially primary data (say from the sample farmers, market intermediaries like commission agents, wholesalers, processors, retailers etc), as such data is based on the recall of memory.
- Repetition problems.
- Outdated and insufficient data may cause problems in analyzing data and in drawing policy implications.
- Resources constraint.
- Paucity of time leads to half-hearted conclusions.
- Non-availability of trained researchers.
- Absence of code of conduct.
- Often times, the researcher choose common 'pet' problems that are interesting to him, but they are not to the scientific community and local population at large. So, the research conducted does not generate new knowledge or insight about the phenomenon being investigated. Some researchers select the popular topics of the present day context for conducting research and such topics may have limited shelf-life.
- Some researchers go for blind data mining. That is, they have the tendency to collect data first (using instruments that are already available), and then figure out what to do with it. Note that, data collection is only one step in a long and elaborate research process and in fact, a series of other activities are needed in a research process prior to data collection. If researchers jump into data collection without such elaborate planning, the data collected will likely be irrelevant, imperfect, or useless, and their data collection efforts may be entirely wasted.

2.9. Issues and Trends in Research

We know research is a dynamic activity and is rapidly changing and growing in importance. The following factors guide the current issues and future trends in research:

- With changing technology, there is increasing competition among the firms and hence, the researchers has to focus more in producing the products that are cost-effective, with differentiation, product distribution strategies etc. These changes call for new trends for research in product development.
- With the advent of internet, there is fast dissemination of information across the countries and in this context, the world has turned into a global village. So, the research gained momentum in the context of studying the export competitiveness of agricultural commodities in the international market, grades and standards preferred by the foreign consumers, market research, marketing management etc. There is increased foreign investment, understanding the tastes and preferences of foreign customers, build networks and forge partnerships etc., and this led to the expansion of research activities in India.
- To capture significant market share, the firms go for long term strategies like horizontal integration/vertical integration/conglomeration etc., and for implementing these market integration strategies, an extensive background research is needed.

- ✰ With increased customization, new products and services as demanded by the customers are needed to be produced. That is, the focus is to be shifted from 'selling' to 'marketing'. This calls for strengthening market research to promote product and service innovations.
- ✰ In the modern era of agri-business, quality consciousness is increasing among both producers and consumers with reference to availability of inputs and products respectively. This will lead to more emphasis on product research, packaging research, distribution research etc.
- ✰ With advances in technologies like biotechnology, production technology of crops, marketing technology, agri-engineering technology etc., there is bright scope in the inter-disciplinary research.
- ✰ There is growing impetus for environment safety in view of the sustainability perspective of future generations. This calls for strengthening of research in producing quality output.

2.10. Typology of Research

Now that you are familiar with, what research is and its guiding parameters, and now you are ready to get into the details of different types of research. It is imperative that, a researcher has to have a broad understanding of the various types of research, in general. This is because, before a researcher begins to conduct a study, he/she must decide on a specific type of research, as a good researcher can easily understand the relative advantages and disadvantages of each type. Research can be classified in many different ways (Figure 2.1.1).

A. Based on the Purpose of Research Study or from the Point of View of Application

This criterion classifies the research as:

a. Basic (Pure or Fundamental) research
b. Applied research
c. Practical research

B. From the Viewpoint of Objectives

According to this criterion, the research can be classified as:

a. Exploratory research
b. Descriptive research
c. Explanatory research or Analytical research
d. Predictive research

C. Based on the Data or Information Collected

The process of the research can be classified as:

a. Quantitative research
b. Qualitative research

D. From the Process Adopted to find Answer to Research Questions (Inquiry Mode)

According to this criterion, there are the two approaches *viz.*,

a. Structured approach
b. Unstructured approach

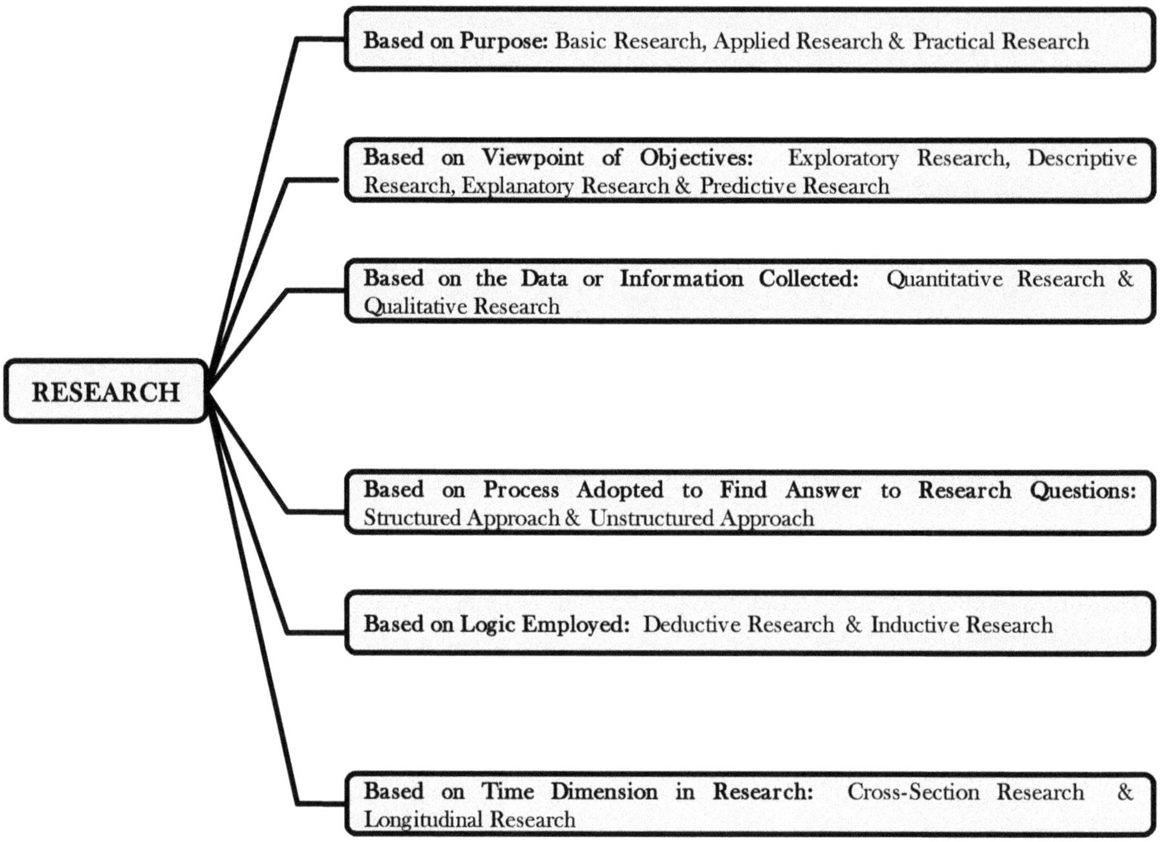

Figure 2.1.1: Different Types of Research

E. Based on Logic Employed

This classifies the research as follows:

a. Deductive research
b. Inductive research

F. Based on the Time Dimension in Research

Based on this criterion, there are two categories *viz.*,

a. Cross section research
b. Longitudinal research

A. Based on the Purpose of Research Study or from the Point of View of Application

a. Basic (Pure or Fundamental or Theoretical) Research

The terms 'basic' or 'fundamental' research indicates that, through theory generation, basic research provides the basis or foundation for further, often applied research. This research is characterized with the following aspects:

- Basic research is an investigation on basic principles and reasons for occurrence of a particular event or process or phenomenon.
- This research aims at investigation on some natural phenomenon or relating to pure science. So, the basic research is only done for knowledge enhancement and not to create or invent.
- This research is original in character. It provides a systematic and deep insight into a problem and facilitates extraction of scientic and logical explanation and conclusion on it. It helps build new frontiers of knowledge.
- It is conducted to satisfy any curiosity such as: what makes things happen, why society changes and why social relations are in a certain way.
- This research involves developing and testing theories and hypotheses that are intellectually challenging to the researcher.
- Can lead to the discovery of a new theory and also can result in the development or refinement of a theory that already exists.
- The main aim of basic research is the discovery of knowledge solely for the sake of knowledge. So, the main motivation is to expand man's knowledge. In fact, it is the source of most new theories, principles and ideas. It is not concerned with practical applicability and most closely resembles the laboratory conditions and controls usually associated with scientific research. It is concerned establishing generally principles of learning.
- Basic research rarely helps anyone directly. It only stimulates new ways of thinking.
- The research does not have immediate use or application. That is, basic research is not concerned with solving any practical problems of immediate interest. Thus, basic research is purely theoretical to increase the understanding of certain phenomena or behavior, but does not seek to solve any existing problem.
- The research is aimed towards human welfare, animal welfare and plant kingdom welfare.
- There is no obvious commercial value to the discoveries that result from this research.
- However, in the long run, basic research lay down the foundation for applied research and for development of commercial products. Researchers working on applied research have to make use of the outcomes of basic research and explore the utility of them.

Highlighting the importance of basic research, Travers opined that, '*Basic research is designed to add to an organized body of scientific knowledge and does not necessarily produce results of immediate practical value*'. This type of research draws its pattern and spirit from the physical sciences. It represents a rigorous and structured type of analysis. It employs careful sampling procedures in order to extend the findings beyond the group or situations and thus develops theories by discovering proved generalizations or principles. Dr. George Smoot says, '*People cannot foresee the future well enough to predict what's going to develop from basic research. If we only did applied research, we would still be making better spears*'. He further says '*people cannot foresee the future well enough to predict what is going to develop from the basic research*'. For example, advances in the practice of medicine are dependent upon basic research in biochemistry and microbiology. The research conducted on issues like 'How did the universe begin?', 'What are protons, neutrons, and electrons composed of?', 'What are the physico-chemical properties of tungsten element', 'At what point in human body did logical thought arise?', Whether virus is a living or non-living organism?' etc., fall under basic research. Similarly, discovery of X-rays, which led to studying bone fractures, psychological studies of decision making that led to important findings in the fields of education, medicine and economics etc., also fall under basic research. However, since there is no guarantee of short-term practical gains, researchers often find it difficult to obtain funding for basic research.

b. Applied Research

Applied research is done to solve specific and practical problems. Its primary aim is not to gain knowledge for its own sake. It can be exploratory or explanatory, but often it is descriptive. It is almost always done on the basis of basic research. This applied research is often carried out by academic or industrial institutions. The academic institution say, State Agricultural University (SAU) will have a specific applied research programme funded by an agency (say, Indian Council of Agricultural Research (ICAR), World Bank etc.). The following are the important characteristics of applied research:

- It is the research performed in relation to actual problems and under the conditions in which they are found in practice. So, the researcher aims at solving research problems by employing well known and accepted theories and principles. Thus, a research, the outcome of which has immediate practical application is also termed as applied research.
- Applied research has most of the characteristics of basic research, including the use of sampling techniques and the subsequent inferences about the target population. Its purpose, however, is improving a product or a process and testing theoretical concepts in actual problem situations.
- Applied research is designed to solve practical problem of the modern world, rather than to acquire knowledge for knowledge sake. Hence, according to Travers, *'Applied research is undertaken to solve an immediate practical problem and the goal of adding to scientific knowledge is secondary'*.
- The goal of applied research is to improve the human condition, as it is problem-oriented. It will focus on analysis and solving social and real life problems.
- Helps in testing the validity of a theory, but under some conditions.
- Provides data that can lead to the acceleration of the process of generalization.
- Applied research is generally conducted on large scale basis and it is expensive. Hence, it often conducted with the support of some financing agency like Government, World Bank, United Nations Children's Fund (UNICEF), University Grants Commission (UGC) etc. The applied research may also be employed by a University or research institute or may be found in private industry or working for a Government agency.
- Applied research should not be treated the same as Research and Development (R&D). This is because, R&D aims at developing new products demanded by the existing clients. Applied research, on the other hand, focuses on uncovering what needs are not being met and use that information in designing products or services that would create their own demand. Thus, applied research brings in new customers and also provides better products and services to the existing customers. In late nineties, in India, the mobile phones were expensive, with more weight and with less features. Applied research foresaw that, this product would have a limited market and thereby, focused on cost-cutting, reduced weight, additional features and long-distance communication. Such measures led to drastic increase in demand both among existing customers and new customers.
- Applied research can be further classified as Problem oriented research and Problem solving research.

i. ***Problem oriented research:*** This research is done by an apex body of the organization for sorting out problems faced by all its members. For example, World Trade Organization (WTO) conducts problem oriented research for analyzing the implementation experience of tariffication among developed, developing and least developed countries. Similarly, in India, Agriculture and Processed Food Products Export Development Authority (APEDA) conduct regular research for analyzing the prospects of agro-processing industries with reference to export of processed food products.

ii. ***Problem solving research***: This type of research is done by an individual firm for the problem faced by it. For example, a pesticide firm (say, Monsanto company) conducts marketing research to promote the sales of its products at the farmers' level.

☆ ***Sub-types of applied research:*** Researchers employ several sub-types of applied research. They include Action Research, Impact Assessment Research and Evaluation research. They are discussed here under.

i. ***Action research:*** This research enables the researcher to investigate a specific problem that exists in practice. The results obtained from this research should be relevant to solve the problem under consideration. That is, the strategies designed through this research should be applicable immediately to address that specific problem. This research is characterized by the following four features:

- Focus mainly in solving a particular problem in a specific situation.
- All the participants in the research (researchers and beneficiaries) form an integral part of action research with the exclusive aim of solving the identified problem.
- Deals with quantitative data and empirical results
- Outcome of research cannot be generalized.

Action research may be Individual action research or Collaborative action research.

❖ ***Individual action research:*** For example, an Agricultural Extension Specialist wants to know, whether group activities will improve the knowledge of farmers regarding the adoption of scientific package of practices on paddy crop. In the training course, he measures the performance of the farmers after giving whole-group instruction for one week. Later, he supplements the instruction with small-group learning activities for one week and again measures the performance of the farmers. He finds that, farmers' awareness and knowledge were increased with the use of small-group activities compared to whole-group instruction. On the basis of these action research results, he changes his approach to impart training to the farmers.

❖ ***Collaborative action research:*** For example, a project is to be implemented in Kurnool-Cuddapah canal command area with the objective of addressing soil salinity problem in the farmers fields. Accordingly, the Principal Investigator and other Scientists (Agricultural Engineering, Agronomist, Soil Scientist, Agricultural Extension Specialist and Agricultural Economist) of the project together examine the farmers fields, collect soil and water samples and identify the specific areas that need improvement. They then execute the need based technological interventions like application of gypsum and installation of sub-surface drainage system in the farmers' fields for about one year. During this one year, the scientific community meets frequently as a group to discuss progress and problems in the execution of the project. At the end of the one year, they compare the yields of paddy crop to those obtained prior to the implementation of the technological interventions. If the results are found significant, the scientists and the farmers-beneficiaries continue monitoring these interventions for sustained output in the crop production programme.

It is to be noted that, the primary goal of action research is the solution to a given problem, not contribution to science. Whether the research is conducted in one area or many areas, the researcher(s) is very much a part of the process. The more the number of times the researcher conducts the action research, the more likely it will produce valid results. However, the major limitation of this action research is that, the results cannot be generalized to other areas. Further, this research studies often lack both internal validity and external validity and generally are not useful for making policy decisions. Despite its shortcomings, it does represents a scientific approach to the problem-solving that is infinitely better than no changes at all.

ii. ***Impact Assessment Research*:** In this research, the researcher highlights the benefits realized through the execution of technological interventions in the project in drawing comparison with the earlier situation. That is, the researcher compares two situations *viz.*, with project situation and without project situation. So, he highlights the consequences due to interventions made through the execution of the project. Such an assessment is essential to suggest various alternative courses of action to address similar problems elsewhere. For example, if a researcher wishes to study the impact of 'Debt Waiver and Debt Relief Scheme' (introduced by the Government of India in 2008-09) implemented by the financial institutions on the indebtedness of farmers and their repayment pattern, it implies Impact assessment research.

iii. ***Evaluation Research*:** This research provides answers to the questions like 'Did the project executed efficiently', 'Did the project realized its goals/objectives', 'Did the project yielded significant results to the beneficiaries' etc. So, in evaluation research, the researcher aims at judging or measuring the project from the point of view of its effectiveness in fulfilling the framed objectives. There are two types of evaluation research *viz.*, formative evaluation and summative evaluation.

- ***Formative or Concurrent evaluation:*** This evaluation is done based on continuous feedback from the project.
- ***Summative evaluation:*** This evaluation is done only after the completion of the project.

According to Hunt, *'applied research is an investigation for ways of using scientific knowledge to solve practical problems'*. The other examples of applied research include: improved production technology to boost crop production, pest management of agricultural crops, dissemination of market information to farmers etc. Most of the experimental research, case studies and inter-disciplinary research are essentially applied research.

Basic Research vis-a-vis Applied Research

It is to be noted that, the procedures and techniques utilized by basic and applied researchers do not differ substantially. Both these methods employ the scientific method to answer the questions at hand. However, there exists important differences between these two and they are discussed here under.

- ☆ Basic research is not problem-oriented, but for the increase in knowledge which may or may not be used in the near future. But, applied research uses basic research, theories, knowledge and methods for solving an existing problem. So, it is problem-oriented. Thus, basic research is academic without paying attention to any practical applications of the real world, whereas applied research aims at practical use in solving researchable problems.

- Basic research is directed towards finding information that has broad base of applications and thus, add new information to the already existing scientific knowledge. But, the applied research is directed towards finding a solution for a practical problem, which warrants solution for immediate use.
- Both basic research and applied research are important to the advancement of human knowledge. Basic research is conducted solely for the purpose of gathering information and building on existing knowledge. But, applied research is geared towards offering solution for a particular problem under consideration. For example, a biologist, who studies about the morphology of plants and animals is doing basic research, while a plant pathologist who is searching for the cause of plant disease is said to be involved in applied research. Similarly, while discovery of laser was through basic research, its subsequent use in surgery is a result of applied research.
- Basic research forms the basis for applied research. So, applied research builds on existing basic research. Thus, basic research could be considered as the foundation of new knowledge. This basic knowledge will be used by the researcher towards solving a particular problem. In the above example, the plant pathologist involved in applied research on the cause of a plant disease will probably use the research findings published by the biologist, who is more generally studying about both plants and animals.
- In basic research, general theories, ideas, and questions are explored and tested. These concepts will be employed by the applied researchers and hence, they don't need to reinvent the wheel every time they start on a new project, because the groundwork has already been done.
- Basic research doesn't always result in an immediate benefit to humans, unlike applied research.
- Basic research is comparatively more costlier and consumes more time compared to applied research.
- Unlike basic research, applied research aims to address and answer real-world problems.
- Basic research has no application on real life, whereas applied research attempts to influence the real world.
- Basic research seeks generalization, whereas applied research studies individual or specific cases without the objective to generalize.
- Basic research aims at basic processes, whereas applied research aims at analyzing cause and effect relationship.
- Basic research attempts to explain *'why things happen'*, whereas applied research tries to say *'how things can be changed, tries to get all the facts, tries to correct the facts which are problematic etc'*.
- Basic research reports will be in technical language of the topic, whereas applied research reports will be in common language.
- Basic research involve discoveries inventions etc., whereas applied research involve innovations, experiments etc.
- Basic research is extensive in nature, whereas applied research is intensive in nature.

- Basic research is research that has no immediate application, whereas applied research is research that does.
- Basic research is driven by a scientist's curiosity or interest in a scientific question. The main motivation is to expand man's knowledge, not to create or invent something. However, applied research is designed to solve practical problems of the modern world, rather than to acquire knowledge for knowledge's sake.
- Basic research is the source of most new theories, principles and ideas. It rarely helps anyone directly. It only stimulates new ways of thinking. There is absolutely no commercial value to the discoveries resulting from such research. However, applied research uses the principles or theories derived from basic research to practical situations and aims at solving researchable problems. Thus, basic research mainly aims at theory building, whereas applied research mainly aims at problem solving. Basic research is more academic and its main aim is to apply this enhanced knowledge in solving practical problems.
- For basic research, the primary consumer is the scientific community, whereas the consumers of applied research include some other than the researchers like professors, managers, committee officials etc. So, the findings of the applied research have to be translated from scientific technical language into the language of decision makers or practitioners.
- The results of applied research are less likely to enter the public domain in publications, unlike the findings of basic research. The findings of the applied research are used only by a small number of decision makers, who wish to put them into practice in practical environment.
- Compared to applied research, basic research emphasize high standards and techniques to try to conduct near-perfect research. Applied research simply aims at getting quick and usable results.
- Compromise is no way in basic research, whereas the applied researcher has to make some comprises in the selected practical environment in conducting the research.

Some scientists opine that, it is high time now to focus more on applied research rather than basic research considering the problems resulting from global overpopulation, pollution, the overuse of the earth's natural resources, ecological imbalances etc. However, the distinctions between basic research and applied research are somewhat ambiguous, as almost all basic research eventually results in some worthwhile application in the long run.

c. Practical Research

This is one type of applied research that goes one step further and applies the findings of research to a specific 'practical' situation. For example, developing a drought resistant jowar variety suiting to dry land tracts of Mahboobnagar district of Andhra Pradesh, economic analysis of consumption pattern of Below Poverty Line (BPL) consumers in tribal tracts of Visakhapatnam district of Andhra Pradesh etc., fall under practical research. The results emanated from this practical research guides in formulating policies for the betterment of people living in these areas. The word 'practical' is perhaps the most significant word in the title. The practical approach means applying the known theories or principles at the grass root level, so as to trace out the solutions to the problem under consideration.

A useful way to look at the relationships among these three research types is illustrated through Figure 2.2. Each of the three different types of research contributes to the other in helping revise and frame the research from each category. For example, practical research may be based on theory that

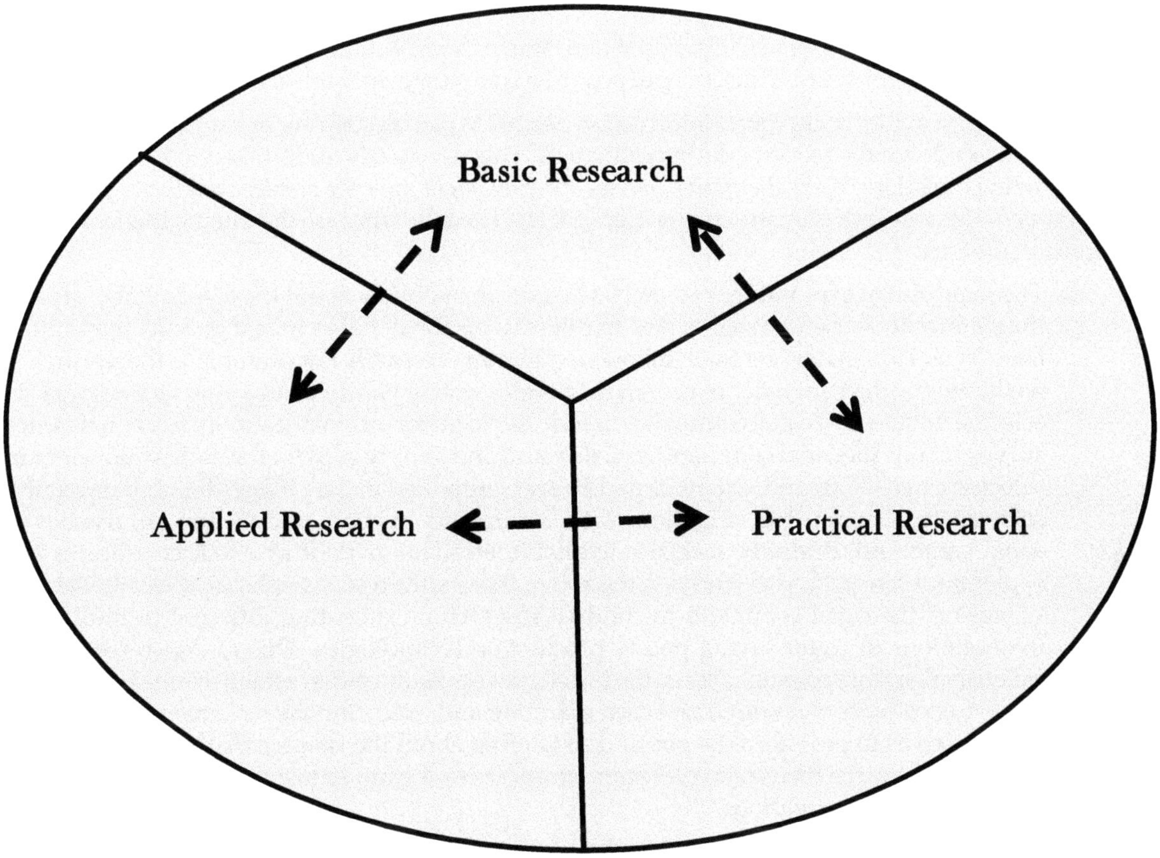

Figure 2.2: Relationships between Basic, Applied and Practical Researches.

came from previously done basic research. Or, theory may be generated by the combination of results from various practical research projects. The same bi-directional relationship exists between applied research and basic research or practical research.

B. From the Viewpoint of Objectives

According to this criterion, the research can be classified as:

a. Exploratory Research

It is also called as Preliminary research or Formulative research. This research is undertaken with a view to 'explore' a researchable area, about which nothing or very little is known or when there is no information on how similar researchable problem have been solved in the past. That is, this research is conducted, when the researcher does not know how and why certain phenomenon occurs or when a problem has not been clearly defined as yet, or its real scope is as yet unclear. In such cases, an extensive preliminary work needs to be done to gain familiarity with the phenomenon in the given situation and understand, identify, prioritize and select the researchable issue for detailed or in depth investigation. Here, the hypothetical solutions or actions are explored and evaluated by the researcher. For example, this research is employed in evaluating the quality of services offered by a bank or hotel or tourism or airline etc. This is because, the benefits offered by these services are intangible and hence, quality of service cannot be assessed directly. Some research projects are entirely exploratory, even though they can go on for years. Thus, the purpose of exploratory research is to know the unknown.

This research provides background information needed to plan descriptive or explanatory research. This also narrows down the scope of the investigation. This research is undertaken to get the answer to the following question: *'What alternative courses of action might solve the problem and thereby reach the final objective?'* This research is unstructured in character. The following are the interesting features of exploratory research:

- ☆ The main aim of exploratory research is to gain an overview about the researchable area so that, it provides useful insights and familiarity for the detailed investigations that follow later. Thus, explanatory research follows exploratory research. For example, if the researcher wishes to conduct an in depth research on water saving paddy production technologies in selected villages of a canal command, first he has to go for an overall survey (reconnaissance survey) about the canal command, *i.e.*, demand and supply gap of canal irrigation water in selected canal command, extent of paddy area cultivated in the villages that fall under the selected canal command, irrigation water availability at head, middle and tail reaches of canal command, drainage facilities available etc. This part of research constitutes the exploratory research. After studying the above aspects, the researcher has to select the sample villages in the canal command to conduct research on executing different technological interventions of water saving paddy production technologies. This in depth research is called explanatory research. Thus, the initial survey conducted to establish areas of concern refers to exploratory research and then planning and executing the research in an in depth manner, so as to provide a deeper understanding about the issues constitute explanatory research. Thus, the focus of exploratory research is on gaining insights and familiarity for later in depth investigations.
- ☆ Exploratory research is conducted to clarify ambiguous problems. This research is generally taken up when a problem is in a preliminary stage of investigation *i.e.*, this type of research is conducted for a problem that has not been clearly defined.
- ☆ The purpose of exploratory research is to define the problem precisely, collect requisite information/data relating to the problem and identify alternative courses of action to address the problem. For example, advertising campaign of a pesticide company has not yielded positive results in terms of increase in sales. Then the company will try to explore the possible reasons relating to media selected, wrong timing of advertising in the media, faulty illustrations, poor performance of the actors acted in the advertisement etc. Thus, in the exploratory research, the possible causes or reasons will be identified, hypothesis will be developed and research activity will be conducted accordingly.
- ☆ The main objective of exploratory research is to gather preliminary information about the researchable problem that will help to identify specific problems and suggest hypotheses. Thus, the principal use of exploratory research is to increase a researcher's understanding of a subject. It will not provide a conclusive evidence to determine a particular course of action for the researchable problem. So, exploratory research is conducted with the expectation that, subsequent research will be required to provide conclusive evidence.
- ☆ In exploratory research, the main focus of the researcher is to explore or discover or search for the ideas/causes. For example, for a pesticide company, the sales have been declining for

the past one year. So, the market researcher of the company conducts a quick survey to elicit the causes or reasons for decline in sales of a pesticide. After prioritizing the causes, relevant strategies will be planned.

- ☆ Another important reason for conducting exploratory research is to analyze the background area for conducting in depth explanatory research in the near future. This is essential because, the technological interventions to be planned by the researcher under explanatory research involve huge costs. So, it is a serious mistake to rush into detailed surveys or investigations or in depth studies without having any preliminary idea about the seriousness of the researchable problem in the study area. Thus, exploratory research helps to crystallize a problem and identify information needs for future research.
- ☆ It is to be noted that, when the purpose of research is to gain familiarity with a researchable issue and to acquire an in depth insight into it by identifying and formulating prioritized issue, the exploratory research come in handy. If the theory happens to be too general or too specific, a hypothesis cannot to be formulated. However, to formulate a hypothesis for the prioritized problem under consideration, it calls for an exploratory research for more definite investigation.
- ☆ As mentioned earlier, exploratory research is conducted on a research problem or issue, when there are very few or no earlier studies pertaining to the selected problem. That is, the researcher had no adequate prior information about the selected problem. So, the aim of this type of research is to look for patterns, ideas or hypotheses, rather than testing or confirming a hypothesis. The quickest and the cheapest way to formulate a hypothesis in exploratory research is by using any of the following four methods:
  - ❖ ***Literature search:*** Here, the researcher develops a hypothesis by referring to literature and secondary data in journals, reports, statistical publications etc. For example, the problem is *'if sales of a pesticide firm is declining'*, it guides the researcher know whether it is an 'industry problem' or 'individual firm's problem'. So, the following are the possibilities to formulate an hypothesis:
    - ❑ The firm's market share has declined, but industry's figures are normal.
    - ❑ The industry is declining and hence the firm's market share is also declining.
    - ❑ The industry's share is going up, but the firm's share is declining.

    From the above hypotheses, if third one is accepted, the researcher has to focus on analyzing the marketing mix variables to improve the sales of the product.
  - ❖ ***Experience surveys:*** In experience surveys, the researcher interacts with the people, who are well informed in the area to be investigated. For example, a food processing company conduct survey among housewives for their opinion about a 'Ready To Eat (RTE)' food. Similarly, if a pesticide firm want to find out the reasons for poor sales of its pesticide introduced recently, the firm conduct interviews with the farmers, scientific community in research stations, Agricultural Officers etc.

    It is to be noted that, no structured schedules will be used in collecting the information from the respondents so that, the respondents can give divergent views. Since the idea of using experience survey is to undertake problem formulation, and not conclusion, probability sample need not be used. Those who cannot speak freely with the researcher should be excluded from the sample.

- ***Focus group (Pilot studies):*** Here, the firm formulates a hypothesis by constituting and drawing conclusions from a number of focus groups. Each focus group is a small number of individuals (8-12 persons) that are brought together to study and talk about some topic of interest. The discussion is coordinated by a moderator. While selecting these persons, care has to be taken to see that, they should have a common background about the topic. This is required because, there should not be a conflict among the group members on the common issues that are being discussed. Firm also takes care to avoid groups, in which some of the participants have their friends and relatives, because this leads to a biased discussion. Normally there are 4-5 groups to discuss on a particular topic say, *'how to boost the sales a pesticide in the given market segment'*. The guiding criteria is to see whether the latter groups are generating additional ideas or repeating the same with respect to the subject under study. When this shows a diminishing return from each group, the discussions will be stopped. The typical focus group lasts for 1 to 2 hours.
- ***Analysis of case studies:*** Analysis of a selected case of a particular firm, which have undergone a similar situation may be available and this guides the present researcher to get an insight into the similar problem, which is being researched. However, the results of investigation of case histories are always considered suggestive, rather than conclusive. For example, if a firm wishes to release a new pesticide into the market, it often critically analyzes the case histories available in the form of previous studies made by its competitors and thereby, designs its pricing, advertising and sales strategies.

- Typical techniques used in exploratory research include case studies, observation and historical analysis, which can provide both quantitative and qualitative data. These techniques are very flexible, as there are few constraints on the type of data collected.
- This research will simply assess, which existing theories and concepts can be applied to the selected problem or whether new ones should be developed. So, this research rarely provides conclusive answers to selected problem, but gives guidance on what future research, if any, should be conducted.
- The results of exploratory research are not directly relevant for formulating need based technological interventions, but guides the researcher to identify and prioritize various researchable issues for in depth analysis in a given agro-ecological situation.
- This research is essential in a given agro-ecological situation, where very few studies might have been conducted in that area. Extensive surveys with farmers and relevant stakeholders will be undertaken besides collecting useful secondary data, so as to handle on the given situation and understand the phenomena. This further facilitates more rigorous (explanatory) research in the future.

Strengths of Exploratory Research

They include:

- It is the starting point in all types of research projects and provides guidance for future research.
- The findings discovered through exploratory research enable the researchers to learn more about the same findings in-detail in the subsequent explanatory studies.
- This research provides useful background information on a particular topic.

- Generate new theories, new ideas, hypotheses etc.
- It helps to determine the feasibility of conducting an in depth study in that area with reference to the selected researchable issue.
- Guides to formulate questions and refine issues for more systematic inquiry.
- Provide techniques and a direction for future research.
- It helps in exploring new ideas. It also helps to determine whether to proceed with a research idea and how to approach it.
- This research helps in gathering broad information about an issue (say, irrigation water management) and this forms the basis to take up a number of researchable issues (say, executing technological interventions like micro irrigation, sub-surface drainage system, gypsum application, water saving crop production technologies etc.).
- This research also helps in knowing the feasibility in attempting an in depth study.
- This research ensures generation of new ideas and assumption, development of tentative theories or hypotheses.
- The execution of exploratory research ensures for more systematic investigation of researchable issues in the study area and this guides to address new research questions. Thus, this research provides direction for future research investigations in the study area and guides in planning need based technological interventions.
- This research is more flexible and can address research questions of all types (WHAT, WHY, HOW).
- This research helps to determine the best research design, data collection methods, identification, prioritization and selection of researchable subjects, selection of technological interventions to address the problem etc.
- This research provides an opportunity to define new terms and clarify existing concepts.
- This research helps in prioritizing researchable problems in the study area.
- This research allows the researcher to familiarize himself with the researchable problem to be studied in depth in the future (through explanatory research). That is, exploratory research is the initial research, before explanatory research is to be taken up. However, it is to be noted that, sometimes the exploratory research may also conclude that, the problem does not exist at the significant level and hence, it is not practically necessary to go for explanatory research.
- Conducting exploratory research can save a great deal of time and money.
- In general, exploratory research is meaningful in any situation in which the researcher does not have enough understanding to proceed with the research project.

Limitations of Exploratory Research

They include:

- The results of exploratory research cannot be generalized to the population at large.
- This research cannot arrive at making definitive conclusions about the findings, as it simply explains the overview about the researchable problem under consideration and due to lack of statistical strength. However, the explanatory research taken up with the background of exploratory research will yield definite findings.

- The exploratory research will yield only tentative results that have limited value in decision-making.
- This research design collects data both from primary and secondary sources, but lacks standard methods of collection and analysis of data, as it simply overviews the researchable problem, unlike explanatory research.
- The results of exploratory research are not usually useful for policy-making by themselves, but they can provide significant insight into a given situation. So, the results of exploratory research cannot be generalized, as they are not representative of the whole population being studied.
- Exploratory research rarely yields definitive answers. It addresses only 'WHAT' question in the problem. For example, this research mainly aims at studying *'what are the problems faced by the farmers pertaining to irrigation water management in Kurnool Cuddapah Canal Command in Kurnool district of Andhra Pradesh'*. Moreover, this research is difficult to conduct because, there are few guidelines to follow.
- Data collected is mostly qualitative.
- Rarely, it can provide conclusive answers to a problem.

b. Descriptive Research

The following are the important characteristics of descriptive research:

- Descriptive research is also known as Statistical research. This research deals with the question 'WHAT IS' of a situation. However, it does not answer questions about 'HOW', 'WHEN', 'WHY' the characteristics occurred, which is done under Explanatory or Analytical research. So, this research describes the phenomena as they exist. The purpose of descriptive research is to observe, describe and document the aspects of a situation as it naturally occurs (Polit and Hungler 1999). The emphasis is to 'describe' rather than to judge or interpret. But, good description of a problem provokes the 'HOW', 'WHEN', 'WHY' questions of explanatory research. For example, a descriptive research explains, *'over the past 25 years, the lending from the financial institutions are more biased towards large farmers compared to small farmers'*. This descriptive analysis, force the researchers to ask `Why is this happening?'
- Descriptive research goes further in examining the selected problem than exploratory research, as it is undertaken to ascertain and describe the characteristics of the problem. This research is followed by one of the two types of research *viz.*, explanatory or predictive, as the researcher is required to go beyond description of the problem to analyze or/and predict the events.
- Descriptive studies are normally factual and simple. However, some studies can be complex and they demand scientific skills on the part of researcher. Compared to exploratory research, descriptive research is well structured. It tends to be rigid (unlike exploratory research which is flexible) and its approach cannot be changed often and again.
- The data collected in descriptive research are often quantitative and requisite statistical techniques are employed to explain or interpret the findings.
- A major difference between exploratory and descriptive research is that, descriptive research is characterized by the prior formulation of the hypotheses. Thus, the information needed is clearly defined. As a result, descriptive research is preplanned and structured. It is typically based on large representative samples. Data collected in exploratory research is mostly

qualitative, whereas data collected in descriptive research is often quantitative and statistical techniques are used to summarize the information.

- In descriptive studies, the researcher has to give adequate thought for framing research questions and deciding the data to be collected and the procedure to be used for this purpose. Descriptive research cannot identify cause and effect relationship. It is designed to describe the present situation or the features of a group or users of a product. In marketing, such research is undertaken to know the characteristics of certain groups or users of a product such as age, sex, education, income etc. Such research studies are based on secondary data or survey research.
- Although descriptive research is factual, accurate and systematic, it cannot describe what caused a situation. Thus, descriptive research cannot be used to create a *causal relationship*, where one variable affects another. In other words, descriptive research can be said to have a low requirement for internal validity. Since descriptive research does not determine cause and effect, it cannot make accurate predictions. That is, researcher has no control over the variables and hence, he can only report what has happened or what is happening.
- Descriptive research generally precedes explanatory research. For example, over time Chemists have described the elements through the periodic table. The periodic table's description of the elements allows researchers to produce new compounds by mixing different elements.
- It concerns with determining the current practices, status or features of situations. The idea behind this type of research is to study frequencies, averages and other statistical calculations.
- Descriptive research is conducted to describe the characteristics of relevant groups, such as farmers, consumers, salesmen etc. For example, the researcher can study the perceptions of the farmers about a pesticide or fertilizer released into the market. This research can also be used to estimate the percentage of sample units in a given population exhibiting a certain behaviour. For example, this research guides to study about the percentage of total number of farmers in an area practicing organic farming and their perceptions on organic agriculture. Similarly, this research also guides to study about the perceptions of consumers about a particular product of a firm. For example, the researcher can find out the preferences of consumers of Hazard Analysis and Critical Control Point (HACCP) labeled products over non-labeled products. A close examination of these examples reveals that, the objective of this research is to describe things as they are. The other examples include:
    - Do farmers have favorable attitudes toward using bio-fertilizers in their crop production programme?
    - How have B.Sc. (Agriculture) courses changed over the last 20 years?
    - What is the best way to provide access to computer equipment in SAUs?
    - Is there a relationship between number of trainings conducted by the Department of Agriculture and increase in managerial skills of farmers?
    - Do farmers actually implement technology in the way they perceive?
    - What is the association between student-professor ratios and student achievement in the colleges of SAUs?
    - What are the qualifications of employees in Agricultural College, Mahanandi of Acharya N.G. Ranga Agricultural University (ANGRAU)?

- What type of packaging for a box of chocolates do consumers prefer?
- What information do consumers prefer to have on food labels?
- What are the sources of supply of irrigation water in Kurnool district of Andhra Pradesh?
- Market studies that describe the size of the market, buying power of the consumers, availability of distributors and consumer profiles.
- Market share studies that determine the proportion of total sales received by a fertilizer firm and its competitors.
- Pricing studies that describe the range and frequency of price changes and probable consumer response to proposed price changes.
- Surveys like population census and educational census.
- A marketing manager plans to formulate a marketing strategy for increasing the sales of pesticides in the given market segment. Before that, he will study the marketing strategies followed by his competitors in that segment *viz.*, prices followed, advertising expenditure, discounts offered, marketing channels etc.
- This research can be used to forecast the market trends of a product *i.e.*, possible market sales of a product after 'n' years.
- It guides to discover the relationship between certain variables. For example, sale of RTE foods among rural population and urban population.

☆ Data collection is either done through asking questions from the selected respondents (through schedules or questionnaires or interviews) or by observation. The data collected are often quantitative, and statistical techniques are usually used to summarize the information.

☆ Existing theories can be easily put under test by empirical observations.

☆ It is not directed by hypothesis. Descriptive research provides 'systematic information' about a social phenomena. The researcher does not begin with hypotheses, but is likely to develop hypotheses after collecting data. Systematic information means careful selection of the respondents and careful measurement of each variable. For example, Indian Medical Association conducts a study on 'Tobacco Use among High School Students' and is a descriptive study. The Association can also suggest policy guidelines to control the tobacco use among the children. So, descriptive research is also a policy research. Similarly, a review paper of previously reported research is also descriptive research. Often new ideas and theories are discovered and presented from this descriptive research.

☆ Although some people dismiss descriptive research as 'mere description', good description is fundamental to the researcher and it adds immeasurably to enhance knowledge about the problem under consideration. Generally, descriptive research encompasses much Government sponsored research such as population census, the collection of a wide range of social indicators, economic indicators such as household expenditure pattern, employment statistics, crime statistics etc.

☆ Descriptive research goes further than exploratory research in examining a problem, since it is undertaken to describe the characteristics of the issue. A research investigation may include descriptive research, but it is likely that, it will also include one of the subsequent two types (explanatory or predictive) as they are required in to go beyond description and to explain or predict.

- ☆ According to Klopper, researchers who use descriptive method for their research usually aim at two aspects:
  - ❖ Selection of representative sample of the population (Test sampling), and
  - ❖ Recording in the form of a written report.
- ☆ Descriptive research can be either quantitative or qualitative. It can involve collection of quantitative information that can be tabulated in numerical form, such as scores on a test or the number of times a farmer chooses to use a certain technology or it can simply describe a information as high or medium or low (say, for example, improvement in managerial ability of farmers practicing mango cultivation after rendering training to them) or described by parameters like their central tendency (mean, median and mode) and their dispersion (range, standard deviation) or even studies the relationship between the variables in terms of correlation coefficient etc. So, the idea behind this type of research is to study frequencies, averages, and other simple statistical calculations. This research often uses visual aids such as graphs and charts to aid the readers in understanding the data distribution. This pictorial representation of data reduces the data to manageable form, as the human mind cannot extract the full import of a large mass of raw data.
- ☆ Descriptive research is unique in the number of variables employed. Like other types of research, descriptive research can include multiple variables for analysis (such as correlation, multiple regression etc), yet unlike other methods, it requires only one variable (Borg and Gall, 1989). For example, descriptive research might simply report the percentage summary on a single variable say, the tally of an Author's citation in different journals publishing articles on Agricultural Marketing'.
- ☆ Since descriptive research is marked by the prior formulation of specific hypotheses, the design requires a clear specification of the six W's of the research. They include:
  - ❖ ***Who:*** who should be considered?
  - ❖ ***Where:*** where should the respondents be contacted to obtain the required information?
  - ❖ ***When:*** when should the information be obtained from the respondents?
  - ❖ ***What:*** what information should be obtained from the respondents?
  - ❖ ***Why:*** why are we obtaining information from the respondents?
  - ❖ ***Way:*** the way in which we are going to obtain information from the respondents.

These questions form the basis for describing the research to be conducted.

- ☆ Descriptive research may be Simple descriptive or Comparative descriptive
  - ❖ ***Simple descriptive research***: This is used, when data are collected to describe persons or organizations or phenomena. For example, a researcher simply describes 'How have B. Sc (Agriculture) courses changed over the last 20 years in the SAUs in the country?'
  - ❖ ***Comparative descriptive research:*** In this case, the researcher describes two or more groups of participants or a same phenomenon in two different situations. For example, a researcher administers a questionnaire to different age groups of professors in three different colleges about their classroom teaching practices. The researcher chooses different age-group professors from different colleges, as both experience and academic environment in the colleges influences their teaching practices and they can be compared.

- ☆ There are three main types of gathering information in descriptive research *viz.*, observational methods, case-study methods and survey methods.
  - ❖ ***Observational method:*** With this method (sometimes referred to as Field observation), human behaviour is closely observed. There are two main categories of the observational method *viz.*, naturalistic observation and laboratory observation. In the naturalistic method, researchers view participants in their natural environments. This leads to greater ecological validity (refers to the extent to which research can be used in real-life situations) than laboratory observation. But, some proponents of laboratory observation method opines that, due to more control in the laboratory, the results found when using laboratory observation are more meaningful than those obtained with naturalistic observation. But, laboratory observations are usually less time-consuming and cheaper than naturalistic observations. Of course, both naturalistic and laboratory observation are important in regard to the advancement of scientific knowledge.
  - ❖ ***Case study method:*** This method involves selection of a unique case purposefully and helps to go for an in-depth study of a rare phenomenon. Note that, these case studies should not be used to determine cause and effect relationship, and hence, they have limited use for making accurate predictions.
  - ❖ ***Survey Method:*** In this method, the researcher selects representative sample and elicit information from them through schedules or questionnaires etc. Later, the researcher compiles the data or information and portrays the results. In order for the survey to be both reliable and valid, it is important that the questions are constructed properly. Questions should be written so they are clear and easy to comprehend.

In addition to the above methods, as discussed earlier, qualitative method is also included in conducting descriptive research.

- ☆ In descriptive research, the following steps should be included:
  - ❖ Problem selection and problem formulation.
  - ❖ Intensive literature search regarding the formulated problem enables the researcher to divide the problem into smaller units.
  - ❖ Hypothesis formulation.
  - ❖ Preparation of schedules or questionnaires for gathering the information
  - ❖ Determination of sample size.
  - ❖ Design of procedure for information collection.
  - ❖ Collection of information.
  - ❖ Analysis of information.
  - ❖ Generalizations and/or predictions.
  - ❖ Final report writing. The report entails the reproduction of factual information, the interpretation of data, conclusions derived from the research and policy recommendations.

Strengths of Descriptive Research

They include:

- ☆ Descriptive research is mainly done when a researcher wants to gain a better understanding of a topic.

- Descriptive research presents an opportunity to fuse both quantitative and qualitative data as a means to reconstruct the "WHAT IS" of a problem under consideration.
- Descriptive research provides a multifaceted approach for data collection. Though this research involves collection of information or data through schedules or questionnaires, it also includes an observational technique for data collection. Because it is so varied in method and usage, researchers favor its use in many different fields including marketing, medical health, science and psychology. A great deal of social research is also descriptive.
- Descriptive studies can yield rich data that lead to important recommendations.
- This research provide an accurate profile of a study group (say, farmers or consumers etc)
- Gives both verbal and numerical picture of the problem under consideration.

Limitations of Descriptive Research

This type of research suffers from the following limitations:

- Confidentiality is the primary weakness of descriptive research. The respondents often do not provide realistic information and they provide such information what they think the researcher wants to hear.
- When the researcher designs a questionnaire, questions are pre-determined and prescriptive. That is, descriptive research can sometimes be skewed to fit the needs of the researcher. For example, if the researcher wants to compare between two products, he might give one product an unfair advantage to get the desired results. Furthermore, the researcher may record what he wants to hear and ignore data that does not conform to the research project's hypothesis. Overcoming a research bias is an extreme difficulty for descriptive research practitioners.
- The results from a descriptive research cannot be used to discover a definitive answer or to disprove a hypothesis.
- Because descriptive designs often utilize observational methods, the results cannot be replicated.
- Although this research is highly accurate, it does not gather the causes behind a situation.

Exploratory Research vs. Descriptive Research

Both these researches include 'non-experimental' studies. That is, both these studies aim at 'observing' their respective research problems. However, the following are the important differences between these two:

- In exploratory research, the researcher simply 'explore' the researchable issue under consideration. That is, the researcher will simply observe the overview of the problem with a view to prioritize the problems relevant to the researchable issue under consideration.
- In descriptive research, the researcher attempts to 'describe' the selected research problem. That is, he will document the characteristics of the problem, selected respondents, sources and tools of collection of data etc.
- Exploratory research is more flexible when implemented than is descriptive research.
- Exploratory research will aim at finding relationship between the variables, whereas descriptive research simply describes the existing situation.
- Exploratory research addresses more open-ended questions than does descriptive research.

- Descriptive research is used more to arrive at statistical tools such as mean, average, median and frequency. On the other hand, exploratory research allows the researcher to develop designs that are more qualitative in nature.
- The amount of information known to the researcher at the start of the research plays an important role in deciding upon the type of research. With only vague ideas in the mind of the researcher, it is better to go for exploratory design. On the other hand, more information such as quantitative data allows a researcher to go for descriptive research that leads to unearthing causal relationships.
- Exploratory research needs to be conducted first to have a platform that allows for collection of data required in descriptive research.

c. Explanatory Research or Analytical Research

This research is a continuation of descriptive research or exploratory research. In this research, the researcher goes beyond both descriptive research or exploratory research, so as to analyze and explain 'WHY' or 'HOW' the phenomenon being studied is happening. Thus, explanatory research aims to understand the cause of the research phenomena and measure causal relations among the variables involved in the problem. So, explanatory research attempts to go beyond what exploratory and descriptive research generates about the research problem. For example, a descriptive research explains, MBBS students have more communication skills compared to under B.Sc (Agriculture) students in SAUs. An explanatory researcher would then be interested in the reasons behind this fact. Similarly, price and quantity demanded relationship, advertisement expenditure and sales performance relationship etc., can be studied through explanatory research. So, the relationships between the casual factors can be studied through this research. In addition, the variables which create effect on other variables can also be studied in depth through this research. Data for this research can be collected through field survey with the help of a questionnaire or schedule or by conducting laboratory experiments. For example, in case of seed testing or testing of new products, laboratory experiments are conducted. This explanatory research also attempts to build and elaborate on theories and add to predictions and principles where possible. This is possible by verifying or testing the causal relations among the different variables and offering relevant explanations.

The way in which researchers develop research designs is fundamentally affected by whether the research problem is descriptive or explanatory. So, the type of information to be collected in explanatory research differs from descriptive research. For example, if the researcher wishes to study the '*marketing channels for transacting paddy in Kurnool district of Andhra Pradesh*', it falls under descriptive study. However, if the researcher wishes to analyze the causal factors for the rise in marketing costs and margins in transacting paddy in Kurnool district of Andhra Pradesh over the past five years, it implies explanatory research. Similarly, if the researcher wishes to study the '*research and extension gaps in paddy cultivation in Kurnool district of Andhra Pradesh*', it falls under descriptive study. However, if the researcher wishes to analyze the causal factors for these gaps in Kurnool district of Andhra Pradesh, it implies explanatory research. Thus, explanatory research involves the study of answering the `why' questions in developing causal explanations. It helps to obtain information about the link between the causes and results of the evidence. It is to be noted that, some causal explanations will be simple, while others will be more complex. The other examples include: *'how an ice cream firm improves the delivery time of its products?', 'how a tractor manufacturing company can improve offering the services to its customers (farmers)?, how a firm try to improve the sales of a pesticide by increasing the advertising expenditure? (here, the pesticide firm want to study the causal relation between advertising and sales)*. The hypothesis that will be tested here is: 'If advertising is increased, then sales will also go up'.

Sub-divisions of Analytical Research

They include:

- Historical research.
- Philosophical research.
- Review.
- Research synthesis (meta analysis *i.e.*, analysis of the review that already published).

Strengths of Explanatory Research

They include:

- Explanatory research goes in depth into the research problems and attempts to build and elaborate on theories and add to predictions and principles where possible.
- This research is problem-solving and thus aims at executing the need based technological interventions at the micro or farmers' level, so as to address the prioritized research problems.
- An important element of explanatory research is identifying and, possibly, controlling the variables in the research activities. This facilitates to analyze the causal links between the variables.
- This research is so designed to achieve predictable results and thus, aims at improving the quality of life for people.
- This research explain things ('WHY') and not just reporting. This enriches the theory's explanation.
- It also helps to compare several explanations about the problem and to identify which analysis is the best way to explain it.
- Extend a theory or principle to new areas of research.

Exploratory Research vs. Explanatory Research

Following are the important differences between exploratory research and explanatory research:

- The basic purpose of exploratory research is to provide broad information that guides to understand, formulate and prioritize various researchable problems in the study area. However, explanatory research goes in depth into the specific problem and traces out cause and effect relationship between the variables.
- Explanatory research is typically more formal and structured than exploratory research. It is based on large representative samples, and the data obtained are subjected to quantitative analysis.
- In exploratory research, the researcher simply 'explore' the researchable issue under consideration. So, exploratory research is 'non-experimental' or 'observational study'. The researcher will not manipulate any of the variables. But, in explanatory research, the primary goal is to understand the nature of the relationship between the independent and dependent variables and to analyze cause and effect relationship.

Descriptive Research vs. Explanatory Research

Both descriptive research and explanatory research differ in terms of the following aspects:

- Descriptive research emphasize on 'what is going on', whereas explanatory research emphasize on 'why is it going on' with reference to the research problem under consideration.

☆ The descriptive research studies associations between the variables, whereas explanatory research studies cause and effect relationship among the variables (Table 2.1).

Table 2.1: Differences between Exploratory, Descriptive and Explanatory (Analytical) Researches.

Item	*Exploratory Research*	*Descriptive Research*	*Explanatory Research*
Purpose	Identify and prioritize the problems and gain insights	Describe things as they are	Determine cause-and-effect relationships between the variables identified
Assumed background knowledge	Minimal	Considerable	Considerable
Research design	Little structured	Highly structured	Highly structured
Flexibility in conducting research	Highly flexible	Moderately flexible	Less flexible
Sample selected for the study	Need not be representative of the population	Sample should be representative of the target population	Sample should be representative of the target population
Cost of conducting research	Low	Medium	High
Findings	Preliminary and forms the basis for descriptive and explanatory researches	Conclusive and even form the basis for explanatory research	Conclusive
Degree of Problem Definition	Key variables not defined	Key variables are defined	Key variables and key relationships are defined
Starts with hypothesis	No	No	Yes
Systematic data collection	No	Yes	Yes
Designed for decision-making	No	Yes	Yes
Example	Reconnaissance survey on demand and supply gap of canal irrigation water in KWD of Andhra Pradesh	Identification of marketing channels for transacting paddy in Kurnool district of Andhra Pradesh	Resource use pattern and efficiency in paddy production in Kurnool district of Andhra Pradesh

☆ In descriptive research, hypotheses are speculative and typically non-directional, whereas in explanatory research, they are specific and directional.

☆ The differences between descriptive research and analytical research can be easily understood from the following examples:

- ❖ Studying the changes in Indian agricultural trade balance (difference between agricultural exports and agricultural imports) during Post-WTO regime (1995-2012) is an example of descriptive research; while explaining '*why*' trade balance move in a particular way over selected reference period is an example of analytical research.
- ❖ If the researcher studies the magnitude of rise in value of US$ against India Rupee during 1995 to 2012, it is an example of descriptive research. On the other hand, if the researcher explains the reasons, 'why', this surge in the value of US$ is occurring, it is a case of explanatory research.

- Both descriptive research and explanatory research starts with the structure already defined and proceeds to actual data collection. Descriptive research simply describes the variables (say, average age of farmers purchasing a pesticide), whereas explanatory research proceeds still beyond descriptive research to infer causal relationships between the variables (say, to study the relationship between advertising expenditure and sales of a pesticide of a firm).
- Explanatory research is appropriate to use, when the objective is to understand which variables are the cause and which variable(s) are the effect, and to determine the extent of the functional relationship between the causal variables and the effect to be predicted. However, in descriptive research, only the characteristics of the variables are simply described and causal effect is analyzed.

d. Predictive Research

Predictive research goes even further than explanatory research. The latter establishes an explanation for what is happening in a particular situation, whereas the former forecasts the likelihood of a similar situation occurring elsewhere. So, predictive research builds on the explanatory research to forecast the future for a similar situation. Thus, predictive research aims to generalize the relationships *i.e.*, if a research finds a solution to a particular problem in an area, the same solution can be generalized elsewhere, if such problem is seen. That is, it aims to generalize from the analysis by predicting certain phenomena on the basis of hypothesized, general relationships. So, the solution to a problem in one area can be applicable to similar problems else where. It provides 'HOW', 'WHY', 'WHERE', 'WHAT IF' answers to current events and also to similar events in the future. For example, we frequently see in the news papers about various research findings say, researchers (doctors) concluded that, '*for every one hour watching TV reduces the life span by 22 minutes*'. So, the viewers lose their life span by about 4.8 years in their life time – a research conducted by the Australian Diabetes, Obesity, Life style Study of National Bureau of Statistics, Australia. This research finding was announced by a website <www.news.com.au>. So, the predictive research might predict a relationship between number of TV viewing hours and life span of human beings ('Eenadu', Telugu daily news paper-Oct-17th, 2012). Similarly, if a researcher conclude that, teens in high schools, aged 13-16, are prone to high dropout rates in Andhra Pradesh, predictive research might assess (predict) a relationship between age or gender and not graduating from high school. Following are the other examples of research questions in a predictive research study:

- Will the introduction of an employee bonus scheme lead to higher levels of productivity in a firm?
- What type of packaging and packing will improve the sales of food products?

C. Based on the Data or Information Collected

It is essential to identify and understand the research approach underlying any given study because, the selection of a research approach influences the questions to be asked, type and size of the sample, statistical techniques used, inferences made, and the ultimate goal of the research. When critically reviewing scientific research, the questions asked, and the answers given, will differ depending upon process of research *i.e.*, whether it is qualitative and quantitative. These two types of research can be easily understood and differentiated from the following example:

For example, a Soil Scientist wishes to conduct a reconnaissance survey in a village to identify saline affected paddy fields to conduct research. If he identifies saline affected paddy fields in the village by means of locating white coloured patches on the superficial layer of soil, it is a case of qualitative analysis of the problem. On the other hand, if the researcher conducts Ec test to analyze salt concentration in the soil and in irrigation water flowing into the fields and conclude Ec is more than 4 dsm^{-1} and selects the saline affected fields for further research, it is a case of quantitative analysis of the problem. The following are the few examples that differentiate qualitative research and quantitative research:

- ☆ If the researcher simply describes the colour of a solution in chemistry lab, it is case of qualitative analysis, whereas if the researcher find out the quantity of solute needed in the solution to turn the color of the solution faint green, it is case of quantitative analysis.
- ☆ If one wishes to investigate, why certain data are random, then it is a qualitative research. If the aim is to study how random the data is, what is the mean, standard deviation and variance of the collected data, then it becomes quantitative.
- ☆ Explaining how digestion of food takes place in human body is a qualitative description. It does not involve any numbers or data and quantities. The detection of changes in physico-chemical properties of intestine, pH in a particular compound is a qualitative analysis. This can be done by carrying out physical or chemical tests.

Thus, it infers that, qualitative research generates non-numerical data or verbal information that cannot be converted into numbers. Quantitative research, on the other hand, generates numerical data. So, let us discuss in-detail about these two types of research.

a. Qualitative Research or Analysis

Qualitative research often involves study of behaviour and the reasons behind it. This type of analysis is more concerned with 'HOW' and 'WHY' about the problem instead of 'WHAT', 'WHERE' and 'WHEN' (which are more focused upon when conducting a quantitative research). So, this research is concerned with qualitative phenomenon. For example, in social sciences, if the researcher studies how farmers feel about using a particular pesticide or how the consumers feel about an RTE food, it falls under qualitative research. The following are the important characteristics of qualitative research:

- ☆ This research merely describes the things as they are, without going into the numerical measurement or quantification of the data. So, it presents non-quantitative type of analysis. It involves collecting, analyzing, and interpreting data by observing what people do and say.
- ☆ This research is much more subjective than quantitative research and uses very different methods of collecting information, mainly individual, in-depth interviews and focus groups.
- ☆ The nature of this type of research is exploratory and open-ended. The sample size is relatively small and in-depth interviews are conducted. The selected respondents were asked to respond to general questions and the researcher probes and explore their responses to identify and define people's perceptions, opinions and feelings about the topic or idea or problem being discussed and to determine the degree of agreement that exists in the group. The quality of the findings from qualitative research is directly dependent upon the skills, experience and sensitiveness of the researcher.
- ☆ This type of research is often less costly than surveys and is extremely effective in acquiring information about people's perceptions about the selected problem and in drawing policy measures on the participatory mode.

- This research is necessary in situations, where the researcher does not have clarity about the different problems pertaining to a particular topic and he may not be able to determine, which data are important and which are unimportant. But, in quantitative research, the researcher exactly knows about the objectives of the study and the nature of data to be collected for in depth investigation. However, in qualitative research, the focus of the study may become more apparent as time progresses.
- The main emphasis of qualitative research is on inductive reasoning, which tends to move from the particular to general *i.e.*, it follows bottom up approach. So, this helps the researchers in developing a theory.
- Qualitative researchers do not base their research on pre-determined hypotheses.
- The approach to data collection and analysis is methodical, but allows for greater flexibility than in quantitative research. Data are collected in textual form on the basis of observation and interactions with the selected respondents and are not converted into numerical form and not statistically analyzed.
- In this research, researchers will give selected respondents a certain degree of freedom in answering the questions. This method of collection of data creates the right atmosphere to enable the respondents to express their views completely to the extent of their knowledge and satisfaction. Thus, the data collection approach is less formal and less rigid than that used in quantitative research.
- Qualitative research can be further classified into three sub-types:
  - ***Phenomenology***: In this research, the researcher attempts to understand how one or more individuals experience a phenomenon. For example, if the researcher collects relevant information from the victims of tsunami tragedy, it falls under phenomenology research.
  - ***Ethnography:*** In this research, the researcher focuses on describing the culture of a group of people. For example, if the researcher studies the culture and habits of the tribal people living in High Altitude and Tribal Area Zone of Andhra Pradesh, it falls under ethnographic research.
  - ***Case Studies***: In a case study, the researcher explores a single entity or phenomenon (the case or event) and collects detailed information through personal discussions with the selected sample. Thus, case study is a descriptive record of an individual's experiences and/or behaviours kept by an outside observer. For example, if the researcher studies about the functioning of ITC's *e-Choupal*, it falls under a case study research.

b. Quantitative Research or Analysis

Quantitative research often involves study of measurement or quantification of the data or information. This type of analysis is more concerned with 'WHAT', 'WHERE' and 'WHEN' instead of 'HOW' and 'WHY' (which are more focused upon when conducting a qualitative research). Thus, a quantitative research method involves a numeric or statistical approach to research design. According to Creswell (2003), *'quantitative research employs strategies of inquiry such as experimental and surveys, and collect data on predetermined instruments that yield statistical data'*. The following are the characteristics of quantitative research:

- This research is an inquiry into an identified problem, based on testing a theory, measured with numerical data, and analyzed using statistical techniques. Statistics is the most widely used branch of mathematics in quantitative research especially in fields such as economics and commerce. The main advantage of using statistical techniques in quantitative research is that, it helps to draw projections for the future. Say, for example, the researcher can study the growth rate in the sales of a pesticide in the market and from this analysis, he can also forecast the sales trends for the future as well. With the advent of sophisticated statistical computer packages, this quantitative research is gaining more significance now, as its permits researchers to discover complex causal relationships and to determine to what extent one variable influences another.

- ✰ The goal of quantitative research is to determine, whether the predictive generalizations of a theory hold true. The findings from quantitative research can be predictive, explanatory and confirming.
- ✰ The interpretations of findings from this research must be with reference to numerical data that are statistically significant. For this, adequate sample size is essential in order to achieve findings with an acceptable degree of accuracy. Generally, researchers seek sample sizes, which yield findings with at least a 95 per cent confidence interval (which means that, if you repeat the survey 100 times, 95 times out of a hundred, you would get the same response), plus/minus a margin error of 5 percentage points. Thus, the analysis in this research is so designed that produce a smaller margin of error.
- ✰ The main emphasis of quantitative research is on deductive reasoning, which tends to move from the general to the specific *i.e.*, it follows top down approach. The validity of findings of this research is compared with premises laid down *i.e.*, prior statements, findings or conditions etc.
- ✰ This research also aims at minimizing the bias in drawing the valid conclusions. Of course, in Astronomy and other hard sciences, this research has a very minimal amount of bias. However, in social sciences research (say, economics), there is a scope for large amount of bias in drawing the information and the researcher tries to minimize it as far as possible through selecting large size samples, repeated surveys, triangulating the information etc.
- ✰ This quantitative research generally comes later in a research project, once the scope of the project is well understood.
- ✰ There are three general types of quantitative methods:
    - ❖ ***Experiments:*** True experiments are characterized by random assignment of subjects or treatments to experimental conditions and the use of experimental controls.
    - ❖ ***Quasi-Experiments:*** Quasi-experimental studies share almost all the features of experimental designs except that, they involve non-randomized assignment of subjects or treatments to experimental conditions.
    - ❖ ***Surveys:*** Surveys include cross-sectional and longitudinal studies using questionnaires or schedules for data collection with the intention of estimating the characteristics of a large population of interest based on a sample that is representative of that population.

Qualitative Research vs. Quantitative Research

Having understood the concepts of these two types of research, it is essential to differentiate them (Table 2.2).

Table 2.2: Differences between Qualitative and Quantitative Research Methods.

Item	*Qualitative Research*	*Quantitative Research*
Philosophy	Qualitative approach research is based on phenomenology. Phenomenology focuses on the processes and experiences one goes through. Experience is a complex concept and cannot externally observable. Hence, it is difficult to get concrete answers, as the problems itself is subjective.	Quantitative research is based on the positivism. According to positivism, reality is stable, observable and can be measured. Knowledge is obtained using the scientific method, which is objective and measurable. To prove that a phenomenon exists, one has to collect numerical data scientifically and that can be analyzed statistically. Also, the research when repeated should yield the similar results.
Aim of research	To provide a complete and detailed description of the research problem. It is usually more exploratory in nature. Researchers are typically open to gather data and information to take them in different directions. So, qualitative research is more open to different ways of interpretations. Hence, qualitative researchers may be more prone to accusations of bias and personal subjectivity.	Focuses more in counting and measuring the numerical data and analyzing the data by statistical means to draw conclusions. This research aims at generalization of the results from the sample to the population of interest.
Research questions	Research questions: How? Why?	What?, When? and Where?
Literature review	Literature review may be done as study progresses or afterwards	Literature review must be done early in study
Suitability	'Soft' science	'Hard' science
Mode of research	It is non-numerical, descriptive, applies reasoning and uses words. Its aim is to get the meaning, feeling and describe the situation.	It is numerical and analytical. Its aim is to trace out cause and effect relationships among the variables. For example, if the statistics show that, excess consumption of alcoholic drinks lead to more cases of heart-attacks, then the researcher concludes that, there is a causal link between these two variables.
Usage	☆ Find out what to investigate. ☆ When research ideas are needed. ☆ Emphasis on understanding the research problem theoretically. ☆ Focus on understanding the problem from respondents' point of view.	☆ The researcher knows what to investigate. ☆ Conducted when research results are needed. ☆ Emphasis on testing and verification. ☆ Focus on facts and/or reasons.
Data employed	No numerical or quantified data pertaining to the variables under consideration are employed in the investigation. The researcher simply describe and analyze a phenomenon using words. That is, qualitative research focuses on gathering of mainly verbal data rather than measurements. Gathered information is then analyzed in an interpretative manner, subjective, impressionistic or even diagnostic. So, in qualitative research, the data	Only measurable data are being gathered and analyzed in quantitative research. Data are quantified in numbers, which are analyzed using statistical techniques. So, in quantitative research, the data are numbers and measurements. Since, the data is numerical, it cannot be easily processed. So, if a researcher conducts a quantitative research, the report comprises of tables containing data in the form of numbers and statistics.

Contd...

Table 2.2–*Contd...*

Item	*Qualitative Research*	*Quantitative Research*
	are in the form of narrative descriptions, observations, words (from interviews), images (videos) etc. Since, the information is verbal, it cannot be easily processed. If a researcher conducts a qualitative research, the report comprises of figures in the form of graphs.	
Development of new theory	Sometimes, qualitative research is used to develop new theory that didn't exist before.	Frequently, quantitative research is used to support or expand a theory that already exists.
Approach	Inductive. This approach is more flexible and informal. The researcher need not know in advance about the topic or problem and about the possible questions he has to ask while interacting with respondents.	Deductive. This approach is more structured, less flexible and more informal. The researcher must know in advance about the topic or problem and about the possible questions he has to ask while interacting with the respondents.
Theory building	Develops theory	Test theory
Objective	Strives for uniqueness	Strives for generalization.
Role of researcher	Researcher is part of process	Researcher is separate
Research design	Research design is flexible and develops throughout the investigation	Research design is specified before the start of the investigation.
Formulation of hypothesis	In qualitative research, a hypothesis is not needed to begin research. Hypotheses are developed during the investigation; questions govern the purpose of the investigation; theories are developed inductively. The hypothesis developed is broad in its scope.	Quantitative research requires a hypothesis before beginning of research. Precise hypothesis is stated before the start of the investigation; theories govern the purpose of the investigation in a deductive manner. This research allows researchers to test specific hypotheses. Depending on research findings, hypotheses are either supported or not supported. The hypothesis framed is narrow in its scope.
Study of a phenomena	Phenomena are studied holistically, as a complex system	Phenomena are broken down or simplified for study
Reliability and validity of findings	Reliability and validity are determined through multiple sources of information (triangulation)	Reliability and validity are determined through statistical and logical methods.
Independent Variable (IV)	There is no specific IV; the concern is to study naturally occurring phenomena without interference	The IV is controlled and manipulated.
Selection of sample	Samples are purposefully selected or single cases are studied. In qualitative research, the sample is small and not chosen randomly. Rather, the choice of a sample is purposeful. These studies usually have fewer respondents than quantitative studies	Samples are selected to represent the population. In quantitative research, the sample should be large enough to be representative of the population. Also, sample size may be determined by the statistics used because certain statistical tools require a

Contd...

Table 2.2–*Contd...*

Item	*Qualitative Research*	*Quantitative Research*
	because, the depth of the data collection does not allow for large numbers of participants.	certain sample size to be effective. It is ideal to have a large number of respondents because, this gives analysis more statistical power.
Study of behaviour	Study of behaviour is in the natural setting.	Study of behaviour is in the natural or artificial setting
Role of researcher	In qualitative research, the researcher can learn the most about a situation or problem by participating and/or being immersed in it.	In quantitative research, the researcher is ideally an objective observer that neither participates in nor influences what is being studied. So, the researcher is detached from the investigation.
Generalizations	In qualitative research, the researcher selects cases or subjects based on their uniqueness. In other words, researcher is not concerned about generalization of research findings.	In quantitative research, the researcher is interested in generalizing on the findings obtained from the sample to the population (as the selected sample is representative of the population).
How to make generalizations of findings	Use of similar cases to generalize the findings	Use of statistical techniques to demonstrate validity of the findings at population level.
Research setting	Qualitative research occurs in more natural and less controlled research settings.	Quantitative research generally occurs in less natural and more controlled research settings.
Focus	Qualitative research studies are focused on differences in quality, rather than differences in quantity.	Quantitative research studies are focused on differences both in terms of quantity and quality.
Representativeness of sample to the population	Qualitative research does not seek to choose samples that are representative of populations.	As discussed earlier, the statistical analysis employed in quantitative research brings more exactness and clarity about the findings and these can be generalized for the population data. So, the goal of quantitative research is to choose a sample that is representative of the population.
Data collection	Qualitative research aims at collecting in depth data that is not possible with quantitative data. This richness of data is possible because, qualitative research focuses on fewer respondents and hence, the researchers generally know more details about each respondent.	Quantitative researchers collect data from more number of respondents and hence, it is quite difficult to have the depth and breadth of knowledge about each respondent.
Data collection instruments	Interviews/Focus Groups	Schedules, Questionnaires etc.
Data Analysis	Non-statistical	Statistical

Contd...

Table 2.2–*Contd...*

Item	*Qualitative Research*	*Quantitative Research*
Presentation of results or findings	Results are in words or pictures rather than numbers.	Results are expressed in terms numbers that too with statistical significance.
Confidence in presenting results	Conclusions are tentative and subjected to ongoing examination	The analytical findings are interpreted with a degree of certainty say, at 1% or 5% level of significance.
Outcome of research	Generation of hypotheses and ideas.	Verify the hypotheses from the results obtained through investigation.
Reliability of outcome of research	The reliability of outcome of research investigation depends upon the researcher in collecting the requisite information without bias.	The reliability of outcome of research investigation depends upon the research design and data gathering instruments
Duration for completing the research	Qualitative research is usually more time-consuming, as it may involve analyzing longer descriptive passages or descriptions instead of numbers. After analyzing these descriptions, the researcher has to investigate about the attitudes, experiences and perceptions of the respondents towards the problem under consideration. Though this research deals with smaller samples compared to quantitative research, it is more exploratory and with detailed explanations. For example, if the researcher wants to find out, why people gets addicted to narcotic drugs, he has to discuss and analyze each respondent's case and has to carefully draw the conclusions.	Quantitative research is less time-consuming, though it addresses larger samples, as it deals with numerical data and use statistical techniques to analyze and interpret the findings. The research is conclusive rather than exploratory.
Phase of research	Qualitative Research is ideal for earlier phases of research projects	Quantitative research is ideal for later phases of research projects, as it provides the researcher a clearer picture of what to expect in his research compared to qualitative research

Strengths of Qualitative Research

They include:

- ☆ The active involvement of the researcher in the investigation makes him to involve deeply in understanding the problem.
- ☆ Facilitate in depth study of the problem.
- ☆ Develop flexible ways to the researcher regarding data collection, analysis and interpretation of collected information.
- ☆ Provide a holistic view about the problem under consideration.

Limitations of Qualitative Research

It is to be noted that, the inherent limitations of the qualitative research also reflect its inherent strengths. For example, the small size sample employed in the research ensures in depth study about the problem under consideration. But, the conclusions derived from small sample research cannot be generalized or make broad recommendations. Further, as the researcher deeply involves in studying the problem, it will increases the opportunity for bias to enter into the way data is gathered, interpreted, and reported. The following are the other limitations associated with qualitative research:

- Researcher may draw different conclusions from the same information collected, as the bias will enter in conducting the research.
- No scope for studying cause and effect relationship among the variables.
- Demand rich expertise from the researcher in drawing realistic conclusions.

Strengths of Quantitative Research

They include:

- Large size sample employed in the research enable the researcher to draw generalizations or broad recommendations.
- There is scope to study cause and effect relationship among the variables.
- Ensure accuracy of analytical findings. Thus, this research can be replicated and compared with similar studies.
- Personal bias can be avoided by researchers by maintaining a 'distance' from participating subjects and employing subjects unknown to them.

Limitations of Quantitative Research

Following are the limitations associated with quantitative research:

- The passive involvement of the researcher in the investigation makes him to involve less deeply in understanding the problem.
- In depth study of the problem is not possible because of large size sample.
- Less flexible or more rigid in conducting the research.
- Do not provide a holistic view about the problem under consideration.
- The development of standard questions by researchers in the schedule or questionnaire can lead to "structural bias" and false representation, where the data actually reflects the view of the researcher instead of the participating subject.

However, referring to a research approach as quantitative or qualitative can be misleading, as a researcher can design a study with a view to collect qualitative data (for example, perceptions about the use of a particular brand of pesticide) and then quantifying them by counting the frequencies of positive answers and negative answers. This allows researchers to analyze their data using statistical methods. On the other hand, a researcher can collect qualitative data with the intention of analyzing them using non-numerical methods, or collect data that are already in numerical form and use statistical methods to analyze them. Some students avoid taking a quantitative research because, they are not confident with statistics and interpretation of analytical findings and hence, opine qualitative research will be easier. Some students also opine that, it is quite difficult to start and decide on an overall design for a quantitative research, but it is easier to conduct the analysis and write up the research because, it is highly structured. It is the general opinion that, qualitative research is normally easier to start, but students often find it more difficult to analyze the data and write up their final report. For example, if a researcher conducts a research to know 'the health problems faced by the farmers in spraying pesticides in their farm fields', he can collect the data/information in two different ways. First, he can go for collecting numerical data such as number of sprayings done, amount spent towards medical treatments, number of farmers died in the village so far due to such health hazards etc. Alternatively, he can go for collecting qualitative information pertaining to how sufferings are experienced by the farmers in terms of their perceptions, problems in their families due to financial burden etc. So, the researcher has to be carefully select the approach that is influenced by the nature of the research problem as well as the philosophical preferences. The choice of which approach to use may reflect the interests of those conducting or benefitting from the research and the purposes for which the findings will be applied. Further, researcher's own experience and preference, time, money, and other resources available also influence the selection of approach (Hathaway, 1995). When both quantitative and qualitative data are collected in the same study, the approach is called as Mixed methods. There are different views pertaining to the use of both qualitative and quantitative researches simultaneously. Some researchers opine that, qualitative and quantitative methodologies cannot be combined because, the assumptions underlying each approach are so vastly different. Other researchers say that, these two can be used in combination only by alternating between methods: *i.e.,* qualitative research is appropriate to answer certain kinds of non-numerical questions in certain conditions and quantitative research is used to answer numerical data and findings. And some researchers think that, both qualitative and quantitative methods can be used simultaneously to answer a research question. Of course, to a certain extent, researchers on all sides of the debate are correct: each approach has its drawbacks. So, rather than discounting either approach for its drawbacks, researchers should find the most effective ways to incorporate elements of both to ensure that, their studies are as accurate and thorough as possible. Thus, these two approaches should be used in conjunction with each other with the aim of presenting the valid information gathered from the respondents.

D. From the Process Adopted to Find Answer to Research Questions (Inquiry Mode)

According to this criterion, there are the two approaches *viz.,* Structured approach and Unstructured approach. These two are discussed here under.

a. Structured Approach

In structured approach, the research process is well-planned and pre-determined *i.e.,* objectives, research design, sample and the questions that the researcher frame to ask the respondents are pre-determined. This type of research aims at analyzing the cause-and-effect relationship among the variables. So, the structured approach of inquiry is usually classified as quantitative research. For example, '*economic analysis of mango cultivation in Chittoor district of Andhra Pradesh*' falls under structured approach of investigation, as the researcher has to systematically follows the research process regarding the framing of objectives, selection of sample regarding mandals, villages and farmers (small, medium and large farmers), data to be collected etc.

b. Unstructured Approach

In this approach, there is more flexibility in adopting the research process. The objective of this approach is to describe the variation in a phenomenon without quantifying it. This unstructured approach to inquiry is usually classified as qualitative research. For example, if the researcher describes about a case study of a farmer regarding the Indigenous Technical Knowledge adopted by him in controlling storage pests of rice, it is a case of unstructured approach of investigation.

E. Based on Logic Employed

We know, social research involve two aspects: testing existing theory by applying it to a situation or problem and developing a new theory or making generalizations by repetitive analysis of the (similar) problems. The former is known as Deductive approach and the latter is known as Inductive approach. This infers that, the relationship between theory and data is two-way. So, in research, conclusions are based on two methods known as the *Deduction* and *Induction methods.* Both these methods are widely used in research projects.

a. Deductive Research

This research follows an approach which is 'top-down' or 'from general to specific'. So, in this approach, the researcher starts from a known theory and try to prove it right with the help of available information at the sample or grassroot level. This deductive method is, otherwise, termed as Analytical or Abstract or Scientific method or Method of logical reasoning. Mill characterized it as *'a priori'* method. Francis Bacon described it as a 'Descending process'. Since, this method of analysis proceeds from 'general to particular on the basis of general truth', it starts with a generally accepted principle and proceeds to the specific. That means, it aims at deriving conclusions at micro level from established general truths. This deductive logic begins with one or more premises. These premises are statements or assumptions that the researcher initially takes to be true. Reasoning then proceeds logically from these premises toward conclusions. To the extent that, the premises are true, the conclusions may also be true. Look at the following example:

- *If all roses are plants, (Premise 1)*
- *And if all plants produce food through photosynthesis, (Premise 2)*
- *Then all roses must produce food through photosynthesis (Conclusion).*

To the extent that, the premises are false, the conclusions may also be false. Look at the following example:

- *If all roses are platypuses, (Premise 1)*
- *And if all platypuses does not respire (Premise 2)*
- *Then all roses must not respire (Conclusion).*

The 'IF-THEN' logic is the same in both these examples. But, the conclusion in second example is erroneous because, both the premises are also erroneous.

According to Wilson Gee, *'By deductive method is meant the reasoning from general to particular or from universal to individual'.* In this method, we accept certain general truths or facts and apply them in specific cases to prove the accepted truths. For example, it is an established fact that, 'man is entirely motivated by self-interest'. The same will be applied at the micro level and its reliability will be established. So, this method involves the application of general principles to a particular situation and draw useful conclusions. This method assumes that, the behaviour of the general public will also be the behaviour of individual person. This method is based on *a priori* reasoning and conclusions are drawn from certain fundamental assumptions. In general, macroeconomic theories are based upon deductive method. Studies of national income, employment, general price level and international trade are made on the basis of deductive method. The main steps involved in deductive logic are discussed here under (Figure 2.3).

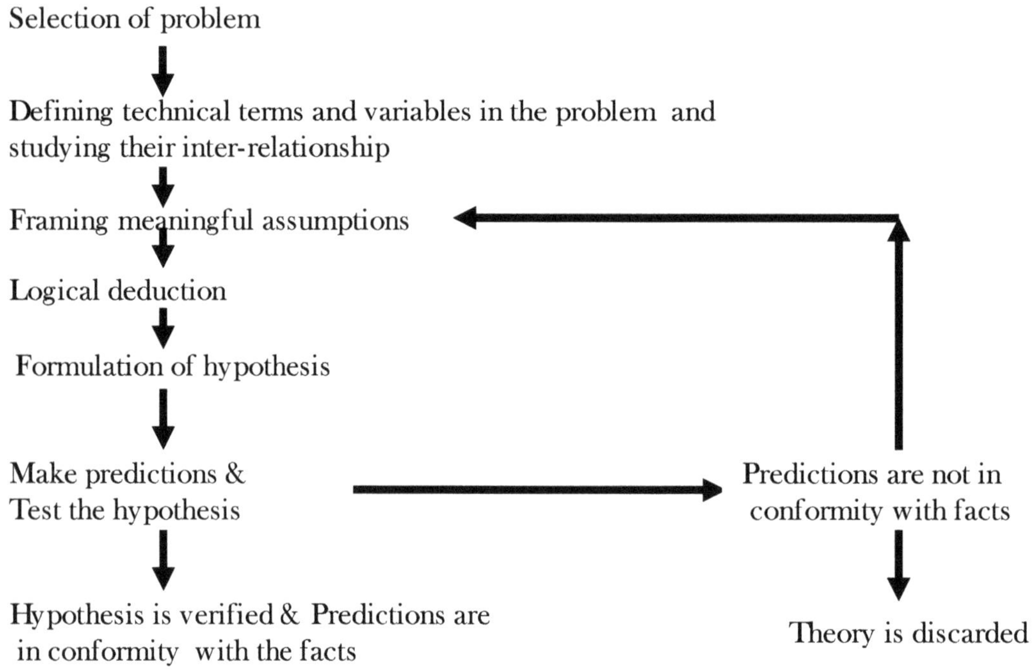

Figure 2.3: Steps Involved in Deductive Method of Economic Analysis.

Step 1–Selecting the Problem

Identification of problem is the basic step. The problem selected for the investigation should be stated very clearly. The problem under consideration may have wide scope such as addressing issues like poverty, inflation, unemployment etc or a narrow scope. But, if the focus or scope of the problem is too narrow, it facilitates the researcher to get accurate results. The problem must be one of significance for the actual world and should have economic significance. While selecting the problem, it is important on the part of researcher to be clear about the objective(s) of the research activity.

Step 2–Defining Technical Terms and Variables in the Problem

After finalizing the problem for study, the investigator should get well versed with the technical terms involved in the study. This helps him to understand about various concepts, identify variables and their inter-relationships.

Step 3–Framing Meaningful Assumptions

The study about technical terms and inter-relationships among variables helps the investigator to frame meaningful assumptions, which further guide him to formulate hypothesis. This method is also known as Hypothetical method, as the assumptions formulated may not correspond to actual facts, but are very near to actual facts and these assumptions are used as premise for starting, reasoning and drawing conclusions. The assumptions may be behavioural relating to the behaviour of variables or technological relating to the state of technology or factors employed in the study. Sometimes, crucial assumptions are also made on the basis of observations or introspection like, farmers try to maximize the profits in the production programme, consumer try to attain maximum satisfaction etc.

Step 4–Process of Logical Deducation

Generally, a hypothesis is formulated from the assumptions framed through logical deduction. This logical deduction may be carried out verbally or symbolically. In the recent days, mathematics is mostly used in the method of logical deduction.

Step 5–Formulation of Hypothesis

Now, the research investigator will formulate the hypothesis duly considering the variables for the study, their inter-relationships, assumptions framed, *a priori* knowledge etc.

Step 6–Making Predictions and Testing the Hypothesis

After formulating the hypothesis, predictions can be made on its basis and they are tested. If the predictions are in agreement with the facts, then the theory is accepted. On the other hand, if the predictions are in conflict with the facts, the theory is discarded in favour of superior alternative and again the process is to be initiated duly modifying the assumptions.

Step 7–Verification of Hypothesis

It is an important step in the deductive method. Verification refers to confirming whether the hypothesis formulated earlier is in agreement with the facts or not. Two possible cases may arise regarding the verification of hypothesis.

- ***Hypothesis is accepted/confirmed:*** For example, a hypothesis is formulated stating that, 'the firm will try to maximize the profits'. This is said be accepted, when the firm under consideration for study is found behave in the same way.
- ***Hypothesis is rejected/not verified:*** If the hypothesis is not verified, it implies two important aspects *viz.*, the hypothesis formulated is correct, but the results are contradictory, as the variables under study behaved in a different way due to special circumstances. Or, the investigator might have formulated a wrong hypothesis without having thorough knowledge about the variables, their inter-relationships, faulty assumptions made, lack of *a priori* knowledge etc.

The above second case *i.e.*, contradictory results not confirming hypothesis is more common in economic sciences, as both human behaviour and economic phenomena are very complex, unlike physical sciences. Moreover, laboratory or controlled experimentation is not possible in Economics.

Merits of Deductive Method

The merits of deductive method are as under.

- This method is nearer to reality. It consumes less time and less expensive compared to inductive method, as useful economic theories can be verified with less data.
- The use of mathematical techniques and logical reasoning brings exactness and clarity in analysis.
- This method is indispensable in social sciences like Economics, where there is limited scope for experimentation. So, this method serves as a substitute for experimentation. Since experimentation is not possible in Economics, this method is based on logical reasoning. As Boulding puts it, *'it is the method of intellectual experiment'.*
- The method is simple because, it is analytical and it has easy application. It involves abstraction and simplifies a complex problem by dividing it into component parts. We know, the Marginal Utility (MU) derived by a consumer goes on diminishing with every

successive consumption of a commodity is a self-evident truth. From this, we can formulate several logical conclusions. For example, rich people have lesser MU for money compared to poor people. This will guide the Government to impose progressive taxation policy. So, the Principle of Progressive taxation is derived from the Law of Diminishing Marginal Utility (LDMU) through deductive reasoning.

- ✰ The inferences drawn through deductive method have universal validity.
- ✰ We try to derive useful conclusions from the well known generalizations, through collecting elaborate statistical information or data and establish the law.
- ✰ This method helps to explain complex economic phenomena through making simple assumptions and draws conclusions.
- ✰ If the premises or hypotheses are true, the conclusions will be true and exact.

Demerits of Deductive Method

No doubt, deductive method is simple and precise only, if the underlying assumptions are valid. There is big 'IF' in the statement. Criticisms were raised against this method and following are the important shortcomings:

- ✰ More often, the assumptions turn out to be based on half truths or have no relation to reality. In such case, the conclusions drawn from such assumptions will, therefore, be misleading. That means, if the assumptions framed are not true, the conclusions will be wrong and the generalizations cannot be accepted as a law. If such generalizations are used in framing economic models and policies, the results would be disastrous. For example, J. B. Say, claimed universal validity for his law, '*Supply creates its own demand*' so that, there will not be any over production in the market. But, this law was discarded, when Keynes proved that, Say's law was wrong and over production would be possible. Such laws or models, which are built with little empirical content, remain as 'intellectual toys' and they have no operational significance.
- ✰ Professor A.P. Lerner describes the deductive method as '*Arm Chair analysis*'. According to him, the premises from which inferences are drawn may not hold good at all times and places. As such, deductive reasoning is not applicable universally. For example, Keynes theory of employment is not applicable to under developed economies.
- ✰ This method is highly abstract and requires great skill or competence in drawing inferences from various premises.
- ✰ This method is based on the assumption that, economic conditions remain constant, But, in practical situation, economic conditions are continuously changing.
- ✰ It is important that, the verification of theories or laws or principles in social sciences like Economics is mainly based on observation. A right observation depends on the accuracy of data. If faulty and inadequate data are employed in the research, it leads to faulty conclusions and thereby, there is a chance of refuting the law/principle under consideration.
- ✰ It is not easy to test the results obtained through deductive method. So, it should be supplemented by induction.

b. Inductive Method

According to Wilson Gee, '*inductive method is the process of reasoning from particular to general or from individual to universal*'. This method is also known as Historical or Concrete or Posterior or

Realistic or Empirical method. It was Francis Bacon, who advocated inductive method in scientific enquiry and described it as an *'Ascending process'*. In this method, first the data are collected about a certain economic phenomenon, they are compiled, systematically arranged and statistical techniques are employed and from these findings general conclusions are drawn. Thus, inductive reasoning does the reverse of deductive reasoning *i.e.,* it begins with an individual problem or question and proceeds to form a general principle based on the evidences observed in the real world. This method involves conductance of micro level study first and this study will be replicated. If same conclusions were drawn from such studies, we generalize the statement as an observed truth and the theory is propounded. The inductive method can be conducted in two ways *viz.,* through experimentation and through statistical means. The former is generally used in biological and physical sciences, and the latter is used in social sciences like economics. This is because, in economics, there is limited scope for experimentation and both economic phenomena and human behaviour are complex in nature. However, some experimentation is possible to validate some economic laws like laws of production, cost principles, price discrimination practiced by monopolist etc. So, the Economists had no other choice to rely upon statistical means in building the economic laws or principles. However, some *a priori* knowledge is also essential, as it gives the researcher some direction in which way he has to proceed. For example, farmer 'A' purchased less quantity of a particular brand of fertilizer when its price was increased in the market. Farmers B, C and D also showed the same pattern of demand due to rise in price of that fertilizer. So, finally, we can generalize their behaviour and an economic theory can be propounded stating that, 'quantity demanded varies inversely with the price of the commodity'. Similarly, assume there are 500 farmers in a village. The data analysis reveals that, nearly 490 farmers are purchasing chemical pesticides manufactured by Multi National Companies, which are highly effective. Out of the ten which remains, nine farmers are purchasing local products just to patronize their own products, while the remaining one farmer is not adopting plant protection measures due to lack of awareness. From this analysis, we can easily conclude that, farmers like to buy the products from Multi National Companies on the criteria of effectiveness, unless they are guided by patriotism or lack of awareness about application of an effective pesticide.

The Law of Diminishing Returns (LDR), LDMU, Law of Equi-Marginal Utility (LEMU), Consumer's surplus etc., have been developed on the basis of inductive method. Malthus formulated the theory of population after analyzing the facts about population trends in different countries. Similarly, Engel formulated the propositions of family expenditure after a factual study of family budgets. Theories of rent, wages, interest and profits are also based upon inductive method. This implies, Microeconomic theories are formulated according to inductive method. However, some of the concepts of Macroeconomics *viz.,* consumption function, the development and use of accelerator principle have been obtained through inductive analysis. The main steps involved in the application of inductive method are given here under.

Step 1–Selection of Problem

In order to arrive at generalization of an economic phenomenon, the problem should be properly selected at micro-level and clearly stated.

Step 2–Defining Technical Terms and Variables in the Problem

The investigator should study about different technical terms related to the study. This helps him to identify different variables and their inter-relationships.

Step 3–Collection of Data

This inductive method can be applied in two distinct ways *viz.,* experimentation and statistical form. The former is used in natural and physical sciences and the latter is used in economics. For the problem under consideration, the relevant data should be collected about the variables and a preliminary thinking will be done to study the possible functional relationships between the selected variables. Once this is done, the researcher can begin to develop theory to explain these findings. Following this, he or she may repeatedly collect more data to ground the developing theory. This means collecting more data to establish whether and to what extent the developing theory is accurate and applicable in different circumstances, and is known as a process of Iteration. Iteration is a step-by-step process of analysis, painstakingly applying a set of data to different related questions or repeating the questions with different data.

Step 4–Processing of Data

The collected data for the selected variables should be compiled and analyzed using appropriate statistical techniques.

Step 5–Development of Theory

From the analytical findings, a theory can be developed, which can be further refined and tested using statistical tests.

Step 6–Making Predictions

Once the theory has been developed, predictions can be made on its basis. If predictions of theory are in agreement with the facts and actual behaviour of the economy, then a new reliable theory has been developed. If a new theory explains the concepts in a better way than the existing ones, it replaces them. On the other hand, if the predictions are in conflict with the facts and actual behaviour of the economy, the theory is discarded and again data collection is to be initiated with much accuracy to make further predictions.

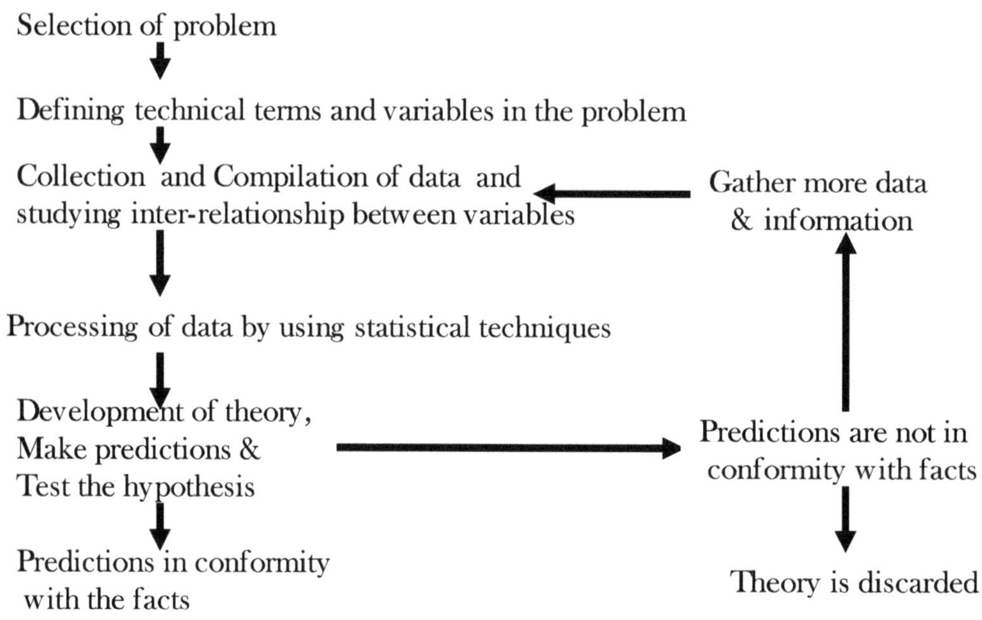

Figure 2.4: Steps Involved in Inductive Method of Economic Analysis.

Merits of Inductive Method

The following are the merits of inductive method:

- This method is more realistic, as it is based on facts.
- It has more practical approach for formulating economic generalizations.
- This method employs statistical techniques to test various economic principles, thereby, it brings exactness, clarity and reliability for the findings.
- Inductive method is dynamic. The changing economic phenomenon is analyzed and on the basis of collected data, solutions and conclusions are drawn from them.
- This method is more useful because, its propositions can be tested and verified easily.
- This method considers the complex relationships found in actual life and examine them carefully.
- The generalizations drawn through inductive analysis are *historico-relative*. Therefore, they are relative to time and place and hence, highly practical.
- Inductive method discovers and proves general principles and this helps in future investigations.
- The conclusions drawn through deductive method can be verified by inductive method.

Demerits of Inductive Method

The following are the demerits of inductive method:

- If insufficient data are used for the analytical study, the conclusions drawn from the study will lead to faulty economic generalizations.
- The collection of data consumes more time and cost and the sources and methods employed in the collection of data differ from investigator to investigator. The results, therefore, may differ even with the same problem. So, there is a danger of investigator bias entering into propositions and this affect the concreteness.
- Since the propositions obtained through this method are based on few facts, they may not have universal applicability and hence, there is limited scope for verification.
- This method has limited scope in economics unlike physical sciences because, controlled experimentation is not possible. This is because, in economics, we deal with complex human behaviour and even economic phenomena are very complex. Every economic event is unique in the sense that, it will favour one segment of the population and disfavour another segment of the population. Further, one economic problem is related to another economic problem and thereby, deduction has its own role.
- In this method, there is a danger of giving hasty and hurried conclusions from insufficient number of facts.

Thus, none of the above two methods provides satisfactory system for solutions of problems. Since, both the methods have their own merits and demerits, we cannot rely exclusively on any one of them. The researchers are of the view that, both these methods are complementary. Deductive conclusions are tested with inductive methods and inductive generalizations are again related to a deductive hypothesis. Deductive method is more suitable to pure theory, while inductive method is more suitable for formulating practical policies. Alfred Marshall has rightly remarked, *'Inductive and Deductive methods are both needed for scientific thought, as the right and left foot are both needed for walking'*.

We can apply any of them or both as the situation demands. Although deduction and induction represents two differing approaches for understanding the economic phenomena, the 19th century American Economist, Henry George observed that, they are related. George noted that, *'induction involves the use of human reason to investigate facts, while deduction is the derivative of the former'*.

Differences between Deductive and Inductive Methods

Table 2.3 and Figures 2.5 and 2.6 shows the major differences between the deductive and inductive approaches of a research problem.

Examples for Deductive Approach

Example 1

- ✰ All planets move on ellipses (Premise 1)
- ✰ Pluto is a planet (Premise 2)
- ✰ Therefore, Pluto moves on an ellipse. (Conclusion)

Example 2

- ✰ Ravi must be either in the college or in the movie theatre.
- ✰ He is not in the college therefore, he must be in the movie theatre

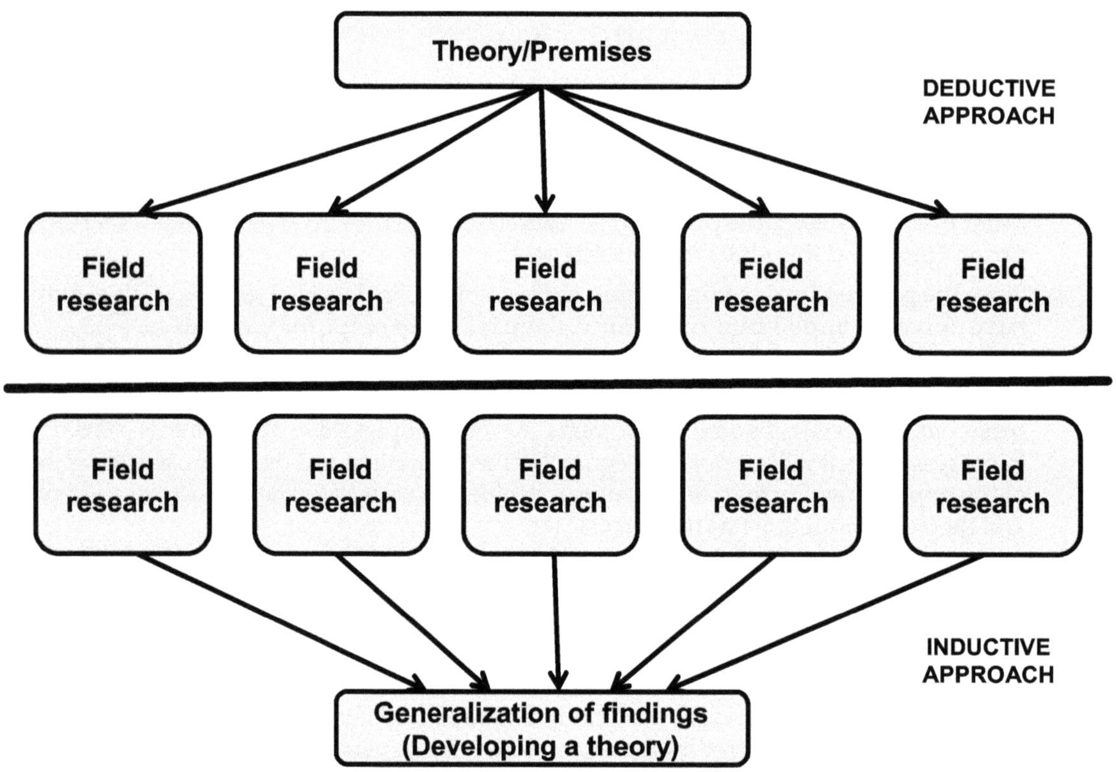

Figure 2.5: Deductive Approach vis-à-vis Inductive Approach.

Table 2.3: Differences between Deductive and Inductive Methods of Economic Analysis.

Deductive Method	*Inductive Method*
Deductive method proceeds from 'general to particular'.	Inductive method proceeds from 'particular to general'.
Deductive approach starts with a general principle and deduces that to apply to a specific case.	Inductive approach takes a specific example or examples, and induces that they can be applied to a much larger group.
Deductive approach is used to try to prove a theory or principle.	Inductive approach is used to try to discover a new piece of information
The deductive process involves a view of the entire picture to arrive at a conclusion.	The inductive process involves a small amount of input observation or information to arrive at a larger conclusion.
In deduction, we start from a theory and try to prove it right with the help of available information. Thus, deductive research aims at deriving conclusions at micro level from established general truths.	In induction, we observe some happening, deduct a pattern and draw conclusion. Thus inductive research examines sample data to make inferences about a population.
This deductive logic begins with one or more premises. These premises are statements or assumptions that the researcher initially takes to be true. Reasoning then proceeds logically from these premises toward conclusions.	Inductive reasoning begins not with a pre-established truth or assumption, but instead with an observation. In inductive reasoning, the researcher observe a sample and then draw conclusion about the population from which the sample has been taken.
Deductive research is characterized by theory and hypothesis driving the collection of data.	Inductive research is characterized by data driving theory development.
This approach aims at hypothesis testing. Deduction is theory-driven, it is based on premises and aims to overcome the limitations of induction.	This approach is hypothesis generating. Inductive theory is being derived from the observations made. This approach cannot test hypotheses, but generates them.
In deduction, it is the hypotheses that are tested by the researchers, not the entire theory. Verification of hypothesis is the outcome of deductive research. The collected data may support or not support the hypotheses. The point is to subject the hypotheses to rigorous testing, which requires the researcher to collect empirical evidence and then to make analytical statements about what the data indicate and what this means for the theory that is being used.	This approach of research begins with observations and findings that are then used to develop (induce) a general explanation, or theory, that accounts for the data. Theory is the outcome of inductive research and emerges out of the data. The point in inductive research is to analyze the data by seeking recurring patterns, themes or topics in the data and to identify generalizable, typical features and properties of the data.
The process of deductive research is often associated with quantitative research and involves starting out with a theory.	Inductive research is generally associated with qualitative research, which typically emphasizes words rather than quantification and measurement.
This is a way of 'fact-building' whereby, general truths are verified at micro level.	This is a way of 'theory-building', whereby 'specific facts are used to create a theory that explains relationships between the facts and allows prediction of future knowledge'.
Mainly aims at verifying the established truths or principles or laws or methods	Mainly aims at establishing the truths or principles or laws or methods as Theory or Generalizations. So, Inductive research is about 'Theory/Model Development'.

Contd...

Table 2.3–*Contd...*

Deductive Method	*Inductive Method*
It is the process of reaching a conclusion that is guaranteed to follow, if the evidence provided is true and the reasoning used to reach the conclusion is correct.	This method is more open-ended and exploratory. This does not give 100 per cent guarantee of truth, but probability of being true. Here is an example: ☆ Socrates was Greek. (premise) ☆ Most Greeks eat fish. (premise) ☆ Socrates eat fish. (conclusion) It is to be noted that, while the premises were true, it is possible that conclusion may be false. This is because, may be Socrates was allergic to fish. Hence, Inductive approach is weaker than deductive approach, as there is a possibility of arriving at false conclusion. That is, the conclusions are not always true.
Deductive systems are usually normative in nature.	Inductive systems are usually descriptive in nature.
It consumes less cost and less time.	It consumes more cost and more time, as it is based on extensive collection of data.
This method is indispensable, as there is limited scope for experimentation in social sciences like economics.	This method helps to make generalizations about laws or principles for which there is scope for experimentation.
In general, macroeconomic theories are based upon deductive method.	In general, microeconomic theories are based upon deductive method.
Limited use to-date.	More frequently used to-date.
It does not give any new knowledge, as it is a method of verification	It gives new knowledge, as it is a method of discovery.
It is a downward process of thought and leads to useful results.	It is an upward process of thought and leads to formulation of economic principles.
It is a method of verification.	It is a method of discovery.
A deductive process requires faith in its premises.	An inductive process requires faith in its conclusion.
It is the method of instruction.	It is a method of teaching.
Researcher gets ready made information and makes use of it.	Researcher acquires first hand knowledge and information by actual observation.
It is quick process.	It is a slow process.
It encourages dependence on other sources.	It trains the mind and gives self confidence and initiative.

Example 3

Writing an 'Abstract' for a thesis is an example of deductive method. This is because, for the selected research problem and obtained research findings (general), the researcher uses certain key points to be included in the 'Abstract' (specific).

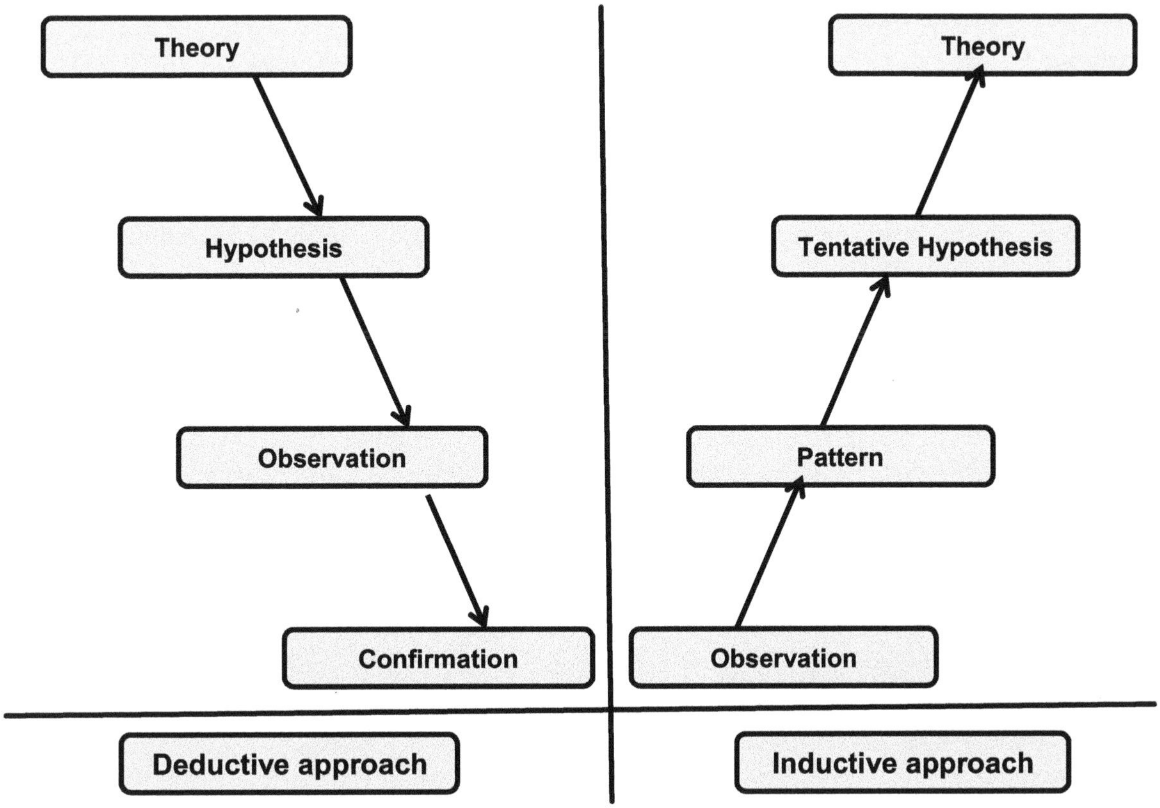

Figure 2.6: Differences in Approaches of Deductive and Inductive Researches.

Example 4

We practically see, students perform salt analysis in Chemistry laboratories. This analysis is an example of deductive method because, first the students know about the general (chemical) properties of given salt (general). Then the students will proceed to perform specific confirmatory test to identify the particular salt (specific).

Example 5

In Entomology, we classify the insects into several orders *viz.*, Coleoptera, Lepidoptera, Diptera, Homoptera, Isoptera etc., on the basis of such differences as metamorphosis, wing structure, and mouth parts. Since, the differences are general in nature, and the classification as mentioned above is particular in nature, it is a deductive approach.

There is, however, one major weakness in deductive reasoning *i.e.*, it relies heavily upon the initial premise(s) being correct. If this premise is incorrect, not only does it jeopardize the deductive reasoning, but the whole process of logic. For example, weather forecasting is an area, where deductive reasoning probabilities are often used. A meteorologist will look at certain parameters and based on his skill and judgement, conclude that, there may be rain today. But, note that, he will never say, it is definitely going to rain today because, the weather is too unpredictable, and he can never be sure that, their initial assumptions are correct.

Examples for Inductive Approach

Example 1

- If Ravi eats mangoes continuously, his MU gained from the consumption of successive mangoes will gradually decrease (Specific, based on a direct observation).
- If Sri Ram eats grapes continuously, his MU utility gained from the consumption of successive grapes will gradually decrease (Specific, based on a direct observation).
- So, the more we consume continuously, there will be gradual fall in the MU gained from the commodity. (General, can be applied to any individual).

Example 2

- Ravi's blood is red in colour. (Specific, based on a direct observation).
- Sri Ram's blood is red in colour. (Specific, based on a direct observation).
- All human beings blood is red in colour. (General, can be applied to any individual).

Example 3

- $6 + 3 = 9$
- $4 + 3 = 7$
- $10 + 7 = 17$
- Therefore, an even number when added to an odd number will result in an odd number.

Example 4

- *'Previous accidents of this sort were caused by brake failure, and therefore, this accident was caused by brake failure'.*

Example 5

'Ravi writes a letter to his father highlighting his selection into Acharya N.G. Ranga Agricultural University'. This event is described as an inductive process because:

- the format of the letter is always specific, as it always starts with 'Respected Father' and ends with 'Yours loving son' followed by name (particular).
- the job selection (*i.e.*, event under description) and the language (use of words) differs from student to student (general).

Example 6

Ravi submitted a thesis *'Performance of Regulated Agricultural Markets in Andhra Pradesh'*. This assignment falls under inductive process because,

- the format of the thesis is always specific, as it always starts with 'Abstract', 'Introduction' followed by 'Review of Literature', 'Materials and Methods', Results and Discussion', 'Summary and Conclusions' and 'Literature cited' (particular).
- the selection of research problem, and the interpretation of analytical findings of the data differs from student to student (general).

Example 7

We know, Dmitri Mendeleev is widely recognized for periodic table of elements. This table is a tabular display of the chemical elements, organized on the basis of their atomic numbers, electron configurations, and recurring chemical properties. The 118 chemical elements in the periodic table are divided into several groups on the basis of electronic configurations, similar chemical properties etc.

Thus, if the properties of individual elements (like melting point, boiling point, ionization energy etc) are known (inductive), these elements can be easily classified into different groups. So, the chemical properties of each group can be easily predicted (deductive) with very few exceptions. Thus, it proceeds from specific to general and so is an example of inductive method.

Now the students are well aware that, deductive logic moves from general to particular and inductive logic moves from particular to general. But, note that, an argument that is deductively invalid can be inductively valid and deductively valid but inductively invalid. Please look at the following examples:

1. Deductively Invalid, but Inductively Valid

- ☆ Ravi's rose is red (Particular).
- ☆ Sri Ram's rose is red (Particular).
- ☆ All roses are red (General).

2. Deductively Valid, but Inductively Invalid

- ☆ If a person possess more of anything, he derives less and less additional satisfaction from each additional unit (General, can be applied to any individual).
- ☆ Ravi derives more and more additional satisfaction for every additional rare stamp he collects (Particular).
- ☆ The more number of books Sri Ram reads, more and more is the additional satisfaction gained (Particular).

3. Both Deductive and Inductive Logics are Not Valid

- ☆ If something is red, it is a Rose (General).
- ☆ Ravi's car is red.
- ☆ So, Ravi's car is also a Rose (Particular).

4. Both Deductive and Inductive Logics are Valid

- ☆ All men are mortal.
- ☆ Ravi is a man.
- ☆ So, Ravi is mortal.

From the Figure 2.7, we can say that, inductive method is a predecessor of deductive method. Deduction is a process particularly suitable for testing a hypothesis, whereas induction is most suitable for hypothesis generating and in exploration of new fields. The probability in induction is raised to certainty in deduction. So, the combination of these two approaches is most appropriate and desirable. For example, when a Professor teaches the students, *'his teaching should begin with induction and end in deduction'*. That is, the Professor has to establish a theory or principle or formula first (induction) and then explain about its application to the practical situation (deduction). Further, note that, *'always understand inductively and apply deductively'*.

F. Based on the Time Dimension in Research

Some research studies analyze the problem at a point of time, whereas some studies the same problem over a period of time. So, considering the angle of time, research could be divided into the following two broad types:

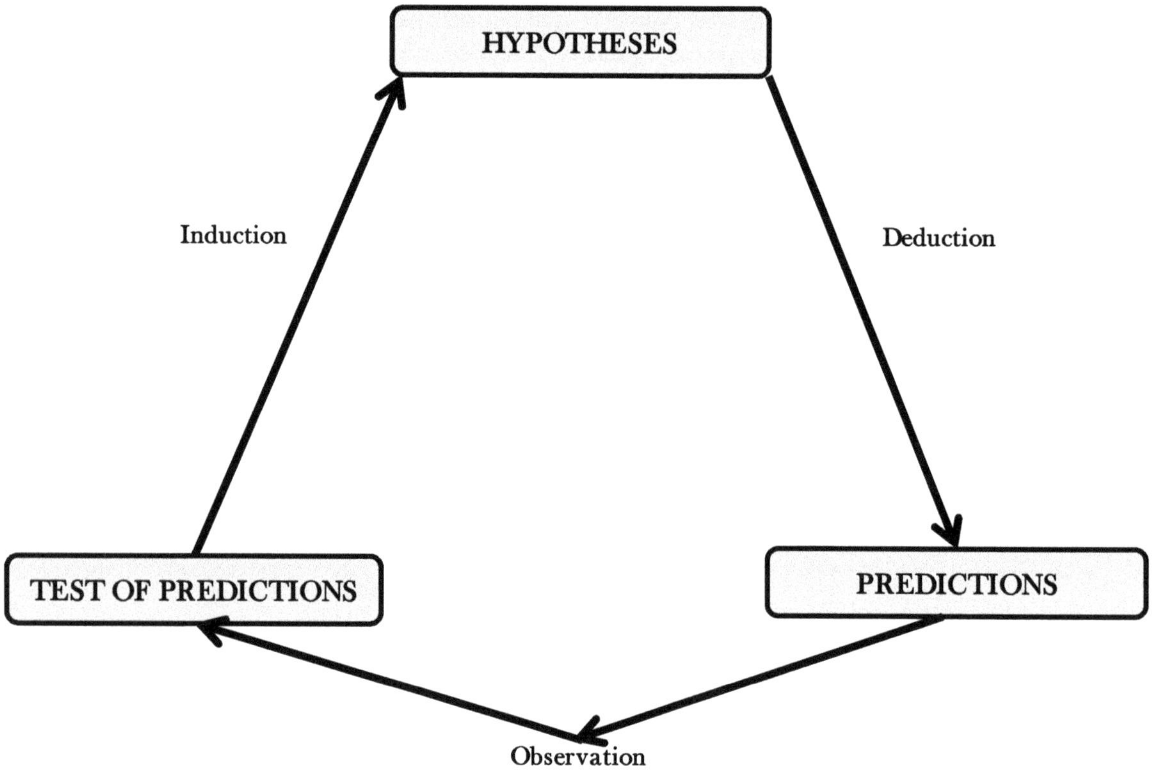

Courtesy: Experiment-Resources.com

Figure 2.7: Inductive Method is a Predecessor of Deductive Method.

a. Cross-Sectional Research

In cross-sectional research, the researcher studies the research problem at a point of time. It is the most frequently used type of descriptive design in marketing research. This type of research is simple and cost-effective. Cross-sectional study is possible through survey. The survey can be conducted from the population (census) or from the sample selected that is representative of the population (sample survey). The time required for sample survey is also less than a census-inquiry. This survey method (large sample) is also economical, as more information can be collected per unit of cost. But, the disadvantage of this method is that, it cannot capture the change processes in the problem over a period of time. This research can be exploratory or descriptive or explanatory, but it is most consistent with a descriptive approach to research. So, cross-sectional research provides a 'snapshot' of the outcome of the research at a specific point in time. For example, economic analysis of paddy in Kurnool district of Andhra Pradesh, studying the perceptions of farmers about the high yielding variety seed cultivated during Kharif season in 2012-13 falls under cross-sectional research.

b. Longitudinal Research

In this research, the researcher analyzes the same problem over a period of time involving a fixed sample of population, which is measured repeatedly. That is, this research captures the change processes in the problem over a period of time (Figure 2.8). He interviews the same sample at regular

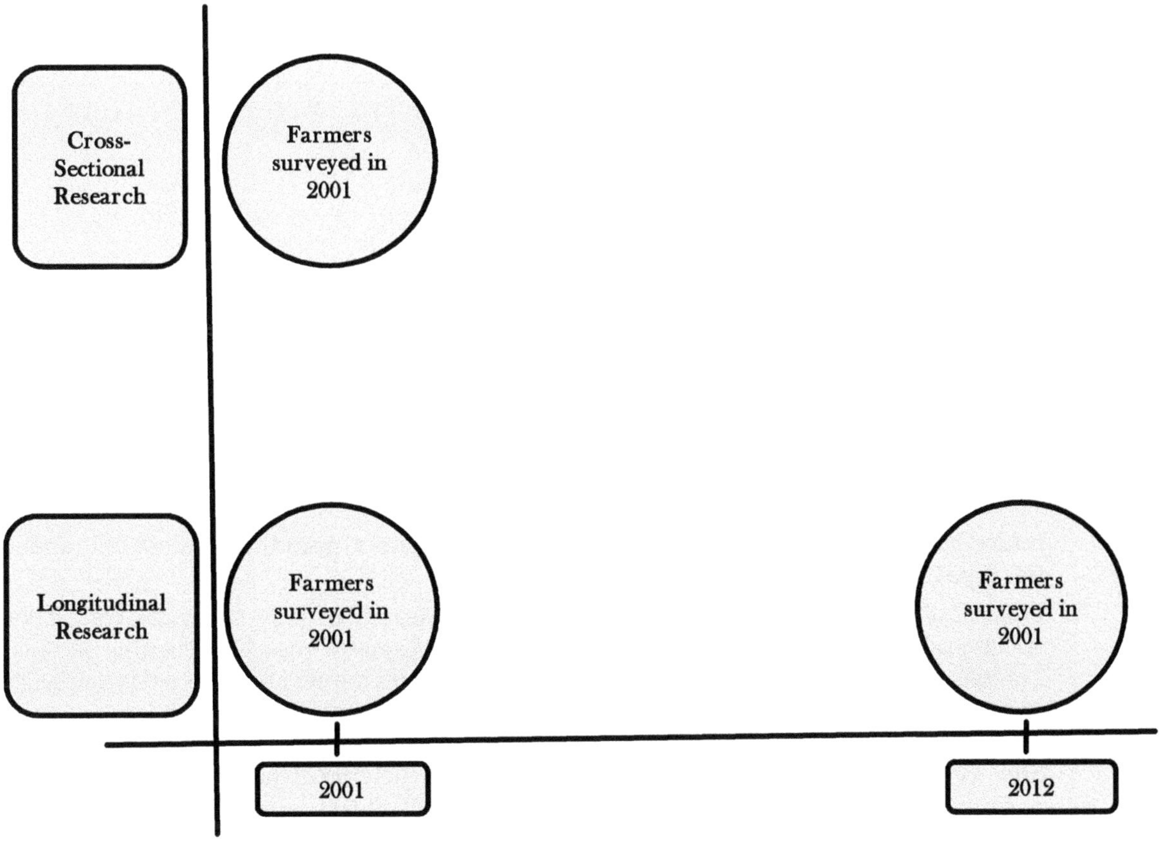

Figure 2.8: Sample Survey in Cross-Sectional Research and Longitudinal Research.

intervals. The longitudinal studies involve collecting the data on the same parameters at different time periods, but the cross sectional research entails collecting the data *at* one point in time. Thus, cross-sectional research is focused on finding relationships between variables at one moment in time, unlike longitudinal research. So, this enables the researcher to fit the trend analysis for the parameters under consideration in longitudinal research. This research also describes the patterns of change of the selected variables and help to establish the direction and magnitude of causal relationships. The measurements are taken on each selected variable at regular intervals and this enable the researcher to measure change in variables over time. Further, longitudinal data tend to be more informative than cross-sectional data because they measure changes in behavior, large amounts of data can be collected, and they are more accurate (Table 2.4). Yet, they suffer from potential bias and non-representativeness. But, the disadvantage is that, this research is more complex and involves huge costs compared to cross-sectional research. Further, it needs a larger sample size (accurate sampling to reach representativeness) and takes a longer period of time to gather results. There are three sub-types of longitudinal research *viz.*, Time series research, Panel research and Cohort research.

- ☆ ***Time series research:*** This is the longitudinal study in which the same type of information is collected on the selected parameters over a period of time. This enables the researcher to analyze the trend in the happening of that event and can also forecast the same for the

Table 2.4: Cross-Sectional Research vs. Longitudinal Research.

Item	*Cross-Sectional Research*	*Longitudinal Research*
Detecting change	Worse	Better
Amount of data collected	Worse	Better
Accuracy of findings	Worse	Better
Representativeness	Better	Worse
Response bias	Better	Worse
Relationship between variables	At a point of time	Over a period of time
Trend analysis	No scope	Scope to analyze trend and forecast the data
Information	Less informative	More informative
Complexity of research	Less complex	More complex
Analyzing the problem	At a point of time	Over a period of time

future. For example, price trends of paddy in Kurnool district of Andhra Pradesh falls under this type of research.

- ✰ ***Panel study:*** This is a powerful type of longitudinal research. In panel study, the researcher selects exactly the same sample (respondents) again and again across different time periods. This helps the researcher to analyze the change processes of the event or selected parameters more exactly over a selected time frame. But, the conductance of this research is not so easy on the part of the researcher because, it is difficult to find out the same respondents every time, as some people may die or cannot be located. Further, panel members may not be co-operative or may leave the panel membership. As a result, the representative character of the original sample may be adversely affected. Secondly, panel members may report wrong data. Their interest may reduce gradually and they may supply information in a casual manner. Their sense of participation/responsibility may reduce. This will affect the quality of data and also of findings. But, if conducted, the results are utmost valuable. For example, studying the perceptions of farmers regarding the effectiveness of a pesticide over a period of time falls under this type of research.
- ✰ ***Cohort:*** This cohort analysis is similar to the panel study, but in the context of non-availability of earlier selected respondents, the researcher will observe and select the other respondents, who share a similar life experience in that specified time period. The main objective of cohort study is to continue the longitudinal (panel) research even in the absence of (some) earlier respondents. Commonly used cohorts include all employees retire in one or two year time frame, all people who graduate in a given year, all farmers who underwent training on a particular technology etc.

G. Other Types of Research

As discussed above, there are a number of classifications of researches based on several criteria. However, a number of researches do not fall squarely on any of the above mentioned classifications; hence, a manner of classifying the said researches to a group is needed by simply listing them as belonging to a typology such as the following:

a. Experimental Research
b. Survey research
c. Causal-Comparative Research
d. Case Study
e. Correlational Research
f. Historical Research
g. *Ex-post* facto research

a. Experimental Research

This type of research is known in literature by a variety of names. Synonyms are, for instance: the Cause and Consequence method, Before and After design, Control group design and the Laboratory method etc. Experimental research is basically the method that can be applied in a research laboratory or even in farm fields. This research attempts to determine *'how'* and *'why'* something happens. It also guides the researcher to know, *'does something cause an effect?'*. For example, does a low student-teacher ratio in Agricultural Colleges cause higher student achievement? It tests the way in which an IV (the factor that the scientist manipulates) affects a Dependent Variable (DV - the factor that the scientist observes). A researcher generally goes for conducting an experiment (say, field experiment on paddy crop in a college farm) when there is time priority in a causal relationship (cause precedes effect), there is consistency in a causal relationship (a cause will always lead to the same effect), and the magnitude of the correlation of the variables is significant. In the words of Landman, *'experiential research is the research designed to study cause and consequence'*. The classic experimental design contains an experimental group and a control group. In case of experimental group, the treatments (different doses of fertilizers, different types of pesticides etc) are applied to the experimental material or plots. Such treatments are not applied to the control plot. This enables the researcher to compare the impact of treatments on the same DV (say, output of paddy) identified in both the groups. Even the treatments are replicated (repetition of treatments) to improve the precision of the experiment.

A number of factors can affect the outcome of any type of experimental research. The number of replications of each treatment directly influences the precision and accuracy of conducting the experiment. Another is experimenter bias, *i.e.*, in which way the researcher is expecting results from the experiment (*i.e.*, what should or should not happen in the study). Still another is controlling for extraneous variables such as room temperature, humidity etc. In case of field experiments, it is difficult control extraneous variables like rainfall, drought that may adversely affect the precision of the experiment and they are collectively called as 'error'. It is to be noted that, only when the experimenter carefully controls for extraneous variables, he can derive valid conclusions from the experiment.

Types of Experimental Research

We know, in experimental research, the researcher manipulates or varies an IV(s) (treatment) and measures its effects on DV. Experiments can be done in the field or in a laboratory. They can involve plant or human or animal subjects. What distinguishes the type of experiment is the *'degree to which the experimenter can assign subjects to conditions'*. There are two types of experimental research *viz.*, True experimental research and Quasi-experimental research. The purpose of both is to examine the cause of certain phenomena.

- ***True experiment research:*** In True experiment research, all the important factors that might affect the phenomena of interest are completely controlled. True experiments are considered the most accurate form of experimental research. They are commonly used in the physical sciences, such as chemistry, geology and physics, to prove or disprove a hypothesis, or theory, but are more difficult to apply to social sciences, such as biology or psychology. The characteristic feature of a true experiment is that subjects (say, experimental material) are randomly assigned. This means that, any difference between them is purely due to chance. For simplicity, just one variable is manipulated and tested. In true experiments, randomization can be achieved by using a computerized random number generator to assign treatments to subjects. The true-experimental design ensures greater internal validity*. However, it suffers from less external validity (not like real world conditions) and not very practical.

- ***Quasi-experimental research:*** The prefix 'quasi' means, in essence, 'sort of'. So a quasi-experiment is a 'sort of' experiment. Since, it is not possible or practical to control all the key factors, so it becomes necessary to implement a Quasi-experimental research design. Considering this, quasi experimental research could be described as a best attempt at an experiment when it is impossible, or not reasonable, to meet all the criteria of a true experiment. Quasi experiments, also known as Natural or *In-situ* experiments. In these experiments, the subjects (say, experimental material) are beyond the control of the experimenter and are not randomly assigned. That is, in this research, the researcher does not randomly assign subjects to the treatments, usually because random assignment is not feasible. These designs are usually constructions that already exist in the real world. A quasi-experimental design will have some sort of control and experimental group, but these groups probably weren't randomly selected. Since, there is a loss of control in the quasi experimental design, it is necessary for the researcher to decide what and when to measure. So, quasi experiments are typically less powerful research tools than true experiments because of the possibility of introducing systematic errors. But, the general goal of quasi experimental research is to investigate cause and effect relationships, like true experiment research. This approach to research allows for greater understanding of programme features and practices. Quasi experiments are often used to investigate subjects other than people, such as plots of land in an agricultural farm or in forest areas (Table 2.5). In quasi experiments, non-random

* The two types of validity of experiments are internal and external. It is often difficult to achieve both, in social science research experiments. When an experiment is internally valid, we are certain that the IV(s) caused the outcome of the study (output). So, internal validity refers to the condition that, observed differences on the DV were a direct result of manipulation of the IV(s). External validity is the extent to which the results of a research study can be generalized to other situations and to other people. That is, external validity refers to how generalizable the results of the study are, or how applicable the results can be applied to groups outside the experimental setting. That is, external validity refers to the extent to which the results of an investigation can be generalized to other samples or situations.

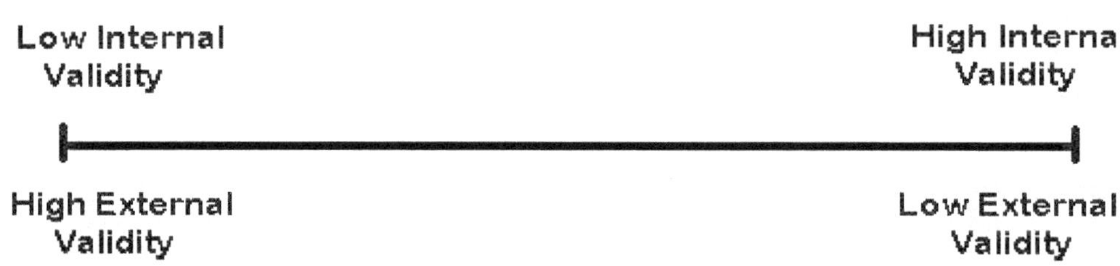

assignment based on first-come, first-served or some other criterion can significantly influence the outcome of the experiment. This research ensures greater external validity* (more like real world conditions) and much more feasible given time and logistical constraints. This research is used extensively in the social sciences and psychology.

Both true experiment research and quasi experiment research attempt to analyze cause and effect relationship. True experiment research accomplishes the same goal through random assignment of subjects (say, experimental material like land or individuals) to interventions or treatments. Quasi experiment research tends to do this through a comparison of existing groups or through non-randomization of subjects.

Table 2.5: Applicability of True Experiment Research and Quasi Experiment Research.

Location	*Type*	
	True Experiment	*Quasi Experiment*
Lab	Common	Unusual
Field	Less common	Common

Similarities between True Experiment Research and Quasi Experiment Research

The following are the similarities between true and quasi-experiments:

- ☆ The experimental material are subjected to some type of treatment or condition.
- ☆ Some outcome of interest (say, output) is measured.
- ☆ The researchers test whether differences in this outcome are related to the treatments employed.

Differences between True Experiment Research and Quasi Experiment Research

They include:

- ☆ In a true experiment, subjects are randomly assigned to either the treatment or the control group, whereas they are not assigned randomly in a quasi-experiment.
- ☆ The term quasi-experiment refers to a type of research design that shares many similarities with the traditional experimental design or randomized controlled trial, but specifically lacks the element of random assignment. With random assignment, participants have the same chance of being assigned to a given treatment condition. As such, random assignment ensures that both the experimental and control groups are equivalent. In a quasi-experimental design, assignment to a given treatment condition is based on something other than random assignment. Factors such as cost, feasibility, political concerns, or convenience may influence how or if participants are assigned to a given treatment conditions, and as such, quasi-experiments are subject to concerns regarding internal validity (*i.e.*, can the results of the experiment be used to make a causal inference?).
- ☆ Since, quasi experiments lack random allocation of subjects or proper controls, a firm statistical analysis can be very difficult, unlike true experiment research. So, the results of quasi experiment research are less powerful compared to true experiment research.
- ☆ Since quasi-experimental designs are used when randomization is impractical and/or unethical, they are typically easier to set up than true experimental designs, which require random assignment of subjects.

- In a quasi-experiment, the control and treatment groups differ not only in terms of the experimental treatment they receive, but also in other, often unknown or unknowable, ways. Thus, the researcher must try to statistically control for as many of these differences as possible.

Importance of Experimental Research

The experimental research guides the researcher in the following aspects:

- It helps the researcher to know the extent of correlation between the variables, cause and effect relationship between the variables, how significantly the causal variable(s) influences the DV.
- Permits the researcher to distinguish placebo (fake) effects from treatment effects.
- Provides highest level of evidence to the researcher regarding the influence of treatments on the DV and in comparison with the control, it further highlight the impact of the treatments.

However, the experimental research suffers from the major limitation that, the results generated are applicable to that practical situation or conditions only, but cannot be generalized. If the conditions change, the results will also change. So, it is not possible to standardize the results generated from the experimental studies.

Descriptive Research vs. Experimental Research

There are several differences between descriptive research and experimental research and they are listed here under.

- The conductance of descriptive research is easier compared to experimental research.
- Experimental research is meant for explaining cause and effect relationship between the variables, unlike descriptive research (which simply explains or describe the characteristics of the variables and their correlation).
- Compared to experimental research, descriptive research is fairly straight forward to design and conduct the research. But, in experimental research, the researcher besides conducting the design and process, he has to look into various aspects like replicating the treatments, minimizing experimenter bias, controlling extraneous variables etc.
- In experimental research, the data (findings) generated are 'original' in the sense that, it is the research conducted for the first time for that given practical situation. However, in descriptive research, no doubt, data is also generated, but it will sometimes incorporate the raw data collected from other previous experimental research.
- Descriptive research is usually conducted in social sciences, whereas experimental research is used in case of biological sciences. However, Psychology, is one example of a social science, where experimental research is often used.

b. Survey Research

The survey is a non-experimental research method. Surveys can be useful when a researcher wants to collect data on phenomena that cannot be directly observed. Surveys are used extensively in social sciences research to assess attitudes and characteristics of a wide range of subjects, analyze the collected data and to draw meaningful conclusions. In a survey, researchers *sample* a *population*. Basha and Harter (1980) state that '*a population is any set of persons or objects that possesses at least one common characteristic*'. Since populations can be quite large, researchers directly question only a *sample* (*i.e.* a small proportion) of the population. Examples include: sample survey among the farmers to study cost of cultivation of paddy in Kurnool district of Andhra Pradesh, sample survey among the farmers and market intermediaries to analyze the marketing efficiency in transacting paddy in Kurnool district of Andhra Pradesh etc. In conducting surveys for agro-economic research, the researchers employ 'schedule' as a tool to collect the information from the sample respondents. The surveys can help to derive both qualitative (*e.g.*, ask open-ended questions in case of questionnaires employed Agricultural Extension studies) and quantitative (*e.g.*, use forced-choice questions) data or information from the respondents. There are two basic types of surveys *viz.*, Cross-sectional surveys and Longitudinal surveys based on the time dimension involved in gathering information or data in the research. These two types are already discussed in the earlier pages.

Advantages of Survey Research

- ☆ Surveys allow researchers to collect a large amount of data in a relatively short period of time.
- ☆ Surveys are less expensive than many other data collection techniques.
- ☆ Surveys can be conducted quickly and administered easily.
- ☆ Surveys can be used to collect information on a wide range of things, including personal facts, qualitative and quantitative data etc.

Disadvantages of Survey Research

- ☆ Poor survey construction and administration can lead to wastage of resources.
- ☆ The answer choices provided on a survey may not be an accurate reflection of how the respondents truly feel.
- ☆ While random sampling is generally used to select participants, response rates can bias the results of a survey.

c. Causal-Comparative Research

Causal-comparative research, like correlational research, seeks to identify associations among variables. The aim of causal-comparative research is to determine the cause of existing differences among groups. The basic causal-comparative design involves selecting two groups that differ on a particular variable of interest and then comparing them on another variable or variables. Correlational research involves collecting data on two or more variables on one group, whereas causal comparative research involves the collection of data on one IV for two or more groups. That is, the basic causal-comparative approach is to begin with a noted difference between two groups and then to look for possible causes for, or consequences of, this difference. When an experiment would take a considerable length of time and be quite costly to conduct, a causal-comparative study is sometimes used as an alternative. As in correlational studies, relationships can be identified in causal-comparative study, but causation cannot be fully established. The basic similarity between causal-comparative and correlational studies is that, both seek to explore relationships among variables. When relationships are identified through causal-comparative research, (or in correlational research), they often are studied at a later time by means of experimental research. For example, people with lung cancer smoke more number of cigarettes than people without lung cancer. There are no other differences in lifestyle characteristics between the groups. So, it can be concluded that, smoking is a possible cause of lung cancer. The following are the merits of causal-comparative research:

- ☆ Used to establish cause-effect relationship when experimental design is not possible.
- ☆ Less expensive and time consuming than experimental research.

The other examples of this causal-comparative research include:

- ☆ A researcher measured the managerial ability of farmers in cultivating paddy who had undergone training at Farmers Training Centre in Kurnool district and compared the scores with a group of similar farmers who had not been to training.
- ☆ A researcher measured the performance of students at Agricultural College, Mahanadi where special coaching to JRF examination was imparted and compared that to other Agricultural College, Tirupati, which did not offer special coaching to JRF examination.
- ☆ A researcher formed three groups of students in Agricultural College, Mahanandi: those who achieved Overall Grade point Average (OGPA) above 8.5, those who achieved OGPA between 7.5 and 8.5 and those who achieved OGPA less than 7.5 – and then compared these three groups on a test to evaluate their knowledge on 'General Agriculture'.

Similarities between Causal-Comparative Research and Correlational Research

They include:

- ☆ In both researches, variables cannot be manipulated (such as age, sex, parent educational level, socio-economic status etc.).
- ☆ Caution is needed in both researches to interpret the results.
- ☆ Both these researches support subsequent experimental research.

Differences between Causal-Comparative Research and Correlational Research

They include:

- ☆ In causal comparative research, attempts are made to understand cause-and-effect relationships, unlike correlational research.
- ☆ At least one variable is IV in causal comparative research, whereas in correlational research, there is no distinction between variables as DV and IV.
- ☆ In causal comparative research, there are two or more groups for comparison, whereas in correlational research, there is only one group.

d. Case Study

A Case study (also known as a Case report) is an intensive analytical study of a particular event or unit or person so that, it guides the researcher to probe into the problem in-detail, analyze and suggest developmental strategies for it. Thus, case study relates to a problem that has to be solved by the researcher. The case-study method was first introduced into social science by Frederic Le Play in studying family budgets in 1829. The findings generated from this case study makes the researcher to suggest effective strategies and this serves to address similar cases. The interesting aspect of case study is that, it makes the researcher to involve deeply into the problem. In case study, the information is arranged in such a way that, the reader of the study is put in the same position as the researcher, when he first faced the problem in his inquiry. So, a good case study is more than just a description. However, a description, on the other hand, arranges all the information, comes to conclusions, tells the reader everything, and the reader really doesn't have to work very hard. The conductance of case study generally involves three basic steps *viz.*, collecting information, analysis, and the actual writing. The researcher start the case study by collecting or gathering the relevant information, but even when he reach the final writing stage, he may still add some more information by going back and research even more information. Thus, case studies provide a systematic way of understanding, why the instance happened, what are the causal factors, what are the strategies to be planned to overcome the problem, what aspects are to be considered to look at more extensively in future research etc. These studies lend themselves to both generating and testing hypotheses. Case studies should not be confused with qualitative research, as the case study may harbor both quantitative and qualitative information. The case study is common in social sciences and life sciences. Case studies may be descriptive or explanatory. The latter type is used to explore causation in order to find underlying principles. They may be prospective (in which criteria are established and cases fitting the criteria are included as they become available) or retrospective (in which criteria are established for selecting cases from historical records for inclusion in the study). An example for case study include: Mumbai's Dabbawalas - Amazing Management Success Story in India'.

Strengths of Case Study

They include:

- ☆ This research enables the researcher to have in depth understanding about the problem.
- ☆ This research guides the researcher to extend the same solutions to other similar problems. In this contexts, Social scientists, in particular, make wide use of this research design.
- ☆ The detailed description of unique cases enriches the body of knowledge.

But, the demerit of this research is that, the findings cannot be generalized, as the selected cases are special or unique in nature. Further, the in depth involvement of researcher into the problem often leads to biased results and interpretations.

e. Correlational Research

Correlational research simply studies the nature of relationship (positive or negative or no relationship) and degree of relationship between two or more variables. For example, we can study the correlation between fertilizer cost and output of paddy, pesticides cost and output of cotton, price of the commodity and its quantity demanded, price of the commodity and its quantity supplied etc. The degree of relationship between two variable is expressed as a Correlation coefficient. The value of correlation coefficient ranges between -1 to +1. That is, correlations are either positive (0 to +1.0), negative (0 to –1.0), or non-existent (0.0). In a positive correlation, the values of the variables increase or decrease ('co-vary') together. In a negative correlation, one variable increases as the other decreases. In a non-existent correlation, no relationship exists between the variables.

Students often confuse between correlation and causation. It is to be noted that, correlation does not mean causation *i.e.,* it won't indicate cause and effect relationship. Rather, it indicates only the relation and degree of association between the selected variables. But, causation infers cause-and-effect relationship. When a correlation exists, changes in the value of one variable reflect changes in the value of the other. But, it does not indicate, which is the 'cause' and which is the 'effect'. (When there exists a cause and effect relationship between two or more variables, we say that causation runs from *cause* variable to *effect* variable. *Causation,* then means changes in one variable brings about changes in another variable. The cause-effect relationship is often called the dependence relationship. But then *how* does a researcher decide on which variable depends on what variable? It is deliberately chosen and imposed by researchers, usually from theory or existing studies on the subject.) To study the effects that variables have on each other, a researcher must conduct a survey or an experimental research. After ascertaining the degree of relationship between the selected variables (*i.e.,* correlation), the researcher can then use the same to analyze the cause and effect relationship (*i.e.,* the rate at which DV changes per unit change in the IV), a technique called Regression analysis. When there is only one independent or predictor variable, the technique is called Simple regression analysis. When there is more than one independent or predictor variable, the technique is called Multiple regression analysis.

Spurious Correlation–The Third-Variable Problem

There is another form of correlation between the variables called Spurious correlation, which refers to the correlation between two variables (A and B variables) that does not result from any direct relation between them, but from their relation to other variable (C). This variable 'C' is an unseen factor referred to as a Confounding factor' or 'Lurking variable'. For example, A causes B, B causes A, or 'C' causes both A and B. In the last case, there is a spurious correlation between A and B. In a regression model, where A is regressed on B, but C is actually the true causal factor for A, this misleading choice of IV (B instead of C) leads to Specification error and variable 'C' is causing third variable problem called Spurious variable. An important example that we commonly notice is that, 'Tea drinking and lung cancer are negatively correlated'. This is because of the spurious variable *i.e.*, 'smoking' *i.e.*, tea drinkers have a lower risk of lung cancer because, they smoke less.

It is to be noted that, when correlation coefficient simply indicate only degree of association between two or more variables and not cause and effect relationship between the variables, it gives the impression that, correlational research has not much important place. But, it has, considering the following reasons:

- ✰ After establishing the relationship (association) and degree of relationship between the selected variables, and if the selected variable have significant positive or negative correlation, then only the researcher shows the interest to conduct cause and effect analysis and he will conduct more studies to determine 'why' such relationship?
- ✰ In case, when the Correlation coefficient indicates no (significant) association between the variables, the researcher need not go for cause and effect analysis. This saves time and resources on the part of the researcher and he can plan for alternative study.
- ✰ When the major alternative to correlational research (*i.e.*, experimental research) cannot be conducted for practical or ethical reasons, in such cases, correlational research might be the best approach available.

Importance of Correlational Research

They include:

- ✰ The results of this research could be used for making predictions.
- ✰ This research forms the basis for an experimental study.
- ✰ Correlational research studies conducted influences the decision making of farmers in application of inputs in crop production programmes.

f. Historical Research

Historical research is the type of research that examines past events or combinations of events to arrive at an account of what has happened in the past. Thus, historical research is conducted to attend the following objectives:

- ✰ To compare the past events with the present.
- ✰ To answer questions about the past events.
- ✰ To analyze the development process from past to the present.

- To study the relationship between past events with the present happenings.
- To evaluate the present events in comparison with the past events.
- To derive new solutions to the present problems through studying the solution measures taken in the past.

We know that, what we see or do today is specifically rooted in the past. So, this research guides the researchers to understand and address the present problems by taking the guidance from the past events. This is because, if the researcher addresses the present problem (if it is prevailed even in the past) by analyzing the solutions taken in the past, he need not reinvent the solutions from the beginning. This saves lot of time and resources on the part of the researcher. The following five steps are involved in conducting historical research:

- Identification and formulation of the research problem that addresses or relates to past events.
- Reviewing the literature and data (both primary and secondary) about the past with reference to the problem under study.
- Critical analysis of literature and data.
- Offering various solutions for the present problem.
- Prioritizing the solutions for the present problems duly considering the present situation.

It is essential to note that, in this research, the researcher devotes much of the time in collecting, reading and analyzing the past research material and preparing the manuscript to address the present problem in tune with the present situation. The researcher often goes back-and-forth between collecting, reading, and writing. For example, when a researcher studies the trends in exports and imports of major agricultural processed products during both pre-WTO and post-WTO regimes, the lending pattern followed for agricultural sector in the 19th century etc., they fall under historical research.

Strengths of Historical Research

They include:

- Historical research guides the researcher to analyze causes and effects of a problem in the past and this helps to derive the trend for the present situation.
- This research may help in explaining present events and anticipating future events. To make such future predictions reliable and trustworthy, the historical researcher needs to identify and clearly describe in which ways the past events differs from the present context and how the present Political, Economical, Social and Technological (PEST) factors could have an impact on the present and the future.
- This research saves time and resources on the part of the researcher, as he need not re-invent solutions for the problem under consideration from the beginning, as he take the guidance from the past happenings.
- It enables the researchers to find out solutions to contemporary problems which have their roots in the past *i.e.*, it serves the purpose of bringing about reforms in solving that problem.
- Studying the past enables a researcher to understand the factors/causes affecting present trends.
- It enables the researcher to understand how and why the theories or principles are developed.

- This research design is unobtrusive *i.e.*, the act of research does not affect the results of the study.
- This research is well suited for trend analysis.
- This research provides requisite background material both in terms of descriptive information and data to address the present problem.
- There is no possibility of researcher-subject interaction that could affect the findings.

However, this research suffers from a major limitation that, it is time consuming (as reviewing the past consumes much of the time of the present research process).

g. *Ex-post Facto* Research

'*Ex-post facto*' is a Latin expression that when translated literally means, 'from what is done afterwards'. In the context of social science research, the phrase means 'after the fact'. In this research, the researcher investigates the possible cause-and-effect relationship by observing an existing condition or state of affairs and searching back in time for possible causal factors. In scientific terms, it could refer to a research conducted into a particular phenomenon conducted 'after' it has taken place. *Ex-post facto* research is the process beginning with a phenomenon and going backward in time to identify casual factors. That is, in *ex-post facto* research, both the effect and the alleged cause have already occurred and are studied by the researcher in retrospect. So, in *ex-post facto* research, the researcher has no control on the variables or he cannot manipulate the variables (IVs), which cause a certain effect (DVs) being measured. Since this type of a research lacks manipulation of variables, the cause-effect relationships measured are only tentative. Generally, *ex-post facto* studies fall into the category of descriptive research. It is a non experimental research method.

The following example will illustrate the basic idea of *ex-post facto* research: Assume that, since 2005, there has been a dramatic increase in the number of suicides of farmers in the State of Andhra Pradesh. So, the Government called an expert to investigate the reasons behind the farmers' suicides. Naturally, there is no way on the part of the researcher to conduct the survey on the farmers, who committed suicides because, suicides have already happened. Moreover, there is no way of using video technology to view the incidents. So, the researcher has to choose the *ex-post facto* research method to elicit the causes for the farmers' suicides. The researcher will collect certain data like acreage of farmers committed suicide, amount indebted to banking institutions, amount indebted to private money lenders, yield of the crop harvested, market prices of commodities, rate of interest changed by banking institutions and private money lenders, increase in cost of cultivation of crops etc., from the respective farm family members. By collecting this data, the researcher can assess the determinants responsible for farmers' suicides. On the basis of these data, the researcher can formulate hypotheses and analyze the data. The findings will be submitted to the Government of Andhra Pradesh with policy suggestions. The point of interest here is, the researcher identified the causes retrospectively by following an *ex- post facto* research method.

The above example infers that, in *ex-post facto* research, the IV(s) have already occurred and the researcher starts with the observation of a DV(s). He studies the IV(s) in retrospect for their possible relationship to, and effects on, the DV(s). The researcher is thus, examining retrospectively the effects of a naturally occurring event on a subsequent outcome with a view to establishing a causal link between them. Interestingly, some instances of *ex-post facto* designs correspond to experimental research in the reverse direction.

The above discussed typology of research infers that, the following questions are to be considered when determining a research 'problem'

- "What is actually happening in this situation"?
- "Would it still happen this way if ..."
- "In this context, if X does this, would Y then ..."
- "What causes X to react to Y in this way"

More concretely, when reviewing possible research "problems," the questions could include the following [points 1-3 would be preliminary to posing a theoretical approach, with points 6-8 required to 'test' the theory. Points 4-5 develop the working theory (or hypothesis) itself]:

1. What bothers the researcher about a particular "problem"? What information seems to be unclear, or incomplete, or missing, or improbable as stated?
2. What are the essential concepts and issues relevant to the problem?
3. Does the description or implementation of the problem vary? If so, how and why?
4. Can you state a relationship between the variables of your problem?
5. Can you hypothesize an answer?
6. Can the researcher collect primary and secondary data to test the hypothesis?
7. Can the researcher collect original data to further test the hypothesis?
8. Is the problem the researcher identify part of a larger problem?

2.11. Academic Levels of Research

We know, research is being conducted at graduation, post-graduation and doctoral levels. Whatever the level at which research is being performed, the basic principles, issues and practicalities remain the same. However, the sophistication of the research design and duration of the project will vary across these levels. The different academic levels of research are discussed here under.

At Undergraduate Level

At this level, the student undertakes a research project as a part of the course or it may even be a complete course. At this level, research is usually exploratory and/or descriptive. The important objectives of the research at this level include:

- To get familiarity with the basic concepts of the course.
- To understand about the applicability avenues about the course/topics.
- To understand about different types of research methods.
- To learn the basics about research methodology.
- To understand how to design a research project, plan the work and work the plan.
- To formulate research objectives and tentative hypotheses.
- To review relevant literature.
- To identify different sources of collection of data and collect requisite data.
- To understand the basics in writing research report and to submit the research report in scheduled time.

At Post-graduate and Doctoral Levels

Having understood the basic concepts about identifying researchable problem, conducting research and writing report, the student at post-graduation and doctoral levels is normally expected to conduct the research on systematic lines by conducting an in depth study on the prioritized researchable problem. The student has to submit a dissertation or thesis to the donor agency (say, Institute or SAU or Directorate etc). In addressing all the basic concepts of research learned at graduation level, further the student has to conduct a critical review of past literature and he should clearly infer that, whether the findings of the present research are in conformity with the earlier studies or generates entirely a new concept or theory. That is, the processes are very similar to undergraduate research, but a more comprehensive approach is needed and higher quality of work will be required at post-graduation and doctoral levels. At doctoral level, the student can even plan to extend the same work performed at M.Sc. (Ag.) level, but the work must be expanded both horizontally and vertically. At postgraduate and doctoral levels, research is always analytical or predictive. It is important to remember that, at these levels, the expectations of the institution or university will have a significant influence on the process and outcome of student's research. This text book offers guidelines to the students pursuing post-graduation and doctoral research programmes to follow the systematic procedures involved in conducting research. The concepts explained in this book also enable the Academic Researchers to write conference papers and journal articles.

2.12. Research Method vs. Research Design

Research method and Research design are two terms that are often confused as one and the same. Strictly speaking, they are not one and the same and they have some differences between them. As discussed in the earlier pages, there are different types (methods) of conducting research say, exploratory research, descriptive research, explanatory research, qualitative research, quantitative research, cross-sectional research, longitudinal research etc. Different research methods are used to address different types of research problems. Thus, research method is broad framework followed to conduct a specific research project. For example, if the researcher wishes to conduct research on 'economic analysis of paddy cultivation in Kurnool district of Andhra Pradesh', he will plan for cross-sectional research method. If the researcher wishes to conduct research on 'Trends in market arrivals and prices of paddy in selected APMCs in Kurnool district of Andhra Pradesh', he will plan for time period study (longitudinal research). So, the researcher has to choose the right research method that suits the requirements of the research project and he is comfortable with.

Research design is a specific outline that is followed within the selected research method to pursue the objectives framed in the project study. While preparing the research design of the project, it is necessary to think of about research method followed in the investigation. That is, research design depends upon the research method followed for the investigation. Research design involves the step-wise procedure that is followed towards the fulfillment of the objectives framed under the study. So, this research design serves as a blue print that the researcher should follow towards attending the problem in a systematic way. That is, the research design guides the researcher regarding what to do at what time. In the above example *i.e., 'Economic analysis of paddy cultivation in Kurnool district of Andhra Pradesh'* , research design indicates the sampling procedure to be followed regarding the selection of mandals, villages and sample farmers; method of data collection (through pre-tested schedules); techniques (tools of analysis) to be followed to analyze the data (Say, Cobb-Douglas production function, Cost Concepts etc.) etc.

The above explanation infers that, research methods are loose framework or guidelines from which one has to select one and then apply a research design on that method to achieve desired results. The researcher had to choose the right research method and well-planned research design to accomplish the research task. If this is not the case, the research findings will not be not satisfactory. Say, in the above example, if the researcher selects 150 farmers in total for data collection of which, 130 are large farmers, 5 medium farmers and 15 small farmers, then the analytical findings will not be accurate. The findings show biased results. In such a situation, because of a flaw in the research design, the researcher has to re-plan the research design (say, selecting 50 farmers from each category). Even in some cases, the researcher has to change the research method (say, employing descriptive research instead of explanatory research method) and this leads to changes in research design as well. The following points further highlight the distinctions between research method and research design:

- ✰ A research method is a general framework guiding a research project, whereas research design is a specific outline detailing how the chosen research method will be applied to answer a particular research question.
- ✰ Research methods are generalized outlines that provide a framework for conducting the research study. Research design refers to the blue print that the researcher uses within the selected the research method and it tells what is to be done at what time. That is, research design tells how the goals of a research project can be accomplished. Key features of any research design are sampling methodology, collection and analysis of data along with procedures and instruments to be used.
- ✰ Research methods are generalized and established ways of approaching research questions. Note that, not all research methods can be applied to the same research problem and the choice of research method is limited by the area and objectives of research that the researcher wishes to explore. Research design involves determining how the chosen research method will be applied to answer the selected research question.

2.13. Research Method vs. Research Methodology

Like research method vis-à-vis research design, the students often confuse another two terms *viz.*, research method and research methodology. As explained above, there are different types (methods) of conducting research say, exploratory research, descriptive research, explanatory research, qualitative research, quantitative research, cross-sectional research, longitudinal research etc. Different research methods are used to address different types of research problems. Thus, research method is broad framework followed to conduct a specific research project. For example, if the researcher wishes to conduct research on 'economic analysis of paddy cultivation in Kurnool district of Andhra Pradesh', he will plan for cross-sectional research method. So, the researcher has to choose the right research method that suits the requirements of the research project and he is comfortable with.

But, the term 'research methodology' seems to be much broader compared to 'research method'. Research method is a type of research that is followed to address a research problem. On the other hand, research methodology may be understood as a science that explains various methods in general by which the researcher may proceed with selected research scientifically. That is, it involves the learning of the various techniques that can be used in the conduct of research and in the conduct of experiments, surveys, case studies etc. Thus, research methods aim at finding solutions to a research problem, but research methodology aims at the employment of the correct procedures to find out solutions. So, to employ the right research method to address the selected research problem, first the knowledge about research methodology is to be known. Thus, research methodology is the beginning, whereas research methods are the end of research activity. It can also be said that, research methodology is the study of methods. Research methods provide the specific details of how a researcher accomplishes a research task. That is, it provides specific and detailed procedures of how to initiate, carry out, and complete a research task by mainly focusing on how to do it. On the other hand, research methodology deals with general approaches or guidelines in conducting research. It provides the general principles for organizing, planning, designing, and conducting research, but it cannot tell the researcher in detail how to conduct a specific, individual research. So, it is essential on the part of the researcher not only the research methods, but also about research methodology. In the above example, the researcher employs Cobb-Douglas production function to study the resource use pattern in paddy cultivation. It is not possible to study the trend in cost of production of paddy in Kurnool district from the above study. The research methodology will make the researcher to be aware of such aspects *i.e.*, what techniques can be employed and what cannot be employed and not the research method. Thus, the scope of research methodology is much wider compared to research method and the research methodology not only advocate the researcher, why to choose a certain research method and why not to choose other methods. So, the researchers/students must be good at both research methodology and research methods in accomplishing the research project successfully. The following points further highlight the distinctions between research method and research design:

- ☆ A research method is a general framework guiding a research project. But, research methodology involves the employment of various techniques in the conduct of research. That is, research method aims at choosing a right method for conducting the research, whereas research methodology aims at the employment of the correct procedures to find out solutions.
- ☆ Every research method will have its own methodology in conducting the research. For example, in case of field or lab experimental research, the researcher conducts a field experiment say, either in a farm or in a laboratory, unlike in case of agro-economic research. Thus, research methodology paves the way for research methods to be conducted properly. So, research methodology is considered as the beginning, whereas research methods are the end of any research.
- ☆ Research methods pave the way or show the outline for conducting the research towards finding a proper solution for the problem under consideration. Research methodology guides the researcher to employ appropriate procedures and techniques in the selected research method, so as to obtain solutions to a problem in hand. So, the researcher must get acquainted with different techniques to be employed across different research methods. Thus, research methodology is considered as the basic or the initial step of carrying out any kind of research. However, to achieve success in conducting the research in a particular field of study or on a particular topic, the researcher must gain excellence in both the research methods as well as research methodology.

2.14. Research Design vs. Research Methodology

These two are often considered as synonyms, but they actually differ. Research design stands roughly for a short outline of the research proposal. For example, if the researcher submits a project report to the funding agency, the outline of the research proposal that deals with nature of research method to be employed, time duration of the project, broad goals and objectives of the project, beneficiaries of the project etc. Note that, like research design, research methodology is also a plan, but it is much more thorough and comprehensive. That is, it clearly defines, how the beneficiaries will be selected, types of statistical techniques to be employed in the project, hypotheses framed for the study, variables to be identified, sources of data collection etc.

2.15. Importance of Research Methodology in Research Studies

It is necessary for a researcher to design a methodology for addressing the research problem under consideration. It is essential to note that, even if the research method considered in two problems are same, the methodology may be different. So, it is important for the researcher to know not only the research methods necessary for addressing the research problem under consideration, but also the methodology. For example, a researcher not only needs to know how to calculate mean, standard error, variance, different econometric models to employ for the set of data collected by him, whether to collect data through schedule or a questionnaire, whether to go for survey research or case study research but he should also know which is a suitable method for the chosen problem?, what is the order of accuracy of the result of a method?, what is the efficiency of the method? etc. Consideration of these aspects constitute a research methodology.

3

Research Process

For conducting research in a systematic manner, it has to pass through sequential steps right from the initial conceptualization stage to the final reporting stage and they collectively constitute Research Process. Thus, the scientific research process is a multiple-step process, where the steps are interlinked with the other steps in the process. If changes are made in one step of the process, the researcher must review all the other steps to ensure that the changes are reflected throughout the process. This research process is similar to undertaking a journey. In this research journey, the researcher takes two important decisions *viz.*, what he want to find out about? (*i.e.*, what research problem he considers for the study) and how to find answers for the problem under consideration? For answering these two questions, the researchers have to pass through the following steps (Figures 3.1 and 3.2):

A. Conception/Identification of Research Problem
B. Formulation of Research Problem
C. Review of Literature
D. Formulation of Testable Hypothesis
E. Preparation of Research Design (Methodology)
F. Execution of Research Project
G. Data Collection

H. Data Processing and Analysis
I. Testing of Hypothesis and Interpretation of the Findings
J. Writing Research Report

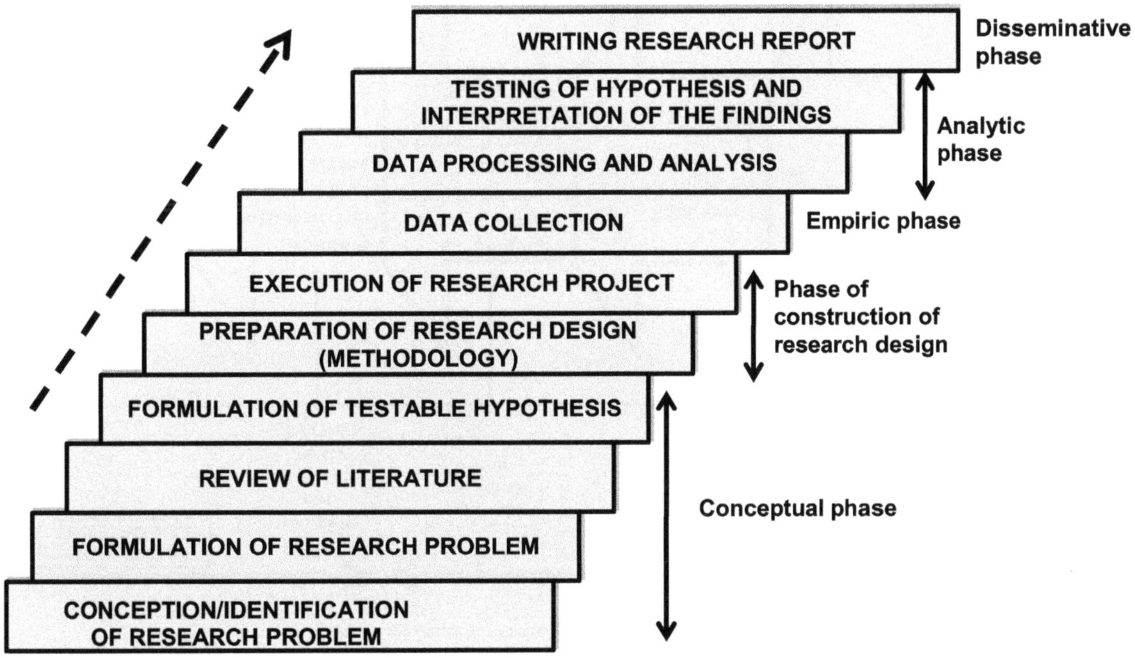

Figure 3.1: Different Phases and Steps in a Research Process.

It is to be noted that, whatever type of research or approach is adopted, these sequential steps in the research process are common to all scientifically based investigations. A close examination of the Figure 3.2 illustrates a highly structured view of the research process with one step leading logically on to the next step. This systematic process enables the researcher to proceed in the right direction right from conceptualizing the research problem till he arrives at conclusions and final submission of the report. This systematic process is documented in such a way that, another researcher can conduct the same study again after a certain period (to analyze the temporal variation) and at a different place. This is referred to as replicating the study. Thus, any research done with proper documentation of results enable others to review the process and this constitute the scientific research process.

However, in practice, research is rarely like that. For example, if the researcher fails at one stage, he has to go back to the earlier stage and thus, many stages will overlap. For example, if the researcher could not get adequate data for the research (for example, cost of production data of paddy from 2001-02 to 2011-12), then he has to rethink about the continuation of the same research problem. So, it may be necessary on the part of the researcher to revise the research problem. But, it leads to wastage of time and resources on the part of the researcher. Hence, this is often a good reason for conducting some exploratory research before commencing a full project, so as to gain better understanding about an overview of the problem, to ascertain whether the problem to be selected falls under prioritized issue

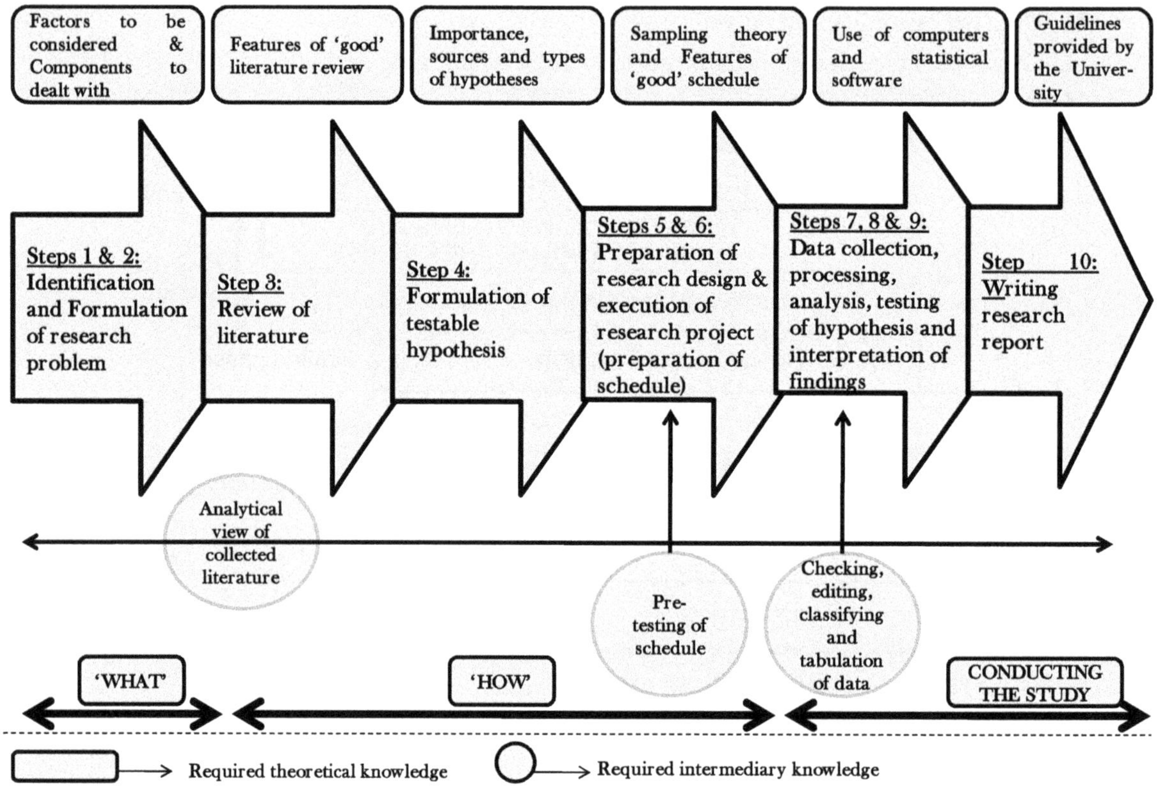

Figure 3.2: The Research Process.

or not in the study area and to ascertain whether data are available or not? The above steps identified in research process are exemplified for easy understanding to the readers (Table 3.1).

A close examination of the Table 3.1 and Figures 3.1 and 3.2 reveals that, research process is commonly referred to as the planning process. It involves step-by-step procedure of developing one's research. It covers a series of steps ranging from identifying or conceptualizing the research problem to deriving conclusions and final submission of report. The investigator has to design this research process in a scientific and logical way. The most important aspect of research process is that, researcher has to incorporate a systematic, planned and logical scheme in his work. These steps are supposed to be intimately linked with each other. It is essential to note that, the above steps generally represent the overall research process and these steps should be viewed as an ever-changing process rather than a fixed set of steps. Most researches begin the research with identification and general statement of the problem. However, in some cases, the researcher first review the earlier studies to identify the gaps in conducting research and then identifies the problem, which provides justification for the study. Thus, the steps in research process need not follow a definite sequence, but they can be changed accordingly depending upon the situation or nature of the problem to be studied. The steps involved in research process are covered in-detail topic-wise in subsequent chapters of this book. However, before we embark on the exhaustive details of these topics, a bird's eye view of the steps is given in following sections.

Table 3.1: Research Steps within the Scientific Method of Inquiry.

Phase	*Research Steps*	*Example*
Conceptual phase	Conception/Identification of Research Problem	Farmers are facing both production and marketing problems of paddy in Andhra Pradesh.
	Formulation of research problem	Economic analysis of paddy production and marketing in Kurnool district of Andhra Pradesh.
	Review of Literature	The earlier studies pertaining to resource use pattern, marketing channels, price spread etc., are reviewed by the researcher.
	Formulation of testable hypothesis	The researcher frame the hypotheses for the problem under consideration say, *'with increase in number of market intermediaries between farmer and the ultimate consumer, the price spread increases'*.
Phase of construction of research design	Preparation of Research Design (Methodology)	Kurnool district is purposively selected for the study, as no studies were conducted earlier on economic analysis of paddy in this district. Three mandals were selected randomly from the district and from each mandal, two villages are selected at random. From each village, 50 farmers, who are cultivating paddy are randomly selected. Thus, one district, three mandals, six villages and 300 farmers forms the sample base. Similarly, 50 commission agents, 50 wholesalers and 50 retailers were also selected to analyze the marketing aspects of paddy in the district. The study is conducted in the year 2012-13.
	Execution of research project	Pre-testing of schedule should be done before actual collection of data from the respondents. The questions framed in the schedule should meet the framed objectives of the study.
Empiric phase	Data Collection	The requisite data (primary data) were collected from the selected sample with the help of pre-tested schedule. Secondary data were collected from authentic published sources.
Analytic phase	Data processing and analysis	The data were subjected to Cobb-Douglas production function to analyze the resource use pattern and efficiency in the production of paddy. Price spread is worked out in transacting the paddy through various marketing channels and Marketing Efficiency Index is worked out.
	Testing of hypothesis and interpretation of findings	Hypothesis framed earlier is accepted that, *'with increase in number of market intermediaries between farmer and the ultimate consumer, the price spread increases'*.
Disseminative phase	Writing Research Report	The researcher has to submit the 'Report' in a systematic form. In case of academic research, the researchers (students) submit a thesis or dissertation to the concerned university.

I. Conceptual Phase

According Brink *et al.* (2006) this phase of research involves activities with a strong conceptual element. Conceptualisation refers to the process of developing refining abstract ideas. During this phase, the researcher categorises and labels his/her impressions. Thus, the activities include thinking, rethinking, theorising, making decision, and reviewing ideas with colleagues, research partners or mentors/supervisors. The researcher also needs to draw on the skills and abilities of creativity, analysis and insight, as well as on the firm grounding of existing research on the topic of interest. Phase of conception is the first phase of original research. In this phase are created content and structure of the planned research. Creation of conception of new research project is structured process. It can be divided into 4 steps as follows:

1. Conception/Identification of Research Problem

The first step of the research process is to identify and state the researchable problem. It is important to clearly state what the problem is to avoid any confusion later. In identifying the problem, the researcher should take into account the purpose of the study, the relevant background information, what information is needed, and how it will be used in decision making. Problem identification involves discussions with the farmers, interviews with the scientific community or experts of the same field, analysis of secondary data, and, perhaps, some qualitative research such as focus groups. 'Identification of research problem' is considered as the 'Heart of the Research Process', as research problem is the axis around which the whole research effort revolves. It is to be noted that, initially the researcher may look into the broader area of inquiry (but it should be an area that is of interest to him) and later, he must begin narrowing the field into a few sub-topics that are of greater specificity and detail. The researcher should prioritize the problems (sub-topics) identified at the farmers' level and should focus on the top most prioritized problem. It is essential to note that, 'research problem' does not necessarily mean that, something is seriously going wrong with the current situation, and needs to be rectified immediately. But, 'research problem' could simply indicate an interest in an issue, where finding the solution(s) might help in improving an existing situation. Thus, 'research problem' may be defined as *'any situation, where a gap exists between the actual and the desired ideal states'*. It is, otherwise, defined as *'a clear, precise, concise, to the point statement of issue that is to be investigated with the goal of finding solution or building understanding'*.

2. Formulation of Research Problem

The formulation of the research problem is the most important issue in the entire research planning. A well defined research problem means half of the journey is done by the researcher. This is supposed to be well thought exercise. This step usually begin with two important questions *viz.*, what problem can be solved? and how can the problem be solved? In general, the identified research problem is in the form of a broad question and it should be narrowed down to be more specific while formulating the research problem. In this context, formulation of research problem is frequently recognized as the most difficult and the most important part of research project. Precisely it was expressed by A. Einstein (paraphrase): *'If I have one hour for solving the problem on which my life depends, than I will devote 40 minutes to study the problem, 15 minutes to analyse the ways how to solve it, and only 5 minutes to solve it'*. So, the most important part of conducting research is formulating a research problem. A well-formulated research problem shows a researcher, where previous researchers have been deficient and identifies avenues of study that have not yet been pursued. So, a researcher must be careful to formulate the research problems properly in order to make sure that, the selected problem is not ambiguous, and the information obtained through the research is of greater significance both in the present and future contexts.

3. Review of Literature

Now that the problem has been identified, the researcher must learn more about the topic under investigation. To do this, the researcher must review the literature related to the research problem. This step provides foundational knowledge about the problem area. The review of literature also educates the researcher about what studies have been conducted in the past, how these studies were conducted, and the conclusions in the problem area. Reviewing the accumulated knowledge about a question is an essential early step in the research process. It is best to find out, what is already known about a research problem before trying to answer it yourself. A literature review is based on the assumption that, knowledge accumulates and that people learn from and build on what others have done. Thus, the goals of a literature review involve:

- To demonstrate a familiarity with a body of knowledge and establish credibility
- To show the path of prior research and how a current project is linked to it
- To integrate and summarize what is known in an area
- To learn from others and stimulate new ideas

So, it is a rule that, research projects begins with collecting literature, which means to identify related research and to set the current research project within a conceptual and theoretical context. It is high time now for the researcher to know, in what way the selected research problem had been addressed in the past by different researchers. For this, he has to go through the earlier studies to analyze the information about the significance of the problem, methodology employed, findings concluded etc. This serves as a bench mark for the researcher to proceed into his selected research problem to carry out the same on the earlier lines and even to improve up on the methodologies. It also makes him to compare his research findings with the findings generated in the past. So, the review of literature serves as an important component in the research process. It guides the researcher to know, what previous investigations have proved and how the present research would fit in the body of works on the subject matter. So, using other words, the aim of literature review is to identify research gaps and to find the current information related to the recognized research problem. In collecting the review, the researcher refer books, journals and internet relevant to the selected problem. By conducting the literature review, it will enable the researcher to find answers to the following questions:

- Did anybody else formulate the same research problem in the past?
- Did anybody else solve the same or similar problem in the past?
- Are the results of the previous solutions are acceptable for us or not?

4. Formulation of Testable Hypothesis

After thorough review of collected literature for the selected research problem, the next step in the research process is to formulate a testable hypothesis (plural is hypotheses) relevant to the framed objectives and identified variables for the study. Formulation of hypothesis requires useful background information about the study and considerable speculation and imaginative guess work. Further, keen observation, creative thinking, thorough knowledge about the phenomenon and its related fields, sound judgement etc., are also essential on the part of the researcher in formulating a good hypothesis. So, hypothesis is the result of researcher's creativity. The formulated hypotheses are more specific predictions about the nature and direction of the relationship between two variables. It is a rational assumption on the possible cause(s) of the observed phenomenon.

II. Phase of Construction of Research Design

The aim of this phase of research is to prepare general plan of real research. This phase is composed of following parts:

5. Preparation of Research Design (Methodology)

A research design is a framework or blueprint for conducting the research project. It details the procedures necessary for obtaining the required information, and its purpose is to design a study that will test the hypotheses of interest, determine possible answers to the research questions, and provide the information needed for decision making. Selecting representative sample, precisely defining the variables, identification of data sources, type of data to be employed in the research etc., constitute research design. In essence, the research design answers the question: *how the researcher is going to get answers to the framed research objectives and to test the formulated hypotheses*? Thus, research design (also called as the research strategy) is the conceptual framework within which the research will be conducted. This is often referred as the 'blue print' of the research. Proper framing of research design ensures greater efficiency on the part of the researcher in conducting the research. Note that, research design is not simply a work plan and it is more than a work plan in the sense that, it guides the researcher how best the collected data and information are subjected to relevant statistical tools and techniques towards answering the framed objectives of the selected problem analytically. Thus, research design is like a roadmap for the researcher, as it makes him clear, what actually he wants to investigate, how he proceeds in the investigation, by what time he has to complete various activities in the design, how efficiently he will utilize the available limited resources etc. Thus, it provides him the most effective and efficient route to accomplish the desired objectives of the research. It also makes him to focus on the ultimate objective constantly in mind and using this road map (research design), he can arrive at desired conclusions from the study.

6. Execution of the Research Project

There are a range of methods to collect data and the right method of data collection can be decided by the researcher keeping in view of the consideration of time, cost, and other resources at his disposal. The primary data can be collected by way of survey or through an experiment. In the case of survey, data can be collected by way of variety of methods *e.g.* by observation, personal interview, telephonic interview, mailing of questionnaires or schedules. In agro-economic research, the researcher mainly employs 'schedule' to collect requisite data and information from the respondents. Thus, a schedule is a formalized written set of questions for obtaining requisite information from the respondents. It carry a set of questions on a given research topic, which are asked by the researcher or interviewer personally from the respondent. It contains direct questions as well as questions in tabular form. If the respondent faces the difficulty in answering the questions, the researcher can explain the same and their importance and can gather relevant information.

III. Empiric Phase

In this phase, the researcher implements all the plans that he/she made in First Phase, Step 3, to collect the data. In many research studies, empirical phase is the most time-consuming part of the investigation. The amount of time spent, however, varies from study to study. This phase is composed of the following activities:

7. Data Collection

After finalizing the schedule, now the researcher will plan to go for collecting the data from the selected sample respondents. Note that, the data or information collected by the researcher is not just meant for performing statistical analysis, but will enable the researcher to use the same at various phases of research right from judging the importance of the study to till drawing the policy implications for the study. So, this phase of research process is very crucial, as it forms of the crux of the research investigation. The collection of data is a critical step in providing the information needed to answer the research question. Every study includes the collection of some type of data, whether it is from the literature or from subjects to answer the research question. The relevant data may be collected from primary sources or secondary sources or tertiary sources. The researcher may go for a census survey or sample survey or collect the data through observation, interviews and case study. Each method has its own pros and cons. When population is large, it is neither necessary nor advisable to cover each and every member, but a sample survey would suffice. It is quicker, cheaper and, if well-designed, it can provide precise information about characteristics of the population. On the other hand, the researcher uses other important method, 'Case study'. The choice of the method depends upon time, money and objectives. Note that, the data collection planning should be done or implemented in a very careful manner with the help of specialist researchers. The data should be good and meaningful in nature should not only be a collection of words or numbers but should provide meaningful information.

IV. Analytic Phase

Once the data collection and surveying activities have yielded sufficient and relevant data, it is time to systematically organize the data so that it can be interpreted and analyzed by researcher. According to Brink *et al.* (2006), the data collected in the empirical phase are not reported in 'raw' form. They must be summarised and subjected to various types of analysis and interpretation. Before starting to analyse or process the data, the researcher must examine them for completeness and accuracy. Incomplete and inaccurate data can be discarded. Than data are organised in an orderly, coherent fashion so that he/she can discern patterns and relationships. So, the content of this phase can be summarising in the following processes:

8. Data Processing and Analysis

Once after completing the data collection, it is not an end for research process, as the researcher has to compile, analyze, interpret and present the findings. By the time the researcher get to the analysis of collected data, most of the really difficult work has been done. The approach to data analysis depends on what methods of data collection were used (*e.g.*, schedules, interviews) and the type of data collected (*e.g.*, quantitative or qualitative). Data analysis can be defined as the process of systematically applying statistical and/or logical techniques to describe and illustrate, condense and recap, and evaluate data. According to Shamoo and Resnik (2003), *'various analytic procedures provide a way of drawing inductive inferences from data and distinguishing the signal (the phenomenon of interest) from the noise (statistical fluctuations) present in the data'*. Before analyzing the data, the researcher has to process the data, which involves four important steps *viz.*, checking, editing, classification and tabulation of collected data so that, it becomes amenable to analysis. Thus, data processing facilitates for statistical analysis of the data and the interpretation of the research findings.

9. Testing of Hypothesis and Interpretation of the Findings

After analysis of data, researcher can move ahead to test the hypothesis. He can employ various statistical tests, like Chi-square (χ^2), '*t*' test, '*Z*' test etc., to find out whether the data generated through study supports the hypothesis or not. If a hypothesis is tested repeatedly, then researcher can move to generalization and construct a theory out of it. This is the real objective of the research. Sometimes, there is no hypothesis to start with. Under such circumstances, the data generated by the study can give clues for constructing hypothesis in future studies. After analyzing the data and testing of hypothesis, the researcher should interpret the findings and this facilitates him to convert the data into information. This, in turn, leads to drawing conclusions about what and why of the problem and thus, helps him to formulate policy implications.

V. Disseminative Phase

The researcher's job is not completed, until he communicates the results of the study to others, who may find it useful. So, dissemination means, process when results of the research are presented or published as Reports, Thesis, Dissertations etc. So, this phase includes the following component:

10. Writing Research Report

The final stage of any research investigation is the preparation of report or thesis. The report should generally carry three sections like Preliminary pages, Main pages and End matter. The Preliminary pages comprises of title page, declaration, certificates, acknowledgements, list of contents, list of tables, list of figures, list of symbols and abbreviations, list of appendices, abstract. Main pages include: Introduction, Review of literature, Materials and Methods, Agro-Economic features, Results and Discussion, Summary and Conclusions. End matter section comprises of Literature cited and Appendices. For the researcher as well as the reader, report writing is very crucial, as it acts as the best way for communication between the two. Report written must be very simple in nature with easy language, high clarity. Report writing cannot be done by everyone and requires a skilled person for this purpose.

Having understood about different phases and steps involved in the research process, it is high time now to discuss them in-detail.

4

Conception/Identification of Research Problem

'Identification of research problem' is considered as the 'Heart of the Research Process', as research problem is the axis around which the whole research effort revolves. Hence, the research problem must be rightly identified and expressed with the utmost precision. It is paramount to the success of the research effort. To see the problem with utmost clarity and to state it in precise and unmistakable terms is the first requirement in the research process. It is to be noted that, initially the researcher may look into the broader area of inquiry (but it should be an area that is of interest to him) and later, he must begin narrowing the field into a few sub-topics that are of greater specificity and detail. The researcher should prioritize the problems (sub-topics) identified at the farmers' level and should focus on the top most prioritized problem. For example, considering the importance being given by the Government of Andhra Pradesh in improving the irrigation efficiency in the canal commands, the researcher must look into this broader area (provided the researcher is also interested in this area). He conducts a reconnaissance survey in different irrigation commands in the State. The survey may reveal major researchable issues like water logging induced soil salinity, water scarcity in tail end areas of canal command, higher percolation and seepage losses in irrigation channels, lack of adequate drainage facilities, salinity of irrigation water etc., in Krishna Western Delta (KWD) of Andhra Pradesh. Thus, the researcher has divided the broader research topic into different sub-topics.

Now, the researcher will attempt to address the top most prioritized problem say, water logging induced soil salinity situation in KWD of Andhra Pradesh.

In the above discussion, dividing the broad research topic into several sub-topics (sub-problems) deserves special mention. Though this is not the right place to discuss this topic in-detail, but it is the right place to mention. It is to be noted that, it is difficult to look into and address the broad research topic by the researcher at macro level (say, at State level). For convenient sake, the broad research topic is sub-divided into different sub-problems (say, at single irrigated command level) and each sub-problem can now be considered as a researchable unit. Thus, sub-problem is the sub-parts of the main problem that are an integral part of the main problem. The sub-problems must add up to the totality of the problem. As far as possible, sub-problems should be small in number. At this juncture, it is also essential to mention about the word, 'Pseudo-sub-problems'. They refer to the procedural issues that involve decisions that must be made before resolving the research problem and its sub-problems say, What's the best way to choose a sample?, What statistical procedures should be used to analyze the data? etc.

4.1. What is Research Problem

It is essential to note that, 'research problem' does not necessarily mean that, something is seriously going wrong with the current situation, and needs to b rectified immediately. But, 'research problem' could simply indicate an interest in an issue, where finding the solution(s) might help in improving an existing situation. Thus, 'research problem' may be defined as *'any situation, where a gap exists between the actual and the desired ideal states'*. It is, otherwise, defined as *'a clear, precise, concise, to the point statement of issue that is to be investigated with the goal of finding solution or building understanding'*. Research problem is also defined as *'an existing problem, where urgent solution is needed, situations in which improvement is desired, or areas where some conceptual clarity is needed for a better theory building'*. Thus, research problem is a problem that someone would like to investigate. It can also be considered as a situation that needs to be changed or addressed.

The symptoms of the research problem should not be confused with the research problem itself. For example, an Agricultural Officer tried to increase the crop productivity in a village by rendering training to the farmers, but with little success. Here, the 'low crop productivity' is the symptom of the real problem, which is the 'low morale and motivation' of the farmers undergone training.

4.2. The Five 'W's of Research Problem

Whether a problem requires research depends on three conditions:

- ☆ There should be a perceived difference or discrepancy between what it is and what it should be;
- ☆ The reason(s) for this difference should be unclear (so that, it makes sense to develop a research question); and
- ☆ There should be more than one possible and plausible answer to the question (or solution to the problem).

The above conditions can be well-understood through the following examples:

Example 1

Problem Situation

In Nandipalli, a remote village of Nandyal mandal in Kurnool district, a report showed that, there are about 500 farmers cultivating paddy and they have to be trained by the Farmers Training

Center, Nandyal on Integrated Pest Management (IPM) on paddy. However, by the end of that year, it was found that, only seven per cent of farmers have under gone training.

Discrepancy

All 500 farmers in Nandipalli village 'should have completed their training, but only seven per cent of them have under gone training'.

Problem (Research) Question

Why only seven per cent of the total number of farmers completed their training?

Definite Answer

There is no adequate staff in the Farmers Training Center and hence, it could not conduct trainings regularly for the farmers in the remote villages. Nandipalli being a remote village, less number of farmers were trained on the said subject.

In the above example, assuming that the given fact is true, there is no need of undertaking a research, since definite answer is obtained to the problem situation.

Example 2

Problem Situation

In Kurnool district, there is one Regional Agricultural Research Station (RARS), Agricultural College, Mahanandi, three Agricultural Research Stations, Agricultural Officers of Line Department of Agriculture, Krishi Vigyan Kendra (KVK), Non-Governmental Organizations (NGO) etc., and these agencies are actively involved in disseminating scientific paddy production technologies to the farmers in the district. However, it was found that, both the research and extension gaps of paddy yields are high to the tune of 1.72 tonnes/ha and 3.58 tonnes/ha respectively.

Discrepancy

Although Kurnool district had adequate scientific man power, both the research and extension gaps of paddy yields are high to the tune of 1.72 tonnes/ha and 3.58 tonnes/ha respectively.

Problem Question

What factors influence the huge research and extension gaps of paddy yields in Kurnool district?

Possible Answers

- The farmers might have problems for not attending in the training sessions.
- The programmes of these agencies in the district might not have been integrated; hence, majority of the farmers might have missed opportunities in getting trained on paddy production technologies.
- Spurious inputs flooded in the market are responsible for huge extension gaps.
- Inadequate extension by these agencies in disseminating the technologies.
- The package of practices disseminated by these agencies is not relevant to the farming situations prevailing in the remote villages *i.e.*, the Farming Situation Based Extension is not practiced by these agencies and they are disseminating only general practices.

Since there are a range of possible answers for the problem under consideration and to ascertain the major causes, the above problem situation is researchable.

Thus, for conducting a research, 'identification of research problem' is very important. Before identifying a researchable problem, the researcher must think about Five 'W's and the possible answers to these Five 'W's will guide the researcher in selecting and executing the researchable problem. The Five 'W's include:

a. What? (What is the research topic?)
b. Why? (Why the researcher selected this topic?)
c. Who? (Who will be the participants in the research?)
d. Where? (Where the research is to be conducted?)
e. When? (When the research is to be started?)

a. What? (What is the Research Topic?)

This is the most important and hardest question to be answered by the researcher. It is essential for the researcher to go for conducting some exploratory research before planning 'what' is the research topic? Most research projects that go wrong do so because, the fundamental problem that has led to the research has not been fully understood or defined. Generally, the developmental programmes and policies emphasized by the Government in the Five Year Plans often guide the researcher to decide 'what' is the research topic? It is to be noted that, the selected research topic should not be broad and it should be 'specific' in addressing a particular problem. That is, it should fall in one of the prioritized areas of the Government towards meeting the welfare of the farming community. For example, welfare of the farming community can be ensured through various means like boosting agricultural production through generating HYV seeds, improving resource use efficiency, improving marketing efficiency in transacting agricultural commodities, regulating marketing practices of agricultural commodities, strengthening agro-processing facilities etc. Among these welfare motives of the Government towards the farming community, one can be selected as a research topic by the researcher (student).

b. Why? (Why the Researcher Selected this Topic?)

This implies the 'purpose' of selecting the research topic. Assume that, the researcher (student) has selected the topic *'Economics analysis of production and marketing of paddy in Kurnool district of Andhra Pradesh'*. The researcher has selected this topic for one or more of the following reasons:

- ✰ If the researcher himself is interested in this topic of analyzing both production and marketing aspects of a commodity (say, paddy).
- ✰ The literature reviewed by the researcher has found this gap of studying production and marketing aspects of paddy in Kurnool district of Andhra Pradesh. That is, not much of work has been carried out earlier towards studying the economic aspects of paddy production and marketing in Kurnool district. This review of literature makes the researcher to find out, what things are already known about the topic of his interest. This is again important because, it also makes the researcher to identify what is to be known and facilitate to identify the gaps in knowledge with reference to the topic of interest. The other advantages of reviewing the literature by the researcher include:
    - ❖ To extend the past research in new areas.
    - ❖ To replicate the past research in a different setting *i.e.,* in a different agro-ecological situation.
    - ❖ To identify and analyze different sub-problems of an already dealt problem.
    - ❖ Challenge the past research findings that seem to contradict, what you know or believe to be true.

- To gain useful knowledge for formulating hypotheses for the present study.
- Guide the researcher to use different methodologies for analyzing and interpreting the data findings.

☆ The Academic Advisor (Professor in Agricultural Economics) had advised him to do research on this topic.

☆ The researcher has conducted the survey among the farmers and identified this topic as the prioritized researchable issue. The following are the criteria that guide the researcher in identifying and prioritizing the research problems:

- How large or widespread is the problem?
- How severe is the problem? (what percentage of population is affected with this problem?)
- Whether any study is conducted earlier on the same lines (Avoidance of duplication of study)
- Whether it is feasible to execute the study?
- Whether that study is politically acceptable? (if it is acceptable, it facilitate the smooth conduct of the research).
- Is it likely that the recommendations from the study will be applied?
- How urgently are the results needed for making a decision?
- Is the research that is going to be conducted will have ethical acceptability?

☆ The researcher may get the funding from the donor agency (say, Directorate of Rice Research, Hyderabad), if he executes this project.

☆ The researcher should closely observe the phenomena that happen around him and try to formulate the research. For example, when young Isaac Newton observed an apple falling to the ground, he began to think about a specific kind of motion and he comes up with the theory of gravity. So, the researcher should try to find out the problems that exists around him and should ask himself: Why does such-and-such happen? What are the causal factors behind such phenomena? and so on.

☆ The researcher should attend the meetings, workshops, conferences etc., so that, he can ascertain "what's hot and what's not" in their field. Further, such attendances will make the researcher to interact with experts, share ideas and exchange information.

Above all, the researcher should think very carefully about 'why' he is doing the research? That is, say, if the researcher is performing research for academic interest (say, for M.Sc (Agricultural Economics) or Ph.D (Agricultural Economics), he should see that, whether the project executed by him will provide the opportunity to reach the required intellectual standard? Whether the research project gets completed in the stipulated period of time? Will the research generate enough material or data base to write a thesis of the required length? Will the research offer suitable strategies for improving both production and marketing aspects of paddy in the study area? Say, if the research is conducted for funding purposes, the researcher is required to know whether the project proposal includes all the information in a specific format as required by the donor (funding) agency?

c. Who? ('Who' will be the Participants in the Research?)

This implies 'who' will be beneficiaries or respondents of the research project? In the above example, 'farmers' are the beneficiaries of this research project, as the research focuses on suggesting the strategies for improving both production and marketing scenarios of paddy in the study area. The term 'who' also implies, with whom the researcher is in interaction in executing the project. In the above study, the researcher interacts with sample farmers, local Agricultural Officer, input agencies, commission agents (in regulated market), wholesalers, rice millers, retailers, consumers, procurement (Food Corporation of India) agency etc. So, all these agencies involved in production and marketing of paddy constitutes 'participants' in this study. At this stage of the research process, the researcher need not worry too much in selecting the number of participants in each agency and this will be decided later in Step number 'E' of the research process. But, the researcher should think about the type of agencies with whom he has to be in contact to get the requisite information. If the researcher (student) is conducting research for academic purpose, he should not include those agencies, who are difficult or expensive to contact. Also, the researcher should bear in mind that, the Internet now provides opportunities for contacting people cheaply, especially if the student is with free internet access.

d. Where? (Where the Research is to be Conducted?)

Before executing the project, the researcher must be clear, where it is going to be conducted. This is because, say, in the above example, paddy is cultivated in all the districts of Andhra Pradesh, as it is the staple food crop. But, the researcher prioritized to conduct the study in Kurnool district of Andhra Pradesh and this might be due to the one or more of the following reasons:

- ✰ The researcher is the native of Kurnool district and hence, he is well versed with the local terminology while interacting with the farmers and other agencies.
- ✰ The review of earlier studies made by the researcher revealed that, no studies were conducted earlier regarding economic aspects of production and marketing of paddy in Kurnool district.
- ✰ The preliminary survey conducted by the researcher revealed that, the agro-ecological conditions prevailing in Kurnool district are highly suitable for paddy cultivation.
- ✰ The preliminary survey conducted by the researcher revealed that, the farmers cultivating paddy in Kurnool district are facing a wide range of problems with respect to both production and marketing.

Thus, the advantage of deciding the place, where the research project is to be executed, will enable the researcher to narrow down the focus of research topic. It also helps the researcher to plan the budget and time frame to conduct the study. In general, in case of academic research, if the students do not get adequate budget or stipend, they will generally select the research topic in their native place or close to their college. In agro-economic research, the researcher has to collect the data with the help of schedules and hence, he has to interact with the sample respondents (farmers, input agencies, commission agents, processors, wholesalers, retailers etc) personally. So, it is essential to decide, where the research project is to be executed. It is not only the convenience on the part of the researcher, but also the responsiveness of selected agencies that provide information to the farmer counts much in selecting the place of the study. Of course, this can be judged well-in-advance by pre-testing the schedules among the respondents.

e. When? (When the Research is to be Started?)

This is essential on the part of the researcher to collect reliable information about the research project. That is, the researcher should initiate the project well-in right time so that, it guides him to collect the requisite data from the selected respondents and to complete the project in the prescribed time frame. In the above example, the researcher should collect the data after the crop (say, Kharif) season, as by that time the farmer will harvest and market the produce. This will enable the researcher to collect the data from all the selected respondents for that season/year. However, if the researcher executes a project say, 'Prices forecasting of major agricultural commodities in Kurnool district of Andhra Pradesh', he has to plan for collecting secondary data on market prices (say, for the past 20 years period) from the selected APMCs in Kurnool district. Based on the trend analysis, he will forecast the prices for the future years (say, upto 2020). Thus, selecting the time period when the project is to be executed and to be completed (period of study) and deciding the time period for which the data are required (say, one season/year in case of cross-sectional data or say, 20 years in case of time series data) are essential for successful execution of the project.

4.3. What Makes a Good Research Problem?

A research problem is some difficulty either of a theoretical or practical nature, which an individual or organization faces and wishes to obtain a solution for the same. As mentioned earlier, research problem is considered as the heart of the research process. This is because, even while conducting the research, researcher always think on the lines that, 'whether the problem undertaken is the right issue in the present day context?', whether he is proceeding in the right direction'?, 'whether the research fulfills the purpose for which it is being executed'?, 'what is future course of direction of the present research?' etc. So, all these questions will move around the present research problem undertaken by the researcher and this indicates the importance of identifying and selecting the prioritized research problem. In view of this, the following are the important guidelines that are to be followed for selecting the 'right' research problem:

- ☆ The research problem (title) must be clearly defined, specific and make sense. That is, the terms included in the research title or problem or question must have clear meaning. For example, *'analyzing the improvement in lives of farmers after rendering training to them on paddy crop'* is a poor research question. This is because,
  - ❖ In the title, there is a vague term 'improvement'. The researcher has not specified the meaning of the term 'improvement' – does it refers to improvement in managerial ability or improvement in socio-economic conditions or improvement in technical knowledge of the farmers etc.
  - ❖ The term 'training' is also not clearly defined – does it refers to training on latest crop production technology or training on Integrated Nutrient Management (INM) or training on weed management or training on IPM etc.
  - ❖ The title has not specified the location of the study – does the study is executed in Kurnool district or East Godavari district or West Godavari district or Visakhapatnam district etc. This specification of 'district' name is important because, say, if the training is given on crop production technology, the dates of sowing, varieties to be cultivated, type of soils, status of N, P and K nutrients in soils etc., vary from district to district and hence, the results achieved from this study cannot be generalized across the districts.

This implies that, the more specific or more clearly defined the research problem is, the better is the clarity in answering the same and in generalization of results. This is because, *'a problem rightly defined is a problem half solved'!* If the problem is not stated with clarity, then the researcher is merely deceiving himself in attending towards that problem. Such self-deception will lead to several problems later on regarding the availability of data, interpretation of findings etc. This is well-exemplified here under.

- ***Too broad/Vague problem:*** 'What are the effects of the All-India Coordinated Research Project (AICRP) on farming systems of the Government?
- ***Specific/Precise:*** What are the effects of the AICRP on farming systems on the livelihood security of small and marginal farmers in North Coastal Zone of Andhra Pradesh?

  The researcher has to develop a unique and specific focus in identifying the research problem. The sequential steps along with an example (Table 4.1) are given below.
  - What is the researcher's interest?
  - In what area the researcher is proficient?
  - What is the focus area of the research?
  - What is the specific focus of the study area?

Table 4.1: Example of Specific Focus of the Study Area.

Category	Agricultural Marketing
Topic	Marketing efficiency
Focus	Operational efficiency and pricing efficiency in transacting paddy
Specific Focus	Study of operational efficiency and pricing efficiency in transacting paddy in Kurnool district of Andhra Pradesh.

- The research question should be more focused and the solutions offered to this problem should enhance the body of knowledge. The solutions of this problem must give some beneficial implications to the society. Further, the selected problem for the study should be the prioritized issue in the study area and it should also have relevance to the time, place, and population of the study. For example, a study on '*Economic analysis of shrimp cultivation in Anantapur district of Andhra Pradesh*' has no meaning at all, as Anantapur district represent dry tract in Andhra Pradesh.
- The research study identified by the researcher should be novel. So, the findings generated from the study must be new and not noticed elsewhere. Otherwise, it leads to wastage of both time and resources on the part of the researcher. So, the study should not be simply replicated. However, if the same study is to be replicated elsewhere say, in other district or other state or on other crop, it is best to change the methodology, so as to increase the novelty of the research.
- The identified problem must be operationalizable or feasible. So, the researcher should have better practical ideas. That is, the problem considered for the study must be executable to get good data base, facilitate statistical analysis and draw useful findings. For example, a study on '*Economics analysis of production and marketing of paddy in Kurnool district of Andhra Pradesh*' is easily operationalizable and generate more data base than a study on '*Assessing the farmers knowledge regarding the importance of training programmes being conducted by the Line Department of Agriculture in the villages of Kurnool district*'.
- The research problem identified must be within a reasonable scope. The more is the specificity of the focus of the problem, the more likely is the success of the project. For example, a study on '*Economics analysis of paddy production in Kurnool district of Andhra Pradesh*' is more likely to succeed than a project on '*Comparative economics of paddy production across different States in India*'.

- ☆ The research problem identified and stated must be precise and have clarity, what researcher actually mean. That is, it should not give scope to others to argue on the title and assume to read researcher's mind for understanding the problem. If the research statement is not precise, it leads to serious problems and others might question about the basic integrity of the researcher in putting the research efforts for solving the problem.
- ☆ As mentioned earlier, handling a broad research problem is very difficult to execute and hence, it will be conveniently divided into different sub-problems. For example, 'Irrigation water management in KWD of Andhra Pradesh' is a broad research problem. For convenient sake, this research problem can be divided into different sub-problems as follows:
  - ❖ ***Sub-problem 1:*** What is the demand-supply gap of irrigation water usage in KWD of Andhra Pradesh?
  - ❖ ***Sub-problem 2:*** What is the strategy followed to overcome water logged induced soil salinity situation in Mutluru canal command in KWD of Andhra Pradesh?
  - ❖ ***Sub-problem 3:*** What are the technological interventions to be taken up in improving the irrigation water use efficiency in paddy cultivation in Modukuru canal command in KWD of Andhra Pradesh?
  - ❖ ***Sub-problem 4:*** What are the economic benefits of executing different water saving paddy production technologies in Modukuru canal command in KWD of Andhra Pradesh?
  - ❖ ***Sub-problem 5:*** Whether the installation of sub-surface drainage system in Mutluru canal command area of KWD in Andhra Pradesh is economically viable?
  - ❖ ***Sub-problem 6:*** How the Water Users Associations in Mutluru and Modukuru canal command areas of KWD are functioning?

  A close examination of the above sub-problems reveals that, the broad research problem is divided across major disciplines such as Agricultural Engineering (Sub-problems 1 and 2), Agronomy (Sub-problem 3), Agricultural Economics (Sub-problems 4 and 5), and Agricultural Extension (Sub-problem 6). From a research standpoint, these sub-problems are easier to address and resolve. The following are the important characteristics of sub-problems:

  - ❖ Each sub-problem should be a completely researchable unit. That is, each sub-problem can be executed by the researcher as an independent research problem. All the sub-problems must fit into the broad framework of the broad problem and thus, if all the sub-problems are executed, it leads to solution against the broad problem. In the above example, each sub-problem is assigned to a different department and all the Scientists work in a collaborative mode towards addressing the broad problem. It is to be noted that, the sub-problems should be stated in the form of a question. This is because, a question tends to focus the researcher's attention more directly in targeting the sub-problem than does a declarative statement.
  - ❖ Each sub-problem must involve collection, analysis and interpretation of the data to analyze the causal relationship between the variables.
  - ❖ The old axiom that, '*the sum of the parts equals the whole*' applies here. That is, the solutions offered under each sub-problem will lead to answering the broad problem. So, the researcher should carefully check about unnecessary problems that do not fit into the framework of the broad problem and these should not be framed at all. Further, the broad problem under consideration must be completely reflected in the sub-problems.

- It is essential to note that, the number of sub-problems should be as low as possible usually in the range between 2 to 6. If the sub-problems are larger in number, it is difficult to look into all these problems and further difficult to consider them collectively to address the broad problem.

Sub-Problems vs. Pseudo-Sub-Problems

From the above example, we know sub-problems are an integral part of the main problem and they are easy to execute at the micro level. However, while executing these sub-problems, some procedural issues may crop up for the researcher and these constitute Pseudo-sub-problems. So, the researcher should take the right decisions to overcome these pseudo-sub-problems in executing the research. For example, in executing Sub-problem 4, the researcher may face pseudo-sub-problems like:

- What method of sampling the researcher should follow in selecting the sample?
- What should be the size of the sample so that, it is representative of the population?
- What instruments should be used to collect data from the selected sample?
- What statistical procedures should be used to analyze the data?

So, the researcher should carefully address all these pseudo-sub-problems and this ensures smooth execution of 'Sub-problem 4' towards the attainment of its objectives. For students, the expertise of Academic Advisor (Professor) is important at this juncture in addressing the possible pseudo-sub-problems they face in conducting the research. This is because, sometimes, the enormous literature available to them in the internet may often confuse them in tackling various pseudo-sub-problems they encounter. Further, it is essential to distinguish sub-problems from pseudo-sub-problems.

- The researcher should ensure that, whether the identified research problem contribute to one or more of following?
  - Develops new knowledge in the selected field of inquiry
  - Expands current knowledge
  - Contributes to theory development
  - Advances methodology
  - Related to a current social or political or economical issue
  - Evaluates specific practice or policy
  - Explores an issue about which little is known
- Note that, the researcher should take both internal (personal) and external factors in the identification of the research problem. External criteria involve such issues as newness and significance for the area, availability of data, administrative and institutional cooperation etc. Personal criteria include interest, training received, cost, time etc. The following are more detailed list of criteria for the choice of a good research problem:
  - ***Novelity***: The research problem identified must be sufficiently original and it must not involve objectionable duplication.
  - ***Importance:*** The identified problem should contribute to a new knowledge area for the future investigators and must at least enhance the existing body of knowledge. If these two merits were not going to be achieved, there lies no meaning in taking up the research project.

- ***Feasibility or Amenability:*** The selected research problem must be 'feasible' to execute under the practical conditions. The term 'feasible' is a very broad word and it may address the following issues:
  - Availability of data
  - Availability of cooperation
  - Availability of guidance
  - Availability of other facilities like computers, transport etc.
  - Researcher should be experienced, creative, competent, knowledgeable, skillful enough to develop, administer and interpret the necessary data.

Considering the above aspects, the researchers should identify the research problems. Above all, 'personal interest' of the researcher in addressing the selected problem should also be given due consideration. This is an essential criterion, as the selected problem must be interesting for the researcher himself to execute it. But, note that, the interest must be purely intellectual and should not be for any reward, material benefit, advancement in position etc. Besides these, the researcher should also pose himself the following questions before selecting the research problem:

- Could this problem can be solved through conducting the research study?
- Is the problem to be selected is significant from the point of view of conducting research?
- Is the problem to be selected will have feasible solutions?
- Whether he is competent enough to handle and tackle the selected problem?
- Whether requisite information and database are available to him for conducting the research?
- Is the researcher can complete the research in the allotted time frame?
- Is funding is available to conduct the research?

4.4. Factors to be Considered, while Identifying a Research Problem

While identifying and selecting the research problem, the following considerations should be borne in mind:

- ***Economic considerations:*** We know, execution of research projects involve spending of time, money and other resources on the part of the researcher. So, considerable costs are involved in the execution of the project. The execution of research project also ensures different kinds of benefits like increase in employment opportunities, increase in output, improved standard of living of beneficiaries, proper income distribution etc. If the value of these anticipated benefits outweighs the costs incurred in the project, it is better to identify, select and execute that project. However, in case of academic research, enhancing the body of knowledge should be considered as the benefit derived from the project and is not comparable with the costs incurred in the project.
- ***Resources criterion:*** The researcher should have adequate resources like time, energy and other resources to fulfill the task efficiently.
- ***Technical considerations:*** The researcher must be aware that, adequate technical information must be available in advance to him before identifying the problem. The exploratory research conducted by the researcher should enable him to gain knowledge on different technical

terms, definitions, concepts etc., and this will guide him in prioritizing and selecting the top most researchable problem in an area. Further, instead of selecting a broad problem, the researcher should aim at selecting the problem that enjoy narrow focus.

- ☆ ***Human considerations:*** This is an important consideration, as the ultimate objective of any research activity, is to benefit the participants in the project. So, the project that contributes much to the participants or beneficiaries must be prioritized, identified and executed.
- ☆ ***Environmental consideration:*** In the context of increasing concerns about environmental safety and restoring ecological balance, this consideration is gaining more and more significance in agricultural research. For example, if a research project entitled, *'Impact of Farmer Field Schools on adoption of IPM technology by Cotton Farmers in Warangal district of Andhra Pradesh'* is identified by the researcher, it highlights the importance of execution of IPM technology in safeguarding the environment.
- ☆ ***Budget criterion:*** In the present day context, as the funds available for agricultural research are getting limited and limited, but the expectations from this research are gradually increasing. This calls for executing the research with more resource use efficiency. Seldom are there unlimited funds for research and more often there are very limited funds. In this case, it is helpful to know what the budget is, and this guides the researcher to plan the research and its objectives.
- ☆ ***Timing criterion:*** It implies, when the findings of the research is to be required? In case of academic research, it is purely time bound, as the researcher (student) has to submit his thesis in a given time period (usually shorter duration *i.e.,* one to two years). However, in case of research that is intended to provide solutions to complex problems, it may take considerable time to generate research findings.
- ☆ ***'Researchability' criterion:*** Though 'problem identification' is the initial force behind conducting the research, not all problems are researchable (Figure 4.1) and a problem to be researchable, it must satisfy at least the following three criteria:
  - ❖ A perceived discrepancy exists between 'what is' and 'what should be'.
  - ❖ A question exists about 'why' there is a discrepancy.
  - ❖ At least two possible and plausible answers must exist to explain the question.

  Among the above three, the last point is important. This is because, if there is only one possible and plausible answer to the question about the discrepancy, then a research situation does not exist.
- ☆ ***Feasibility criterion:*** As per this criterion, the researcher must consider the following important aspects:
  - ❖ How much time does the researcher takes to execute and find solution for the problem to be considered?
  - ❖ Whether adequate data are available?
  - ❖ How much will it cost?
- ☆ ***Expertise of the researcher:*** The researcher should have adequate level of expertise before taking up the research task, as he himself has to work on the project.

Above all these criteria, the researcher should select the topic of great interest to him.

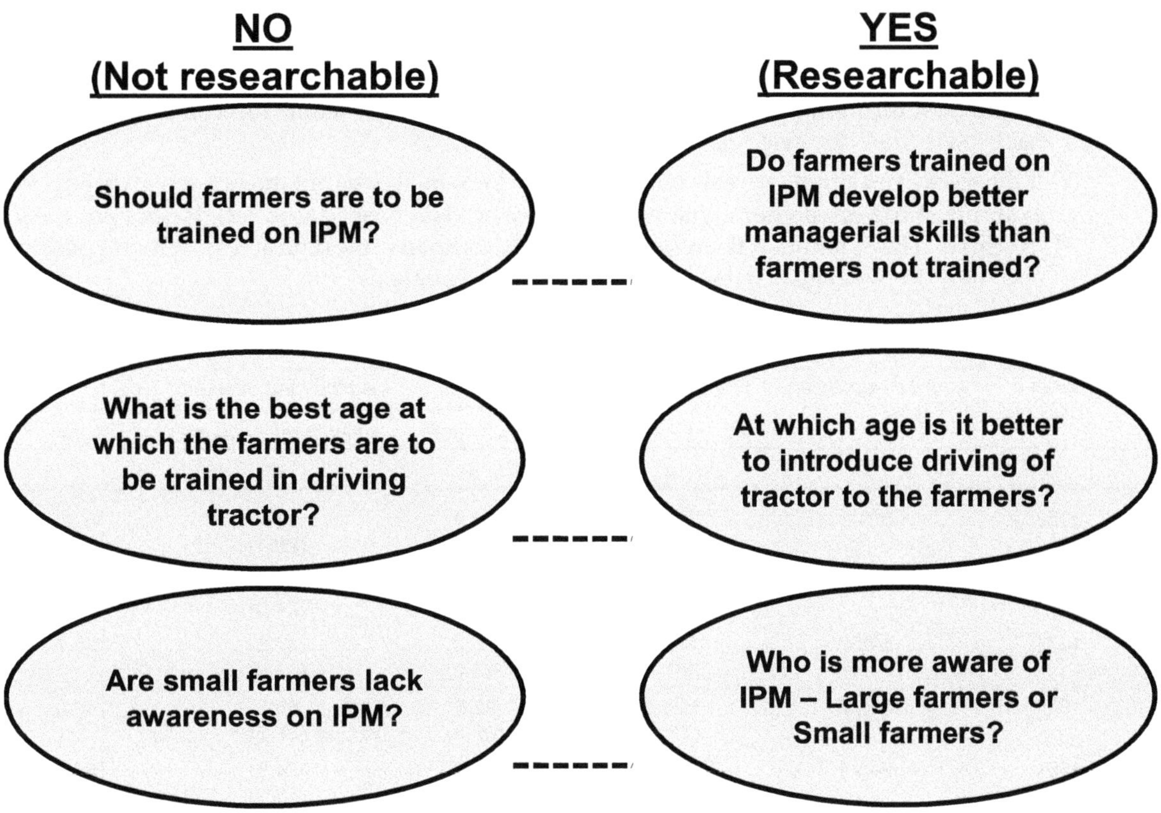

Figure 4.1: Researchable vs. Non-Researchable Problems.

4.5. Situations to Avoid when Identifying a Research Problem

It is to be noted that, when a research problem is identified, that problem will take a significant amount of time, energy and resources on the part of the researcher for its completion and hence, the selected problem should be worth that time, and resources and energy involved. So, while identifying the problem itself, the researcher has to focus on the following two important aspects:

- First, the identified research problem should address an important and prioritized question (say, prioritized problem faced by the farming community) and the answer (solutions) the researcher generates from the problem can actually 'make a difference' in some way *i.e.*, in solving the problem.
- The knowledge gained through addressing the selected research problem should open new frontiers through advancing the knowledge, leading to new ways of thinking and increased applications in the relevant fields thus, paving the way for further research in that field.

To accomplish both of these ends, the research project selected and executed must involve not only the collection of data, but also the interpretation of analytical findings. As mentioned earlier, some problems are not suitable for conducting research because they lack the 'interpretation of data' and hence, they do not contribute towards enhancement of knowledge. In this context, the following situations are to be avoided, when identifying and considering a research problem:

- ☆ Simply gaining more knowledge about a particular topic does not mean conducting research. For example, if a student wants to know, how a knapsack sprayer works or how a two stroke engine works, it does not mean research. This is because, the student simply gained more satisfaction through increasing his knowledge about a particular topic, but it would not have led to 'new' knowledge.
- ☆ If the researcher simply reveals what is already known, it does not mean a research. Say, for example, if the researcher explains the following data base (Table 4.2), it does not mean 'research'. This is because, the researcher simply compared the data across different periods and 'new' knowledge is gained through this trend analysis.

Table 4.2: Population and Agricultural Workers (in millions).

Year	*Total Population*	*Rural Population*	*Agricultural Workers*		
			Cultivators	*Agricultural Labourers*	*Total*
1	*2*	*4*	*5*	*6*	*7*
1951	361.1	298.6 (82.7)	69.9 (71.9)	27.3 (28.1)	97.2
1961	439.2	360.3 (82.0)	99.6 (76.0)	31.5 (24.0)	131.1
1971	548.2	439.0 (80.1)	78.2 (62.2)	47.5 (37.8)	125.7
1981	683.3	523.9 (76.7)	92.5 (62.5)	55.5 (37.5)	148.0
1991	846.4	628.9 (74.3)	110.7 (59.7)	74.6 (40.3)	185.3
2001	1028.7	742.6 (72.2)	127.3 (54.4)	106.8 (45.6)	234.1
2011*	1210.2	833.1 (68.8)	NA	NA	NA

Note: NA – Not Available. Figures in parentheses indicate percentage to respective total population in that year.

Source: www.agricoop.nic.in

- ☆ If the researcher simply studies 'what' is the relationship between the variables, without assessing 'why' is that relationship, it does not mean research. For example, if a researcher simply studies the correlation between various quantities demanded of a commodity at various prices of the same commodity, it does not mean research. This is because, the basic requirement for research is ignored here *i.e.*, there is no mind struggle on the part of the researcher to analyze and interpret the data. The researcher can simply compute the correlation coefficient between the selected variables with the help of a computer, which can perform this statistical operation infinitely faster and more accurately than a researcher can. Further, a correlation coefficient is nothing more than a statistic that expresses 'what' is the relationship and 'how' closely the two variables are related or associated with each other. But, it will not explain about 'why' is that association. So, simply calculating a correlation coefficient between two sets of data to show a relationship between them is not acceptable as a problem for research.

- Research problems that result in a 'yes' or 'no' answers are not suitable for research. For example, if the researcher enquires 'whether radio broadcasts are useful to the farmers in disseminating technology?', it is not a researchable question. This is because, for the same reason explained above *i.e.*, merely finding a correlation coefficient is unsatisfactory. But, if the question asked in a different way, 'which agricultural programmes of radio broadcasts are useful for them', then it is a researchable question. This is because, the researcher tries to find out, whether the programmes related to package of practices or IPM or INM or weed management or irrigation water management or marketing technologies etc., are useful to the farmers and why?

4.6. Applying for Funding

After identifying the research problem, the researcher (say, Scientist (Agricultural Economics)) can apply the project to a funding agency to get funds for its execution. Different agencies like NABARD, ICAR, World Bank, The Royal Netherlands Embassy etc., will provide funds for agricultural research projects, if they meet the requisite criteria prescribed by them. The application of identified project should give the names of the Principal Investigator and the other researchers to be involved, as well as describing their relevant experience and qualifications. As a rule, applications also have to state that the same research report is not applied to any other funding agency. It is to be noted that, while applying for funds to the funding agency, the applicant should be aware that the identified research problem must satisfy certain important criteria like specificity, fall in the prioritized researchable area, fits into the broad framework of the Government's programmers and policies, alleviate poverty, promote livelihood security of the beneficiaries etc. This is because, the funding agencies will receive hundreds of applications every year and they evaluate the case by case based on certain (above) criteria. It is, therefore, important to ensure that the research proposal submitted must be relatively easy to compare across other applications. Recipients of research grants are expected to submit regular progress reports, describing how the funds have been spent and what has been achieved during the period covered. The scientific reports submitted in this context should give a clear account of the work for which the grant was sought and has been used during the funding period.

4.7. Delimiting the Research

When the research problem is identified and clearly stated by the researcher, it implies the researcher now precisely knows what do and what not to do. For example, if the researcher wishes to conduct a survey on analyzing the '*Marketing efficiency in transacting paddy in Kurnool district of Andhra Pradesh*', the researcher clearly knows about the following 'What to do' in his research:

- Collect the information pertaining to different marketing channels involved in transacting paddy in the district.
- Collect the data pertaining to marketing costs and marketing margins realized by each agency in the identified marketing channels.
- Analyze the Marketing efficiency Index for each marketing channel and compare them.

Similarly, the researcher must also know, 'what not to do' in the research and this is exemplified below:

- The researcher need not collect the data pertaining to costs and returns of paddy enterprise.
- The researcher need not analyze the growth rates in paddy area, production and productivity in the selected district.

- The researcher need not collect information pertaining to the input subsidies offered by the Government in cultivating paddy.

Thus, the above aspects can be easily avoided by the researcher, as they are not at all related to the research problem under consideration. So, what the researcher intends to do is to be stated in the problem and what the researcher is not intended to do is to be stated in the '*delimitations*'. Thus, the delimitations of a research problem include the unnecessary aspects that are not related to the current research problem. It is helpful to narrow down the topic by telling what the study is not intended to do. This mention of 'delimitations' is important in the context that, the researcher need not waste his precious time, energy and other resources towards attending unnecessary aspects that are not related to the stated problem. The researcher, who critically thinks about the identified problem, will be able to distinguish 'what to do' and 'what not to do' in the research. By doing so, he can carefully eliminate all irrelevancies in the statement of delimitations. It is also essential to note the difference between delimitation and limitation. Delimitation is a premeditated limitation that the researcher narrows down, whereas 'Limitation' is a weakness in the research study which becomes apparent during or after its completion. The categories of delimitations may include:

- The number of observations, subjects or cases
- Time and geographic location
- The selection of sources (especially in historical-documentary research).

The following examples highlight the differences between delimitations and limitations in a research:

Delimitations

- In order to assure manageability of the collected data, survey instruments used only multiple-choice items and did not include open-ended response items.
- Due to the large number of potential participants in the study population, the population involved in the current study focused only on members located in KWD in Guntur district of Andhra Pradesh.

Limitations

- Due to the small and unique sample available for the study, results may not be generalizable beyond the specific population from which the sample was drawn.
- Due to poor maintenance of farm records, the respondents fail to express answers with clarity and the results might not accurate enough to be generalizable for the entire target population.
- Due to the length of the study, a significant number of respondents available in the preliminary testing may be unavailable or unwilling to participate in the final stage of testing.

4.8. Practical Problem vs. Research Problem

From the above discussion, we can infer that, research problem is the heart of the research process and the entire context of research revolves around this problem for which the researcher is trying to get the answer. Before discussing what is a practical problem, let us go through the following example:

'We know that, cotton boll worm menace has adversely affected many numbers of farmers. The Entomologists before resolving this practical problem, first tries to understand and solve the research question about the life

cycle of boll worm and how it damages the cotton crop. Then, after identifying the solutions for this research problem, the same can be applied to control boll worm menace on cotton crop in the farmers' fields. After resolving this practical problem, the Scientists will try to apply that knowledge in solving other practical problems of the farmers'. Thus, before solving a practical problem, the researcher must think about solving the research problem of that topic and based on the solutions achieved for research problem, the same can be applied to solve the practical problem. Thus, we can infer that, by thinking about a practical problem, it motivates the researcher to think and solve the concerned research problem. The solutions we learn for answering the research problem can be applied to resolve the practical problem. So, thinking about a practical problem motivates the researchers to identify and answer the research problem of that concerned topic and the solutions generated in answering that research problem will helps to solve the practical problem. That is, the knowledge gained from solving the research problem is applied toward finding the solution to the practical problem.

The above explanation clearly shows that, a practical problem happens in the real world, but a research problem starts in researcher's mind when he does not understand about a practical problem. Unless the researcher faces the practical problem first, he may not have thought of it for conducting the research. So, to solve the practical problem, first the researcher have to think up a relevant research problem of the same topic and then solve that research problem. Learning the answers about solving that research problem, will let the researcher understand how to resolve the practical problem. Thus, before solving any practical problem, a research problem has to be posed. But, it is to be noted that, solving the research problem does not automatically solve the practical problem. Rather, the solutions identified in solving the research problem will make the researcher to learn more and understand about solving the practical problem.

4.9. Research Problem vs. Topic

Before discussing this, it is essential to mention that, the word 'problem' in research problem is to be taken in the positive senses *i.e.*, it is good piece of work to act upon and not in the negative sense to mean something is 'bad'. That is, the researcher is aiming at a 'good' research problem to solve many practical problems in the world.

Regarding the comparative picture between 'research problem' and 'topic', the former is something to solve and the latter is some information to read. That is, regarding 'research problem', the researcher has to put all the efforts to identify and analyze the causal relationship about the problem and to suggest solutions to overcome the same. Thus, research is a search for a solution to a research problem and makes the researcher to focus on it to bring it to an appropriate end. On the other hand, topic simply involves information and data that makes the reader to enrich his knowledge in that particular field. Further, in analyzing a research problem, the researcher can focus on a particular topic that is useful to solve practical problems in the society. On the other hand, the reader will have elaborate information while going through a topic and hence, he cannot focus on a particular aspect.

5

Formulation of Research Problem

After identifying, selecting and signifying the importance of the selected research problem, now the researcher will formulate the research problem in such a specific manner that, it can be subjected to empirical verification. This is essential because, once after identifying and selecting the research problem, it is very difficult on the part of the researcher to go for direct empirical attack on it. Thus, it calls for formulating the selected research problem and it usually begin with two important questions *viz.*, what problem can be solved? and how can the problem be solved? In general, the identified research problem is in the form of a broad question and it should be narrowed down to be more specific while formulating the research problem (Figure 5.1). The identified research problem can be narrowed down by:

- a particular aspect (say, economical, social etc.)
- a particular time period
- a particular technological intervention
- a geographical area (district or state etc.)
- gender
- age group

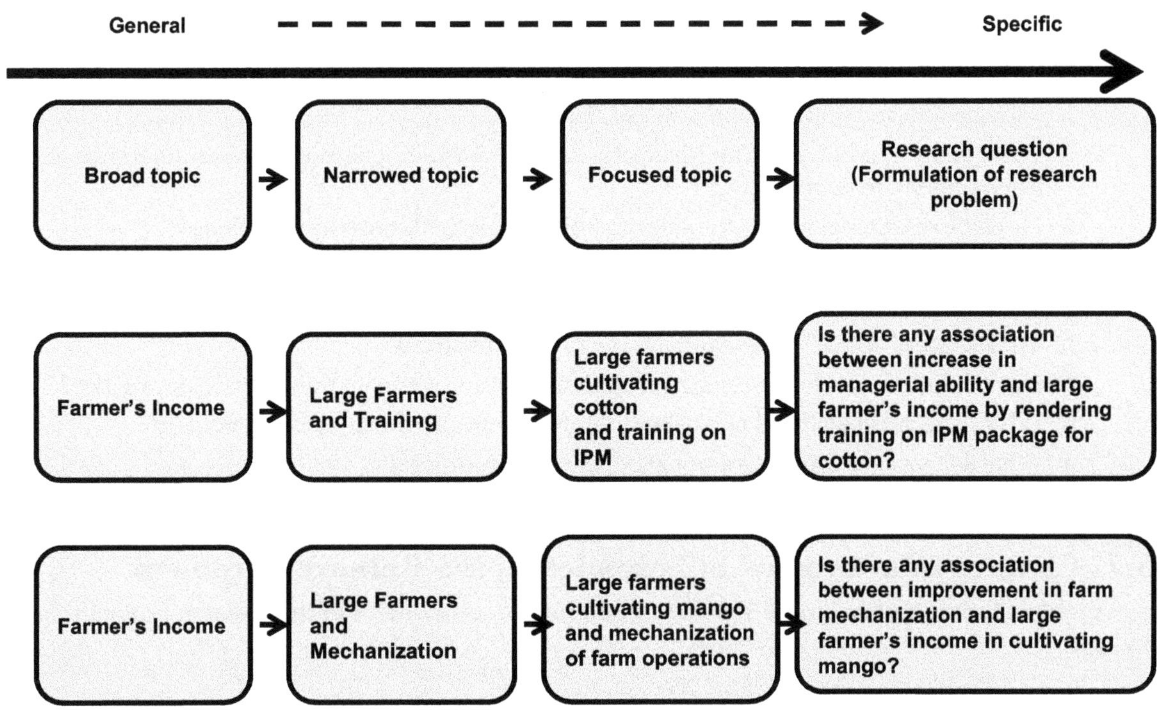

Figure 5.1: Formulating Research Problem from a Broad Researchable Area.

This formulation of research problem is considered as the most crucial step in the research process. Its main function is to decide what the research wants to find out about. The way the researcher formulate a problem determines almost every subsequent step in the research process.

Perhaps, the most important part of conducting research is formulating a research problem. A well-formulated research problem shows a researcher, where previous researchers have been deficient and identifies avenues of study that have not yet been pursued. So, a researcher must be careful to formulate the research problem properly in order to make sure that, the selected problem is not ambiguous, and the information obtained through the research is of greater significance both in the present and future contexts. There is a popular saying that, *'a problem well put is half- solved'*. This implies that, problem formulation is probably an important step towards overcoming a problematic state than problem-solving. A good definition of 'what the problem is', is believed to be more than half of the way towards its eventual elimination. The formulation of research problem can also be viewed from the managerial concept *i.e.*, a well-formulated research problem fits the level of researcher's research skills, needed resources, and time restrictions. Further, it ensures the accomplishment of desired objectives of the study duly considering both the economical and ethical aspects of the participants. So, problem formulation is like an identification of a destination before undertaking a journey. We know, identification of a research problem serves as a foundation of a research study *i.e.*, like a foundation for a building. If the problem is well-formulated, the researcher can expect a good study to follow.

Thus, once the researcher has decided, what problem he has identified and selected, he is now ready to formulate the research problem. For formulating the research problem, the researcher has to go through the thinking process with several questions in his mind *viz.*,

- ☆ Why he selected that particular research problem?
- ☆ What is the gap between 'an ideal situation' and 'existing situation' with reference to the issue under consideration?
- ☆ How the researcher justify the selection of that particular research problem?
- ☆ Who are the participants in the execution of research problem?
- ☆ What is the purpose of conducting this in depth study?
- ☆ What is the scope of the research on the present problem?
- ☆ How the answer(s) to the research problem will contribute to theory and/or practice?
- ☆ What aspects (objectives) he wishes to pursue through the identified problem?
- ☆ What aspects he wants to learn from the research problem?
- ☆ What changes he can bring about after solving the selected research problem?

5.1. Components Involved in Formulating the Research Problem

Problem formulation is simplified once we define the components of the research problem. They include:

A. Statement about the problem (Problem statement)
B. Justification of study of the selected research problem
C. Scope of the selected research problem
D. Framing the objectives for the selected problem
E. Identification of key terms and concepts of the selected problem
F. Identification of variables relevant to the selected problem
G. Assumptions, Limitations and Delimitations of the research study

A. Statement about the Problem (Problem Statement)

For formulating the research problem, first the researcher has to state it in precise terms. Writing the problem statement for the selected problem is one of the critical parts of the research process, as it provides focus and direction for the subsequent steps. Thus, it describes the selected research problem, which needs to be addressed. It provides the context for the research study and generates the questions, which the research aims to answer. It is defined as a clear, precise, concise, to the point statement of issue that is to be investigated with the goal of finding solution or building understanding. The type of statement to be employed depends on the preference of the researcher and the nature of the problem. There are two alternative ways of stating a problem.

- ☆ Posing question(s)
- ☆ Making declaration statement(s)

As mentioned earlier, question form has an advantage in sharpening and focusing the issue. But, the declarative form is perhaps more common. However, both the ways may be combined easily.

Question Form

The following are the examples of an interrogative statement, or a statement that raises a question:

- ☆ Is there a relationship between the number of hours that B.Sc (Agriculture) students have studied and their anxiety levels before the Semester-Final examinations?
- ☆ Is there any influence of IPM technology on cost-effective production in cotton?
- ☆ Is there any relationship between effective supervision and business performance?
- ☆ Is sub-surface drainage system is an efficient mechanism to reduce waterlogging induced soil salinity?

Declarative or Interrogative Form

The following are the examples of research problem in declarative form:

- ☆ To determine the relationship between the number of hours that B.Sc. (Agriculture) students have studied and their anxiety levels before the Semester-Final examinations.
- ☆ Study of performance of regulated agricultural markets in Andhra Pradesh.
- ☆ Production and marketing of mangoes in Chittoor district of Andhra Pradesh – An Economic analysis.
- ☆ Impact of IPM technology on cotton and paddy production in Kurnool district of Andhra Pradesh – An Economic analysis.

Compared to declaration form, the use of a question format to state the research problem seems to be the clearest way to identify the problem area of a study. If the research problem is in a question form, it is called as a research question. If it is in a declarative sentence form, it is called as a problem statement. But, both these forms are equivalent and it is up to the researcher to use his desired form.

Many University Professors believe that, problem statement is not a single question that a researcher tries to pursue, but it serves as a springboard for generating and presenting research questions in the schedule to gather requisite information and data from the respondents. Thus, problem statement is a concise description of the issues that need to be addressed by the researcher in solving the research problem. It makes the researcher to clearly focus on the research problem. It clearly indicates the worthiness in taking up the research project. This is because, it describes the research problem faced by the researcher that does not have adequate solutions available in the past. It (research-worthy problem statement) addresses all six questions: what, how, where, when, why, and who of a research problem.

Features of a 'Good' Problem Statement

A good problem statement need not be lengthy and it should incorporate the following features:

- ☆ It starts with a question.
- ☆ The statement of the problem should clearly indicate, what is to be investigated.
- ☆ It should indicate the central focus of the study
- ☆ It should be as specific as possible. For example:
    - ❖ ***Precise:*** 'Impact of IPM technology on production of cotton and paddy in Kurnool district of Andhra Pradesh – An Economic analysis'.
    - ❖ ***Too broad/vague:*** 'Impact of IPM technology on production of cotton and paddy'.

- The question raised in problem statement must be ethically neutral.
- It should not contain the word 'should' or 'ought'. For example: 'Should the training rendered to the cotton farmers on IPM technology is beneficial'? It could be restated scientifically as: 'Impact of IPM technology on cotton production in Kurnool district of Andhra Pradesh – An Economic analysis'.
- It should communicate the focus and importance of the problem.
- It should generate a variety of viewpoints from the readers' side.
- It should provide the necessary framework for reporting the results.
- A research problem should always be stated in a complete and grammatically correct sentences so that, the research consumer (reader) can read, understand, and respond to.
- It should indicate the feasibility of taking up the study.
- It should be unbiased.
- The solution(s) derived from the problem should make a difference.
- It should be justified with historical examples.

Functions of a Problem Statement

They include:

- It serves as the focal point of any research activity.
- It will justify the utility, significance, or interest inherent in the pursuit of the problem.
- It lays the foundation for further development of the research problem like research objectives, methodology, work plan etc.
- It helps the researcher to find information and reports of similar (earlier) studies from which the present study can benefit.
- It enables the researcher to systematically point out, why the present research is to be undertaken and the possible achievements of the study.
- It will bring clarity in the conceptual, theoretical and methodological issues of the research.
- It defines the problem precisely and helps to identify the variables that will be investigated in the study.
- It focuses only on the relevant variables and not on others. It also provides an opportunity for the researcher to demonstrate, why the selected variables are important.
- It is more specific than a topic and it limits the scope of the research problem.
- It communicates the focus and importance of the problem.
- It enables the readers to understand the importance of the selected problem.
- It serves as a framework for reporting the results. These indicate, what is probably necessary to conduct the study and explain how the findings will present the information. The researcher could relate the findings to prior research, or it could be presented as a comparison, or he could draw implications from the findings to generalize to other related conditions.
- It places the problem into a particular context that defines the parameters of what is to be investigated and indicate the scope of the research.
- It defines the interaction among various causes and thus, leads to a problem solving strategy.

Problem Statements Across Different Research Methods

- ***Correlation statement***
  - ***Format:*** Is there any correlation between two variables X and Y in the sample?
  - ***Example:*** Is there any correlation between fertilizers cost and gross returns of paddy enterprise?
- ***Descriptive study (comparative study)***
  - ***Format:*** *Is there any difference in Y (DV) between people in the population, who have X characteristic (IV) and those who do not have X characteristic?*
  - ***Example:*** *Is there any difference in the managerial ability of the farmers, who have more farming experience than the farmers who have no farming experience?*
- ***Experimental study (Comparative study)***
  - ***Format:*** *Is there any difference in Y (DV) between group A who received X treatment (IV) and group B who did not receive X treatment?*
  - ***Example:*** *Is there any difference in the yield levels of plots that receive 'fertilizer' treatment compared to those that do not receive fertilizer treatment.*

How to Write a Problem Statement

A good problem statement is just one sentence (with several paragraphs of elaboration). For example it could be: *'Market chain analysis of honey production in tribal areas of Visakhapatnam district in Andhra Pradesh'*. While this problem statement is just one sentence, it should be accompanied by a few paragraphs that elaborate on the problem. The researchers should also brief the elements so that, the reader does not get lost. One page is enough for a statement problem. This is given below:

World-wide, Governments have recognized the importance of liberalizing agriculture. The heart of the agriculture reforms lies in the 'market place'. With the advent of trade liberalization in Indian agriculture, there had been marked increase in the investments for strengthening the agricultural marketing sector. In spite of these reforms, symptoms of poorly functioning markets in India especially in the tribal areas are evident in the form of low investments in the market infrastructure, the persistence of high margins and marketing costs, unscientific grading, storage and processing facilities, poor market information network etc. These inefficiencies are especially true with respect to tribal farmers of Andhra Pradesh in general and Visakhapatnam district in particular. Hence, enhancing the ability of tribal farmers to reach markets is one of the most pressing development challenges. As these farmers are located in the interior areas, the remoteness results in reduced farm-gate prices, returns to labour and capital, and increased input and transaction costs. This, in turn, leads to subsistence rather than market-oriented production systems. Lack of cooperative marketing organizations, poor awareness on modern marketing techniques, higher transportation costs, lack of negotiating skills etc., are other important impediments to market access. This calls for an efficient, integrated, and responsive market mechanism, which is crucial for optimum allocation of resources in agriculture and for stimulating farmers to increase output. However, without having basic and convenient marketing conditions, the desired results both from production and marketing of agricultural produce cannot be expected. This calls for strengthening the agricultural marketing network on the desired lines, so as to safeguard the interests of farming community.

Honey production in Visakhapatnam district is mainly characterized by its seasonality of production. Besides this seasonality of production, the marketing pattern of honey in this district is highly unorganized and this calls for systematizing its marketing mechanism, so as to make the tribal farmers to derive higher share in the consumer's rupee. No studies have been carried out so far, to identify the existing pattern of marketing of honey, to identify the loopholes and to suggest suitable measures to improve the marketing efficiency in transacting honey in this district.

Market chain analysis is supposed to be the current approach working in studies of such type of production and marketing problems. Analysis of profitability of both honey production and marketing in the tribal areas counts much in identifying the bottlenecks and in planning the requisite strategies. Even though both honey and honey by-products are economical and socially important, no adequate study has been made in the tribal areas of Visakhapatnam district to improve the sector. This study, therefore, has attempted to contribute to filling the information gap by investigating the honey marketing chains and factors affecting honey supply in Visakhapatnam district of Andhra Pradesh.

In order to investigate the aforementioned problems, the study undertook the following research questions:

- ✰ *What is the profitability of honey production in tribal areas of Visakhapatnam district?*
- ✰ *What are the factors influencing supply of honey in the market?*
- ✰ *What is the Structure-Conduct-Performance Paradigm with reference to honey marketing in Visakhapatnam district?*
- ✰ *What are the channels involved in marketing honey in Visakhapatnam district?*

Common Deficiencies of Problem Statement

The problem statements with the following features are not valid:

- ✰ Failure to establish the existence of a problem
- ✰ Do not describe actual current condition or problem condition
- ✰ Problem is not measurable
- ✰ Problem is unclear and vague
- ✰ Problem area is not specific
- ✰ Indicate the areas that lies outside the scope of actual problem
- ✰ Possible solution expected that makes no difference when compared to the earlier studies
- ✰ Does not carry justification about the problem

B. Justification of Study of the Selected Research Problem

After stating the problem, it is necessary next to justify the importance of the problem. This implies that, 'why the selected research problem is important' say, in the context of participants or finding solution to the new problem or expanding the body of knowledge etc. This is essential because, we know, conducting research often is expensive and time consuming. Unless the problem is important, there is no meaning in wasting the time, money, energy and other resources on the part of the researcher. The justification of research problem is not only from the view point of the researcher, but also from the readers' side and other researchers, who wish to apply the same knowledge in other practical situations. Thus, this section describes or explains the potential value of the research study. As mentioned above, since, research is conducted to enhance the existing body of knowledge and/or to solve a practical problem and in this context, how the solution(s) of the selected research problem will do this, should be articulated in this section. While justifying the study, the researcher should pose the following questions on himself and then try to answer each of them:

- Is the problem considered for the study refers to the current and timely one?
- How widespread is the problem?
- Are many areas and many people affected by the problem?
- Does the problem affect key populations such as farmers, farm-women, rural youth etc?
- Whether the researcher had adequate time, resources, and skills to carry out the research?
- Does this research study advance knowledge?
- Does this research study contribute to practice?
- Will this study fill a gap or void in the existing literature?
- Will this study replicate a past study, but examine different participants and different research areas?
- Will this study extend past research or examine the topic more thoroughly?
- Whether the Government launched any programmes earlier to rectify the same problem?
- Does the problem relate to ongoing developmental programmes and policies of the Government?
- Does the problem relate to broad social and economic issues such as education, unemployment, income distribution, poverty etc.?
- Who else is concerned about the problem?
- Whether Government intervention is necessary and possible to address this problem?
- How the findings or results of the study will help the participants?
- How will the understanding of the selected topic be improved?
- What contribution to existing theory will be made?
- Why will the research be worth the effort?
- How will it benefit people?
- Why does the problem matter?
- How might the study improve policy?

For the above questions, the researcher has to prepare the answers to justify the selection and execution of the selected research problem. While answering the questions, the researcher first has to discuss the broad issues and gradually focus on the specific issues related to a particular population group or geographical setting. The main objective of answering the above questions is not the purpose to describe the whole context of the research problem in-detail. Rather, it aims to describe the severity of the problem in the study area so that, both the researcher(s) and readers will appreciate its selection and execution. In this context, a brief introductory discussion is essential to find out, which part of the total population is being addressed through this research. The justification for the selected research problem can easily be generated, if the researcher believes that, his study offers relevant solutions for the problem. That is, the solutions designed by him will alleviate the consequences faced due to the existence of the problem. Such consequences and possible solutions can be listed out by the researcher in a paragraph or two. It should also highlight, who are the beneficiaries of this study and how they are benefitted as a result of successful completion and dissemination of the study. In dissertations, this is a section typically labeled as 'Significance of the Study'.

In writing the justification for the study, the researcher must note that, at each every step, he has to convince the readers about the significance of the study and why it has been taken up for the investigation. In dissertations, this topic is mentioned in about one to three paragraphs and in explaining the justification of the study, the researcher has to take help of the reviewed articles. He has to closely look, how the earlier researchers had convinced the readers about the importance of their topics. He has to deduce the same logic in making his research study, signifying its importance in solving the problem and convincing the readers. While justifying the study, the researcher should remember that, his objective is not to show or gather evidence to prove something about the idea, but to make the readers to convince about the severity of the problem and to learn what they should know.

How to Write a Justification for the Selected Problem or Significance of the Study

In the dissertations, justification about the selected researcher problem will be written in a paragraph or two duly answering the above questions. For the above dissertation on '*Market chain analysis of honey production in tribal areas of Visakhapatnam district in Andhra Pradesh*', the following is the justification for the study:

'This study is essential because, both production and marketing constraints of honey are showing significant influences on the socio-economic well-being of the tribal farmers in Visakhapatnam district. Hence, this study would generate useful information in order to formulate honey marketing development projects and guidelines for interventions that would improve the efficiency of honey marketing system in the study area. The potential beneficiaries of this study include producers of honey, marketing agencies, Governmental and non-governmental organizations, that have an interest to intervene in honey marketing system. Researchers, who want further investigation on honey marketing, would use the results from this study. The finding of this study could be gainfully utilized by the administrators, policy makers, scientists and extension workers to know about the status of honey producers and to make both production and marketing of honey more profitable and economical'.

C. Scope of the Selected Research Problem

'Scope' describes the areas covered in a research *i.e.*, it explains how far the research area has been explored. It provides factual statements on the extent of the problem in the past, current or future. Thus, the scope of a research problem refers to breadth or depth of it being investigated. It lays out the plans for a research project and the hopes for what will be achieved. A well-written scope statement is crucial because, it makes the researcher to plan and take fruitful decisions, while executing the subsequent stages of the project. This also enables the researcher to be more adaptable to deal with the obstacles that may arise during the execution of the project. The writing of scope statement for the research problem establishes a solid agreement between the researcher and the participants towards the achievement of the project objectives. The two parties reach an agreement by explaining the need or issue to be resolved in studying and analyzing the problem. For example, if a research project on '*Economic appraisal of sub-surface drainage system to reclaim waterlogged induced soil salinity in Mutluru pilot command of KWD in Andhra Pradesh*', the research staff and the participant farmers come to an agreement regarding the potential costs involved in the execution of the project. Thus, scope statement becomes a useful resource for everyone involved in the research because, it will answer any questions that arise.

The advantage of writing the scope statement for studying the research problem is that, it excludes other areas beyond the research topic in collecting the information and hence, the researcher is not criticized for not covering it. Thus, the scope statement refines the research topic from a general to a specific level.

How to Write 'Scope of the Research Study'

In dissertations, this information can be provided in a paragraph or in a bullet point form. For the above selected dissertation *'Market chain analysis of honey production in tribal areas of Visakhapatnam district in Andhra Pradesh'*, the scope statement is given below:

'The area covered in this study was limited to tribal areas of Visakhapatnam district. The study mainly focused on the functioning of the honey market and relationships among the marketing agencies with reference to performance of different marketing functions. Profitability of honey production and efficient marketing were also addressed. The results of the study will be helpful to identify the bottlenecks both in the production and marketing of honey at farmer's level.'

D. Framing Objectives for the Research Study

Having decided what to study (problem statement), and knowing why the researcher wants to study it (justification of the research problem), the researcher now has to frame the objectives to address the problem (Figure 5.2). Before going into the details, it is essential to define 'objectives'. They are the set out goals of a research study through which the researcher describes and analyzes the

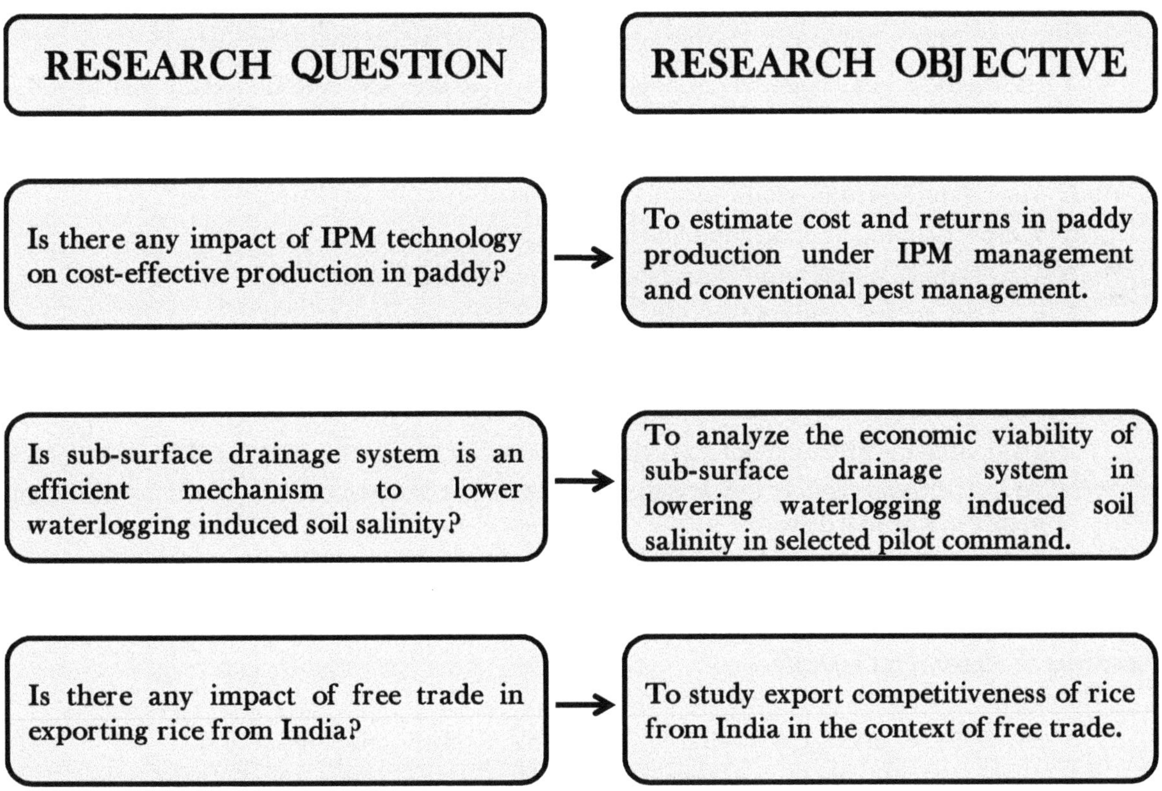

Figure 5.2: Framing Research Objective from its Research Question.

problem to draw useful conclusions and to plan strategies to address the same. According to Churchill, *'research objectives are the specific components of the research problem that you'll be working to answer or complete, in order to answer the overall research problem'*. Thus, objectives refer to the questions to be answered by the researcher in addressing the research problem. They indicate the possible outcomes from the research study. They are essentially a transformation of researcher's questions made as statements. It is to be noted that, objectives should be closely related to the problem statement. For example, if the problem identified is 'lower marketing efficiency in transacting paddy in Kurnool district of Andhra Pradesh', the general objective could be to assess the reasons for this lower marketing efficiency. If we break down this general objective into smaller and logically connected parts, then we get specific objectives. Thus, there are two categories of objectives to deal with the research problem: *viz.*, General (Main) objective and Specific (Sub) objectives.

a. General Objective

This objective is generally stated and rather widely focused. It refers to an overall statement of the thrust of the research study. It gives information about the main associations and relationships that the researcher wishes to seek to establish. For example, in the research problem, 'Marketing efficiency in transacting paddy in Kurnool district of Andhra Pradesh', the general objective was 'to assess reasons for lower marketing efficiency in transacting paddy in Kurnool district of Andhra Pradesh'.

b. Specific Objectives

General objective is followed by the specific objectives – which are more focused and direct. These specific objectives are the measurable statements framed to answer the specific questions within the main framework of research study. The important features of specific objectives include:

- ✰ Unlike the general objectives, the specific objectives are more specific and are related to the research problem situation.
- ✰ They indicate the variable(s) to be examined and measured.
- ✰ They should be numerically listed.
- ✰ Wording should be clear, complete and specific.
- ✰ They should communicate the readers about researcher's intention.
- ✰ Each objective should contain only one aspect of the study.
- ✰ Use action oriented words or verbs, when writing objectives.

In the above example, the specific objectives include:

- ✰ To identify the predominant marketing channels for transacting paddy in Kurnool district.
- ✰ To assess the marketing costs and marketing margins of all agencies involved in transacting paddy in Kurnool district.
- ✰ To compute pricing efficiency in transacting paddy in Kurnool district.
- ✰ To analyze spatial market integration in transacting paddy in Kurnool district.

Features of Research Objectives

The objectives framed by the researcher must be:

- ✰ SMART – *i.e.*, Specific, Measurable, Achievable, Realistic and Time bound.
- ✰ Address all the aspects of the problem in a coherent way and in a logical sequence.

- Meet the purpose of the study
- Generally, they are written as statements, using the word '*to*'. Include action verbs such as:
  - To find out ...
  - To determine ...
  - To examine ...
  - To identify ...
  - To analyze ....
  - To study ....
  - To ascertain ...
  - To compare ...
  - To verify ...
  - To describe ...
  - To establish ...
  - To explore ...
- Avoid the use of vague non-action verbs such as;
  - To appreciate
  - To understand
  - To believe
  - To grasp the significance of
  - To really understand
  - To have an awareness of
- Research objectives can be stated as:
  - ***Questions:*** The objectives of the research are to answer the following questions .... For example: Where should the pesticide firm market its new product?, Where should the fertilizer firm market its new product?
  - ***Positive sentence:*** The objectives of this study are to find out, to analyze, to determine, ... For example, *'to analyze the factors influencing farmers' indebtedness in Srikakulam district of Andhra Pradesh'*.
- Clear without ambiguity
- Precise in nature (micro in scope)
- Actionable
- Flexible
- It is normal to write in the passive voice. Use the future tense in the proposal, present tense when the study has been completed.
- Ethical (would not cause physical and psychological injury to anyone, or to the environment).
- Feasible (could be studied and would not take too much time, energy and cost on the part of the researcher).
- Sequentially appropriate.
- Attainable within a reasonable amount of time.

Thus, research objectives with the above features give a clear picture about, how to proceed with the research investigation. It is essential to note that, framing of research objectives in a logical sequence is very important, as the researcher will present the Results and Discussion chapter in the Dissertation in that sequence only. So, for the research study to be more logical and in a systematic manner, framing the objectives in a logical sequence counts much. It is further important that, the researcher (student) should limit the number of objectives for about four to six, as it is not uncommon to come across a situation in which some of the specific objectives are not addressed in the 'Results and Discussion' chapter at all. So, it is, therefore, advisable to limit the number of specific objectives.

Objectives Across Different Research Methods

The framing of objectives (*i.e.*, how the researcher state the objectives) also determines the nature of research method say, Descriptive or Correlational or Analytical research etc. The following are the examples of framing of objectives under different research methods:

Examples of 'Descriptive' Objectives

In these objectives, the emphasis is to describe the existing phenomenon as it is.

- ☆ To find out farmers' opinion about training programmes conducted at Krishi Vigyan Kendra, (KVK), Yagantipalle in Kurnool district of Andhra Pradesh.
- ☆ To find out farmers' opinion about the functioning of paddy transplanter.

Examples of 'Correlational' Objectives

In these objectives, the emphasis to study the degree of association between the variables.

- ☆ To ascertain the influence of bio-fertilizers application on paddy yield.
- ☆ To compare the effectiveness of different methods of technology dissemination in improving the technical-know-how of the farmers.

Examples of 'Analytical (or Hypothesis Testing)' Objectives

In these objectives, the emphasis to study the direction of the relationship between the variables being correlated.

- ☆ To analyze the factors influencing market arrivals of jaggery in Anakapalle regulated market of Visakhapatnam district in Andhra Pradesh.
- ☆ To analyze the factors influencing farmers indebtedness in Srikakulam district of Andhra Pradesh.

The above discussion can be summarized through Figure 5.3 in relation to the features of objectives:

Importance of Framing the Research Objectives

The framing of research objectives will help the researcher to:

- ☆ Narrow the focus of the study
- ☆ Avoid collection of data that are not necessary to address the problem
- ☆ Systematically organize the study (in a logical sequence) in clearly defined parts
- ☆ Guide the researcher to address the problem in a more systematic way
- ☆ The researcher can plan the analytical tools to address the framed specific objectives

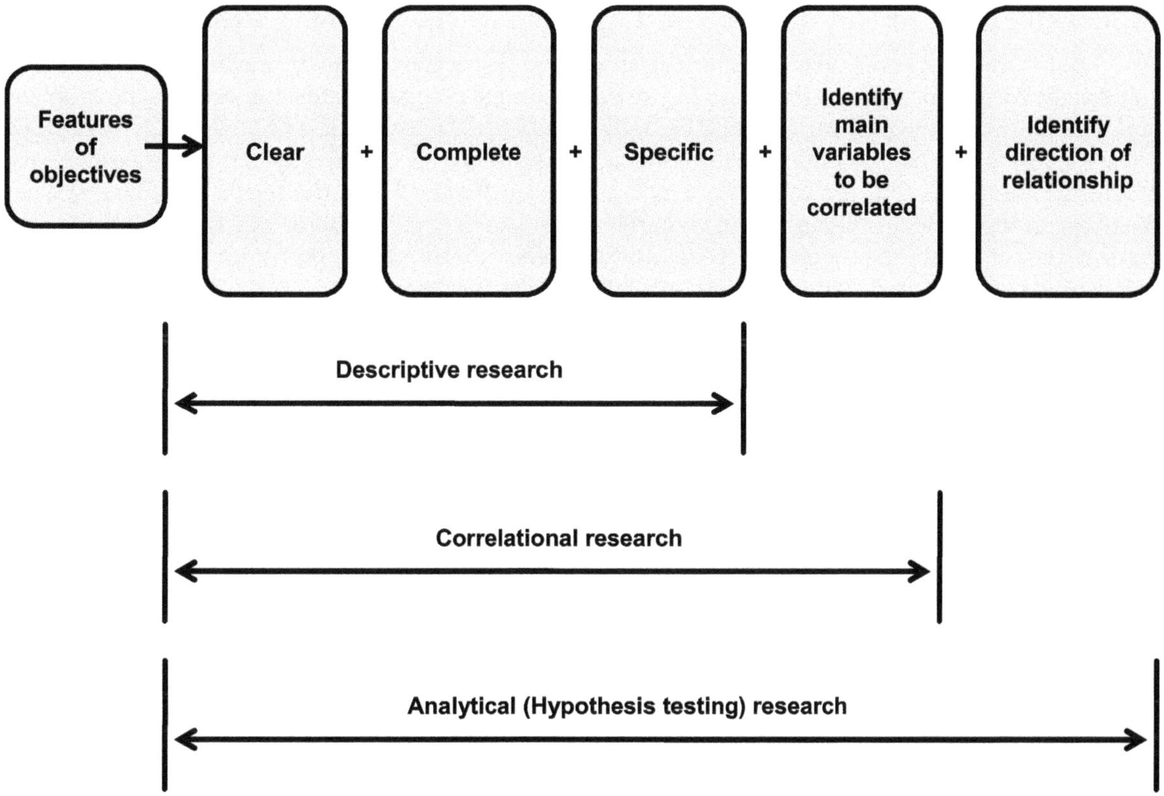

Figure 5.3: Features of different Types of Objectives Across different Research Methods.

How to Write 'Objectives for the Research Study'

In dissertations, this information can be provided in bullet point form. For the above dissertation *i.e., 'Market chain analysis of honey production in tribal areas of Visakhapatnam district in Andhra Pradesh'*, the following are the general and specific objectives framed:

General Objective

The overall objective of this study was to analyze honey market chain in tribal areas of Visakhapatnam district.

Specific Objectives

The specific objectives of the study include:

- *To analyze the structure of production costs of honey and determine its profitability*
- *To analyze the determinants of honey supply in the tribal areas of Visakhapatnam district*
- *To analyze the structure, conduct and performance of honey market*
- *To compute the marketing efficiency in transacting honey in tribal areas of Visakhapatnam district*
- *To identify the major constraints both in production and marketing of honey and offer relevant suggestions.*

E. Identification of Key Terms and Concepts of the Selected Problem

After framing specific objectives for the study, the researcher has to go through key terms and concepts related to the selected problem. The researcher must precisely define the terms in the problem and the sub-problems. Each term should be defined, as it will be used in the final report. In defining a term, the researcher has to define it in the context of the research problem only and not as a dictionary meaning. The terms that are unusual or not widely understood and the terms that have special meaning in the context of the selected research study should be defined in this section. Acronyms (except those in common usage) used frequently also require definition at this point. The identification of key terms and concepts of the selected problem enables the researcher:

- To gain overall understanding about the issues he has to deal in meeting the framed objectives.
- To frame meaningful questions.
- To collect the literature on the aspects related to the selected problem.
- To identify the variables relevant to the selected problem.
- To plan tools of analysis for the study.
- To interpret the analytical findings.
- To present the information in a logical sequence.
- To frame meaningful suggestions for addressing the problem.

Thus, studying key terms and concepts enable the researcher to conduct meaningful inquiry for all the questions he raise in the study and help to cut away unimportant details. It also makes the researcher to verify, whether any of the terms have been used in a vague or unclear manner. If so, he will substitute with a more precise word. So, it ensures the presentation of report in a professional way.

For the above dissertation on '*Market chain analysis of honey production in tribal areas of Visakhapatnam district in Andhra Pradesh*', the researcher will identify and define the following terms and concepts:

- *Total costs*
- *Gross returns*
- *Net returns*
- *Marketing efficiency*
- *Market structure*
- *Market conduct*
- *Market performance*
- *Price spread*
- *Marketing channel*
- *Gross marketing margin*
- *Net marketing margin*
- *Marketing costs*
- *Marketing agency*
- *Wholesaler*
- *Retailer*

F. Identification of Variables Relevant to the Selected Problem

After framing specific objectives for the study and identifying the key terms and concepts, these guide the researcher to identify key variables, their method of measurement and units of measurement in relevance to the selected problem. The researcher has to carefully define and quantify these variables. A variable generally measures a characteristic of the survey or study that changes, like the fertilizers cost, irrigation costs, market prices etc. Some variables change easily (as mentioned above), while other variables are almost constant, like the name of the respondent. Researchers are often seeking to measure variables. This will enable the researcher to plan the analysis to perform in the future. Of course, the nature of variables identified differs across the type of research methods employed in the research. For example, in descriptive research, the researcher simply identify and describes the variables as it is, without any manipulation. However, say, in case of analytical research, there is scope for the researcher to manipulate the variables. For example, in experimental research, the researcher tries different doses of 'N' fertilizer to analyze their influence on the output of the crop (paddy). Thus, identifying and understanding about the key variables of the study will guide the researcher to plan the data collection and analysis in the future. The researcher, in the next step of the research process *i.e.*, 'Review of literature' will further go in-detail about understanding the behaviour of these selected key variables. For example, in analyzing the factors influencing gross returns of paddy, the researcher identifies different variables like gross returns (DV), human labour costs, fertilizer costs, pesticides costs, irrigation costs (IVs) etc.

Thus, during the problem formulation stage, it is necessary to identify the key variables of the study. In general, four types of variables are important in research *viz.*, IVs, DVs, Confounding or Intervening variables and Background variables. The characteristics that define these variables must be clearly identified for the purpose of the study. The variables may also be quantifiable (fertilizers cost, irrigation costs, market prices) and non-quantifiable (sex, managerial ability etc). Non-quantifiable variables are also called as Dummy variables. The variables that are often stated in experimental research or analytical research aim at studying the direction of relationship between the variables (*i.e.*, which variables 'cause' and which variable(s) is the effect). However, in correlational research, the researcher aim at studying the degree of association between the variables. In explorative research, (qualitative research), the independent and the dependent variables might not be identified beforehand. This is because, the researcher does not have a clear idea yet on what is really going on. The ability to control variables is an important key to the success of a research study; however, some variables are more difficult to control than others. These variables are discussed in-detail in the ensuing Chapter 8.

The following are the independent and dependent variables identified across different research studies:

- ☆ An Agricultural Economist collected the data and analyzed to find out the determinants of farmers' indebtedness in Srikakulam district of Andhra Pradesh.
  - ❖ **DV:** Magnitude of debt
  - ❖ **IV(s):** Land holding, Cost of cultivation, Family expenditure, Family income
  - ❖ **Type of study:** Survey research
- ☆ An Entomologist has conducted a field experiment to test, whether a new brand of pesticide is effective in controlling boll worm on cotton. To test this, he followed Randomized Block Design experiment and applied the treatment randomly in some plots and other plots are kept as control.

- **DV:** Increase in output of cotton
- **IV:** Dosages of pesticide applied
- **Type of study:** Field experiment

☆ An Agricultural Economist opined that, if the farmers grade their produce before transacting their produce, they get higher profits in the market, than the farmers who do not grade their produce. To test this hypothesis, he randomly asked 100 farmers, how many are grading their produce before selling the produce in the market.

- **DV:** Gross returns
- **IV:** Grading cost
- **Type of study:** Survey research

☆ An Agronomist conducted an experiment on 'Response of kharif maize to nitrogen levels and plant population densities'.

- **DV:** Maize output
- **IV:** Plant population densities, different nitrogen levels
- **Type of study:** Field experiment

☆ A Plant Physiologist conducted an experiment on 'Nitrate reductase and Nitrate reductase activity in cotyledons and embryonic axis of germinating *Phaseolus vulgaris* L. seeds after GA_3 and MH treatment'.

- **DV:** Germination of seeds
- **IV:** Different levels of GA_3 and MH
- **Type of Study:** Lab experiment

Criteria for Selection of Variables

They include

☆ The selected variables should be relevant to the study.

☆ The variables selected should be in accordance with the objectives planned for the study

☆ The variables should be selected depending upon the available analytical techniques to the researcher.

☆ The variables selected should be mutually exclusive and not overlapping

☆ The selected variables should be limited in number, so as to avoid confusion on the part of the researcher

☆ They should be easily measured

☆ The data sources for these variables should be easily accessed by the researcher.

For the above dissertation on '*Market chain analysis of honey production in tribal areas of Visakhapatnam district in Andhra Pradesh*', the researcher will identify the following variables (To analyze the determinants of market supply of honey production):

☆ **DV:** Quantity Supplied

☆ **IVs:**

- Quantity of honey produced
- Distance to nearest market

- Price of honey
- Age of the household head
- Experience in beekeeping
- Access to market information
- Access to credit

G. Assumptions, Limitations and Delimitations of the Research Study

a. Assumptions

Assumptions in a research study implies the things that are somewhat out of researcher's control, but if they disappear, the study would become irrelevant. Assumptions may also be used to state whether or not limiting factors are likely or unlikely to affect the outcome of the study. Generally, assumptions are made about the theory under investigation, the phenomenon under investigation, the methodology, the analysis, the power to find significance, the participants in the study, and the results. For example, Linear Regression analysis makes several key assumptions like linear relationship, multivariate normality, no multicollinearity, no auto-correlation, homoscedasticity etc. After the analysis is completed, the researcher assume that, the results are generalizable beyond the sample being studied. Finally, it is assumed that, the results of the study will be relevant to stakeholders. This is the most compelling assumption that the results will be meaningful. Similarly, if the researcher wants to analyze the importance of Agricultural Experience Learning Programme (AELP) for the B.Sc (Agriculture) students, he first assume that, entrepreneurial skills will continue to be important in the agricultural education. If he conducts a survey, he need to assume that, the students will answer truthfully. If he chooses a sample, he needs to assume that, this sample is representative of the population for drawing the inferences. Generally, it is assumed that, in survey research, all respondents will answer all questions in the schedule honestly. According to Leedy and Ormrod (2010), *'assumptions are so basic that, without them, the research problem itself could not exist'*. It is essential to note that, simply listing out the assumptions will not mean the purpose, but the researcher must justify that each assumption is 'probably' true, otherwise the study cannot progress. In the above example, the researcher assumes that, the students (participants) will answer the questions honestly and this enables the researcher to present his findings with more confidence. Thus, assumptions are those things, which the researcher takes granted in the study. These assumptions are to be explicitly stated in the body of the dissertation, usually in Chapter 1 (Introduction) of the Main text in a dissertation or thesis.

b. Limitations

These are the factors or elements that indicate the weakness in the research and are beyond the control of the researcher in conducting the research study. In most instances, any assumption the researcher make becomes a limitation. We know, while conducting the research, the researcher may encounter several difficulties say, in collecting the requisite data from the farmers. For example, if the researcher makes an attempt to study *'Economic analysis of paddy production and marketing in Kurnool district of Andhra Pradesh'*, the researcher tries to elicit primary data from the sample farmers. We know, the farmers generally do not maintain farm records and whatever data they provide to the investigator is based on the recall of their memory. This is a serious limitation, as certain bias will definitely enter in providing the data by the respondents. Further, the generalizability of the analytical findings is limited by how well the sample represents the population. The stating of such type of limitations, may be very useful for the readers because, they provide a method to acknowledge possible errors or difficulties in interpreting results of the study. Sometimes, the limitations of conducting research may not be apparent at the time of beginning of research, but they become apparent during the course of investigation. These limitations can be considered as 'excuses' made in conducting the research and also enable the readers to understand the inferences of the results of the study in the realistic context. The researcher should list out the possible limitations in conducting the study and they should be in relevance to the selected problem under consideration. The limitations encountered by the researcher can be conveniently categorized as below:

Methodological Limitations

They are related to sample size, lack of available and/or reliable data, lack of prior research studies on the topic, measure used to collect the data etc.

Researcher's Limitations

They are related to access, time availability, resources availability, biasedness, fluency in a language etc.

The following are the examples of limitations:

- ✰ The respondents were not in the habit of maintaining records of their income and expenditure. The entire information was by recollecting past events by the farmers.
- ✰ District wise growth rates in relation to area, production and productivity was unable to carry out due to time and manpower limitations.
- ✰ Since the sample selected for this in depth study is small (n=90), the analytical findings generated from this study alone cannot be generalized at the population level.
- ✰ The growth rates of the area under paddy could not be calculated for Kurnool and Anantapur districts, as for some mandals, the data were not available with their respective Chief Planning Offices.
- ✰ Production function analysis was limited to three districts because, the Department of Agriculture has not completed the processing of data.
- ✰ Time series data on physical and financial indicators required for the study is not readily available for the past fifteen years. Hence, data on physical and financial performance of regulated markets in Andhra Pradesh state as a whole and in selected markets for the factor analysis through principal component technique were limited to five years only.
- ✰ Primary data collection was not undertaken to ascertain impacts of crop diversification on rural agrarian economy due to financial constraint.
- ✰ Literature review on export competitiveness of paddy was limited to few studies in India due to nonexistence.
- ✰ Comparison of market integration between pre and post liberalization periods was not made due to absence of market-wise price data during pre-liberalization period.
- ✰ Marketing research faces several constraints like budgetary constraints, time constraints, reliability of the data, legal and ethical constraints etc.
- ✰ Since a large majority of the farmers are not educated therefore, the degree of reliability of their response need not be always correct.
- ✰ The quality of primary data collected is subject to the level of cooperation of the respondents.
- ✰ The study was limited to the farmers trading in five Rythu Bazaars of Hyderabad and one each from Vijayawada and Visakhapatnam.

It is to be noted that, if the researcher do not mention this section in Chapter I (Introduction) of Main text in his dissertation, it implies that, the research study has accounted for all (or nearly all) variables, the findings of the sample can be generalized to the population and could be replicated accurately under all conditions. Hence, the possible important limitations encountered by the researcher in his study have to be clearly mentioned in the research report. When discussing the limitations of the study, the researcher should keep in mind two important aspects *viz.*, each limitation should be explained clearly along with the reasons and he should also explain the reasons, why he could not overcome such limitations. This enables the readers to plan to overcome such limitations in their future investigations. Generally, conditions that have already been stated as assumptions for the research study should not be included in this section. For example, if respondents' honesty in answering the questions of the schedule/questionnaire has been listed as a possible assumption, there should not also be any limitation that the correctness of findings are subjected to respondents' honesty. Generally, researchers confuse between scope and limitations of the research and they assume both are one and the same, but they are actually quite different. As discussed earlier, scope describes the areas actually covered in a research. This section will indicate what was done and where the information that was used specifically came from. However, 'limitations' include those factors that not are actually under the control of researcher, while conducting the research study. This section highlights the excuses made in conducting the research and thus, prevents other people from criticizing research methods or results or conclusions. Thus, this section reduces the likelihood of the reader raising those issues to criticize the data used or methods employed or analytical findings. For example, scope can be, what are the marketing problems in selling a pesticide in a locality, whereas the limitations include the inaccuracy of data, sampling errors, data gathered from a specific locality only etc. Every research report should include both scope and limitations of the research. Without these two sections, the reports would be meaningless and hence, it would not benefit anyone in the long run.

For the above dissertation on '*Market chain analysis of honey production in tribal areas of Visakhapatnam district in Andhra Pradesh*', the following are the important limitations:

- ✰ *Due to time and resource constraints, the study is restricted to only Visakhapatnam district in the Andhra Pradesh state.*
- ✰ *This study being the first in the district, it lacks detailed investigations.*
- ✰ *The respondents were not in the habit of maintaining records of their income and expenditure. The entire information was by recollecting past events by the producers.*

c. Delimitations

As mentioned earlier, delimitations are those elements the researcher can control and these elements are not included in the research study because, they are irrelevant for the study. Setting limits on the sample size, extent of the geographic area selected for collecting the requisite data, time frame considered for the study etc., fall under delimitations of the research stuidy and this makes the study feasible for the researcher. Technically, delimitations (factors which the researcher controls) are distinct from limitations (over which the researcher has little or no control) and it is discussed earlier under Section 4.7 of Chapter 4. But, some researchers (students) indicate the delimitations within the section on limitations, but they should be indicated as a separate section. The following are the important examples for delimitations of a research study:

- ✰ In order to assure manageability of the collected data, questionnaire included only multiple-choice questions and did not included open-ended response items.

- Due to the large number of potential participants in the study population, the population involved in the current study focused only on farmers located in KWD of Guntur district in Andhra Pradesh.

The above discussion infers that, it is essential to convince the readers through clearly indicating the assumptions, limitations and delimitations of the research study. Once a statement of limitations and delimitations has been prepared, the question about where in the research report to place it arises. A logical place to present delimitation is after listing out the limitations for the study in Chapter I (Introduction) of the main text in the thesis.

6

Review of Literature

So far, the researcher has focused on the identification and formulation of selected research problem. It is high time now for the researcher to know, in what way the selected research problem had been addressed in the past by different researchers. For this, he has to go through the earlier studies to analyze the information about the significance of the problem, methodology employed, findings concluded etc. This serves as a bench mark for the researcher to proceed into his selected research problem on the earlier lines and even to improve up on the methodologies. It also makes him to compare his research findings with the findings generated in the past. So, the review of literature serves as an important component in the research process. It guides the researcher to know, what previous investigations have proved and how the present research would fit in the body of works on the subject matter. Previous researches may indicate unanticipated consequences, which can be validated in the present investigation. It is to be noted that, 'review of literature' is also done prior to the selection of research problem. For selecting the researchable problem, the researcher proceeds in different ways like conducting a survey among the famers to identify and prioritize the problems, may review earlier literature to identify the gaps in the research areas, identifying the priority areas of Government's programmes and policies etc. So, 'review of literature' can be defined as '*an extensive survey of all available past studies relevant to the field of investigation*'. It is, otherwise, defined as '*the process of reading, analyzing, evaluating, and summarizing scholarly materials about a specific topic*'. The collection of literature review is based on the assumption that, knowledge accumulates, and that we learn from and build on what others have done. It gives an indication to the researcher, what others have found out in the related field of study in the past and how they have done so. It is a careful examination of a body of literature pointing toward the answer to the formulated research question. This enables the researcher to acquaint himself with the available body of knowledge in the area of his interest and to test the research question against what already is known about that subject. Hence, this step (review of literature) is considered as an integral part of entire research process and makes valuable contribution to every operational step.

6.1. Why to Collect Review of Literature?

It is essential to note that, there should be clear links between the aims of the research and the literature review. The literature review makes the researcher to search, whether the same problem has answered earlier or not. If the same problem has addressed earlier, then the researcher has to change or modify the problem. For example, if the researcher has attempted a research study on '*Economic analysis of sub-surface drainage system in Mutluru canal command of KWD of Andhra Pradesh*', then he should thoroughly analyze himself about the following questions:

- ☆ What is known about sub-surface drainage system?
- ☆ Has anyone else done anything similar in the past?
- ☆ What techniques have been used to appraise the economic aspects of sub-surface drainage system?
- ☆ What is the chronology of the development of knowledge about economic appraisal of sub-surface drainage system?
- ☆ Are there any gaps regarding the methodology and tools of analysis in analyzing the research problem? How the researcher intend to bridge the gaps, if any?
- ☆ In what way the execution of this project is significantly different from the earlier projects?
- ☆ What is the knowledge gained through reviewing the literature? How this knowledge will help the researcher to plan the direction of work in the research project?
- ☆ Whether the execution of activities in the present research project are in consensus with the earlier works or they deviate?
- ☆ Is the present technological intervention is relevant to the problem?
- ☆ How might the present research add to the understanding or challenge existing theories?

The researcher must be ready to answer the above questions, if any one poses at him. It should be noted that, the literature reviewed by the researcher must guide him in conducting the present research work through exposing him to various tools and techniques. This makes the researcher to generate the output from the present research and that must be significantly different from the past findings. Mere repetition of the research that too in the same agro-ecological situation will not serve the basic purpose of conducting the research.

Thus, colleting the review of literature sets the background framework for the research problem identified by the researcher and enables him to address it in a planned and systematic manner. It also gives indications to the researcher the ideas that are to be explored. While studying the research problem, it is essential on the part of the researcher to clarify the relationship of previous research to the justification of the selected problem, even he writes it before initiating the data collection. So, it is meaningful to define literature review *'as a survey and discussion of the literature in the selected area of study. It is a concise report on the review made regarding the past studies that are in relevant to the selected research topic for investigation'.* The reviews should be made from various sources like journals, internet, books, reports, conference proceedings etc., with reference to the selected topic and are arranged chronologically. It is to be noted that, literature review is both a summary and synthesis of past research studies so that, it gives a new interpretation of old material or combine new with old interpretations. The review is to be organized in a systematic pattern dividing the selected research topic into different sub-topics in accordance with the objectives and key terms identified in the study. Since, the reviews are arranged chronologically, it discusses the trends and developments of a theory or model or technique etc., rather than focusing on one item at a time. Thus, review is not a mere summary of the past research findings, but evaluates previous and current research in regard to how relevant and/or useful it is and how it relates to selected research topic. The format of a review of literature may vary from discipline to discipline and from assignment to assignment, but in general, it contains introduction about the review, main body and conclusions (Figure 6.1).

Sometimes, the researcher collects a significant volume of literature even before selecting the research problem, as it helps him in identifying the research problem that was not executed earlier in the same study area. However, if the researcher has to work out research in the same study area, the literature review provides an option to modifying the selected research work by adding a new perspective or altering some of the methods of research to obtain a perspective that will be different from earlier works and thus more valuable. In general, researcher usually goes through a large volume of literature than that is actually incorporated in the report. This is because, different literature may have differing relevance for the selected research topic. Some literature collected are directly relevant to the selected problem and hence, it serves as a backdrop for the study to write problem statement. The

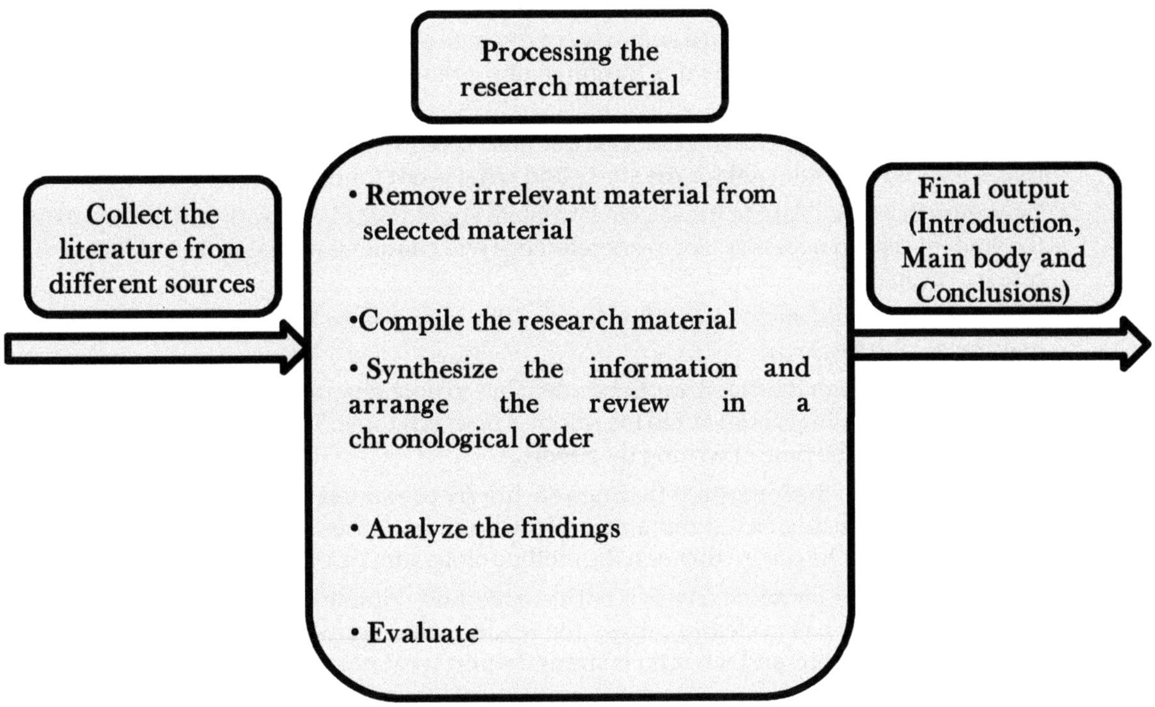

Figure 6.1: Stages of Effective Literature Process.

other literature reviewed may enable the researcher to plan methodology, formulation of hypotheses, interpretation of findings, suggesting policy implications etc. It is to be noted that, there is hardly any present research project which is totally unrelated with past research. So, literature review serves as one of the important components of research process.

This literature review benefits the students (post-graduate and doctoral students) in pursuing the academic research in the following ways:

- ✰ It proves the Advisory Committee that, the student has put all his efforts in understanding the concepts (both theory and methodology) about the problem under consideration.
- ✰ It makes the Advisory Committee to know that, the student had successfully related the past research methodology to the present selected problem.
- ✰ The Advisory Committee will understand about the gaps in the available literature for the selected problem and they get convinced about the originality of the problem selected for the study.
- ✰ The Advisory Committee gets convinced that, the originality of the topic selected for the study by the student can be an extension of past research or a modification of existing methodology in a new agro-ecological situation.

Thus, review of literature serves as a springboard for conducting the research. It aids the student in giving an idea to identify the research gaps in the past research, select the problem that is relevant to the study area, forms a frame work both for the theory and methodology of conducting the study etc. The students should further keep in mind the following aspects in collecting the review of literature:

- ✰ Literature review is more than a summary of research publications. It provides the evidence that, selected research problem is 'original' and relevant to the agro-ecological situation under consideration.
- ✰ Writing the literature review is not a simple task. It requires the researcher to focus on the issues that, what is relevant for the study and what is not (or irrelevant).
- ✰ Literature that is relevant to the selected problem should only be included in the report and the student should note that, not every reference will contain material that is relevant to the research problem.
- ✰ The student should include the findings of the past research *'as it is'* and should not manipulate the research.
- ✰ While going through the literature, the student has to note separately the assumptions made and important findings relevant to the selected research topic. This will save lot of time for the researcher at the time of writing the report.
- ✰ In journal articles, the important findings are briefly presented as 'Abstract'. The students should go through this, as it summarizes all the findings. If these findings are relevant with the selected topic, he can go through the methodology and details in that article.
- ✰ It is to be noted that, literature review is not just a mere description of the available information, but the researcher has to clearly convey the reader, what knowledge and ideas have been established on a topic, and what their strengths and weaknesses are.
- ✰ The researcher has to spend a significant amount of time for literature, as it forms the basis right from identification of the problem and submission of final report.

- ☆ Weak literature review signifies the fact, the student has not adequately reviewed the past researches in relevance to the selected research problem. This also infers that, the researcher has not properly explained the research findings *i.e.*, whether the results are in line with the past findings or gives altogether a different picture.
- ☆ Every statement in a literature review must be supported by the source listed in the Literature cited chapter of the dissertation.
- ☆ In a literature review, the student need not present all the details found in the reference. For example, a student from the Department of Agricultural Economics can choose a model from the past research for using it in his research topic. But, he need not show the proof, where that model was actually used and all the inferences derived from that model. Simply, the student can indicate the name of the model, its specification and its source. The Advisory Committee can go to the original reference for all of the details.
- ☆ This topic of 'Review of Literature' should be sufficiently long enough (at least 20 pages) to convince the Advisory Committee that, the researcher has thoroughly explored the research topic. The student should not merely summarize the findings of the past research studies. He must also evaluate and comment on the past studies regarding their findings. Sometimes, the researcher may find that, past research findings are not valid and this may counter the hypothesis formulated for the present study. In that case, the researcher should comment on past study's worth and validity, rather than completely ignoring that study. After conducting the present study, the researcher should clearly mention, how the findings of the present study deviate from that past study and how his findings are more valid compared to that of past findings.
- ☆ The student need not include all the research articles that fall in the area of his selected research topic. Only those studies that are most relevant to his selected topic with reference to background information, tools and techniques etc., are to be included.
- ☆ After prioritizing the literature (say, articles) to be included in the review, now the researcher has to follow a proper sequence in presenting them in the dissertation. In general, most students present the list of summarized articles in a chronological order. But, the most effective way of presenting the review is that, the researcher should follow the objectives framed for the study as a sequence. That is, entire topic of review of literature should be broadly categorized into different sub-heads in accordance with the objectives framed for the study and under each category, the most relevant articles have to be placed. These sub-heads should flow together covering all the major key issues of the selected research topic. For example, a researcher wishes to conduct a research on '*Production and marketing of pomegranate in Visakhapatnam district of Andhra Pradesh* : An economic analysis'. The following are the specific objectives framed for this study:
  - ❖ To study the investment pattern in pomegranate orchard
  - ❖ To compute the cost and returns in pomegranate cultivation
  - ❖ To estimate the resource use efficiency in pomegranate production
  - ❖ To assess the efficiency of marketing channels

  For collecting the relevant literature review, the researcher has divided the chapter 'Review of Literature' into the following sub-heads:
  - ❖ Investment pattern of plantation crops
  - ❖ Cost and returns in the cultivation of perennial fruits

- Resource use efficiency of plantation crops
- Marketing costs and margins in plantation crops

- At the beginning of writing the topic of review of literature, the student should highlight the importance of collecting literature review and division of research topic into different categories in accordance with the objectives framed for the study. After placing the review under each category, the student should write in-brief about the important (common) points highlighted from the reviewed articles, possible hypotheses that can be framed, tools and techniques employed and how these articles will guide the researcher to accomplish or fulfill the objectives of selected research topic.
- The student should keep on updating the literature reviews. This is because, if he reads the articles, he can get some cross references and it enables him to go through such original articles. Most of these new references added by the researcher are part of the reference list of the articles or books.
- The student must note that, the subject concepts in a literature review must be related and do not mix unrelated concepts.
- Wherever possible, the student should use original source material rather than summaries or reviews made by others.
- The student should note that, the purpose literature review is to form the backdrop for his research study and to initiate and execute the research process. So, after reviewing the literature, the student should refine the problem statement, objectives and formulate the hypothesis for the study. When this is done, the researcher is ready to proceed with next step *i.e.*, 'Formulation of hypothesis' of the research process.
- The researcher should not merely present, what other authors have written relating to the present selected topic. Rather, he should explain and evaluate, how the past information is relevant to the present topic selected for investigation. The researcher should use present tense for relating what other authors say and for discussing the literature, theoretical concepts, methods, etc. For example, the researcher should write, 'In her study, Meena (2012) stipulates that, —————'. Similarly, use the present tense when the researcher discusses about his observations on the literature collected. For example, the researcher should write, 'On explaining the reasons for dismal performance of Kurnool regulated market, Meena remains silent ……..'. The researcher should use past tense for the results that have been found in the past. For example, the researcher should write, 'Meena concluded that,……………..'. Look at the following example:
    - Kumar *et al.* (2004) show a positive correlation between price and quantity supplied of chillies in Warangal regulated market.
    - Kumar *et al.* (2004) showed a positive correlation between price and quantity supplied of chillies in Warangal regulated market.

  The difference is that, the former review uses the present tense 'show', whereas the latter uses the past tense showed.' Most of the time, use the past tense to discuss papers that were published in the past. Present tense can make sense in the context, if the researcher think of the previously published paper existing in the present and contributing to the current discussion of the topic. If the researcher think of it that way, it works perfectly fine, albeit it may sound different to ears, if we're not used to it.

- ☆ The student should clearly define the sub-topics in a few words or a sentence in accordance with the framed specific objectives. The literature collection should be limited to the defined areas of sub-topics so that, no time is wasted on peripheral topics.
- ☆ The student should immediately list the key words while referring the literature, indexes, databases etc., to identify relevant publications. As he read, take notes on the key information, putting it into his own words will help him to understand and remember the information better than if he simply copy the author's exact words.
- ☆ An excellent approach to identifying further sources is to review the list of references included in the books and articles you find. These will typically lead you to many other valuable publications. These cross references are important and can be used to download the articles from the internet. This is important because, even large university libraries will not carry all the books, journals etc.
- ☆ Once after the student located an article or book, he has to quickly skim the abstract and the conclusions to see, whether it includes the information relevant to selected research topic. Identify those sources that harbor the material closely related and important to the selected topic and then begin reading the most recent publications of these sources. Recent sources are most likely to be valuable because, they have the benefit of earlier work and research as a foundation.
- ☆ The student should record the bibliographic information immediately on the index cards after writing the review. Note that, if the student is collecting different articles from the same journal, he need not write the particulars of name of the journal, volume number, issue number, year of publication etc., as they are common for all the reviewed articles. Only title of the reviewed article and page numbers will differ from article to article. Hence, on one card, he has to note down all the above particulars *i.e.*, name of the journal, volume number, issue number, and year of publication and on the remaining cards, he has to just indicate the source only, as year of publication will be given within the brackets after the author's name in the review. Of course, on each and every card, he has to note down title of the article and page numbers. This procedure will save lot of time on the part of the researcher.
- ☆ When writing the dissertation, the student need to take a caution about 'Plagiarism'. This plagiarism is a serious and dishonest offense and it may involve any of the following:
    - ❖ *Turning in someone else's work as your own.*
    - ❖ *Copying words or ideas from another work without giving credit.* That is, any idea that does not come directly from student's mind must be referenced.
    - ❖ *Giving incorrect information about the source of a quotation.*
    - ❖ *Failing to put a direct quote in quotation marks.* All words in a paper must be your own. Any direct quotes must be put in quotation marks.
    - ❖ *Changing the words, but copying the sentence structure of another work without giving credit.*
    - ❖ *Copying so many words or ideas from another work that it makes the majority of the report.*
- ☆ The student should give last priority for 'Google' search, as it provides huge information provided by many people, not only by the academicians. So, this is not an appropriate source for collecting review for academic research. However, if a student finds the title of a research article in the 'Google', he can use the same in the Google search box to find, whether the author has put the research material on the internet. Similarly, do not cite 'Wikipedia' as a source.

6.2. Checklist for Selection of Literature to be Included in Literature Review

In post-graduate and doctoral researches, the collection, analysis and reporting of literature review is often considered as an important academic skill. This is often considered as a sub-project of the selected research project, as it forms the main background or framework to select, formulate, analysis, interpretation and report writing for the selected research project. The main source of researcher's contributions (applying own thoughts) towards achieving solution(s) for the research project will be generated from literature review only. As mentioned earlier, while collecting the literature and going through the same, it facilitates the researcher to identify the gaps and to select the research problem. At this juncture, the researcher has to borne in mind that, he should not be too ambitious in solving the problem. It would be safer and probably more realistic to say that, the solution(s) generated from the research will 'address a gap', rather than that it will 'fill a gap'. Once after reviewing the literature, the researcher should aim at framing the objectives of the study. Note that, the literature reviewed also helps the researcher in refining the problem statement written earlier and justification of selecting that research problem. After collecting the relevant literature, the researcher should question himself on the following grounds:

- ☆ Whether any important aspect or concept or key term is missing from review?
- ☆ Have you defined key terms?
- ☆ Is the purpose of the review clear?
- ☆ Is the scope of the review reasonable?
- ☆ Is the selected literature relevant to the selected research problem?
- ☆ Are the references or bibliography is complete, accurate and up to date?
- ☆ How effectively he linked all the elements in the review?
- ☆ Is the material presented in the most effective order?
- ☆ Are there places, where the researcher is left with unanswered questions?
- ☆ Is every element of the literature review support the objectives framed in the study?
- ☆ Whether he explained adequately about the justification for selected research topic?
- ☆ What type of literature review the researcher is conducting *i.e.*, about a theory or methodology or policy?
- ☆ Whether the collected literature is sufficient enough to answer all the objectives of study?
- ☆ Whether the researcher critically analyzed the concepts discussed in the reviews?
- ☆ Whether the researcher reviewed the studies that are contrastingly different from the views of the researcher?
- ☆ Who will be the readers of the review?
- ☆ Will the readers find the literature review relevant, appropriate and useful?
- ☆ Whether the researcher included any irrelevant literature?
- ☆ Have you used mostly primary sources with some use of secondary sources?
- ☆ Have you organized the main body of the review report in accordance with the framed objectives for the study?
- ☆ Is your criticism of design and methodological issues adequate?
- ☆ Are most of the sources from reputed, refereed journals?

- Are recent developments in the literature emphasized in the review?
- Is the review organized by topics or ideas, not by author?
- Is the review organized logically?
- Are studies compared and contrasted and conflicting or inclusive results noted?
- Does the review of major studies relate explicitly to the research problem and methods?
- Do the methodological analyses provide a rationale for the design to follow?
- Does the review of the literature help establish the significance of the research?

If the researcher goes through answering the above questions, he can avoid may mistakes in writing the review report and bibliography. At each and every stage, the researcher should question himself, 'why am I including this?'. Further, the researcher should keep the focus of research in mind and should collect the review accordingly. In the words of Rudestam and Newton (1992:49), the aim of literature review is to *'build an argument, not a library'*. According to Wellington *et al* (2005:80), writing of review should also be considered as a part of research process and not something that happens after the researcher have finished reading the literature. So, he suggested that, *'writing while you collect and collecting while you write.'* Thus, the review made by the researcher should be comprehensive. The references should be critically analyzed and the results of the various studies should be compared and contrasted. Finally, the review should conclude with a brief summary of the literature and its implications for the problem investigated.

6.3. Characteristics of a Good Literature Review

The role of a good literature review is to find and present the pertinent work from the collected literature in a logical, organized manner and to bring the reader as up-to-date as possible with reference to the research project under consideration. The following are the characteristics of good literature review:

- The review must be as recent as possible so that, it enables the researcher to understand the recent concepts and methodology about the selected problem
- The review must be relevant to the selected research problem
- The review must have been selected from a genuinely original and reliable sources
- The available review must not be too few nor too many
- The review must have clarity in explaining the methodology
- The review must not have bias in explaining the concepts
- The review must explain, how the methodology of past researches must integrate into the proposed research problem.
- The review of literature should not only confined to mere description rather it should also have some evaluation of the quality and findings of the research.
- The review should justify the reasons for selecting the research problem
- The review should allow the researcher to establish theoretical framework and methodological focus.
- The review should enable the researcher to know:
  - What has and has not been investigated so far?
  - What knowledge and ideas have been established on the selected research topic?

- What are the strengths and weaknesses of research on the selected topic?
- How the selected research project is related to the work of others?
- Whether it provide a solid background for present research investigation?

6.4. Importance of Collecting Review of Literature

There is often a reciprocal relationship between a literature review and the research project for which it is written. That is, there needs to be a reciprocal relationship between the ongoing literature review, analysis, and writing of the exegesis and the development of the creative research project. It is to noted that, the purpose of the literature review remains the same irrespective of the research method the researcher employs. The following are the ways in which reviewed literature will help or guide the researcher:

- To identify and select the right research (prioritized) problem for in depth study. That is, it form a basis for determining the significance of your study.
- To understand the research topic better through clarifying vague points, if any.
- To define and limit the study area of investigation.
- Prevents the researcher from doing duplicating work that has been done before.
- To integrate and summarize what is known in an area. A good review points out areas where prior studies agree, where they disagree, and where major questions remain.
- To show, how the current project links to past research. A review outlines the direction of research and places a research project in a context, demonstrating its relevance by making connections to a body of knowledge. Thus, it helps to build on existing knowledge and ideas, to provide the intellectual context for the selected work, to identify methods that could be relevant to the selected project.
- Gives confidence to the researcher in pursuing the research problem.
- Makes the researcher familiar to understand and execute different research methods, research design and methodology.
- The researcher gain more knowledge about the problem and this guides him to refine the 'problem statement'.
- Enable the researcher to interpret the findings accurately and in new ways and try to fill the gaps noticed in previous researches.
- To resolve conflicts that raised in the previous studies.
- The knowledge gained through literature reviewed helps the researcher to formulate researchable hypotheses.
- To identify data sources used by other researchers.
- To learn, how earlier researchers structured their reports.
- To conceptualize the research problem clearly and precisely. It also helps the researcher to understand the relationship between research problem and the body of knowledge in the study area.
- To contextualize research findings *i.e.*, it enables the researcher to compare the answers of selected problem with past findings, whether the present findings enhance the existing body of knowledge?

- ☆ It demonstrates researcher's preparedness to complete the research.
- ☆ To show the path of prior research and indicate, how a current project is linked to it.
- ☆ To learn from earlier research studies and thus, stimulate new ideas.
- ☆ Avoid *'reinventing the wheel.'* That is, it helps to carry on from, where others have already reached. So, it saves time, resources, energy etc., on the part of the researcher.
- ☆ Compare the views of earlier studies and enable the researcher to note the areas of disagreement and gaps in the knowledge.
- ☆ To formulate questions that need further research.
- ☆ Helps the researcher to discover, what statistical knowledge exists related to the selected research topic and to improve his statistical knowledge in the study area.
- ☆ Help the reader with an up to date account and discussion of the research findings on a particular topic and to justify the findings of selected research study.
- ☆ It makes the researcher to understand what others have done and thus, enable him to execute the current assignment. That is, it guides the researcher to study the methodology followed in analyzing and interpreting the related problems in the past and makes him to follow on their footsteps and copy their methodological approach.
- ☆ It enable the researcher to detect conflicting points of views expressed by past researchers. This makes him to analyze the indicators of diverging theories within the same topic. Besides conflicting theories, the review also helps him to gain information on theories that support the topic.
- ☆ Gives credit to those, who have laid the groundwork for the selected research.
- ☆ Convinces the readers that, the selected research will make a significant and substantial contribution to the literature (*i.e.,* in resolving an important theoretical issue or filling a major gap in the literature).
- ☆ It serves as a stepping stone for conducting the present research. Its depth and breadth emphasizes the credibility of the researcher in his or her field.
- ☆ The review of literature serves as a link between current research topic and past researches already done. Thus, the review of past research studies forms the basis of current research topic and it also forms the basis for the scope and purpose of the present study. It tells the readers about aspects that have been already established or concluded by earlier researchers and thus, projects the current research work in the proper perspective.
- ☆ It helps to identify other people working in the same field *i.e.,* it strengthens researcher's network.
- ☆ It helps to identify positive and opposing views about the selected research topic.
- ☆ Helps the researcher to develop a better understanding of key variables to be employed in the study. It also provides information about, how the key variables should be measured.
- ☆ It demonstrates a familiarity with a body of knowledge and establish credibility. That is, a review tells the reader that the researcher knows the research in an area and knows the major issues. A good review increases a reader's confidence in the researcher's professional competence, ability and background.
- ☆ It helps to know the path of prior research and how a current research project is linked to it. Thus, a review outlines how the background of the selected research problem has been developed over a period of time.

- ✰ Review helps to learn from previous researchers and thereby, stimulate new ideas.
- ✰ Review helps in developing theoretical framework for the research problem under consideration.

Highlighting the importance of literature review, Bruce (1990) opined that, *'as far as the literature review process goes, ultimately the goal for students is to complete their review in the allocated time and to ensure they can maintain currency in their field of study for the duration of their research'*.

6.5. Four Ws before Writing Literature Review

The aspects to be considered before writing literature review can be conveniently studied in terms of four Ws. They include:

a. Where to read
b. What to read
c. Whom to read
d. What to search (how to find what you need to read)

a. Where to Read

For the student, University library is the best place to conduct literature review. This is because, it harbours all kinds of sources *viz.*, internet, journals, books, reports, conference proceedings etc. The student can easily get the photocopy material of the selected material and this enable him to read it at home. The advantage of using photocopies is that, it makes the student can underline the material and make notes.

b. What to Read

As discussed above, the researcher has to go through different sources like internet, journals, books, reports, conference proceedings etc., for collecting literature relevant to the selected problem.

c. Whom to Read

While reviewing the research articles and data base for compiling literature, the researcher should ascertain how authoritative the author(s) is in explaining the methodology, concept, interpretation of findings, suggesting policy implications etc. That is, the researcher should be able to ascertain the authors' motivation in writing those articles. In case of data base, the researcher should assess how authentic the source is? For example, the data derived from www.indiastat.com, www.agmarknet.nic.in, www.agricooop.nic.in etc., are considered reliable data sources for agro-economic surveys.

d. What to Search (How to find what researcher need to read)

The researcher will collect review from different sources like journals, internet, books, reports, conference findings etc. But, if the researcher goes through all these sources completely, it consumes lot of time. So, the researcher has to decide 'what to read' to know, whether the information he requires is available in the selected sources or not? The researcher has to go through the contents of the reports or books or journals, 'Abstract' of the research articles, looking at the introduction and conclusions of the research articles etc., to collect the relevant literature. These hints should be followed by the researcher for a speedy selection of the literature. However, it is to be noted that, these hints should be followed only for identifying and collecting the literature, but the researcher has to go through full length of the articles critically for writing the review. This way of literature collection comes through experience on the part of the researcher. In the context of 'what to search', there are different types of reviews like Self-study reviews (increase reader's confidence in the researcher as well as the researcher's confidence in his/herself), Context reviews (review that place the current project in the big picture and helps to create a link in developing a body of knowledge. This will guide the researcher, how the current research continues as a developing line of thought or it can point to an unresolved question or conflict in prior research to be addressed), Historical reviews (trace the development of an issue over time), Theoretical reviews (compare how different theories address a particular issue), Integrative review (summarizes what is known at a point in time *i.e.*, it presents the current state of knowledge) and Methodological reviews (point out how methodology varies by study).

6.6. Sources of Literature Review

Noticing the importance of literature review in conducting research for the selected topic, now the researcher has to identify different sources to collect the requisite data or information that forms the background for the entire research process. The researcher has to write the summary of the studies that are most relevant to the selected problem. This will help to refine the research question, framed objectives, contextualize the research topic etc., at the same time, enable the readers to understand the current research problem and its solutions. Sources of information are generally categorized as primary, secondary or tertiary depending on their originality and their proximity to the source or origin.

a. Primary Sources

These are the sources that give first hand information to the researcher about the literate review. According to Borg and Gall (1979), '*Primary sources are direct descriptions of research studies or other events written by an individual, who actually conducted the study or witnessed the event*'. Primary sources present original materials on which others research is based. They are usually the first formal appearance of results in print or electronic format. That is, they present original thinking, report a discovery, or share new information. For example, the articles written and published in peer* reviewed research journals and reporting research studies conducted by the researcher (and his team) represent primary source of literature. Similarly, the information collected from the internet (say, electronic data bases) or a book or a research report authored by a researcher also represents the primary source. Thus, primary sources of literature harbor the information that is original, represent original thinking or experiences of the researcher in the past and have not been filtered through interpretation or evaluation by others. So, the primary sources offer a direct view of the researcher or thinking of the researcher for the study without the intermediate interpretation of others. They offer the depth and veracity required for an in-depth study. The other important primary sources include: information gathered from radio programmes; TV telecasts; interviews; news paper articles; proceedings of the conferences, workshops, symposia; survey findings; original documents like farm particulars, bank pass book, date of birth certificate etc.

Some examples of Primary Sources

They include:

- ☆ Scientific journal articles reporting experimental research results.
- ☆ Proceedings of Meetings, Conferences and Symposia.
- ☆ Technical reports.

---

* Peer review: Process by which articles are submitted to experts in the scientific discipline for comments and criticism before publication in a scholarly journal.

- Dissertations or theses (*may also be secondary*)
- Patents
- Sets of data, such as census statistics
- Works of literature (such as poems and fiction)
- Interviews, surveys and fieldwork
- Newspaper articles (*may also be secondary*)
- Government documents
- Photographs and works of art

b. Secondary Sources

These sources provide the information or discussion about a topic that is originally presented elsewhere. So, secondary sources gives interpretations and evaluations of the topics of primary sources. Thus, secondary sources are not evidence *per se*, but rather, a commentary or a discussion of evidence published through a primary source. According to Borg and Gall (1979), *'secondary sources are documents written by an author, who did not directly observe or participate in the events described or who was not the originator of the concepts outlined'*. So, secondary source contains commentary on or discussion about a primary source. The most important feature of secondary sources is that, they offer an interpretation of information gathered from primary sources. These sources can be articles in newspapers or popular magazines, book reviews, or articles found in scholarly journals* that address someone else's original research. So, secondary sources are 'second-hand information,' analogous to human conversation. *If I tell you something, I am your primary source. If you tell someone else what I told you, you are a secondary source.* Majority of the sources in a literature review are secondary sources that present research findings, analysis, and the evaluation of other researchers works. It is important to note that, secondary sources represent various opinions, interpretations and arguments about the selected research problem such that, the researcher may agree or may not agree with them while synthesizing the literature for his own analysis of the selected problem. The researcher can use the information in the review article and he can update the information with the literature it quotes for primary literature. Thus, secondary sources are those, which are published about the primary literature, that generalize, analyze, interpret, evaluate or otherwise 'add value' to the original information or which simplify the process of finding and evaluating the primary literature.

Some Examples of Secondary Sources

They include:

- Bibliographies (*may also be tertiary*)
- Dictionaries and encyclopedias (*may also be tertiary*)
- Dissertations or theses (*more usually primary*)
- Handbooks and data compilations (*may also be tertiary*)
- Journal articles, particularly in disciplines other than science (*may also be primary*)
- Newspaper and popular magazine articles (*may also be primary*)
- Review articles and literature reviews

---

* Scholarly journal is the journal that publishes mostly scientific, primary articles; exists to disseminate knowledge (Examples: *Science*, *Nature*, *Cell*, *Ecology*).

- ☆ Textbooks (*may also be tertiary*)
- ☆ Monographs
- ☆ Internet sites (also considered primary)

c. Tertiary Sources

These sources consist of information, which is a collection of primary and secondary sources. So, these sources provide rich information to the researcher compared to the above two sources. This tertiary source serves as a good starting point for the researcher to collect voluminous literature. It is difficult to distinguish between secondary sources and tertiary sources of information, as both these sources do not provide 'original' (primary source) information. However, for the purposes of reviewing the literature, tertiary source serves as an important one, as it distills lot of information compared to each of the primary and secondary sources. The importance of tertiary sources is that, they compile the information from primary and secondary sources at one common place. They guide the researcher to access to other sources through indicating the bibliography. Online site <*www.wikipedia.org*> is the best example for tertiary source. Other examples include: text books, manuals, databases, search engines, manuals, statistical compendiums etc. Note that, these tertiary sources does provide a broad overview of a topic or data, already proven facts, and definitions etc., but does not provide any new information. These include 'reference' types of works such as textbooks, encyclopedias, fact books, guides and handbooks, and computer databases.

The researcher, while conducting the research has to depend upon one or more of different types of sources for collecting the requisite data and information. This is so because, all good research and writing is guided by a review of the relevant literature. As mentioned earlier, primary sources provide the information about the research problem, as perceived or viewed by the researcher who actually studies the problem in-detail. Secondary and tertiary sources are also useful because, they integrate information from many different sources into one document and provide an easy method of developing an overall understanding of the topic. The importance of literature review can also be noted from the point that, its purpose remains same irrespective of the research method the researcher follows in conducting the research. This highlights the aspect that, 'review of literature' is an integral component of the research process. The different sources of review referred by the researcher in general across different steps of research process is shown here under (Table 6.1).

6.7. Search for Existing Literature

A range of information sources might be used to collect the literature review. The evaluation of these sources is a very real problem. Typically in professional disciplines, like Agricultural Economics, there will be both academic and professional literature. Both these sources are important for the student in collecting the review. While the professional literature guide the researcher in the identification of research gaps and formulating the research problem, the academic literature provides theoretical background to analyze and interpret the research findings. In Colleges, libraries serve as a major place for accessing to different research materials like books, journals, reports, internet access etc. The students should spend considerable time in the library to access to different research materials. The students should note that, all the research materials (say, books, reports, conference proceedings, articles etc) cannot be downloaded from the internet. Hence, the student has to refer various research materials to collect the requisite information for the research study. He may even obtain the information from the primary author through using email. To produce and submit a high-quality dissertation, the student must learn to access different research materials that provide requisite information to conduct

Table 6.1: Different sources of literature reviewed across different steps in research process

Steps in Research Process	*Sources of Literature Referred*	*Purpose of Review*
Conception/Identification of Research Problem	☆ Journals ☆ Proceedings of Meetings, Conferences and Symposia ☆ Technical reports ☆ Previous dissertations or theses ☆ Newspaper articles (*may also be secondary*) ☆ Government documents	☆ To gain preliminary and background information about the research problem. ☆ To know, what research has already done on the same topic ☆ To identify research gaps ☆ To avoid duplication
Formulation of Research Problem	☆ Journals ☆ Proceedings of Meetings, Conferences and Symposia ☆ Technical reports ☆ Previous dissertations or theses	☆ To become familiar with the objectives to be formulated, discuss the problem statement and scope of the present investigation.
Review of Literature	☆ Journals ☆ Proceedings of Meetings, Conferences and Symposia ☆ Technical reports ☆ Interviews, surveys and fieldwork ☆ Previous dissertations or theses ☆ Sets of data, such as census statistics ☆ Newspaper articles (*may also be secondary*) ☆ Government documents	☆ To analyze the progress of work on the same field of investigation·To identify the progress in using different methodologies in analyzing the problem
Formulation of Testable Hypothesis	☆ Journals ☆ Proceedings of Meetings, Conferences and Symposia ☆ Technical reports ☆ Previous dissertations or theses	☆ To gain familiarity about different ways of finding solution to a research problem
Preparation of Research Design (Methodology)	☆ Journals ☆ Technical reports ☆ Previous dissertations or theses	☆ For gaining knowledge on the methodological aspects of addressing the research problem
Execution of Research Project	☆ Journals ☆ Technical reports ☆ Previous dissertations or theses	☆ For gaining knowledge regarding the methodology to be followed for selecting a representative sample, variables to be identified, data to be collected etc.
Data Collection, Data Processing and Analysis	☆ Journals ☆ Proceedings of Meetings, Conferences and Symposia ☆ Technical reports ☆ Interviews, surveys and fieldwork ☆ Previous dissertations or theses ☆ Newspaper articles (*may also be secondary*) ☆ Government documents	☆ For preparing the data collection instruments like schedule or questionnaire to be employed in investigating the research problem. ☆ To identify the right statistical techniques for analyzing the data

Contd...

Table 6.1–*Contd...*

Steps in Research Process	*Sources of Literature Referred*	*Purpose of Review*
Testing of Hypothesis and Interpretation of the Findings	☆ Journals ☆ Technical reports ☆ Previous dissertations or theses	To gain knowledge on how to interpret the analytical findings
Writing Research Report	☆ Journals ☆ Proceedings of Meetings, Conferences and Symposia ☆ Technical reports ☆ Interviews, surveys and fieldwork ☆ Previous dissertations or theses ☆ Newspaper articles (*may also be secondary*) ☆ Government documents	To gain familiarity in writing the thesis

the research. In the present day context, the important research materials include *viz.*, books, journals and internet.

☆ ***Books***: Books comprise a central part of any bibliography. It comprises good quality material and serves as a major backdrop for a research report. But, the disadvantage is that, it is not completely up to date, as considerable time gap is indispensable between the completion of a work and its publication in the form of a book. Note that, all the books and all the contents in each book may not be relevant for the research topic and hence, the researcher has to be vigilant in selecting those topics that are found relevant. For example, if the student is conducting a research study on '*Marketing efficiency in transacting paddy in Kurnool district of Andhra Pradesh*', the students will refer books on Agricultural Marketing'. For example: Acharya, S.S. and N.L. Agarwal, (2001), Agricultural Marketing in India, New Delhi: Oxford and IBH Publishing Co. and Kumar K.N.R. (2014), 'Agricultural Marketing', Astral International (P) Ltd., New Delhi. It is essential on the part of the researcher to select standard textbooks. These books also harbor good number of references to other useful sources. The following are the important aspects should be borne in mind by the researcher, while reviewing the standard books:

- ❖ The selected topic in the book should be relevant to the research topic;
- ❖ The book should be written by an authoritative author and published from an authentic source (indicated by the presence of ISBN number)
- ❖ Good source of bibliographical details should be given in the book
- ❖ Preferably latest publication should be referred (refer to the publication date)
- ❖ Well-structured and well presented, and easy to read.

☆ ***Journals:*** Unlike books, journals provide the most up-to-date information, even though there is a gap of one to two years between the completion of a research project and the publication of research article in a journal. Compared to books, the articles published in the journals have a narrow focus and hence, they will be more relevant for the research study. These journals also provide data base that are essential to include and analyze in the research report. So, the researcher has to identify the journals, from which he could get literature relevant to the selected research topic. In each journal, the student has to go through the index, identify the article that is relevant with the selected topic and read the articles. If the student wishes to use a relevant article, he has to take a photocopy of it or prepare a summary and record it for reference for later use. As far as possible, the student should select the latest issue of the journal to collect the latest information about the selected topic. For example, in the above research study, the student refer journals like 'Agricultural Marketing', Indian Journal of Agricultural Marketing' etc.

Articles published in research journals offer an invaluable source of review to the student. This source should form the core of the literature review. This is because, the published articles are also written by the researchers, who are doing research in the same field. These articles are presented in a systematic fashion *i.e.,* Title of the article, Name(s) of the researcher(s), Abstract, Introduction, Review of literature, Materials and Methods, Results and discussion, Summary and conclusions, Policy implications and Bibliography. Moreover, these articles are peer reviewed before their publication. These journals also publish review articles that provide a review of all of the recent work in an area. Both research articles and review articles will include a significant bibliography that may serve as an invaluable sources of reference to the researcher in accomplishing the selected research topic. Professional journals include research articles about a particular topic or area and each article may run about 3 to 4 pages in length. These articles also guide the researcher in collecting the review pertinent to his research topic in general and framed objectives in particular.

☆ ***Internet:*** The Internet is relatively a new tool for conducting literature reviews. It provides enormous collection of research materials such as articles, books, journals, databases, graphic images, video and audio files etc. The student can also obtain relevant research material by accessing through various websites on the internet. The advantage of internet is that, the student can get updated information and data base. Further, he can access the relevant information from different corners of the world, where similar research was conducted in the past. For example, for the above research study, the student refer different sites like www.agmarknet.nic.in, www.agricoop.nic.in, www.fao.org, www.indiastat.com etc. 'Google' search engine provides 'too much information', but it does not distinguish or judge the quality of information. Further, it cannot compare the quality of information across the downloaded articles or reports etc. So, it is better for the researcher to start reviewing the literature from academic library through referring books, journals etc. After referring them, the next step is to go to the internet to collect the information. Whatever the gaps left over after collecting the requisite information from books and journals can be easily accessed through internet. But, as mentioned above, it is very difficult to judge the credibility of an online paper. So, after downloading the vast information available from the internet, the student has to categorize it in accordance with the framed objectives for the study and gaps noticed after reviewing books and journals. However, the student has to be careful on the following aspects, while referring the internet source for collecting data base and literature:

- What is the frequency of update?
- Any copyright protection is there for the database or article?
- Which website is providing the relevant information?
- Is any payment is needed to download information or data base from the website (For example, to download data base from www.indiastat.com, it requires payment or membership on the part of the researcher or organization).
- Are there links or references to other relevant web sources?

Noticing the relative advantages of one source over the other in gathering relevant information, the student should follow the easiest way to collect and place the information in a systematic order. The student should first go through the 'Abstracts' of the articles available either in journals or internet and this helps the student in a big way in eliminating the unwanted literature or non-relevant works. After selecting the literature through judging 'Abstracts', next the researcher should go through each article critically (objectives, methodology, findings etc) to evaluate it, on what aspects it will be relevant for the selected topic. Note that, the student need not include a long list of references for the study, but a quality review is sufficient enough to conduct good study. That is, a smaller number of quality sources is far preferable than a long list of irrelevance. Further, the student should check into the credentials of literature source. In this context, the reputation of the 'University' or 'researcher' counts much in selecting the article or book or report for reference. If the selected review or article for the study is replicated several times by other researchers (say, for different commodities or in different agro-ecological situations) with similar conclusions, then it is safe on the part of the student to include those articles in the review. In case of contrasting results, the student may approach the Advisory Committee for help and for judging the quality articles depending upon the reputation of journals (for example, based on NAAS ratings, the journals can be ranked: Agricultural and Forest Meteorology (NAAS Rating 2011 :7.9), Agricultural Economics Research Review (NAAS Rating 2011 :4.7), Indian Journal of Agricultural Sciences (NAAS Rating 2011 :6.6), Indian Journal of Weed Science (NAAS Rating 2011 :2.9), Journal of Agricultural Economics (NAAS Rating 2011 :7.4 etc) in which the articles have been published. So, the student should not fall into the trap of using a poor evidence and all these will be gained in due course by gaining experience in the research. This discussion infers that, literature review is one of the challenging tasks in the research process and it requires the culmination of many skills such as library research, logical arrangement of collected information, synthesis of literature and scientific writing.

6.8. Review the Literature Collected

After collecting the literature from different sources, now the student has to go through the information and data to put together themes and issues that are associated and relevant to the selected research topic. The student has to prepare a notes regarding the background information about the research topic, compile requisite data from different sources at one place and note the findings. After doing this, he has to classify the collected material objective-wise framed in the research study. In general, the student gets two types of information about the selected topic *viz.*, general information and more specific (*i.e.*, information pertaining to local area or same agro-ecological situation or same crop etc). So, it is important on the part of the student, first to start with the general information and gradually narrow down to the specific information (Figure 6.1.1). Note that, at the time of collecting the information and data, he should immediately note the bibliography for those references.

As mentioned above, the student has to classify the collected information into different themes in accordance with the objectives framed for the study. This is because, the literature available to the student may cover a wide range of topics and hence, the student should categorize the information. The student should remember that, review of the literature should not be a loose collection of articles, citations, facts etc. Hence, for each theme (framed in accordance with the specific objectives), the student should write its importance before starting the review and at the end of the theme, he should highlight the concepts, facts, brief conclusions etc., from different reviews made. Further, a critical analysis of the literature enable the student to identify several aspects that may have direct and indirect bearing on the selected research topic. The student should use these aspects to develop background theoretical framework. Note that, until the student go through the literature, he cannot

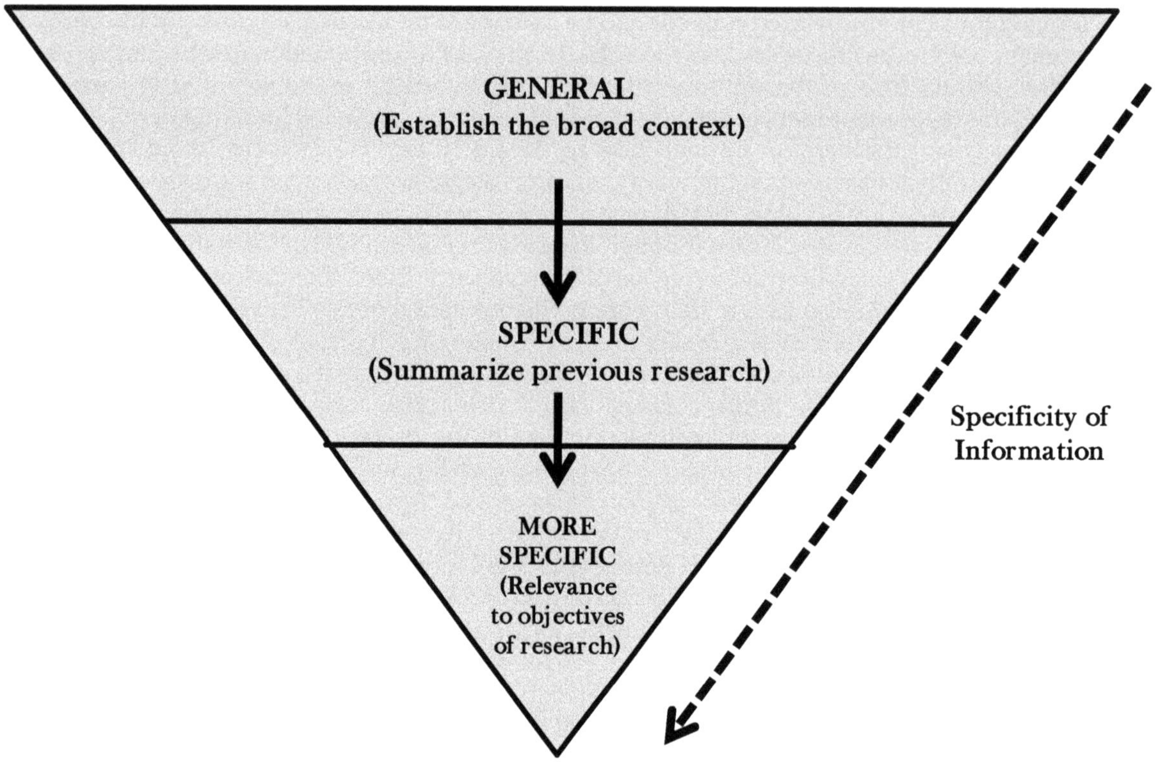

Figure 6.1.1: Outline of the sequence of information that needs to be presented in the Review of Literature.

develop a theoretical framework and until he developed a theoretical framework (identifying key terms and concepts), he cannot effectively review the literature.

As the student reviews large number of research materials, he should critically go through the concepts, analytical findings, conclusions etc., and compare these items to analyze their strengths and weaknesses. He should focus on those materials that have close relationship with the selected research topic. Then, in writing the summary at the end of the review, he should explain to the readers clearly, how the past researches have addressed the selected problem, how the present study is attempting to address the same problem and how the present study is superior compared to the past studies. Note that, writing summary to the literature review is like a summation to a jury. We know, a person want the jury to bring in the verdict as desired by him (a preferred verdict). So, he must submit the facts to the jury in a nut shell to get a preferred verdict. Note that, a literature review is not judged by its length or by the number of references it includes. Rather, it is judged by the inclusion of relevant information and critical analysis made through comparing different reviews. Thus, the student should pose himself two important questions, while reviewing the literature *viz.*, 'whether I am including the relevant literature that is supportive to the selected research topic?' and 'whether I am critical about the review made for the study?'

6.9. Steps in Reviewing the Literature

There are six steps in writing the literature review and they are discussed here under.

a. Identifying the Topic for Review (Problem Statement)

As discussed earlier, the formulated problem statement guides the researcher to plan for literature review. The first step to a literature review is a clear statement of the research problem. Both the problem statement and the literature review are mutually supportive. The problem statement directly indicates the subject area for collecting literature review. On the other hand, the literature review shows the selected research problem has received prior attention before and indicate the gaps to the researcher to plan the further research to resolve the problem. By integrating a literature review with the formulated problem statement:

- It makes the focus clear for the researcher to plan for literature collection.
- It makes the readers worthy reading the research topic to know, how far the conducted research has filled research gap(s).
- It guides other researchers to plan future research works in that area.

It is essential that, the researcher should aim at collecting the literature that are relevant to the selected topic. While collecting the literature, the researcher should be cautious whether he can manage the literature available to him or not. If sufficient (manageable) literature is available, he has to note down the important terms, concepts, methodology, findings, and conclusions separately. These items will guide for further review after framing the specific objectives for the study. In case, if the researcher finds difficulty in knowing the terms and concepts of literature review, he can seek the advice from the Advisory Committee.

b. Deciding on the Scope of Literature Review

After identifying the problem statement for conducting the research, the researcher should know the scope of the review. That is, he should be certain regarding how many number of relevant articles he require to conduct the research study smoothly, how comprehensive the literature should be, for how many number of years the data are to be collected, from which year the review is to be collected etc. It is important that, the researcher need not exhaustively cover each and every aspect of the selected topic, but to cover it thoroughly. That is, the researcher should critically go through the literature and he has to strive for maximum coverage for a minimal word count.

c. Searching the Literature from Different Sources

Now, the researcher has to search different sources (primary, secondary and tertiary) to collect the relevant literature for the selected research topic. Generally, journals are regarded as most important sources (Figure 6.2), as they are more up-to-date than books as sources of information. Books do not contain latest information, as it takes considerable time for their publication. This does not mean that, they should be excluded, but they also contain an acceptable and valuable source of information. The researcher should also consider, how much time he has to allocate to collect the literature review in the total time frame available for him in conducting the research.

d. Gathering, Reading and Analyzing the Literature

After collecting the literature, now the researcher has to compile the information obtained from different sources. He should critically note duplications, if any. He should separate out those materials, which contain the best relevant material for the selected topic and which make the greatest contribution to the understanding and development of his area of research. He should also use the bibliographies and references of relevant research studies to locate other studies. The researcher should consult the Advisory Committee from time to time, to verify whether he collected the information pertaining to all

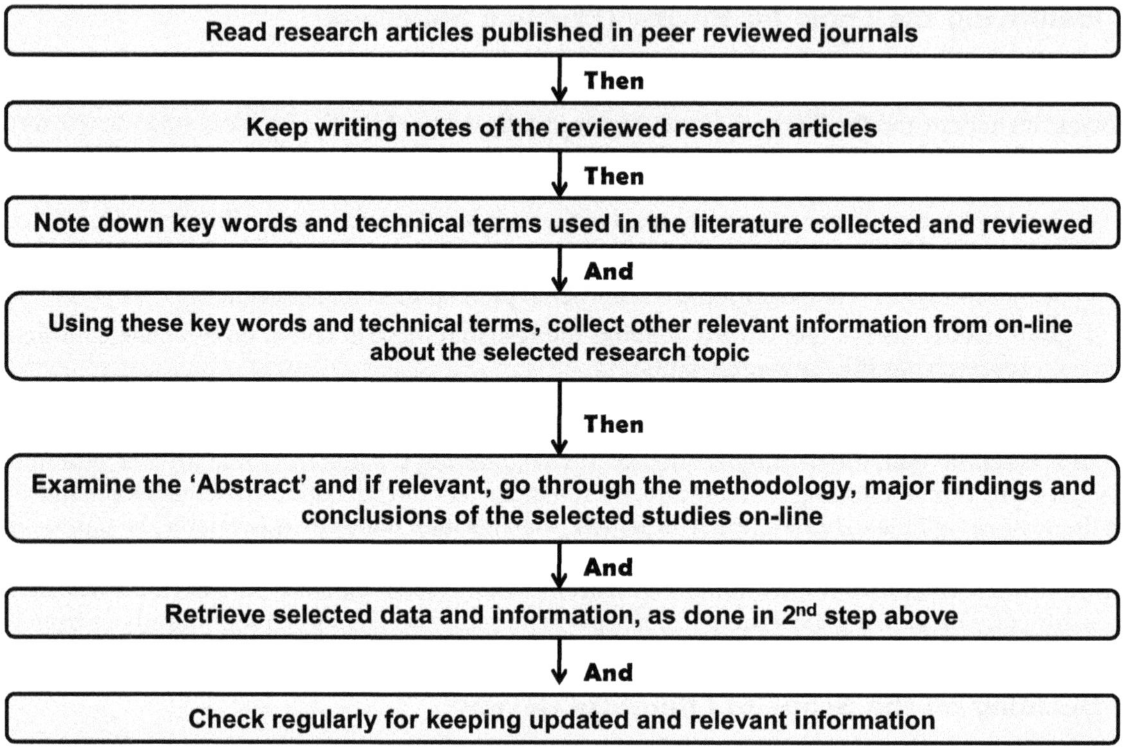

* Researcher while writing the notes for review, should always keep in mind that, how well each item will be relevant to the selected research problem and contribute to the body of literature.

Figure 6.2: Steps in Collecting Review of Literature.

the key words and concepts and whether he has to collect any additional information. When the Advisory Committee satisfies with the information gathered, now the researcher has to analyze and synthesize the collected literature. Initially, it is advisable to undertake a first read of all the articles that have been collected to get a sense of what they are about. We know, most published articles contain a summary or abstract at the beginning of the paper. The researcher should carefully go through the 'Abstract' of each research study, as it will make him to understand in brief about the objectives aimed, methodology followed, important findings and conclusions and this saves time on the part of the researcher. It also guides the researcher to take a decision, whether it is worthy of further reading and inclusion in the review. Once after finalizing the articles that are to be considered and relevant for the research study, it is necessary to undertake a more systematic and critical review of the articles. In this process, the researcher has to note key terms, concepts, analytical findings and their interpretations etc., and this information is essential in the future, while describing the research and interpreting the current research findings. This enables the researcher to write a short summary of each article in the context of its usefulness to the present study indicating, where its information will be useful in the current research topic. The researcher may also indicate the key findings, comments, strengths and weaknesses of the publications. Note that, the researcher should write this brief review in a simple language, as it facilitates him to write the review of literature for that article and can use that material at a later phase.

e. Writing the 'Review of Literature'

After synthesizing and analyzing the gathered literature, it is essential on the part of the researcher to structure each article or publication in a systematic form such that, it demonstrates the key concepts in a clear and consistent way. As far as possible, the researcher should avoid using long sentences and confusing words. Each sentence must have clarity in explaining the meaning without any spelling and grammatical mistakes. In general, the chapter on 'Review of literature' comprises of three sections namely, introduction, main body and conclusion *(Burns and Grove, 2007).*

- ☆ ***Introduction:*** This is the first section for presenting the review of literature. This section provides the information regarding a brief overview about the selected problem. The researcher should indicate the reason (point of view) for reviewing the literature; organization of the review (sequence) and, when necessary state why certain literature is or is not included (scope). The researcher should also indicate the key terms considered for reviewing the articles. The researcher may also comment, whether there was a dearth or wealth of literature on the selected topic. This gives the reader some insight into the breadth and depth of the literature sourced and also facilitates some judgement, so as to analyze the validity of the claims being made. The researcher can also state about the developments made about the particular theory or model, conflicts in theory and methodology, gaps in research etc.
- ☆ ***Main body:*** The main body of the report presents and discusses the findings synthesized and analyzed from the collected literature. For convenient sake, the researcher divides the main body into different sub-heads in accordance with the objectives framed for the study. Under each sub-head, he should include the main findings of the relevant articles. The articles should be arranged in the chronological order *i.e.,* from the past to the latest year possible, as it indicates the development of the concepts over a period of time. Note that, the reviews should not be arranged in the alphabetical order of the researchers, as it does not assist in understanding the growth in concept(s) to be studied. The researcher should not write the same sentences, what authors have written in their articles and should not make broad conclusive statements about the reviewed articles. According to Polit and Beck (2006), *'when describing a study's findings, it is best to use language that indicates the tentativeness of the results rather than making definite statements about the research'*. Note that, the researcher should not give his personal opinions about the findings of the reviewed articles. The review should be written in such a way that, the reader should judge that, the researcher has understood and synthesized the relevant information, rather than merely describing what other authors have found. That is, the review should be like a critical evaluation of the information available on the topic, highlighting and comparing results from key sources (Beyea and Nicholl, 1998). If there are any consensus or differences among the findings between the selected articles with reference to the topic in each sub-head should be clearly mentioned. This section also indicates the gaps in the research area and how the present research will be directed towards filling those gaps.
- ☆ ***Conclusions:*** The researcher should summarize and write the conclusions of main findings emerged from the literature reviewed in the main body, maintaining the focus established in the introduction. This serves as a backdrop to proceed with the current research. That is, the main conclusions reviewed from the literature highlight the progress of research in the study area and help the researcher to identify the research gaps. This guides the researcher to frame the objectives for the present study. The researcher should also summarize key

ideas, inconsistencies and gaps identified in the collected literature. It also offers suggestions for future research.

Note that, the purpose of writing the literature review is to document the state of the art (science), with respect to a particular question or problem. The literature review is itself a research method. It takes raw data (the annotated bibliography (discussed in Chapter 13)) and converts it into information (a critical appraisal). The written review should:

- ☆ be organized and directly related to a research problem/question that the researcher is thinking of developing.
- ☆ organize and synthesize the findings of previous researchers into a summary of what is and is not known.
- ☆ identify areas of controversy in the literature.
- ☆ formulate questions and identify requirements for further research.

The review is 'PROSE', not a list describing or summarizing one citation after another. Organize the review into sections that represent themes or sub-topics and identify trends in the development of a theory or a model etc.

f. Literature Cited or Bibliography

It is essential to note that, the researcher should acknowledge each and every source in the 'Literature cited', provided if he referred and used that work or information or data in his current research. That means, every citation that is used in the report must appear in the 'Literature cited'. All sources like books, journal articles, reports and other media are to be included in the 'Literature cited'. This references list will serve as a source of literature to other researchers, who are interested in attempting the same or related topic. So, every effort made to prepare the references list should be accurate. The most suitable method of recording literature is the 'Card' system. This recording system involves two sets of cards *viz.*, Source cards and Note cards.

- ☆ ***Source cards:*** They are of the size: 3"x 5". They are used for noting bibliographic information. The entire information from all the score cards will be pooled and arranged in an alphabetical sequence to prepare Bibliography or Literature cited at the end of the dissertation. This score card also facilitates to record the bibliographic information in a proper bibliographic format. The following are the examples:
  - ❖ ***Format for citing a research article:*** (*i*) Name(s) of author(s). (*ii*) Year of publication. (*iii*) Title of article. (*iv*) Name of Journal in italics with volume number, Issue number in brackets and pages of the article.

    Example: Kumar, K.N.R. 1995. Marketing efficiency in transacting paddy in Kurnool regulated market of Andhra Pradesh. *Indian Journal of Agricultural Marketing*, 2(2): 45-49
  - ❖ ***Format for citing a book:*** (*i*) Name(s) of author(s). (*ii*) Year of publication. (*iii*) Title of book in italics. (*iv*) Name and city of publishers. (*v*) Pages referred to in the book.
  - ❖ ***Example:*** Kumar, K.N.R. 2012. *Microeconomic Analysis in Agriculture*. Daya Publications, New Delhi. 207-217.
- ☆ ***Note cards:*** They are of the size: 5"x 8". They are used for noting down detailed information from a printed source.

It is essential to note that, the researcher should note a single fact or idea on each card, on one side only. The student should write the reviews on the cards and this works well because, he can easily shuffle the cards into categories, when he finished reading all the sources. Some students prefer to photocopy articles and portions of books rather than taking notes on index cards, but this approach does not generally allow for internal cognitive processing and tends to result in a more superficial understanding of the material. A useful alternative to handwriting notes on index cards is the use of bibliographic software such as Endnote (www.endnote.com), which serves as a relational database of bibliographic information that can be input by hand or downloaded from the Internet.

The above discussion infers that, the chapter on review of literature in the dissertation should have the features given in Table 6.2.

At the end of writing the chapter on Review of Literature, the researcher should verify the content of review with the following checklist:

- ✰ Have you indicated the purpose of the review?
- ✰ Have you emphasized recent developments?
- ✰ Is there a logic to the way you organized the material?
- ✰ Does the amount of detail included on an issue relate to its importance?
- ✰ Have you been sufficiently critical of design and methodological issues?
- ✰ Have you indicated when results were conflicting or inconclusive and discussed possible reasons?
- ✰ Has your summary of the current literature contributed to the reader's understanding of the problems?

6.10. Characteristics of a Good Research Question for a Literature Review

As discussed earlier, selecting and formulating a good research question is essential for collecting the literature review relevant to it. If the problem statement is not clear, it is quite difficult on the part of the researcher to collect requisite information and data base. So, unless the researcher formulate a good research question, it is difficult to collect review and hence, it is difficult find apt solutions to the selected problem. So, for a good literature review, the research question should have the following characteristics:

- ✰ Should have narrow focus
- ✰ Should assume the possibility of different outcomes
- ✰ Should be clear
- ✰ Built on sound assumptions
- ✰ Should have a single question

6.11. What the Researcher Should do Before Writing the Literature Review?

The researcher should go through the following aspects before writing the literature review:

a. The researcher should have clarity on the following aspects:
   - ❖ How many sources the researcher has to refer to collect the requisite information?
   - ❖ What type of sources he has to refer to collect the requisite information (Journals, Internet, Books, Reports, Conference proceedings etc.)

Table 6.2: Features of Review Literature to be included in the Dissertation.

Section	*Content*
Sub-heads	Should be in accordance with the framed objectives for the study
Introduction	This section occupies 10 per cent of the chapter. It provides the following information: ☆ Define or identify the general topic to provide the context for reviewing the literature ☆ Outline why the topic is important ☆ Brief overview about the selected problem ☆ Identify overall trends in what has been published about the topic ☆ Importance of writing the review of literature ☆ How the review is arranged ☆ Identify conflicts in theory, methodology, evidence, and conclusions ☆ Identify gaps in research ☆ Explain the criteria to be used in analyzing and comparing the literature ☆ Highlight the scope of review collected ☆ Indicate the key terms considered for reviewing the articles ☆ Comment on dearth or wealth of literature on the selected topic ☆ Indicate the developments made about the particular theory or model, conflicts in theory and methodology, gaps in research etc.
Main body – Review is divided into different (usually 3 to 6) sub-sections.	This section occupies 80 per cent of the chapter. It provides the following information: ☆ Organize the review collected into paragraphs that present themes and identify trends relevant to your topic. ☆ Each paragraph should deal with a different theme. The researcher need to synthesize the collected readings into each paragraph in such a way that, there is a clear connection between the sources. ☆ Presents and discusses the findings synthesized and analyzed from the collected literature*. ☆ The articles should be arranged in the chronological order *i.e.,* from the past to the latest year possible. ☆ Identifies, analyses and evaluates the selected article. ☆ The researcher should not list all the materials he identified in literature search.
Conclusion	This section occupies 10 per cent of the chapter. It provides the following information: ☆ The conclusions of main findings emerged from the literature reviewed. ☆ Summarize the main agreements and disagreements in the literature. ☆ Summarize the general conclusions that have been drawn. ☆ Summarizes key ideas, inconsistencies and gaps identified and establish where your own research fits in the context of the existing literature. ☆ Summarizes suggestions for future research.
Literature cited or Bibliography	This is presented at the end of the dissertation or thesis or report as a separate chapter. It provides complete, consistent, accurate and updated list of cited research material.

*Note: In writing the overall review (sub-head wise) after arranging all the relevant collected material, the researcher has to brief upon the findings of one author followed by other author and then by another .... It is now essential to group authors and link ideas across the authors. To group authors, who draw similar conclusions, the researcher can use linking words such as: '*also*', '*additionally*', '*again*', '*similarly*' *etc.* However, when the authors disagree on pa particular aspect, linking words that indicate contrast will show how the researcher have analyzed their work. Words such as: '*however*', '*conversely*', '*on the other hand*', '*nonetheless*' *etc.,* will indicate to the readers, how the researcher have analyzed the reviewed material. At other times, the researcher may want to qualify an author's work (using such words as *specifically, usually,*or *generally*) or use an example (*thus, namely, to illustrate*). In this way, the researcher ensure that, he has synthesizing the material and not just describing the work already carried out in the selected field. Always note that, there needs to be a clear relationship between the literature review and the methodology to follow and do not include the review that stand alone without any links to the rest of the thesis.

- ❖ Should he synthesize, analyze and evaluate the information collected from different sources?
- ❖ Should the review report is to be divided into different sub-topics in accordance with the framed objectives?

b. The researcher should aim at collecting the information mostly from the primary sources and he should not depend exclusively on secondary and tertiary sources. Journals are the best source to get the primary information about the research articles. Even internet is also an important source to get primary information provided, the researcher has to download the soft copies of the original articles or documents published by the writers. Thus, the researcher has to go through a comprehensive examination of all sources to collect the requisite information. However, the number of sources vary with the nature of the study. For example, if the researcher aims at conducting the benchmark survey among the farmers with reference to irrigation water management, he may wish to limit the sources to professional literature within that single field (Irrigation water management). Of course, if the researcher aims at analyzing the *'Economic efficiency of irrigation water resource in Kurnool district of Andhra Pradesh'*, he has to collect review from various sources (research materials related to Agricultural engineering, Agronomy, Agricultural Economics etc).

c. The researcher should narrow down the focus of the study and this facilitate him to collect the research articles relevant to the selected topic. We know, the objectives framed for the investigation are highly specific and they serve as useful guide for the researcher to proceed for collection of literature. This is so because, the researcher may get access to hundreds or even thousands of articles and books on the selected areas of study. But, to proceed with the research on the selected topic, it is high time for him to narrow the focus of the study through framing the objectives of investigation. This facilitate him to limit the number of sources to refer in order to get a good quality review material.

d. The researcher should decide upon the scope of the review. That is, from which period to which period, the information has to be collected. It all depends upon the type of research investigation. For example, if the researcher is conducting research study on *'Comparative analysis of trends in prices of major agricultural commodities in India during pre and post-WTO regimes'*, he has to collect literature with reference to both in pre and post-WTO regimes. On the other hand, in some disciplines like medical research, the researcher has to use information that is as current as possible. We know, the treatments for medical problems are constantly changing according to the latest studies. Information even two years old could be obsolete. So, depending upon the discipline, the researcher has to fix the time frame for collecting the literature. The time may include all documents published on the selected topic from the current day back through the first publication on the topic. Alternatively, and much more commonly the review is done from current day back to a certain defined period. Generally, in Agro-Economic research, the researcher will collect the information from the latest source back to the past 15 or 20 years. For a topic that is relatively new, the time horizon may be further limited (for example, economic analysis of SRI technology of paddy production in Kurnool district of Andhra Pradesh). The common mistake made by the researcher is, especially when he is new to a field, to assume a current topic that emerged recently, when in fact, it has long roots in the past that continue to have a major impact on current practice. For example, if a student conducted a research study on 'e-learning' and collected the data for the recent 10 years period, it implies, the researcher has considered a significant time period for collecting the requisite data. But, if the concept of 'e-learning' actually had the roots over the past 50 years in the State, it implies that, the student was missing a large amount of important information on this topic.

e. After deciding focus and scope of the review, the next important job is to organize the review in a systematic fashion in the report. As discussed earlier under Section 6.1, the literature review is to be presented in three different sections *viz.*, introduction, main body and conclusions. After framing these sections, it is essential on the part of the researcher, how to present the review in the main body section. There are several ways like chronological review, trend review, thematic review (organized around a topic or issue, rather than the progression of time), methodological review etc., to present the reviews on the topic.

f. After deciding one of the above ways for the body of reviews, it is quite easier for the researcher to organize the review or begin composition in a systematic order. In writing the body, the researcher should select only the most important points in each source to highlight in the review. But, remember, the researcher should not present the review as it is from the research material and he should synthesize, analyze and evaluate the review material before including it in the body of review. That is, though the literature review presents others' ideas, the researcher's voice should remain front and center. The researcher should spend considerable time for revising the review because, the main objective is to present the material, not the argument. So, he should review again and again to make sure it is relevant to the selected research topic and presented in the most concise manner possible. He should be sure to use terminology familiar to the readers.

6.12. Common Mistakes the Researchers do in Reviewing the Literature/ Problems Faced when Writing a Literature Review

The ability to carry out a literature review is an important skill for any student. For writing good review, the researcher has to spend considerable time for reading and understanding the literature relevant to the selected research topic. This understanding will make the researcher to prevent from committing errors, redoing work which has already been done and gaining insights to explore the present research. The above discussion infers that, literature review is done to accomplish three important aspects *viz.*, whether there are any previous studies related to the selected research problem for investigation, to obtain guidance from the past research studies to conduct present research and to build a background or platform to execute the present investigation (Mertler and Charles, 2005). However, many students do not understand why they have to write a literature review!. While collecting literature, the researcher may perform some mistakes and they are listed here under:

- ✰ The researcher may simply go through the information without noting down the important findings. Writing the findings will help the researcher to identify how the past research findings are related to the selected research problem and also helps him in interpreting the current research findings.
- ✰ The researcher may simply note the findings of the past studies in the review without maintaining the bibliographic information. In some cases, reviews are not cited in text, but their sources are listed in reference list. These inconsistent references usually erode away the credibility of the literature reviews.
- ✰ The researcher may not indicate, whether the current findings are related to or deviate from the past research findings.

- ☆ The researcher may not find sufficient time to identify and locate best findings. He carry out the literature review in a hurried way and this makes him in overlooking previous studies containing ideas that could improve the project.
- ☆ The researcher may not locate primary sources of information and simply depend upon secondary and tertiary sources for gathering the information.
- ☆ The researcher may simply write the past findings 'as it is', without critically reviewing the available literature.
- ☆ The researcher mostly rely on internet sources for collecting literature without looking into the journals, as it saves lot of time on the part of the researcher.
- ☆ The researcher may simply concentrate on the findings of the past research studies and overlooking different methodologies followed for the studies.
- ☆ The researcher may collect information covering broad topics or too narrow topics about the selected problem. The former makes the researcher fail to understand about the topic limits of the literature review, while the latter can cause to overlook many articles.
- ☆ The researcher may collect too much information about the selected topic and this makes difficult for him to separate important from unimportant information.
- ☆ The researcher generally use other authors' literature reviews in his research report. This makes him to do less review job and hence, he could not understand in-detail about the selected project.
- ☆ The researcher, while collecting the review may not focus on the selected topic (*i.e.*, the review include many disconnected ideas within one paragraph). Hence, the selected literature reviews are unsatisfactory.
- ☆ The research may not synthesize the literature review. That is, the review consists of many separate paragraphs, but they are not integrated.
- ☆ The researcher may not organize and structure the literature review.
- ☆ The researcher may include repetitive articles.
- ☆ The researcher may fail to cite influential papers.
- ☆ The researcher fail to keep up with recent developments.
- ☆ The researcher may include the article reviews in the main body that do not clearly relate to the research problem;
- ☆ The researcher may uncritically accepts another researcher's findings and interpretations as valid, rather than critically evaluating them.

To avoid the above common mistakes in writing Review of Literature, the researcher must note the following tips for good writing of review:

- ☆ Plan your writing *i.e.*, clearly understand what you are going to write about, in what order, and what you're going to say.
- ☆ Write an outline of the review and note that, it may not be the final review. But, it just help you think about what you are going to say.
- ☆ Ignore the language! When you plan your writing, don't worry about the language. Concentrate on what you are going to say and not about verb agreement.

- Don't waste time worrying about spellings, prepositions and conjunctions. This should be taken care at the time of writing the final review. If you concentrate at the beginning, it will kill most of your time, as at the final stage of writing review, you may change the content substantially in accordance with studying the trends in review, willing to cut out paragraphs or sentences that you later realize aren't necessary.
- Similarly, don't put a lot of effort into proofreading (punctuations and spelling) until you are sure that final review is prepared from an article or any other source.
- Write and rewrite the review duly incorporating the new material in accordance with the content, trend etc., you are willing to present in the final review. Note that, more experienced writers rewrite more times and more substantially than less experienced writers. Repeated writings improve your thoughts and in each successive case, the presentation of review will get improved.
- You should find readers after writing the review to read what you've written. Ask friends, ask professors etc., for proof reading the literature. Don't finalize the review before giving it to the readers, as they may offer valuable suggestions for improvement.
- Incorporate the feedback suggestion offered by the readers (if found relevant) into your content and now finalize the review.

6.13. Advantages of Literature Review

In addition to the importance of literature review discussed earlier, the following points further highlights its meritorious aspects:

- The literature reviews are essential irrespective of the type of topic considered by the researcher, as they provide both over view and in depth knowledge about the selected research topic.
- The sources of literature collection like internet, journals etc., are relatively inexpensive and huge amount of information and data base can be collected efficiently *i.e.*, large amount of data can be collected quickly at minimal cost.
- The researcher can conduct this exercise on his own.
- The literature review can be considered as a first step in the identification of research problem and formulation of problem statement.
- The literature review provide a conceptual framework for further planning and study.

6.14. Disadvantages of Literature Review

They include:

- The literature review commands high level of skill on the part of the researcher in selecting the sources of information, identifying the relevant material, synthesizing and analyzing the relevant information and writing a meaningful summary.
- A poorly collected review will adversely influence the subsequent steps of the research process.
- Accessing to primary sources of information is sometimes a costly affair on the part of the researcher.
- In some cases, the researcher could not get the up to date information or data base.

6.15. What Do Examiners Look for in a Literature Review?

According to Holbrook *et al.* (2007), in general, Examiners look into three basic aspects while commenting on the review in the dissertation or thesis or report. They include:

- They comment on the coverage of literature. That is, they verify the period for which the review has covered. As mentioned earlier, in Agro-Economic research, the reviews should at least cover 15 to 20 years period. Any review less than this period, can be subjected to comments by the examiner. This is because, it gives the clear cut indication that the researcher has not built sufficient theoretical and methodological backgrounds to explain the selected research topic.
- The examiner identify errors in the report. That is, he will critically examine, how the review is introduced, how the main body is divided into different sub-topics that too whether these sub-topics cover all the specific objectives framed for the study or cover the key concepts of the study and how the review is concluded. He also identify typographical and grammatical errors, locate references omitted from the bibliography though mentioned in the main body of literature, locate references in the bibliography but not mentioned in the dissertation etc.
- The examiner may also comment on the presentation style of the body of the literature regarding mentioning of author's names, years etc. He may comment the literature review as below by using the following techniques:
    - ***Incomplete:*** The collected review is considered 'incomplete', if there exists large gaps in understanding key terms and concepts.
    - ***Inadequate:*** The collected review is not sufficient enough, as it provides only overview about the topics and could not provide deeper understanding about sub-topics.
    - ***Incommensurate:*** This infers that, the collected literature is not actually related to the topics included in the dissertation so that, it could not develop and sustain interest on the part of the reader.

7

Formulation of Testable Hypothesis

After thorough review of collected literature for the selected research problem, the next step in the research process is to formulate a testable hypothesis (plural is hypotheses) relevant to the framed objectives and identified variables for the study. This formulation of testable hypothesis is usually considered as the principal instrument in research and it goes hand in hand with the formulated research problem. The formulation of hypothesis helps to explain the facts and to guide in the investigations of others and hence, hypothesis is frequently called as a Working hypothesis. That is, a working hypothesis is a provisionally accepted hypothesis proposed for further research. The word hypothesis is derived from the Greek word *'hypotithenai'* meaning 'to put under' or 'to suppose'. Etymologically, 'hypothesis' is made up of two words, *'hypo'* (less than) and *'thesis'*, which mean less than or less certain than a thesis. The following are the important definitions of hypothesis:

'A hypothesis is a tentative generalization, the validity of which remains to be tested'.

(Lundberg)

'Hypothesis is a proposition, which can be put to test to determine its validity'.

(Goode and Hatt)

'Hypotheses are single tentative guesses, good hunches –assumed for use in devising theory or planning experiments intended to be given a direct experimental test when possible'.

(Eric Rogers)

'A hypothesis is a conjectural statement of the relation between two or more variables'.

(Kerlinger, 1956)

'Hypothesis is a formal statement that presents the expected relationship between an independent and dependent variable.'

(Creswell, 1994)

'It is a tentative prediction about the nature of the relationship between two or more variables.'

'A hypothesis can be defined as a tentative explanation of the research problem, a possible outcome of the research, or an educated guess about the research outcome.'

(Sarantakos)

'Hypotheses are always in declarative sentence form, and they relate, either generally or specifically, variables to variables.'

'A hypothesis is a statement or explanation that is suggested by knowledge or observation but has not, yet, been proved or disproved.

(Macleod Clark J and Hockey L)

'A research question is essentially a hypothesis asked in the form of a question'

'Hypothesis is a proposition made as a basis for reasoning, without the assumption of its truth, a supposition made as a starting-point for further investigation from known facts'.

(The Concise Oxford Dictionary)

'A hypothesis is a statement capable of being tested and thereby verified or rejected.'

(Rummel and Balline)

'A hypothesis is a tentative solution posed on a cursory observation of known and available data and adopted provisionally to explain certain events and to guide in the investigation of others.It is, in fact, a possible solution to the problem.'

(M.H. Gopal)

The above definitions infer that, hypothesis is a tentative statement or assumption or guess made, which serves as a guide in the search of real truth about the investigation. But, these guesses are not merely accidental. Moreover, mere collection of factual information alone about the selected topic does not lead to successful formulation of hypothesis. So, formulation of hypothesis requires both *i.e.,* useful background information about the study and considerable speculation and imaginative guess work. Further, keen observation, creative thinking, thorough knowledge about the phenomenon and its related fields, sound judgement etc., are also essential on the part of the researcher in formulating a good hypothesis. However, there are certain conditions or sources that are essential for the successful formulation of a hypothesis and they are discussed in-detail in the ensuing pages. This formulation of hypothesis indicates the researcher's background knowledge and his intellectual guess about the probable outcome of the research. To validate the findings, the framed hypothesis is to be tested by statistical means. For example, consider statements like the following ones: *'there exists an inverse relationship between price of rice and quantity demanded by the consumers in the market'* or *'farmers, who receive training on IPM practices will have a greater managerial ability to control pest incidence on cotton than farmers, who do not received training'* etc. These hypotheses are capable of being objectively verified and tested (Figure 7.1). Thus, it is clear that, hypothesis serves as a tentative statement, based on which the researcher verifies the findings and it can be put to a test to determine its validity. It is to be noted that, hypothesis can be proved or disproved. By formulating the hypothesis, it guides the researcher to proceed in a particular direction in selecting the right variables for analysis purpose, whether to prove it or disprove it.

7.1. Characteristics of 'Good' Hypothesis

A 'good' usable hypothesis is the one, which satisfies many of the following criteria.

- ✰ Hypothesis formulated should be clear and precise. If the hypothesis is not clear and precise, the inferences drawn on its basis cannot be taken as reliable.
- ✰ Hypothesis should be specific. A specific hypothesis leaves no ambiguity about the subjects and variables, or about how the test of statistical significance will be applied. It should indicate the nature of IVs and DV and how they will be measured. Often these details may be included in the study proposal and may not be stated in the research hypothesis. However, they should be clear in the mind of the investigator while conceptualizing the study.
- ✰ Hypothesis should identify the independent and dependent variables to be studied. That is, it should clearly specify the nature of the relationship that exists between the selected variables for the study.

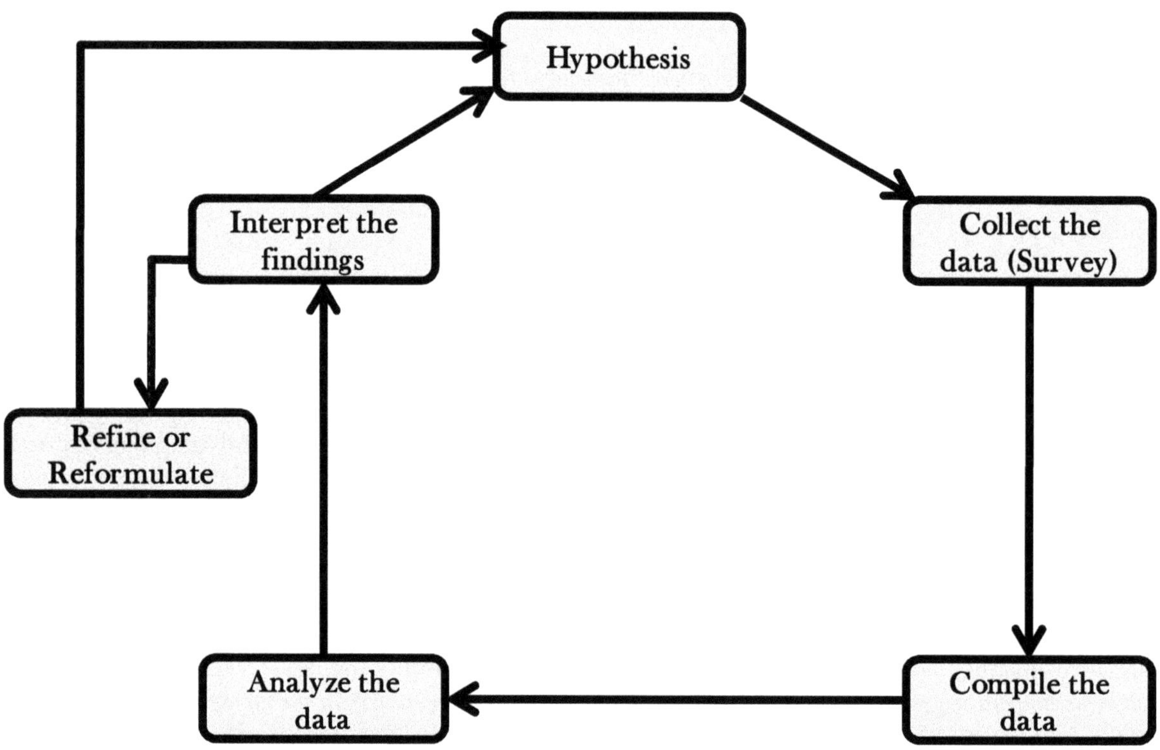

Figure 7.1: Steps in Research Process after Formulating Hypothesis.

- It should be empirically testable. That is, it should be verified. If it is not testable, there lies no meaning in interpreting the analytical findings as significant or not. That is, hypothesis does not automatically provide authentic answer to the problem under investigation, until it is tested by statistical means. Further, hypotheses formulated should be amenable to testing within a reasonable time. That is, those problems which can be resolved in time and that too which involves hypotheses that can be tested within a reasonable and specified time must be selected for the investigation. So, the researcher should not aim at complicated problems and frame non-testable hypotheses.
- Hypothesis should be formulated in simple terms so that, it can be easily understood. A simple hypothesis contains one IV and one DV. That is, the formulated hypothesis should be specific with a narrow scope. It is essential that, instead of having one complicated hypothesis, it is better to have several simple hypotheses, as they can be easily understandable. A complex hypothesis contains more than one IV or more than one DV. Complex hypothesis cannot be easily tested with a single statistical test and should always be separated into 2 or more simple hypotheses. It is better to limit the number of hypotheses to six or fewer. If the researcher assumes more hypotheses than six, it will make him to collect extensive data in the investigation and at the same time it is difficult to keep participants interested. So, by limiting the number of hypotheses, it makes the researcher to collect the requisite data in time, analyze and interpret the findings with more clarity. But, note that, there is no relation between the simplicity of formulated hypothesis and its testability for its significance.
- Hypothesis does not indicate specific statistical procedures to be used in the analysis.
- Hypothesis indicates the population that the researcher is going to study in the investigation.
- They must be capable of verification or rejection within the limits of the research resources.
- Hypothesis formulated should be relevant for the selected topic under investigation. If an irrelevant hypothesis is formulated, it will lead to an invalid research.
- They must be stated in a manner, which provides direction for the research.
- They must provide for an acceptable level of confidence in the results, but at the same time economize the use of scarce resources.
- Hypothesis should be consistent with most known facts *i.e.*, it must be consistent with a substantial body of established facts.
- It should provide tentative answer to the proposed problem. This can be in the form of a declaration statement or in the form of a directional statement or in a null form.
- It must be operational, *i.e.*, the variables to be dealt with must be measurable
- A good hypothesis does not conflict with any law of nature which is known to be true.
- It should be so designed that, its test will provide an answer to original problem which forms primary purpose of the investigation.
- Hypothesis must be stated well-in advance at the time of writing the research proposal. This will help to keep the research efforts focused on the primary objective and create a stronger basis for interpreting the study's results as compared to a hypothesis that emerges as a result of inspecting the data. The habit of post hoc hypothesis testing (common among researchers) is nothing but using third-degree methods on the data (data dredging), to yield at least something significant. This leads to overrating the occasional chance associations in the study.

- The hypothesis should be related to a body of theory or some theoretical orientation.
- A hypothesis has a forward or future reference *i.e.*, it is future oriented. It relates to the future verification and not to the past facts and information.

Some Examples of 'Good' Hypotheses

Impact of IPM Technology on Cotton Production in Kurnool District of Andhra Pradesh – An Economic Analysis.

Specific Objectives of the Study

The following are the specific objectives outlined for the above study:

- *To estimate costs and returns in cotton production under IPM and conventional pest management.*
- *To analyze resource use efficiency in cotton production under IPM and conventional pest management.*
- *To identify and prioritize the constraints faced by the farmers in the adoption of IPM in cultivation of cotton in the study area.*

Hypotheses

To study the above stated specific objectives, the following hypotheses are formulated:

- *The IPM technology is more economical than conventional method of pest management in cotton cultivation.*
- *Resource use efficiency is higher under IPM compared to conventional method of pest management in cotton cultivation.*
- *Farmers encounter several problems in the practice of IPM technology compared to conventional method of pest management.*

7.2. Importance of Hypothesis

Though formulation and testing of hypothesis constitute an insignificant part in the thesis or dissertation, but it significantly directs the researcher in accomplishing the framed objectives of the research study. This can be highlighted from the following points:

- It provides the right direction for the researcher to conduct the research *i.e.*, it guides the researcher to identify what is relevant and what is irrelevant. Thus, it prevents the researcher in collecting unrelated literature, unrelated variables, unrelated data, excess data etc. So, the formulation of hypotheses makes the researcher to move on the right track and stay intact to the specific objective for which it is formulated. Thus, research becomes well-organized and haphazard actions are minimized. That is, it identifies facts that are relevant to the study and those that are not: Who shall be studied (men farmers, women farmers, traders etc), in what context they shall be studied (their decision making) and what shall be studied (their individual perceptions or their roles).
- A well worked up hypothesis is half the answer to the research question. For this, both knowledge of the subject derived from extensive review of the literature and working knowledge of basic statistical concepts are desirable.
- It brings clarity to the researcher in conducting the research by placing clear and specific goals before him.

- It helps the researcher in selecting the representative sample and in planning the tools for the analysis of data.
- It helps to make predictions about some outcome of the problem under consideration.
- It sensitizes the investigator on certain aspects, which are relevant from the standpoint of the problem to be dealt by him. Thus, it facilitates the extension of knowledge in an area and provides tentative explanations about the problem to be investigated.
- It enables the researcher to relate logically known facts to intelligent guesses about unknown conditions. Thus, it serves as a guide in the thinking process and in the process of discovery of solution to the problem under consideration.
- It enables the researcher to draw meaningful conclusions for the study after testing the hypothesis. Thus, it serves as a background framework right from identifying the relevant variables and up to interpreting and drawing conclusions for the study.
- Since, it makes the researcher's focus narrow and specific, it saves his time and resources in conducting the research study.
- A hypothesis may enable the researcher to add to the formulation of a theory. It enables him to specifically conclude, what is true or what is false.
- It transforms research questions into testable propositions.
- A hypothesis adds to the available body of knowledge. For example, in selecting and formulating a research problem, the researcher will collect literature to find out what work has already been done on this topic in the past and based on the research gaps, he will select a research problem and formulate objectives for the study. To accomplish the framed objectives, the researcher will formulate hypothesis that helps him to be in right direction in the research work.
- It suggests, which form of research design is likely to be the most appropriate *i.e.*, survey design/case study/participation observation study/Focus Group Discussions.
- It provides a framework for organizing the conclusions of the findings.
- Hypothesis is a temporary solution of problem concerning with some truth, which enables the researcher to start his work.
- Each hypothesis may lead the formulation of another hypothesis
- A preliminary hypothesis may take the shape of final Hypothesis
- Hypothesis serves as the investigator's 'Eyes' in seeking answers to tentatively adopted generalization (Good, 1963).
- The use of hypothesis prevents a blind search or research and saves the researches from the gathering of masses of data which may later prove irrelevant to the study (Young, 1965).

7.3. Sources of Hypothesis

There are several ways of identifying a hypothesis and they are discussed here under.

- ***Rich background experience about the selected topic:*** This guides in formulating the hypothesis inductively. For example, in a Farmers' Training Centre, the trainer based on his close observation analyzes the farmers' knowledge about the modern techniques of crop production and may attempt to relate this with his teaching methods. From this observed relationship, the trainer may inductively formulate a hypothesis that attempts to explain the relationship between training rendered and improved managerial ability of the farmers. So, hypothesis may be formulated correctly by researchers, who have rich experiences and academic background, but they can never be formulated by those who have poor background knowledge.

- ***Background theory:*** This is one of the main sources of formulating hypothesis. This source makes the researcher to formulate hypothesis deductively. The researcher may select the known theory in a particular area of interest and with this background, he can formulate a hypothesis. For example, we know the ultimate objective of a private enterprise is 'profit maximization'. This known assumption guides the researcher to formulate hypotheses like: 'the capital turnover of a business firm is an index of business success', 'higher the proportion of non-equity capital in total firm's capital increases risk in the business' etc.
- ***Analyzing the past studies:*** This may give the clues to the researcher so that, he can formulate a hypothesis for finding solutions to the present selected problem. For example, a researcher wishes to analyze the factors influencing market arrivals of paddy in a market. If the researcher knows from previous related studies that, market prices of maize and production of maize influenced market arrivals of maize in a market, he reasons that, perhaps in the present study also the same factors *viz.*, market prices of paddy and production of paddy also influences market arrivals of paddy in the market. However, such analogies should be used by the researcher with caution.
- ***Discussions with Professors:*** Sometimes, conversations and consultations with Professors or Academic Advisors also help the researcher in formulating meaningful and testable hypothesis.
- ***Observation:*** Hypothesis can be derived from observation. If a researcher observes several farmers suffer from cancer due to application of chemical pesticides without taking necessary precautions, then he can easily formulate a hypothesis stating that, 'application of chemical pesticides without taking necessary precautions leads to cancer'.
- ***Intuition and personal experience:*** These may also contribute to the formulation of hypothesis. The story of Newton and the falling apple is a classical example for this source of hypothesis.
- ***Continuity of research:*** The continuity of research on a particular aspect (say, marketing of agricultural produce) itself constitutes important sources of formulation of hypotheses. The rejection of some hypotheses in the past studies leads to the formulation of new hypotheses, if the researchers continue to work on the same subject.
- ***Analogy:*** Sometimes, certain commonalities between various things can lead to formulation of hypothesis. For example, if a researcher find that, many farmers from Northern parts of India are more progressive in practicing modern techniques of agricultural production compared to Southern parts of India, he can hypothesize that, 'farmers from Punjab state are more progressive in the adoption of crop production technology than in Andhra Pradesh' (because, the former belong to Northern part of the country). Such a hypothesis based on a phenomenon is called Analogical reasoning.
- ***Concomitant Variation:*** At times, hypothesis can be formulated about two phenomena, by correlating relationship between them. For example, if a researcher find that, graded produce fetch more income to the farmers than selling ungraded produce, then he can formulate a hypothesis stating that, 'grading of produce fetches higher price to the farmers than non-graded produce.

☆ ***From problem statement:*** A hypothesis can be derived from a problem statement. For example, a researcher wishes to conduct a study on *'An Economic Analysis of Farmers' Indebtedness in Srikakulam District of Andhra Pradesh'*. The problem statement include: *'In the economic context, several studies have been taken up regarding the performance of different financial institutions, assessing the credit needs of farmers, demand-supply gap of credit requirements of farmers etc., in Srikakulam district. But, the economic analysis pertaining to the extent of indebtedness of farmers, factors responsible for farmers' indebtedness, impact of Agricultural Debt Waiver and Debt Relief Scheme announced by different financial institutions etc., is of immediate concern. This is because, such analysis will help the researcher to formulate strategies to address the farmers' indebtedness in the study area. In this context, the present research investigation is certainly a contributing one, both from the farmers' side and also from the financial institutions' side'*. Considering the above problem statement, the following are the hypotheses formulated for this study:

- Cost of cultivation and family expenditure do not exert any influence on farmers' indebtedness.
- Land holding of farmers does not show any relation with average debt per farmer household.
- Family size do not show any influence on average debt per farmer household.

7.4. Classification of Hypotheses

The following are the different types of hypotheses:

A. Main Hypothesis and Sub-hypotheses

Depending upon the stage of conducting research, the following are the different types of hypotheses formulated.

a. Main Hypothesis

When the researcher(s) generally goes for conducting an exploratory research or in the stage of collecting literature for the selected research topic, he frame main hypothesis for the study. This main hypothesis is related to the ultimate goal of the study. Say, for example, if the researcher wishes to take up an exploratory research on analyzing the marketing efficiency in transacting agricultural produce in Kurnool district of Andhra Pradesh, he will conduct preliminary survey (Reconnaissance survey) in the study area to collect information from farmers and various stakeholders regarding marketing pattern of major agricultural commodities in the study area. From the survey and collected data, he has found that, there is low marketing efficiency in transacting major agricultural commodities due to various malpractices in the marketing system. This guides the researcher to formulate the Main Hypothesis *'malpractices by the traders are the main reason for low marketing efficiency in the agricultural marketing system'*. Now, the researcher to support Main hypothesis, will develop many Sub-hypothesis.

b. Sub-Hypothesis

From the Reconnaissance survey, the researcher selects the top most prioritized research problem in the study area for conducting an in depth study through framing well-specified objectives. To answer these objectives and keeping the Main hypothesis under view, now the researcher will frame Sub-hypothesis and they may be as follows:

☆ *Marketing channels of selected commodities are efficient.*

☆ *With increase in number of intermediaries in the marketing channel, the price spread increases.*

- *With increase in the length of marketing channel, the marketing efficiency decreases.*
- *The farmers face several marketing constraints in transacting agricultural produce.*

The Sub-hypotheses will be either accepted or rejected based on the statistical analysis conducted like percentage analysis, trend analysis etc. Consolidating the Sub-hypotheses, the researcher arrives at a conclusion whether to accept or reject the Main hypothesis stated earlier.

B. Descriptive Hypothesis

This hypothesis simply describes the characteristics of a variable such as farmers, agricultural labour, organization, situation, event. For example, 'the rate of unemployment among unskilled agricultural labour is more than skilled agricultural labour', 'Private firms dominate the Indian seed sector' etc.

C. Relational Hypothesis

As the name indicates, this hypothesis indicates the relationship between two variables in a study. The relationship between the variables may be positive or negative correlation. For example, 'the increase in production of groundnut resulted in increase in its arrivals in Kurnool regulated market', 'Farm families with higher incomes spend more on comforts and luxuries' etc.

D. Causal Hypothesis

This hypothesis clearly demarcates the causal relationship between two variables *i.e.*, one as DV and another as IV. So, the researcher is interested in analyzing the cause and effect relationship in the study *i.e.*, which is the cause and which is the effect. For example, 'the increase in prices of groundnut in the regulated market resulted in increase in its market arrivals', 'the increase in fertilizer application resulted in increase in output of paddy' etc.

E. Tentative Hypothesis

The researcher generally goes for framing tentative hypothesis, when a phenomenon cannot be fully understood because of some technical difficulties. So, the framed tentative hypothesis will guide the researcher to know, how far it will answer the research question. Sometimes, the researcher simultaneously frames two or more such hypotheses. For example, while explaining the price and output determination under oligopoly market competition several models have framed different hypotheses, but none of these are final and they are only tentative. Thus, tentative hypothesis is the statement made by the researcher, when there is no adequate supportive information in formulating it. In conducting practical research, this hypothesis is framed when the magnitude of the problem is not that serious or when the problem is from a very general area. So, the researcher is not interested to waste his time, money and other resources in conducting exploratory research to collect adequate supporting information about the hypothesis to be framed. He formulates hypothesis on a tentative manner from the data which he obtained easily and hence, it is termed as Tentative hypothesis. This is just like a person instead of proceeding in total darkness without direction proceeding with a torch light, which may be sufficient as long as his speed is very minimal. So, the researcher proceeds from no direction situation (no hypothesis research condition) to some direction situation. For example, the famous hypotheses about propagation of light namely waves theory and corpuscular theory of light and both explain the phenomenon of light, but none of them are final. They are tentative.

F. Working Hypothesis

At the beginning of the study of the research problem, researcher formulates the hypothesis. But, initially, the hypothesis so formulated may not be very specific and this hypothesis is subjected to revision towards more specificity, as the investigation proceeds. Such hypothesis is called as Working hypothesis. In some cases, first the researcher arrives at the Main hypothesis keeping the primary objective (goal of research) in mind and develops many more Sub-hypotheses (Working hypotheses) as the investigation proceeds. Similarly, when the nature of the problem is complex and demands using more number of statistical tools, it makes the researcher inevitable in formulating several working hypotheses to proceed with the research process. For example, in addressing the research issue on irrigation water management in KWD of Andhra Pradesh, the researcher(s) has to divide the topic into several sub-projects addressed by the Scientific Personnel of departments like Agricultural Engineering, Agronomy, Soil Science, Agricultural Economics and Agricultural Extension and then formulate working hypotheses under each sub-project.

G. Research Hypothesis vs Statistical Hypothesis

a. Research Hypothesis

It is also known as Scientific hypothesis. This hypothesis indicates the expected relationship between the variables (IVs and DV(s)). Thus, it indicates what the outcome of the study is expected to be. If the analytical findings are statistically significant, then the research hypothesis is supported. So, the research hypothesis is formulated in a testable form. It is usually expressed using an *'if-then'* structure such as: '*if* Hinosan is applied, *then* rice blast is controlled'. Research hypothesis may be again classified as Directional research hypothesis or Non-directional research hypothesis.

- ✰ ***Directional hypothesis:*** This is also known as one-tailed (or one-sided) hypothesis. This hypothesis indicates the expected direction of the relationship between the independent and dependent variables. For example, the research hypothesis: *'Farmers with high managerial ability will take more risk in the farm business than the farmers with low managerial ability'* is a directional research hypothesis because, it stipulates the direction of the difference between two categories of farmers. This hypothesis includes terms such as 'positive', 'negative', 'more than', 'less than' etc.
- ✰ ***Non-directional hypothesis:*** This is also known as two-tailed hypothesis. This is a research hypothesis, which simply indicate the existence of relationship between independent and dependent variables, but does not specify the anticipated direction of the relationship between these variables. The researcher may not know what can be predicted from the past literature. It may read, *'.there is a difference between.'*. For example, the research hypothesis: *'There is a difference in the managerial ability of farmers in practicing paddy cultivation'* is a non-directional hypothesis because, though it stipulates there will be a difference, the direction of the difference is not specified. So, the research hypothesis can take either statistical form, declarative form, the null form, or the question form.

The proponents of non-directional hypothesis state that, this form of hypothesis is non-biased (*i.e.*, impartial) and objective compared to the directional hypothesis. But, the proponents of directional hypothesis opine that, this format will generate guesses or expectations about the outcome of the research study and this initially guides the researcher to speculate about the selected research problem. In research studies, directional hypothesis are much more commonly used than non-directional hypothesis.

As mentioned above, directional hypothesis is also known as one-tailed (or one-sided) hypothesis and non-directional hypothesis is also known as two-tailed hypothesis. The word tails refers to the tail ends of the statistical distribution such as the familiar bell-shaped normal distribution curve that is used to test a hypothesis. In a two-tailed hypothesis (Figure 12.3), one tail represents a positive effect or association; the other, a negative effect. A one-tailed hypothesis (Figures 12.1 and 12.2) has the statistical advantage of permitting a smaller sample size as compared to that permissible by a two-tailed hypothesis. Unfortunately, one-tailed hypotheses are not always appropriate; in fact, some researchers believe that, they should never be used. However, they are appropriate when only one direction for the association is important or meaningful.

b. Statistical Hypothesis

This hypothesis does not indicate the expected relationship between the variables (IVs and DV). This hypothesis is to be proved statistically, *i.e.*, through statistical inference the researcher can accept or reject the hypothesis. This hypothesis is an assumption about a population parameter. This assumption may or may not be true. The best way to determine, whether to accept or reject the formulated statistical hypothesis is by examining the entire population. Since conducting census study (dealing with entire population units) is often impractical, researcher select a representative sample from the population. Note that, the statistical hypothesis often is given in quantitative terms and is numerically measurable. For example: 'Group A farmers recorded higher paddy yields than Group B farmers'. In the dissertation or thesis, the statistical hypothesis should be discussed with the results, whereas the research hypothesis should be addressed in the discussion or conclusions. There are two types of statistical hypotheses *viz.*, Null hypothesis (denoted as H_o or H_N) and Alternative hypothesis (H_1 or H_A). These two are discussed here under.

i. Null Hypothesis (H_O or H_N)

The null hypothesis is a statistical hypothesis, which is basically the opposite of research hypothesis. That is, if the researcher gets statistically significant findings of H_O, the H_O is rejected, but the research hypothesis is accepted. That means, rejection of null (statistical) hypothesis is equivalent to the acceptance of the research hypothesis. The H_O is a term that was invented by an English geneticist and statistician called Ronald Fisher. In a mathematical formulation of the null hypothesis, there will be an equal sign. So, in the H_O, the researcher makes a statement that no relationship exists between the selected variables. H_O is normally formulated by the researcher in the negative direction with a sole aim of rejecting it. So, the H_O is a hypothesis, which the researcher tries to disprove or reject or nullify. The statistical analysis of the data is subjected to significant tests ('t', 'Z' etc) to reject or accept H_O (rather than H_A). So, H_O is also called as Testing hypothesis. This H_O serves as the opposite of the Alternative hypothesis, H_A. In H_O, the term 'null' often refers to the common view about the variables or 'no difference' or 'no correlation' or 'no association' of the variables, while the H_A is what the researcher really thinks is the cause of a phenomenon. That is, H_A of indicates 'there is a difference' or 'there is a correlation', or 'there is an association' between the variables. If the H_O is rejected by the statistical test, then only the H_A remains valid (because, H_O and H_A are mutually exclusive and exhaustive). The H_O is formulated to be tested for possible rejection. If statistical analysis leads the researcher to reject the H_O, then H_A could be supported. That is, if the H_O is not accepted, then the H_A is used. Thus, two conclusions generally arise for the investigator in the formulation of H_O *viz.*, accept the H_O and reject the H_O in favour of an H_A. It is to be noted that, the researcher should carefully use the 'wording' of the null and ensure that, it is as specific as possible. Say, for example, a researcher has formulated the following H_O and H_A:

H_O: There is no difference between the efficacy of two pesticides, on average

H_A: The new pesticide is effective than the current pesticide, on average

In the above hypotheses, there is a major error. Say, if the farmers actually receive low crop yields due to new pesticide than applying current pesticide, here both H_O and H_A formulated are not valid. That is, if the H_O is rejected, even the H_A is also not accepted or it is rejected or it is also invalid. The researcher rejects the H_O, when the statistical tests generate the results such that, they do not fit the H_O at 1 per cent or 5 per cent level of significance. In such case, he has to accept the H_A. But, when the statistical tests generate the results such that, they fit the formulated H_O at 1 per cent or 5 per cent level of significance, the researcher has to accept the H_O. Note that, not rejecting H_O (or accepting H_O) does not really mean that, formulated H_O is true. This is because, there might not be enough evidence against H_O. Though the H_O is accepted, it is still a hypothesis, and must conform to the principle of falsifiability*, in the same way that, rejecting the H_O does not prove the H_A. Further, even if the H_O is not rejected (*i.e.*, accepted), then the researcher must be careful to interpret the findings. The thinking on this is similar to a legal verdict. Just because a person has been declared 'not guilty' in a case, it does not mean that, he is innocent. In the same way, just because a H_O is not rejected, it does not mean that, the statement is true.

Even H_O can be formulated in the form of a negative statement. Say, for example, the H_O: *'There is no relationship between farmer's income and expenditure on luxuries'*. This H_O can be tested for statistical significance and the statistical tests would aim at 'nullifying' this hypothesis and hence, it is called as 'Null hypothesis'. The formulation of H_O will have two advantages. First, they have to be formulated in a more exact way. Second, it is quite easier to 'reject' the hypothesis than to confirm it with complete certainty. Hence, the concept of H_O is found to be very useful. In general, when the researchers accept the H_O, they presume that, the conducted survey or experiment is a failure. But, note that, when a researcher accepts the H_O, it simply implies, the initial assumption is only incorrect and not the entire survey or experiment. Hence, accepting or rejecting the H_O should be seen as a positive result.

ii. Alternative Hypothesis (H_A)

This is also known as Directional hypothesis. Although this H_A is generally paired with the H_O, this was not actually developed by Fisher and was actually introduced by Jerzy Neyman and Egon Pearson. As discussed above, the researcher will formulate both H_O and H_A in his research study, and if H_O is rejected, he consider the H_A formulated. Since, H_O is formulated with the objective of its rejection after conducting the statistical significance tests, the H_A indicates the relationship or difference that exists between two variables or groups and goes further to state the nature or direction of that relationship or differences. Note that, H_A can be stated in directional form (Directional H_A) or non-directional form (Non-directional H_A) and this can be easily understood from the following example:

Assume the researcher formulated the following H_O and H_A in conducting a research problem to study the differences in managerial abilities of men and women farmers in practicing paddy cultivation:

---

* 'Falsifiability', as defined by the philosopher, Karl Popper, defines the inherent testability of any scientific hypothesis. In its basic form, falsifiability is the belief that for any hypothesis to have credence, it must be inherently disprovable before it can become accepted as a scientific hypothesis or theory. For example, if a scientist asks, "Does God exist?" then this can never be science because it is a theory that cannot be disproved. The idea is that no theory is completely correct, but if not falsified, it can be accepted as truth. Falsifiable" does not mean "false". For a proposition to be falsifiable, it must be possible in principle to make an observation that would show the proposition to be false, even if that observation has not been made. For example, the proposition "All crows are black" would be falsified by observing one white crow.

H_O: There is no difference in the managerial abilities of men (M_M) and women (W_M) farmers in practicing paddy cultivation *i.e.*, H_O: M_M=W_M or H_O: M_M-M_W=0

H_A (Non-directional)

There is difference in the managerial abilities of men and women farmers in practicing paddy cultivation *i.e.*, H_A: $M_M \neq W_M$.

H_A (Directional)

- ✰ The managerial ability of men farmers in practicing paddy cultivation is higher compared to the managerial ability of women farmers *i.e.*, H_A: M_M>W_M (or)
- ✰ The managerial ability of men farmers in practicing paddy cultivation is lower compared to the managerial ability of women farmers *i.e.*, H_A: M_M<W_M.

Thus, H_A is a statement that suggests a potential outcome that the researcher may expect after rejecting H_O. The sources of H_A comes from earlier research studies. It is established only, when a H_O is rejected. This H_A serves as the desired conclusion of the researcher.

Null Hypothesis vs. Alternative Hypothesis

These two hypotheses are used by the researcher to validate the research findings. The researcher conduct the statistical significance tests to verify the H_O (*i.e.*, to reject the H_O) and to accept the H_A. That is, by conducting statistical tests, the researcher will try to reject the H_O or to prove the H_O as wrong and will not try to prove its H_A as true. This is where the name 'null' comes from, as the testing aims to 'nullify' the H_O. The following are the interesting differences between H_O and H_A:

- ✰ In a mathematical formulation of the H_O, there will be an equal sign (=), while in H_A, there will be 'not equal to (≠)' or 'less than (<)' or 'greater than (>)' signs.
- ✰ The H_O states the exact opposite of what a researcher predicts or expects from the investigation. The H_A makes a statement that suggests a potential result or an outcome that a researcher may expect.
- ✰ H_O is a statistical hypothesis and is the default or original hypothesis, while an H_A hypothesis is any hypothesis other than the null. That is, if the H_O is not accepted (*i.e.*, rejected), then the H_A is used.
- ✰ H_O indicate only one prediction regarding the relationship between the selected variables, whereas H_A predict all other possible outcomes. For example, if the researcher formulate H_O: A=B, then H_A: A≠B (Non-directional) or H_A: A>B or A<B (Directional).
- ✰ Note that, although the H_A is oppositely framed to the H_O, it is not necessarily a negation of the H_O. This is because, H_A serves the purpose to the researcher to analyze the research outcome, especially when the H_O is rejected. Since, the statistical significance tests are conducted only with reference to the H_O, its rejection makes the researcher to consider H_A. So, the conclusion is always stated in terms of the H_O. In brief, the possible conclusions include: reject H_O in favor of H_A, or not reject (or accept) H_O. As mentioned earlier, a failure to reject H_O (or considering H_O is true) does not prove it is true, it just suggests that, there is not enough evidence to reject it in favor of H_A. So, the conclusion is never written in terms of accepting or rejecting H_A. That is, if the researcher concludes 'do not reject H_O or accept H_O', this does not necessarily mean that the H_O is true, but it only suggests that, there is no sufficient evidence against H_O in favor of H_A. But rejecting the H_O suggests that, the H_A may be true.

- ☆ H_O is only the prediction of outcome of a research, while an H_A is all other outcomes other than stated under H_O.
- ☆ H_A is the reverse of the H_O. That is, H_A is accepted only, when H_O is rejected or not accepted.
- ☆ The H_O is formulated in the sense of rejecting it after analyzing the statistical findings. The H_A is the opposite one.
- ☆ The H_O is the formal basis for testing statistical significance. The H_A cannot be tested directly and it is accepted by exclusion, if the test of statistical significance rejects the H_O.
- ☆ In hypothesis testing, the researcher gives consideration only for H_O, whereas the H_A relates to the statement to be accepted if /when the H_O is rejected.
- ☆ The final conclusion after hypothesis testing is always given in terms of the H_O. The researcher either 'reject H_O in favor of H_A' or 'do not reject H_O'; he never conclude 'reject H_A', or even 'accept H_A'.
- ☆ The following set of negations may help the students, while differentiating H_O and H_A.
  - ❖ Null hypothesis: 'X is equal to Y.' Alternative hypothesis 'X is not equal to Y.'
  - ❖ Null hypothesis: 'X is at least Y.' Alternative hypothesis 'X is less than Y.'
  - ❖ Null hypothesis: 'X is at most Y.' Alternative hypothesis 'X is greater than Y.'

In spite of the above differences, both the H_O and H_A are necessary in statistical hypothesis testing in social science research and other research studies. The following are the examples for H_O and H_A across different statistical tests conducted by the researchers:

Comparison Tests (For the Z-test, t-test and One-way Analysis of Variance (ANOVA) for Two Samples)

H_O: There is no difference between the means of the two samples

Or

H_O: There is no difference between the mean plant heights of sugar cane at two different locations

The H_A would be:

H_A: There is a difference between the means of the two samples

Or

H_A: There is a difference between the mean plant heights of sugarcane at two different locations (non-directional)

After framing these hypotheses, now the researchers are in a position to collect the data for the selected variables and able to test the hypothesis based on the statistical findings.

Multiple Comparison Test when there are more than Two Samples

H_O: There are no differences between the means of the samples

H_A: There is a difference between the means of the samples

For two-way ANOVA:

H_O: There are no significant differences between the effects of plant growth regulators on the paddy yields

H_A: There are significant differences between the effects of plant growth regulators on the paddy yields.

Correlation Analysis

H_O: There is no correlation between the two variables (There is no correlation between the samples)

H_A: There is a correlation (relationship) between the two variables.

Chi-Squared (χ2): A Comparison of Observed Data to Expected Data

H_O: There is NO difference between the (observed frequency) and the (expected frequency)

H_A: There IS a difference between the (observed frequency) and the (expected frequency).

Multiple Regression

An analysis of the relationship between multiple variables (say, one DV and more than one IV) within a sample to determine the relationship (strength and nature) of those variables. There are usually more than one set of hypothesis statements needed to complete the problem when performing a multiple regression analysis. The first set is used when performing a 'global test' to see, if there is a relationship between any of the IVs and the DV. The Null and Alternate Hypothesis statements use the symbol β to represent the regression coefficients in the population. There is a β for each independent variable in the problem

H_O: $\hat{\beta}1 = \hat{\beta}2 = \hat{\beta}3 = 0$

H_A: Not all $\hat{\beta}$ are 0

H. Declarative Hypothesis

This hypothesis is in the form of a positive statement about the outcome of the study. For example, the hypothesis: 'Resource use efficiency in cotton production is higher among IPM farms than the conventional farms' is stated in the declarative form.

I. Simple Hypothesis and Complex Hypothesis

Simple hypothesis is a hypothesis that expresses an expected relationship between one independent and one dependent variable. Say, for example, the hypothesis is: '*There is no influence of fertilizer use on paddy output*'. On the other hand, Complex hypothesis is a prediction of a relationship between two (or more) IVs and two (or more) DVs. It is, otherwise, referred as Multivariate hypothesis, as it deals with multiple variables. For example, in Data Envelopment Analysis (DEA) Model, the researcher employs multiple inputs and multiple outputs to analyze the functional relationship of selected Decision Making Units.

J. Parametric and Non-Parametric Hypotheses

A hypothesis, which specifies only the parameters of the probability density function, is called parametric hypothesis. For example, the hypothesis: 'Mean of the population is 2 feet' is a parametric hypothesis. If a hypothesis specifies only the form of the density function in the population, it is called a non- parametric hypothesis. For example, the hypothesis: 'population is normal' is non - parametric.

K. Simple Hypothesis and Composite Hypothesis

A Simple hypothesis is one, where the value of the parameter under H_O is a specified constant and the value of the parameter under H_A is a different specified constant. For example, if we test:

H_O: $\mu = 0$ vs. H_A: $\mu = 10$

Then, it is a simple hypothesis test. In the above formulated hypothesis, we have a particular value for H_O and a different particular value for H_A.

A Composite hypothesis is one, where the value of the parameter under H_O is a specified constant, but the value of parameter under H_A is not a specified constant *i.e.*, the value of parameter under H_A can assume any value. For example, if we test:

H_O: $\mu = 0$ vs. H_A: $\mu \neq 0$

Then, it is a case of composite hypothesis test and is a two-sided H_A. A one-sided H_A can be written as:

H_O: $\mu = 0$ vs. H_A: $\mu > 0$

or

H_O: $\mu = 0$ vs. H_A: $\mu < 0$

All of these hypotheses are composite because, they include more than one value for H_A.

Based on the above explanation, hypotheses may be catregorized as 'exact' or 'inexact'. In case of 'exact' hypothesis, it selects a unique value for the parameter such as H_O: $\mu = 0$ vs. H_A: $\mu = 10$. A hypothesis is called an inexact hypothesis, when it indicates more than one possible values for the parameter such as H_O: $\mu = 0$ vs. H_A: $\mu \neq 0$; H_O: $\mu = 0$ vs. H_A: $\mu > 0$; H_O: $\mu = 0$ vs. H_A: $\mu < 0$. Thus, a simple hypothesis must be an exact one, while an exact hypothesis is not necessarily a simple hypothesis. An inexact hypothesis is a composite hypothesis. Further, a simple hypothesis specifies the population distribution completely, unlike composite hypothesis.

7.5. Limitations or Difficulties in the Formulation of Testable Hypothesis

Formulating a hypothesis is not an easy task and the researcher may face the following difficulties:

- Lack of adequate and clear theoretical background
- If the researcher could not logically use the available theoretical background, then also it is difficult to formulate the hypothesis.
- The researcher may not have clear understanding about the key variables to be considered in the study.
- If the research problem is not clear and precise, it is difficult to formulate the hypothesis.
- If the objectives of the problem statement are not so clear and specific, it is difficult to formulate the hypothesis.

7.6. Parametric Tests and Non-parametric Tests

After formulating the hypotheses, the next step is to choose an appropriate statistical test. There are basically two types of statistical test *viz.*, Parametric and Non-parametric.

- ***Parametric tests:*** If the information about the population is completely known by means of its parameters (say, population mean, population standard deviation etc), then statistical test is called Parametric test. If we assume, for example, that the distribution of the sample mean is normal, then we require to use a parametric test. Examples include: t- test, F-test, Z-test, ANOVA etc.
- ***Non-parametric tests***: Nonparametric tests are also called as Distribution free tests or Distribution free statistics. If there is no knowledge about the population parameters, but

still it is required to test the hypothesis of the population, then it is called Non-parametric test. Examples include: Wilcoxon signed rank test, Whitney-Mann-Wilcoxon (WMW) test, Kruskal-Wallis (KW) test, Friedman's test etc. Thus, non-parametric test do not require any assumption about population parameters and they relate mainly to that branch of statistics known as 'Order Statistics'. In these tests, we discard actual numerical values and focus on the way in which things are ranked or classed. Thereafter, the choice between alternative types of test is determined by 3 factors:

- whether we are working with dependent or independent samples
- whether we have more or less than two levels of the independent variable, and
- the mathematical properties of the scale which we have used, *i.e.* ratio, interval, ordinal or nominal.

Differences between Parametric tests and Non Parametric Tests

Following Table 7.1 highlight the differences between these two tests:

Table 7.1: Differences between Parametric tests and Non parametric tests

Parametric Tests	*Non Parametric Tests*
Information about population parameters is completely known.	No information is available about the population parameters.
Specific assumptions are made regarding the population	No assumptions are made regarding the population
H_0 is formulated with reference to the population parameters.	H_0 is free from population parameters.
Parametric tests can lead to eliciting inferences about the population.	Because non-parametric tests do not make strong assumptions about the population, a researcher could not make an inference that the sample statistic is an estimate of the population parameter.
Test statistic is based on the distribution	Test statistic is arbitrary
These tests are applicable only for numerical variables.	These tests are applied to both variables and attributes.
No parametric test exist for nominal scale data. This test is employed for statistical procedures that use interval or ratio scaled data and assume populations or sampling distributions with normal distributions.	Non parametric test do exist for both nominal and ordinal scale data and make no assumptions about the distribution of the population.
This test is more powerful	This test is less powerful
Parametric tests assume the population parameters and hence, more complicated compared to non-parametric tests.	Non parametric test are simple and easy to understand, as it will not involve complicated sampling theory. This is the only test available for nominal scale data.

8

Preparation of Research Design (Methodology)

The next step after stating the research problem, research purpose, research objectives and hypotheses, is to formulate a research design. The starting point for the research design is, in fact, the research objectives and hypotheses that have been so carefully developed. In essence, the research design answers the question: how the researcher is going to get answers to the framed research objectives and to test the formulated hypotheses? Thus, research design (also called as the research strategy) is the conceptual framework within which the research will be conducted. This is often referred as the 'blue print' of the research. Proper framing of research design ensures greater efficiency on the part of the researcher in conducting the research. The importance of research design can be easily understood from the following example: When constructing a farm storage structure, there is no point in ordering construction materials or setting critical dates for completion of project stages, until the farmer knows about which type of storage structure he has to construct. Similarly, when constructing a building, first the builder should know, whether he is planning an office building or an independent house or an apartment or a school building before placing the order for purchasing construction materials. The same is the case in conducting social sciences research, as it demands a proper research design before the researcher actually proceeds for collecting requisite data for the formulated research problem. Note that, research design is not simply a work plan and it is more than a work plan in the sense that, it guides the researcher how best the collected data and information are subjected to relevant statistical tools and techniques towards answering the framed objectives of the selected problem. Thus, the research design aims at, given the research problem, what type of evidence is needed to answer the problem in a convincing way? As discussed in the above example, the farmer before ordering the construction materials to construct a storage structure, he must establish earlier, what type of storage structure is needed, what type of produce is to be stored etc. Then the work plan follows from this. Similarly, in social sciences research, the work plan begins after framing the research design *i.e.*, selection of sample respondents by following a suitable sampling design, variables to be identified, type of data to be collected, sources of data collection, statistical techniques to be employed etc. The students, too often, start collecting the data and information without actually framing the proper research design. The statistical findings from such data and information normally be weak and unconvincing and fail to answer the research question(s). At this juncture, it is essential to differentiate between research method and research design and the same is already discussed under Section 2.12 of Chapter 2. The research design to be followed by the researcher depends upon the research method and the objectives framed for the investigation. Irrespective of the research method, the research design aims at collection of relevant data and information with minimal expenditure of effort, time and money. The above discussion infers the following highlights or significance of research design:

- It provides a systematic approach in answering the formulated research problem and also facilitates the researcher to undertake various research operations smoothly.
- It ensures efficiency in the pursuit of research work through generating maximum information with minimum effort, time and money.
- It serves as a blueprint for the researcher to collect the relevant data and information and also to use best suited statistical techniques for analyzing the collected data and information.
- It makes the researcher cautionary in each and every aspect of conducting research, as any error may upset the entire project. So, the reliability of results for which the researcher is looking in the research project all depends upon the well-laid out research design.
- It constitutes a firm foundation for entire body of research work.
- It provides answers regarding what, where, when, how much, by what means etc., about the research study.
- It ensures project time schedule.
- Helps the researcher in planning the various activities of research and in procurement of various resources in time.
- It ensures proper documentation of the various activities, while the project work is going on.
- It provides satisfaction and confidence, accompanied with a sense of success from the beginning of the work of the research project.
- It specifies the methods and procedures for acquiring the requisite data and information for solving the problem. For example, schedules, questionnaires, samples etc., needed for investigation are decided, while framing the research design.
- It enables the researcher to arrive at certain meaningful conclusions at the end of proposed study.

Thus, research design is like a roadmap for the researcher, as it makes him clear, what actually he wants to investigate, how he proceeds in the investigation, by what time he has to complete various activities in the design, how efficiently he will utilize the available limited resources etc. Thus, it provides him the most effective and efficient route to accomplish the desired objectives of the research. It also makes him to focus on the ultimate objective constantly in mind and using this road map (research design), he can arrive at desired conclusions from the study. The selection of research design for a particular research problem is not like solving an algebraic problem, where there is only one correct answer and infinite number of wrong answers, but it is just like selecting an ice cream, where some are better than others and there is no one which is universally accepted as 'best'. Thus, the researcher may have several options in selecting the right research design for addressing the research problem. For example, if the researcher wishes to conduct research among the farmers for eliciting information on 'Appraisal of installation of sub-surface drainage system in Mutluru canal command of KWD in Andhra Pradesh', he can plan the study in terms of various disciplines like Agricultural Engineering, Agronomy, Soil Science, Agricultural Economics, and Agricultural Extension. Thus, the activities of research design may differ to address the same problem and it is up to the researcher to select the right activities and right design. Much of the choice will depend upon the fundamental objective implied by the research question and also influenced by other factors as given below.

- ✰ Availability of scientific information
- ✰ Availability of sufficient data
- ✰ Time frame of the research study
- ✰ Access to different sources for collecting requisite data and information
- ✰ Size of research budget
- ✰ Availability of manpower
- ✰ Ability, knowledge, skill and technical background of the researcher
- ✰ Researcher's attitude toward risk
- ✰ Urgency to solve the problem
- ✰ Nature of the research objectives

8.1. Definitions of Research Design

The following are the important definitions of research design:

'A research design is the determination and statement of the general research approach or strategy adopted/ or the particular project. It is the heart of planning. If the design adheres to the research objective, it will ensure that the client's needs will be served.'

(David J. Luck and Ronald S. Rubin)

'Research design in the plan, structure and strategy of investigation conceived so as to obtain answers to research questions and to control variance.'

(Kerlinger)

'A research design is the specification of methods and procedures for acquiring the information needed. It is the over-all operational pattern or framework of the project that stipulates what information is to be collected from which source by what procedures.'

(Green and Tull)

'Research design is the process of making decisions before a situation arises in which the decision has to be carried out. It is actually a process of deliberate anticipation directed towards bringing an unexpected situation under control'.

(Russell Ackoff)

'The research design is a plan of action indicating the specific steps that are necessary to provide answers to those questions, test the hypotheses, and thereby achieve the research purpose that helps choose among the decision alternatives to solve the research problem or capitalize on the market opportunity'.

'Research design can be better defined considering three important terms viz., plan, structure and strategy. The 'plan' is the outline of the research scheme on which the researcher is to work. The 'structure' of the research work is a more specific scheme and the 'strategy' suggests how the research will be carried out i.e., methods to be used for the collection and analysis of data'.

8.2. Steps in Planning the Research Design

For planning the research design, the researcher to consider the following steps:

- ☆ The researcher should have thorough understanding about the objectives and hypotheses considered for the study. This guides the researcher in various aspects like type of data to be collected, identification of sources for data collection, identification of variables, selecting the representative sample for the study, questions to be included in the schedule/ questionnaire etc. This planning avoids confusion, misdirection and wastage of time, money and efforts at later stages of research work. The whole research project moves smoothly due to initial planning of the research project.
- ☆ The researcher should have thorough knowledge about the costs involved in executing the research project. This is because, depending upon the money resource available with him, he can plan the resources usage in the project. Say, for example, if the researcher wishes to conduct research study on '*Economic analysis of sub-surface drainage system in Mutluru command area of KWD in Andhra Pradesh*', the researcher has to spend significant amount of money resources towards preparation of large number of copies of schedules, appointing interviewers for data collection, staff will be required for tabulation and analysis of data collected, tour expenditure for the staff in moving to the study area, purchase of requisite materials to install sub-surface drainage system etc. These costs should be estimated accurately.
- ☆ Besides cost structure, the researcher should be well-planned in preparing the time schedule for various activities in the research project. This is an essential aspect to be considered at the initial stages of the project. The time calculations are essential for each and every aspect of research design like preparation of schedule and its pre-testing, training of interviewers, actual survey work, compilation and analysis of data and finally report writing. The time requirement at each stage needs to be worked out systematically. Such study will indicate the time requirement of the whole project. It is essential to note that, too long period for the completion of research work is undesirable, as the conclusions and recommendations may become outdated, when actually available. Similarly, time-consuming research projects are not useful for solving urgent problems (say, soil salinity problems faced by the farmers, marketing problems faced by the pesticide dealer). While executing these activities of the research design, the researcher should critically note that, they should be completed well-in right time. This also brings certainty as regards the completion of the whole research project in right time.

- ☆ As discussed above, planning the activities involved in the research design both in terms of costs and time enable the researcher to keep effective check on the whole research work during the implementation stage. This makes him to accomplish the research tasks both effectively and efficiently.
- ☆ The various sources of the data or the information should be very clearly stated by the researcher.
- ☆ The researcher should employ right techniques for collecting the requisite data and information from the selected sources.
- ☆ The researcher should select the sample for the investigation and it should be adequate and representative of the population.
- ☆ Besides the above steps, the researcher should have thorough know-how about the influence of various risk factors that are detrimental for the study and also the measures to overcome the same.

8.3. Components Dealt in Research Design

The following are the components dealt in research design of a survey research method in an agro-economic analysis of a research problem:

A. Preparation of sampling design
B. Identification of variables relevant to the objectives framed in the study
C. Types of data to be collected
D. Identification of the sources of data

A. Preparation of Sampling Design

In social sciences research, the researcher conducts survey research method to elicit requisite data and information from the respondents for the objectives framed for the problem. Sampling design refers to the methodology followed for selecting a reliable sample for conducting the research study. This is essential because, it is very difficult on the part of the researcher to elicit information from all the units of the population (called Census survey) and hence, he goes for sampling procedure to select a sample representative for the population so that, the analytical findings of the sample can be attributed and related to the overall population units. Before discussing the different types of sampling techniques, it is essential to understand the following basis concepts of sampling design:

i. Universe or Population

The aggregation of all units of any research project will constitute Universe, whereas, the aggregation of all units of a particular research topic constitute population. That is, a finite group is called Population, whereas a non-finite (infinite) group is called Universe. However, these two terms are often used interchangeably. A population may be defined as the totality of elements in any specified group of individuals or objects. It is, otherwise, defined as the entire aggregation of items from which samples can be drawn. 'N' represents the size of the population. Basha and Harter (1980) state that *"a population is any set of persons or objects that possesses at least one common characteristic."* For example, all the farmers cultivating paddy in Kurnool district of Andhra Pradesh, all the books in a library, all the workers in an agro-processing factory etc., are examples of population. Note that, a population may be finite or infinite. A population is said to be finite, if it consists of a limited number of elements so that, it can be easily counted. For instance, the number of workers in an agro-processing factory, number of farmers cultivating paddy in Kurnool district of Andhra Pradesh, number of books in a library etc., fall under finite population, as these elements of population can be easily counted. But, an infinite population is one, whose size is unlimited and therefore, its elements or members cannot be counted. For example, the population of fishes in a river, the number of microbes in the soil etc., are examples of infinite population. Further, population can be categorized as Existent and Hypothetical. An Existent population is one, which comprises of tangible items or units that can be counted. For example, the number of books in a library, the number of pesticides firms in Andhra Pradesh etc., as these items can be easily counted. A Hypothetical population is one, which is not existent *i.e.*, it is purely imaginary. The examples include number of stars in the sky, number of moves on a chess board etc.

ii. Sampling

It refers to the statistical process of selecting a subset (called a 'sample') of a population of interest for purposes of making observations and statistical inferences about that population. Sampling process is inevitable in social sciences research, as it is very difficult to address all the units of the population by the researcher considering time and cost constraints. In selecting a sample, the sampling process comprises of several stages. In the first stage, the researcher has to identify the target population. For example, if the researcher wishes to conduct a study on *'Economic analysis of paddy production and marketing in Kurnool district of Andhra Pradesh'*, obviously the target population is farmers cultivating paddy in Kurnool district. It is to be noted that, the target population (unit of analysis) may comprise of farmers, organizations, State, country or any other entity that the researcher wishes to draw scientific inferences about. In some cases, study about entire population units (Census survey) is obvious. For example, if there are only 50 farmers in Kurnool district practicing organic cultivation of paddy, then it is essential on the part of the researcher to consider all these 50 farmers for the survey to analyze about the economic aspects of organic paddy cultivation. Sometimes, it is quite difficult to understand about the target population. For example, if the researcher wishes to analyze the motivation factors that influence the farmers in adopting modern technologies in paddy, then what is the target population: farmers, training personnel, private input agencies, local agricultural officers, scientific community? The right answer in this case is the farmers cultivating paddy, as the researcher is interested in knowing their performance in adopting modern technologies of paddy production rather than the performances of training personnel, private input agencies, local agricultural officers, scientific community etc.

The second stage in the sampling process is to choose a 'Sampling frame'. It is essential to note that, in the above research study on *'Economic analysis of paddy production and marketing in Kurnool district of Andhra Pradesh'*, the researcher knows 'paddy cultivating farmers' constitute the units of population or analysis. But, the researcher may not get the information or access about all the paddy cultivating farmers. He may even contact the local agricultural officers or Gram Panchayat Offices to collect the information about the paddy cultivating farmers and still he may not get the complete information about population units cultivating paddy in Kurnool district. Moreover, some of the farmers for whom the information is available may not be accessed in the village or may not be interested in providing the requisite information to the researcher. So, this infers that, only a part of the population units can be accessed by the researcher for eliciting the requisite information and this part of the population constitute 'sampling frame' (Figure 8.2). But, it is essential to note that, sometimes the sampling frame may not be representative of the population and in such case the inferences derived by such a sample may not be generalizable to the population. Sometimes, it also happens in such a way that, the population from which a sample is drawn may not necessarily be the same as the population about which the researcher actually want information. For example, if a researcher wants to study the *'Economic analysis of organic paddy production in Nandyal region of Kurnool district in Andhra Pradesh'*, assume that, the population comprises of 70 small farmers, 35 medium farmers and 20 large farmers. But, after considering the sampling frame and if it contains only 10 small, 5 medium and 20 large farmers, then the analysis may not give reliable results, as the sampling frame comprises less number of small farmers as against its population units and hence, the results are more biased towards large farmers.

After identifying the sampling frame, the last step in the sampling process is to select a representative sample from the sampling frame (Figure 8.2). For this, there are various types of sampling techniques and they are discussed in-detail in the ensuing pages. So, now, sampling can be defined as the method or the technique consisting of selection for the study of the so called part or the portion or the sample, with a view to draw conclusions or the solutions about the universe or the population. According to Mildred Parton, *'Sampling method is the process or the method of drawing a definite number of the individuals, cases or the observations from a particular universe, selecting part of a total group for investigation.'* Sampling ensures various advantages over census survey as listed below:

- ☆ It enables researchers to make estimates of some unknown characteristics of the population in question.
- ☆ Requires less budget and consumes less time compared to census survey.
- ☆ High degree of accuracy and reliability (if sample is representative of population).
- ☆ Sampling may sometimes produce more accurate results than taking a census, as in the latter, there are more risks or errors due to the high volume of persons contacted and the number of census takers, some of whom may not be well-trained.

iii. Sample

The subset of total population or sampling frame that is selected for in depth investigation constitute sample. Since, the observations and statistical findings obtained from the sample are to be generalized back to the population, the selected sample should obey two important aspects *viz.*, it should be adequate in size and it should be representative of the population. Note that, as long as the sampled (surveyed) population is a high proportion of the target population, the results obtained should also be true for the target population. So, improper and biased sampling are to be avoided in deriving true information about the population units. Thus, a sample is made up of some of the units of the population. A representative sample of any population should be so drawn that, every unit of that population has a specified non-zero probability of being included in the sample. In other words, every unit of the population has a statistically equal chance of being selected into the sample. Sample size depends mainly on the size of the research study and the size of population. Although sample is a subset of population, it is suitable for research in terms of cost, convenience and time. The sample can be selected from the population based on a probability or a non probability approach. The size of the sample is represented by 'n'.

We know, the world is full of potential data, but the researcher is interested only on such data that are relevant and specific to the selected research topic. Even the amount of data available on the selected research topic is enormous, it is essential on the part of the researcher to limit the collection of data relevant to his study and for this, he has to select a representative sample from the population. When the research problem is specified and an appropriate research design is developed, the next step in the research process is to collect data and information relevant to the research topic under consideration. There are two ways of getting reliable information about the population. One way is to collect data from each and every unit of population. Of course, it is more time consuming and costly on the part of the researcher. Another important way is to draw a representative sample from the population (Figure 8.1) and collect information from each and every unit of the sample. The findings derived from the data analysis of sample can be generalized back to the population, as the selected sample is representative of the population. This method of addressing the research problem is more feasible, less time consuming and less costly on the part of the researcher. In agro-economic research, this sampling is more important. For example, if the researcher wishes to conduct research study on ''*Economic analysis of paddy production in Nandyal region of Kurnool district in Andhra Pradesh*', it is quite difficult on the part of the researcher to contact each and every farmer (marginal, small, medium and large farmers) to collect data about the cost of cultivation and returns from paddy crop. Hence, he will go for a sampling technique for selecting a sample (representative of total population (total number of farmers cultivating paddy)). However, the researcher requires special skills in drawing a representative sample from the population and to elicit relevant data and information from the sample so collected. So, in collecting the information from the sample, the researcher has to bear in mind an important question, how representative is the information collected about the selected sample to the whole population?

Features of Good Sample

A good sample must have the following characteristics:

- ☆ Chosen at random.
- ☆ Adequate in size to meet the needs of the research.

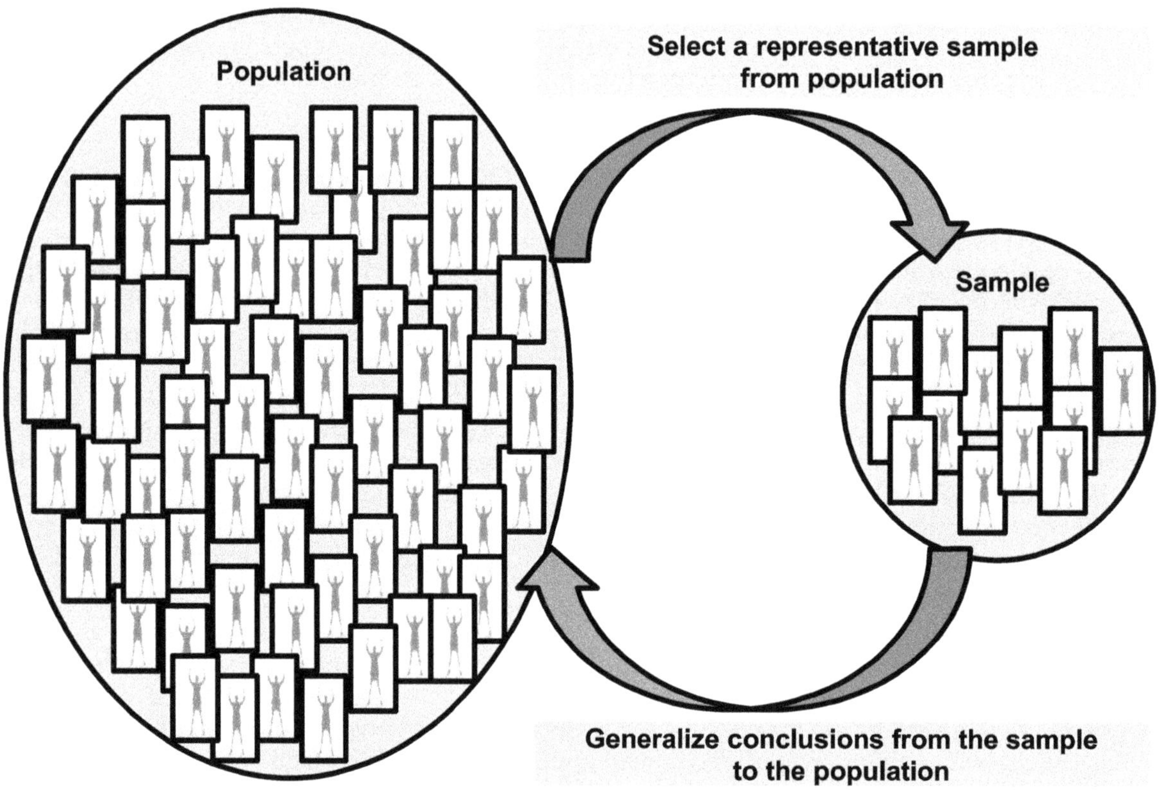

Figure 8.1: Selection of Representative Sample from the Target Population.

- ☆ Representative of its population so that, the sample statistics can be generalized.
- ☆ Unbiased.
- ☆ Units of sample should be independent and relevant.
- ☆ Units of sample should be complete, precise and up to date.
- ☆ Free from random sampling error.
- ☆ Avoid substituting the original sample for convenience.
- ☆ Every member of sample belongs to the population. That is, every individual in the sample bears the characteristics of the population.

iv. Sample Unit

It refers to the single respondent of selected sample from which the required data and information are collected.

v. Sample Design

The technique or scheme followed for selecting the sampling units from the population is called Sample design.

vi. Statistic and Parameter

A Parameter is a characteristic of a population, but a Statistic is a characteristic of a sample. For example, the mean income of all farmers in Kurnool district constitute a Parameter. The mean income computed from a sample of farmers of Kurnool district constitute a Statistic. Similarly, the word Parameter is used to indicate other technical measures like median, mode, standard deviation, correlation etc., as these measures describe the characteristics of a population. When we work out the same measures for samples drawn from the population, then they are known as Statistics, as they are based on the characteristics of a given sample. Thus, sample estimates are called Statistics (a 'statistic' is a value that is estimated from observed data of a sample). Since, it is difficult to study the entire population, population characteristics are always unknown, and are called Parameters (and not 'statistic' because, they are not statistically estimated from data). Sample statistics may differ from population parameters, if the sample is not perfectly representative of the population; the difference between the two is called Sampling error. If the researcher gradually increases the sample size so that, the sample approaches representative to the population, then sampling error will decrease and a sample statistic will increasingly approximate the corresponding population parameter. So, if a sample is truly representative of the population, then the estimated sample statistics should be identical to corresponding theoretical population parameters.

vii. Sampling Frame

As discussed earlier, that part of the total population from which the researcher selects a representative sample constitute Sampling frame. Thus, sampling frame is a list of all those population elements that will be used in drawing the sample. Often, the list does not include the entire population. The discrepancy is often a source of error associated with the selection of the sampling frame (Sampling frame error, Figure 8.2A). For example, the list of farmers (say, cultivating paddy) available in Gram Panchayat Office, list of names available in a telephone directory etc., constitute a sampling frame. From this sampling frame, the researcher has to draw the sample. So, it can be inferred that, population size>Sampling Frame size>Sample collected (Figure 8.2).

So, the researcher has to take the necessary care to keep the sampling frame up to date and also it should be free from the various errors of the omission and duplication of the sampling units. In many

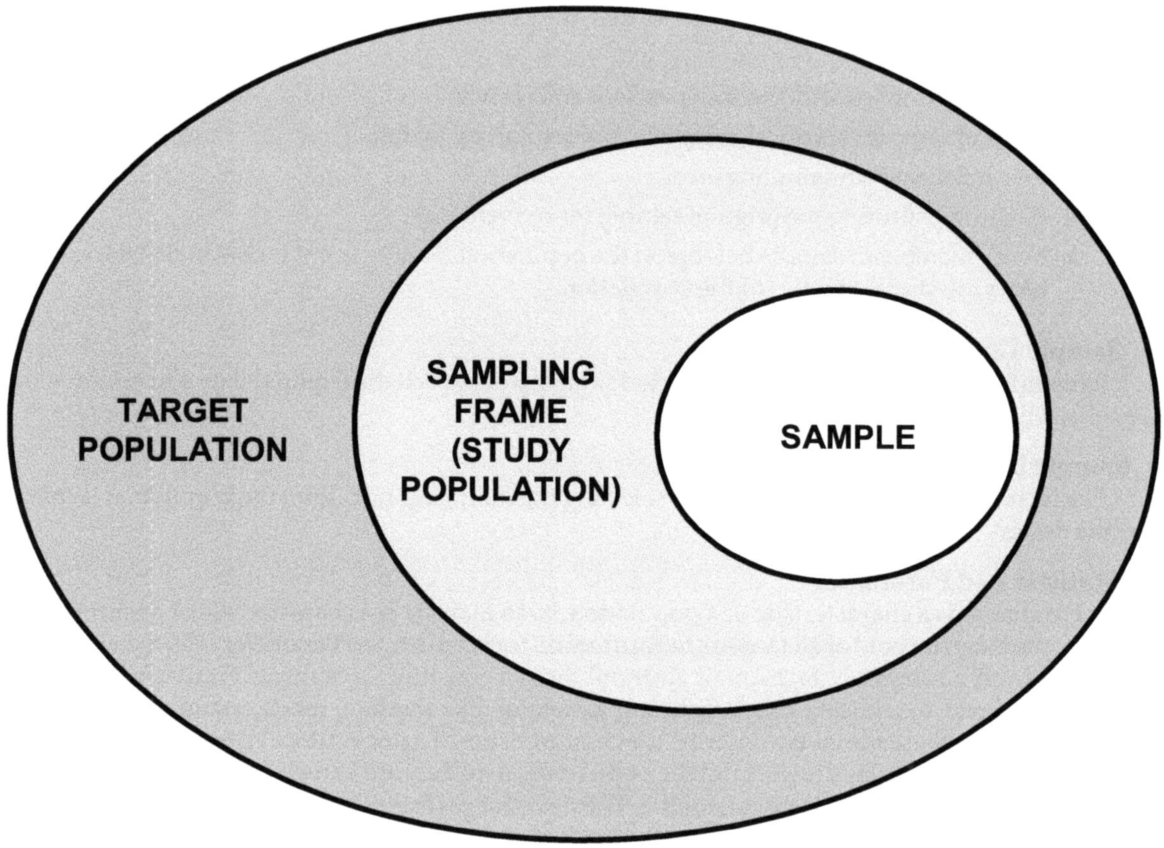

Figure 8.2: Selection of Representative Sample from the Sampling Frame.

cases, it has been observed that the preparation of the sampling frame sometimes becomes a very critical practical problem. A perfect sampling frame is the one that is able to identify each element only once and these frames are very rarely available in the real life. In general, a sampling frame may consist of different types of defects and these defects can be classified on the following lines:

- ***Incomplete frame:*** In this case, some of the sampling units of the population are omitted.
- ***Inaccurate frame:*** In this case, some of the sampling units of the population are inaccurately listed or sometimes, such units are involved which actually are not existing.
- ***Inadequate frame:*** In this case, the sampling frame does not include all the units of the population by its structure.

viii. Sampling Fraction

This is given by f = n/N. For example, a researcher wishes to select 40 sample farmers from a population of 800 farmers cultivating maize in a remote village. But, what does it mean to select 40 out of 800? Which proportion of the population are the researcher is selecting? To calculate the proportion of farmers that the researcher will interview can be obtained by dividing the sample size by the population size *i.e.*, 40/800 = 0.05, and this means that, the researcher selected 5 per cent of the population. So, sampling factor is the quotient or ratio between the size of the sample and the size of the population *i.e.*, n/N and is expressed in percentage. If this quotient is multiplied by 100, we get the percentage of the population represented in the sample. Sampling fraction is also called as Sampling intensity.

ix. Elevation Factor

Now the researcher is going to estimate, how many farmers in the population are represented by each unit sample farmer. This is given by, Elevation factor = N/n = 800/20 = 40. So, this mean that, each of the farmer in the sample represents 20 farmers in the village. So, 'Elevation factor' is the quotient between the size of the population and the size of the sample (*i.e.*, N/n). It represents the number of elements existing in the population for each element of the sample.

x. Sampling Error

As discussed earlier, the total difference between population parameter and sample statistic constitute Sampling error or Total sampling error. It is, otherwise, defined as the absolute value of the difference between an unbiased point (sample) estimate (statistic) and the corresponding population parameter. It arises because, the researcher collected the data for a sample rather than from all the units of a population. The total sampling error can be conveniently divided into two types *viz.*, Random sampling error (or Error variance or Non-systematic error) and Non-sampling error (Systematic Error).

- ***Random Sampling Error:*** Random sampling error is the difference between the population parameter and sample statistics due to random selection of sample from the population. So, random sampling error arises only due to errors made, while selecting the sample from the population. Random sampling error is otherwise defined as the 'difference between the sample result and the result of a census conducted using identical procedures' and is the result of chance variation in the selection of sampling units. If sample is selected properly (for *e.g.*, through the technique of randomization), the sample is usually deemed to be a good approximation of the population and thus capable of delivering an accurate results. Usually, the random sampling error arising from statistical fluctuation is small, but sometimes this error can be significant. In general, the sample selected from the population is not true representative of the population and hence, deviations arise between population parameter (say, population mean) and sample statistic (say, sample mean). So, random sampling error is evident. As the random sampling error is the outcome of chance, the laws of probabilities are applicable to it. So, the greater the sampling error, the lesser is the representativeness of the sample to its corresponding population and *vice versa*. So, the researcher often aims at minimizing the random sampling error in the research study. If this error is considerably high, then the researcher has to increase the sample size, so as to make the sample representative to its population and to minimize the error. That is, the random sampling error is inversely proportional to the sample size. Although this error cannot be avoided altogether, it can be controlled through careful sample designs, large samples, and multiple contacts to assure representative response. It is essential to note that, no two samples drawn from the same population will be identical, as they will be composed of different units or individuals and hence, they will have different sample statistics. Thus, random sampling error varies across different samples drawn from the same population. Thus, a random sampling error occurs when sampling is done from the population and it increases with increase in non-representativeness of the sample to its population. Random sampling error occur randomly and are equally likely to be in either direction (positive or negative). The magnitude of the random sampling error depends upon the nature of the population. The more homogenous the population, the smaller is the random sampling error and as discussed above, it is also inversely related to the size of the sample.

☆ ***Non-sampling error:*** Besides random sampling error, selection of incorrect research (sampling) design, improper collection (sometimes interviewers may not pose the questions properly) or measurement of data, no measurement of data or collection of data from only some sampling units, induced biasedness in getting responses etc., also contribute to the deviation between population parameter and sample statistic. The error component that is contributed by these factors other than due to non-representativeness of sample to its population (*i.e.,* Random sampling error) constitute Non-sampling error or Systematic error or Measurement error. That is, Non-sampling error includes non-observation errors and measurement errors. Non-observational errors occur when data cannot be collected from some of the sampling units and when improper data are collected from the sample units. Measurement errors include improper tabulation and processing of collected data. Non-observational errors can be further sub-divided into Non-coverage errors and Non-response errors. In probability sampling, each unit in the population has a fixed probabilistic chance of being selected into the sample. So, Non-coverage error occurs, when a unit in the population has no chance of being selected into the sample. Non-response error occurs, when no data are collected from the unit that is entered into the sample. This may be due to the refusal of the respondent because of his non-availability during the survey period, language barriers, health problem, non-cooperation from him etc. So, selection of faulty sampling frame result in a Non-response error (Non-sampling error). The faulty sampling frame exists, when some non-responsive respondents are included and responsive respondents are excluded from the sampling frame.

Both Random sampling errors and Systematic (non-sampling) errors reduce the representativeness of a sample and consequently the value of the information which is derived by the researchers from the sample units does not reflect the true population (Figure 8.2A). Among these two errors, the most frequent cause of the sampling error is due to non-representative sample selection from the population *i.e.,* biased sampling procedure (Random sampling error). So, every researcher must seek to establish a sample that is free from bias and is representative of the entire population and thus, be able to minimize or eliminate random sampling error. When two researchers conducts similar studies with the same population and same sampling design, the study with a larger sample size will have less random sampling error compared to the study with smaller sample size. This is because, as the sample size increases, it approaches the size of the entire population and therefore, it almost represents the characteristics of the population, thereby, random sampling error decreases (Figure 8.3). Thus, total sampling error includes:

❖ ***Over coverage:*** Inclusion of data from outside of the population.

❖ ***Under coverage:*** Sampling frame does not include elements in the population.

❖ ***Measurement error:*** *e.g.* when respondents misunderstand a question, or find it difficult to answer.

❖ ***Processing error:*** Mistakes in data coding.

❖ ***Non-response***: Failure to obtain complete data from all selected individuals.

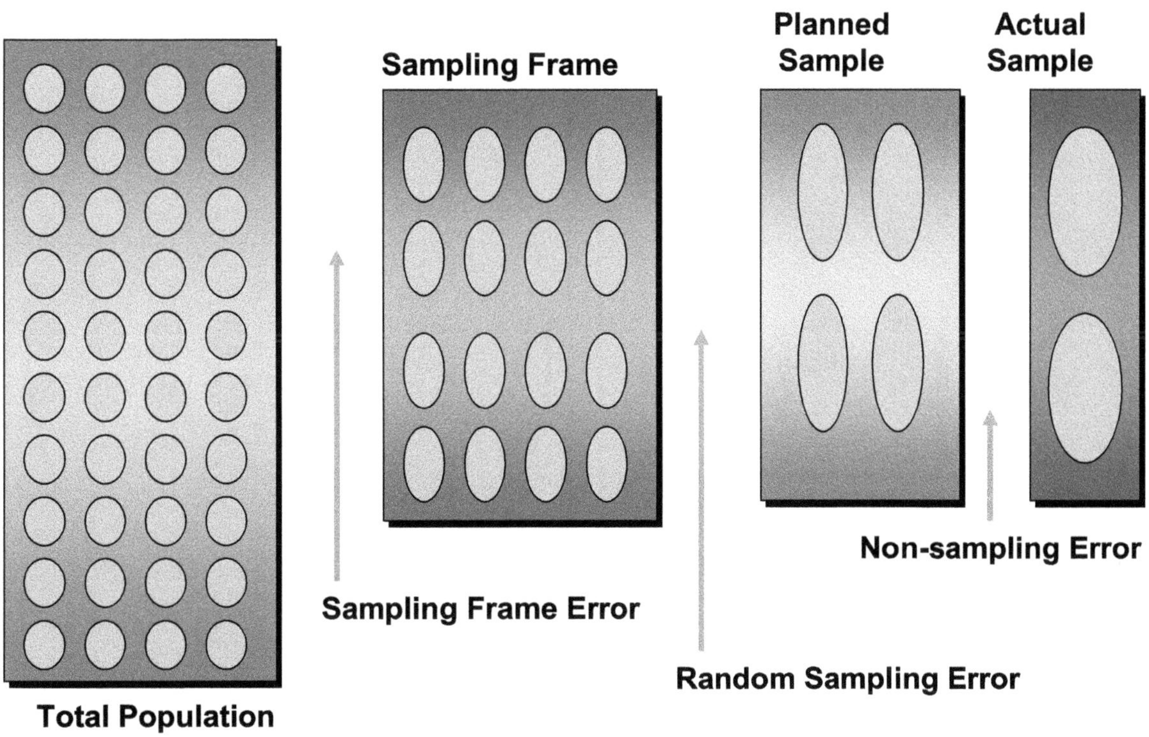

Figure 8.2A: Different Types of Errors in Sampling Process.

Since sample surveys take into account only a part of the total population for in depth study, there is always a certain amount of inaccuracy in the information obtained. This inaccuracy is said to be Sampling error or error. The concept of sampling error can be understood from the Figure 8.3A.

The following is the formula employed to compute sampling error for a known population size (N):

Sampling error = $(1/\sqrt{N}) \times 100$

Methods to Reduce Sampling Error

The following are the two ways by which sampling error can be reduced:

- ***Increasing Sample Size:*** If the researcher increases the sample size for a given population, the chance of occurrence of the sampling error will be less and there will be no sampling error, if the researcher goes for complete enumeration study (*i.e.*, if the sample size and the population size coincide). Thus, sampling error is in inverse proportion to the sample size.
- ***Stratification:*** When all the population units are homogeneous, the researcher can draw a small sample and this can be representative of the population. But, when the population is heterogeneous, drawing a perfect sample becomes impossible. In such a situation, the researcher has to go for stratification of heterogeneous population units into different homogeneous population units (called Strata) and from each stratum, he can draw a random sample representative of the population in each stratum. This method of sampling (Stratified random sampling) is discussed in-detail in the ensuing pages.

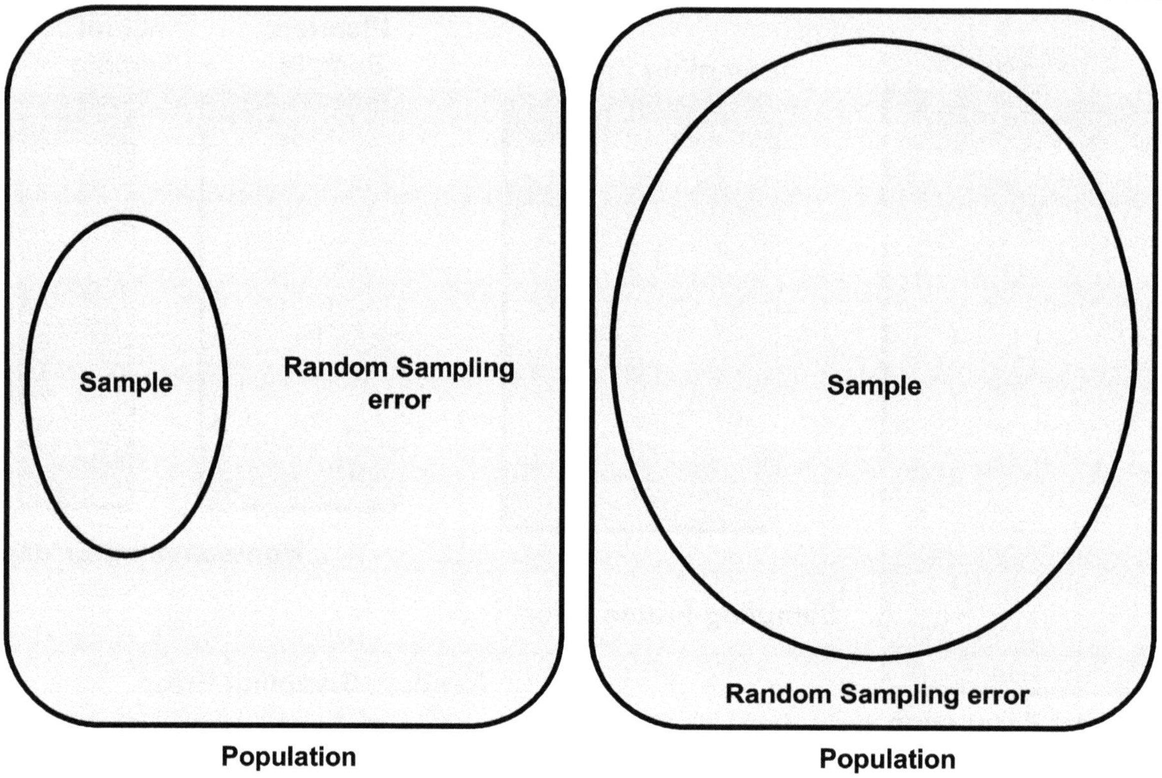

Figure 8.3: Relation between Sample Size and Random Sampling Error.

Relation between Random Sampling Error and Variance

Variance is the mean of the squared deviations of observations from their mean value. Therefore, if the sample has high variance, it follows that sample also has high random sampling error. We know, the higher the variance, the greater is the standard deviation and hence, the higher is the random sampling error. So, we can relate sample size and standard deviation of the sample and note that, with increase in sample size, the standard deviation decreases. For example, the researcher has taken a sample of only 10 plants from a population of 1000 units. So, with this very little sample size, there is more deviation across the observations and hence, standard deviation will be high. On the other hand, if the researcher increases the sample to 300 plants, the tendency of observations is to cluster and hence, the standard deviation decreases. When the researcher confronts with the situation of high random sampling error, the only way to eliminate this error is by conducting an enumeration study. Of course, this is not at all feasible considering the time, energy, money and other resources on the part of the researcher. So, the best way to limit the random sampling error is by taking a representative sample from the population by following suitable sampling technique.

xi. Census Study

An inquiry in which the entire population is considered for study is referred to as a Census study or Inquiry or Enumeration study. 'Population' here refers to all the units or items under consideration. However, it is a time consuming and costly process and is, therefore, not usually conducted by the

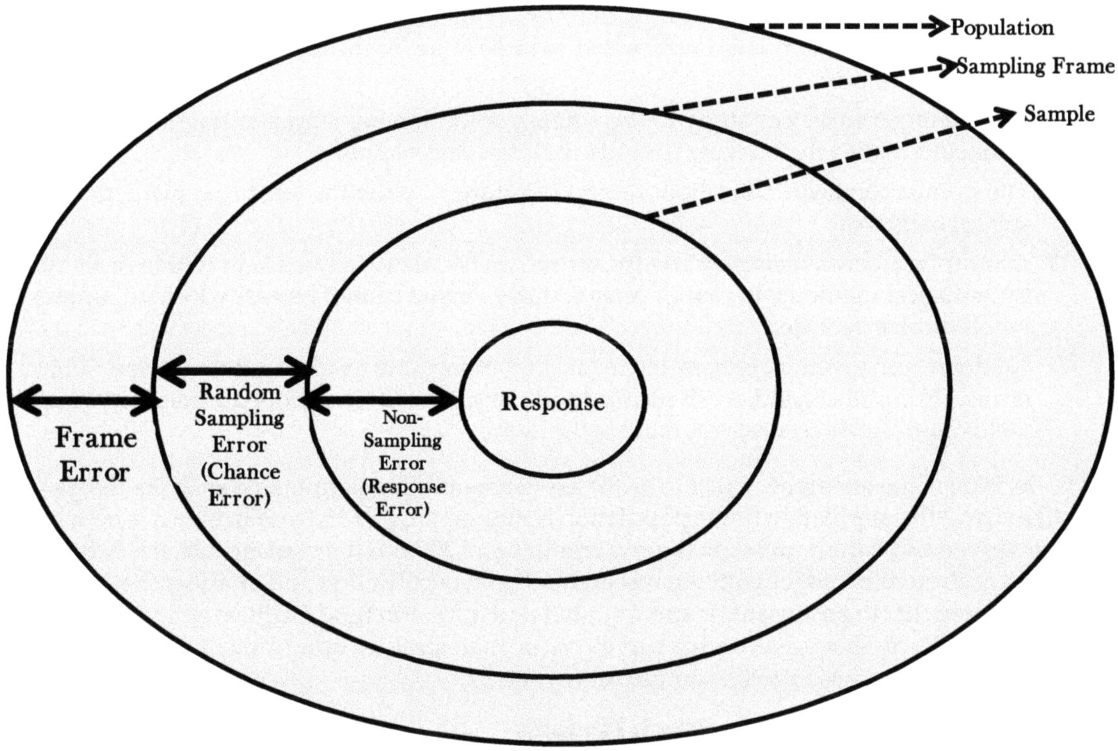

Sampling Error = Frame Error + Chance Error + Response Error
Total (survey) error = Sampling error + Non sampling error

Figure 8.3A: Components of Sampling Error.

researchers in social sciences research in agriculture. The other limitations include failure in generating a complete and accurate list of all the members of the population and some of the members refute to provide requisite information to the researcher. In such cases, the researcher often resorts to sampling. A sample is that part of a population or sampling frame that is actually considered for in depth study. While selecting the sample, efforts should be made by the researcher such that, the selected sample is not biased, it should be adequate and is as representative of the population as possible. The general notion that, a census study generates more accurate data than sampling is not always true. The national population census is an example of census survey.

Census Study vis-à-vis Sample Study

There are major differences between census study and sample study and the choice between these two studies depends upon:

- ☆ When a large area are to be covered, sample study has to be done to save time, funds and labour.
- ☆ In some areas like forests, there will be varieties of crops composition and density and in such case, census study have to be carried out.
- ☆ To estimate population parameters, census study have to be carried out.

- If resources like budget, time and labour are available in adequate quantity, census study have to be carried out. This implies that, sample study is cheaper and less time consuming than census study.
- Conducting a survey or study to the whole population has a greater possibility of having erroneous results than having to study only selected sample.
- The population bears the whole subject of interest, while the sample is just a part of the subject of interest.
- In sampling, checking is easy and the errors can be calculated and kept within desire limits by statistical methods. In case of census study, errors cannot be easily located, unless the whole work is repeated.
- No doubt, census study is expected to yield most accurate result; it is nevertheless subject to errors. On the other hand, when planned and carried out to an appropriate intensity, sampling yields sufficiently accurate estimate of the items.

The decision of the researcher either to go for a census survey or sample survey depends upon the factors like size of the population (if the population is limited around 150 to 200, then it is better to go for Census survey and if the population is very large beyond 200 and not manageable, it is better to for sample study), available budget with the researcher for data collection and analysis (lower budget compels the researcher to go for sample survey), staff and infrastructural facilities with the researcher (when these facilities are extensive, census survey can be manageable), time element (if the time period is shorter, the researcher has to go for sample survey) etc.

xii. Precision and Accuracy of Research Findings

Accuracy is a measure of how close a sample estimate or statistic is to the population parameter. But, Precision is a measure of how repeatable the result is. For example, if the sample mean is 98 and population mean is 100, the researcher can infer about the accurate measurement of population mean with the help of drawing a representative sample. So, he can make a statement about the sample accuracy, as the sample mean is accurate to population within two units. Sampling accuracy is usually expressed as a relative index in percentage form (*i.e.* between 0 and 100 per cent). When sample size increases and becomes representative to the population, sampling accuracy also increases.

On the other hand, Precision refers to how close the sample estimates (statistics) are, when different hypothecated random samples are taken from the same population. For example, the standard error is a measure of precision. When the standard error is small, estimates from different samples will be close in value and *vice versa*. So, precision is inversely related to standard error. When the standard error is small, sample estimates are more precise; when the standard error is large, sample estimates are less precise. When sample size increases, precision also increases as a result of decreasing variability (Figure 8.4).

Thus, 'Accuracy' describes 'close to true value' or 'trustworthiness of the sample statistic', whereas 'Precision' describes 'reproducibility' or 'repeatability'. It is to be noted that, a sample estimate is considered 'valid', if it is both accurate and precise. The sample estimates may not have both precision and accuracy (Panel A of the Figure 8.5) or the sample estimates may have accuracy but may not have precision (Panel B of the Figure 8.5) or have precision but may not have accuracy (Panel C of the Figure 8.5) or the sample estimates may have both precision and accuracy (Panel D of the Figure 8.5. This can be well explained through the Figure 8.5 illustrating the performances of above four cases at the shooting gallery.

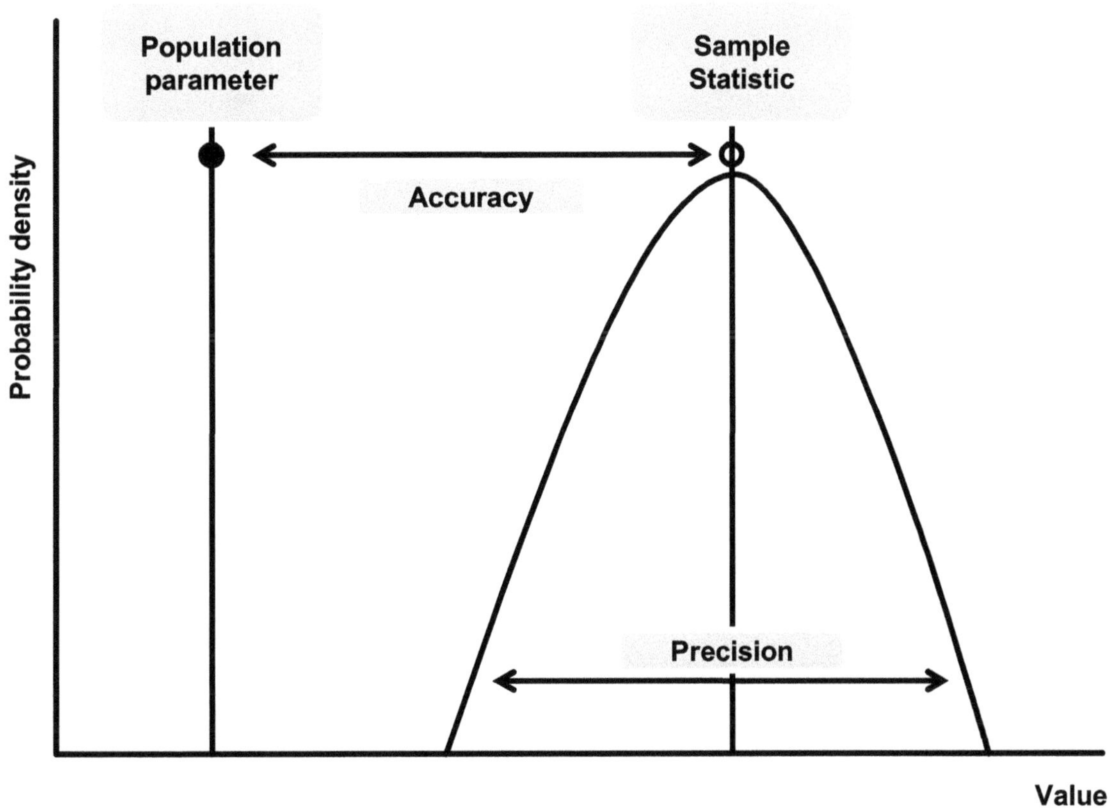

Figure 8.4: Accuracy and Precision of Research Findings.

xiii. Bias

Bias is the term refers to, how far the average statistic estimated lies from the true parameter. When a researcher conducts a survey repeatedly by drawing several random samples from the same population, he will get different sample statistics with each replication. We know that, each of the different sample statistic would be an estimate for the *same* population parameter. The individual sample statistic may differ from the population parameter, but the average of all the statistics from all possible samples may get equal to the true population parameter. If that is the case, the bias is said to be zero or the average sample statistic is unbiased. If there is any difference between true population parameter and average sample statistic, then the sample is said to be biased. Thus, bias refers to the difference between population parameter and average statistic of different samples. This bias can take different forms. Bias often occurs when the survey sample does not accurately represent the population. The bias that results from an unrepresentative sample is called Selection bias. This selection bias may arise due to under coverage of some of the respondents, non-response of some of the respondents etc. Poor measurement process can also lead to bias. In survey research, the measurement process includes the environment in which the survey is conducted, the way the questions are asked and the state of the survey respondent. From this discussion, now, it is quite easy to differentiate between bias and sampling error and they are listed below:

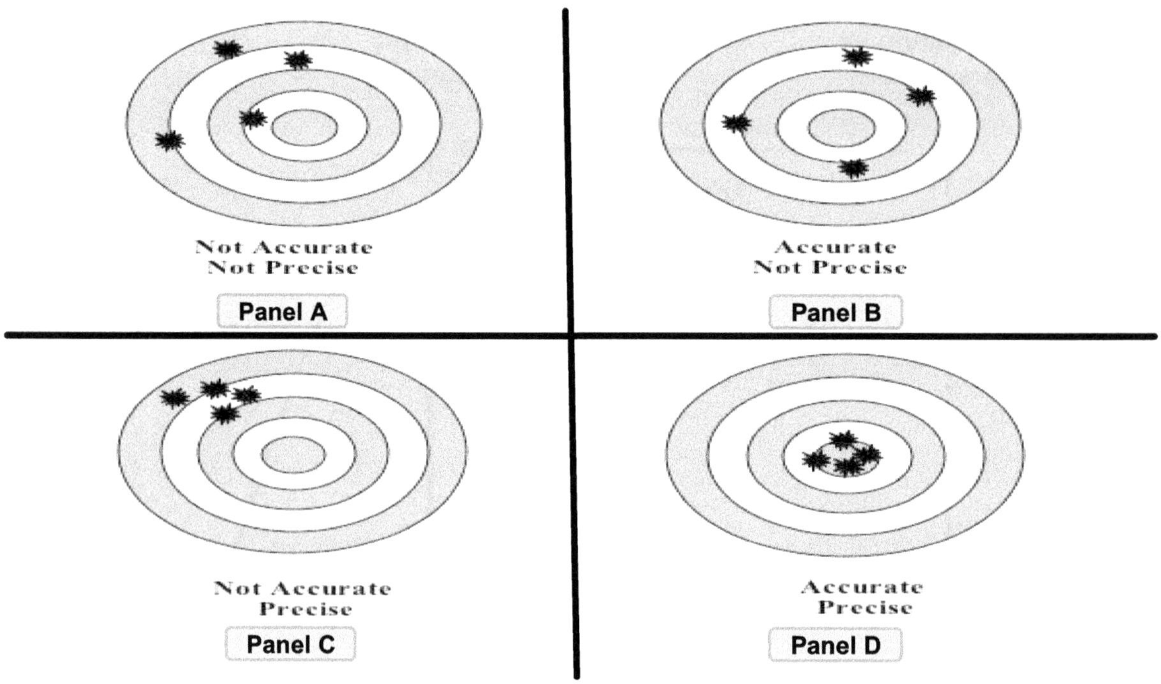

Figure 8.5: Different Possibilities of Accuracy and Precision of Research Findings.

- The difference between sample statistic and its corresponding population parameter is called Sampling error. But, the difference between average sample statistic and its corresponding population parameter is called Bias.
- Increasing the sample size tends to reduce the sampling error *i.e.*, it makes the sample statistic less variable. However, increasing sample size does not affect survey bias. That is, larger sample size increases precision, but does not guarantee absence of bias. Therefore, the researcher may be better off with smaller sample size, less precision, but much less bias.
- Bias is due to factors other than sample size, unlike sampling error. Therefore, a large sample size cannot guarantee absence of bias.
- Bias cannot be precisely measured, whereas sampling error can be precisely measured after the fact. So, control and prevention of bias requires careful attention, whereas sampling error can be easily controlled by selecting a proper sampling technique and by selecting a representative sample size.
- Bias is due to mistakes made by the researcher himself and hence, can be avoided, whereas sampling error is unavoidable unless the researcher goes for enumeration study.

The following are the important sources of bias that should be considered when selecting a sample, by whatever method:

- Any changes from the pre-agreed sampling rules can introduce bias.
- Bias is introduced if distant respondents are purposively omitted from the survey.
- Replacing selected individuals with others, for example if they are difficult to contact, also introduces bias.

- Low response rates can introduce bias.
- If an out of date list is used as the sample frame, it may also introduce bias, if it includes the people who are not currently residing in the study area.

xiv. Response Rate

The response rate is the percentage of people in the sample selected for the study, who actually participate in the study.

xv. Samples and the Validity of a Study

A valid research study is one that finds the truth. There are two components of validity that have been identified viz., External validity and Internal validity. We can use these components of validity as criteria to see if a particular study is valid.

- ***External validity:*** This validity is dependent on the adequacy of the sample. If the sample is representative of the target population, then the results of the study will generalize the population. This is called Generalizability. To have a generalizable sample, first define target population, then randomly select a large sample. With a random sample of sufficient size, research findings can generalize to the larger population. As a rule of thumb, random sample sizes of twenty subjects per group are minimally sufficient.
- ***Internal validity:*** This refers to the adequacy of the study design and the degree of control, the researcher have exercised in data gathering. Good internal validity is ensured by controlling all the variables except the dependent variable. In this way, if the dependent variable changes during the study, then that change is due to the changes the researcher made in the independent variable(s). The concept of control has six major parts: In evaluating

External Validity and Internal Validity

Population	→ Random sample	→ Good external validity
	→ Non-random sample	→ Reduced external validity
Sample	→ Random assignment	→ Good internal validity
	→ Subjects matched	→ Good internal validity
	→ Non-random assignment	→ Reduced internal validity

Consequences of:

Good external validity	Reduced external validity
Results will generalize	Results will NOT generalize

Good internal validity	Reduced internal validity
Groups are equal at start, and you can attribute changes to the independent variables.	Groups are not equal at the start, and thus changes could be due to initial differences, NOT the independent variables.

Source: http://www.sahs.utmb.edu/pellinore/intro_to_research/wad/pop&samp.htm

the internal validity of a research study, the researcher should pose himself the following question: *'was the manipulation of independent variables is the only possible cause of a change in the dependent variable?'* The researcher should also control all the extraneous variables so that, it will allow him a good chance of detecting any change in the dependent variable by the manipulated independent variables, and will have internal validity. Note that, a study can have good internal validity and NOT find any changes in the dependent variable due to the independent variable. Also, a study can have good internal validity, but without a generalizable sample, it may have no external validity. Finally, remember that a study with no external validity still found true relationships for the sample that was studied. Do not confuse internal validity with the validity of the method by which the dependent variable is measured, called Test validity. Internal validity refers to the overall degree of control exercised. Test validity refers to the suitability of the measuring instrument used.

a. Steps Involved in Sampling Process

Having understood the basic concepts of sampling, it is high time now to discuss about the steps to be followed by the researcher to conduct sampling process and to derive a representative sample from the population (Figure 8.6).

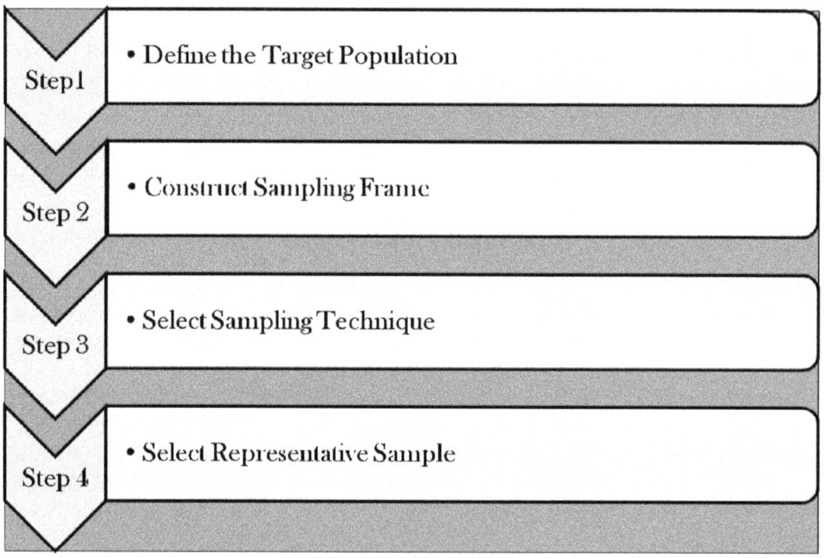

Figure 8.6: The Sampling Design Process.

Step 1. Define the Target Population

First, the researcher should define the target population from whom the requisite data and information are to be collected and for whom the research study is intended to. For example, if the researcher wishes to conduct a research study on *'Economic analysis of paddy production in Kurnool district of Andhra Pradesh'*, it may define the target population as 'farmers who cultivate paddy in Kurnool district'. This makes the research more focused, as the researcher will deal with the farmers cultivating paddy or paddy based cropping system and not other farmers cultivating crops other than paddy. The population so defined can be further refined by considering the farmers cultivating paddy in major paddy cultivating mandals of the district. Thus, defining the right target population reduces the probability of including the respondents, who do not fit well into the objectives framed for the research study.

Step 2. Construct the Sampling Frame

After deciding the target population (*i.e.*, paddy cultivating farmers), now the researcher has to focus on the sampling frame *i.e.*, he has to collect the list of farmers cultivating paddy or paddy based cropping system in the major paddy cultivating mandals of Kurnool district. So, a sampling frame contains the list of farmers from which the researcher has to collect the representative sample. However, in practice, it is quite difficult to get an exhaustive sampling frame that exactly fits the requirements of a particular research. In general, researchers use easily available sampling frames with the Department of Agriculture, Gram Panchayat Offices etc. Even private input agencies also provide the sampling frame, as they will be in touch with the farmers through supplying credit, inputs like seeds, fertilizers, pesticides etc. Sometimes, maps and aerial pictures are also used as sampling frames. Whatever may be the case, an ideal sampling frame is one that have a complete and detailed list of the respondents so that, it facilitate the researcher to select the representative sample. If the sampling frame does not accurately represent the total population, it may lead to selection of non-representative sample and this finally leads to Random sampling error.

Step 3. Select Sampling Technique

After collecting the sampling frame, now the researcher has to choose a sampling technique (discussed in the subsequent section) to draw a representative sample from the population. A range of sampling techniques are available to the researcher (Figure 8.7) and the choice of the sampling method is influenced by the objectives of the research study, availability of financial resources, time constraints, and the nature of the problem to be investigated. All sampling techniques can be grouped under two broad heads *viz.*, probability and non-probability sampling techniques.

Step 4. Select Representative Sample

With the selected sampling technique (say, Simple Random Sampling (SRS)), the researcher has to select the representative sample from the sampling frame. This representative sample or sample size plays a crucial role in the sampling process. This is because, after the analysis of collected data and information, the sample statistic should be generalized about the population parameter. In deriving the representative sample, the researcher should carefully outline the sample design. That is, he should outline the way through which the final (representative) sample units will be selected, so as to elicit requisite data and information from them. He should also outline the *modus operandi* of the sampling plan *i.e.*, he should pose himself various questions like: how to contact the sample farmer when he is not readily available in the village, how to cross check the data and information provided by the sample farmer and many other such questions need to be answered for the smooth execution of data collection process. A well-designed sampling process will enable the researcher to contact the selected sample units and draw requisite data and information from them in right time and thus, ensure the research process to move on the right track.

b. Determining Sample Design

We know, going for a census study is very difficult on the part of the researcher in conducting social sciences research in agriculture. Hence, the researcher has to select a sample segment from the population to represent the population as a whole. Ideally, the sample should be adequate in its size and should be representative of the population such that, the sample estimates correspond to the targeted population. Designing the sample calls for three decisions:

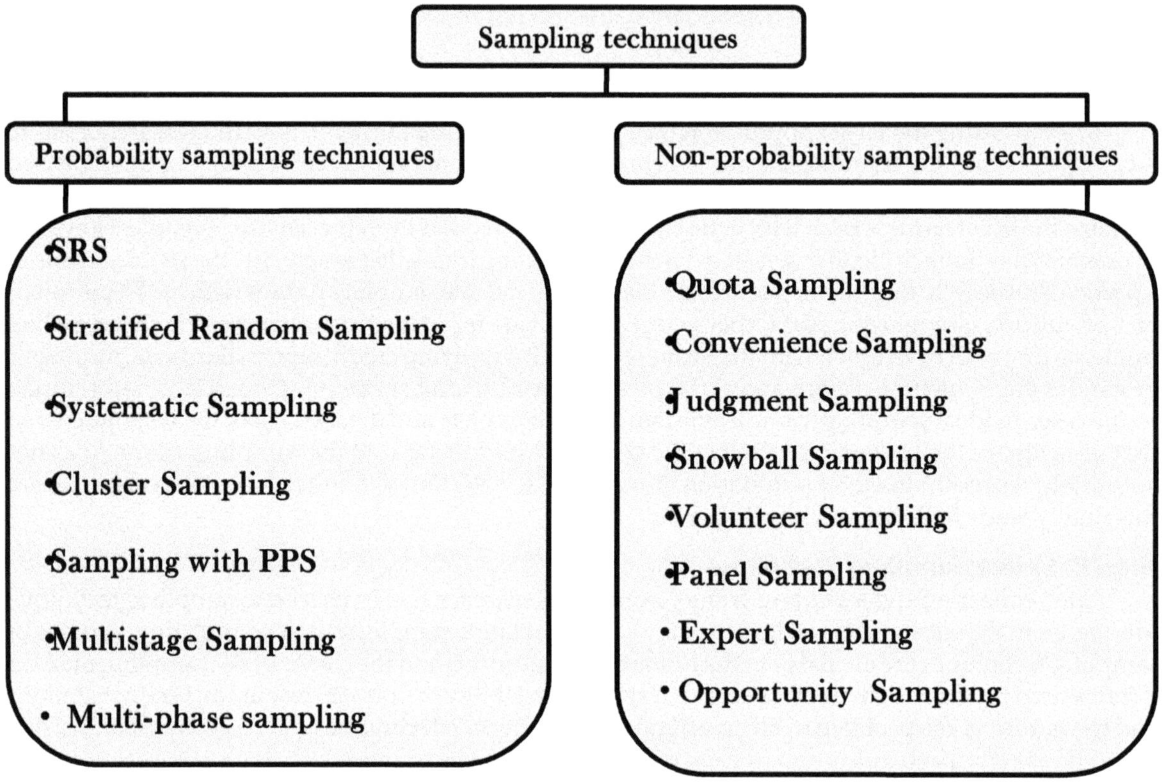

Figure 8.7: Different Types of Sampling Techniques.

- ✰ ***Who will be surveyed? (The Sample):*** The researcher must determine what type of information is needed and who is most likely to have it.
- ✰ ***How many people will be surveyed? (Sample Size):*** Large samples give more reliable results than small samples. However, it is not necessary to collect data and information from the entire target population.
- ✰ ***How should the sample be chosen? (Sampling):*** Sample units may be chosen at random from the entire population (called probability sampling methods) or the researcher might select the sample (say, farmers), who are easier to obtain information from (Non-probability sample).

Note that, depending upon the nature of the project and the goals and objectives of the project, the nature of sampling design varies. The above discussion infers that, sample design can be defined as a definite plan followed for drawing a sample from a target population. It also refers to various techniques followed in selecting the representative sample from the target population. Note that, the sample design must be determined before the data collected.

Characteristics of Good Sample Design

They include:

- ✰ The sample design must ensure representative sample for target population.
- ✰ The sample design must ensure that the sampling error is small.

- ☆ The sample design must be selected keeping in view of the availability of funds with the researcher.
- ☆ The sample design must ensure to apply appropriate statistical techniques, so as to draw valid conclusions.
- ☆ The sample study must ensure that the results have validity and can be generalized to the target population.

c. Types of Sampling Techniques

Sampling technique refers to the methodology or procedure followed for selecting a representative sample from the population. According to Mildred Parton, *'Sampling method is the process or the method of drawing a definite number of the individuals, cases or the observations from a particular universe, selecting part of a total group for investigation.'* As discussed above, the 'best' sample design selected depends upon the objectives framed for the study, resources available with the researcher, time frame for conducting research etc. Whatever the design the researcher follows, he ultimately aims at getting accuracy and precision of the sample statistics. He will aim at selecting a sample that is representative of the target population. That means, the information he derives from the analysis of the sample is the same (allowing for inevitable variations in the estimates due to imprecision) as he would find, if he carried out a census of the target population. So, when selecting a sampling method, some prior knowledge about the target population is essential and based on this and with certain reasonable assumptions, he can plan a sample size required to achieve a reasonable estimate (with acceptable precision and accuracy) of population characteristics. Sampling methods can be categorized according to the approach they take to the probability of a particular unit being included in the sample. Accordingly, there are two broad methods of sampling used by researchers *viz.*, Probability (or Random) sampling and Non-probability (Non-random or Judgment) sampling.

I. Probability (or Random) Sampling

As the name indicates, this method of sampling depends on the theory of probability*. The selection of the units into a sample from a population is done depending on the probabilities that are known. Probability sampling is a technique in which every unit in the population has a chance (non-zero probability) of being selected in the sample, and this chance can be accurately determined. Sample statistics thus produced, such as sample mean or standard deviation, are unbiased estimates of population parameters, as long as the sampled units are weighted according to their probability of selection. Since the units are selected at random, every unit or element of the population has a known probability of being included in the sample. That is, every unit in the population has a chance (greater than zero) of being selected in the sample, and this probability can be accurately determined. Note that, the probability of selection need not be equal for units of the population. If the purpose of a research is to arrive at conclusions or make predictions affecting the population as a whole, then the choice of a probabilistic sampling approach is desirable. For example, if we have 100 units (N) in the sampling frame, and if the researcher wishes to collect 10 units into the sample (n), then the probability of each unit being selected as sample is one in ten, or 0.1 (assuming each unit has the same chance). There are various probability sampling methods that we can use to select the sample units. The important aspect here is, in case of probability sampling, each time we apply the same method to the same sampling frame, we will generate a different sample. So, from a finite population (N), we can

---

* Probability may be defined as the relative frequency of an event occurring usually reported as a percentage or fraction. The probability of a rolling a 5 on dice is 1/6 because there are six possibilities on lines. The probability of a tail on a single toss of coin is ½ or 0.5. Probability of 0.05 means that, an event is likely to occur 5 times out of every 100 times, if it takes place in the same situation.

draw NCn different samples, each of which contains n sampling units. In fact, in probability sampling, we are concerned with the probability of each sample being chosen, rather than with the probability of choosing individual units. If each sample is as likely to be selected as every other sample (assuming equal probabilities), then each sampling unit automatically has the same chance of being included as every other sampling unit. In this method, it is possible to determine, which sampling unit belong to which sample and the probability of each sampling unit to be selected. The other merits of this method of sampling include: it yields a representative sample (so that, the findings of the sample can be generalized back to the population) and facilitate estimation of sampling or error (and hence, randomization can be employed to overcome this sampling error). The following are the important types of probability sampling (Figure 8.7):

A. SRS
B. Stratified Random Sampling
C. Systematic Sampling
D. Cluster Sampling
E. Sampling with Probability Proportional to Size (PPS)
F. Multistage Sampling
G. Multi-phase sampling

All the above probability sampling methods have two things in common *viz.*, every unit or element in the population has a known non-zero probability of being sampled and involves random selection at some point. The method chosen depends on a number of factors, such as the available sampling frame, scatter of elements in the sampling frame, cost of the survey, analytical techniques to be employed, time frame for the study etc. When choosing a probability sample design, the goal of the researcher should be to minimize the sampling error of the estimates, while simultaneously minimizing the time and cost of conducting the survey. The above probability sampling techniques differ in terms of sampling efficiency, which is a concept that refers to trade off between sampling cost and precision. Precision is inversely related to sampling errors, but directly related to cost. The greater the precision, the greater the cost and there should be a tradeoff between sampling cost and precision. The researcher is required to design the most efficient sampling design in order to increase the efficiency of the sampling.

II. Non-probability (Non-random or Judgment) Sampling

In this method, the researcher cannot specify the probability of selection of each unit or element of the population to be included in the sample. That is, some elements of the population have no chance of selection (these are sometimes referred to as 'out of coverage'/'under covered') into the sample. The researcher selects items to be drawn from the population based on his or her judgement about the characteristics of these units and how well these units represent the whole population. The researcher may follow on or more of the following criteria in selecting the sample units:

- the units are self-selected.
- the units are the most easily accessible (in geographical terms).
- the units are selected on economic grounds (say, only farmers from local village are selected compared to distant villages).
- the units may be selected in a 'typical' manner say, on first fifty farmers, who enter the regulated market).

It is clear that, in this method of sampling, the researcher depend on unreliable and unquantifiable factors. The chance of an item being included in the sample is not influenced by the probability or randomness of selection, but by the characteristic of the item being judged by the researcher. So, this method is considered as 'inferior' to probability method because, the selected sample is not representative to the target population. So, through this method, the analytical findings cannot be generalized back to the population. Further, this method suffers from sampling error or bias and this will distort results. But, this method of sampling is simple and less expensive to use compared to probability sampling. This method also assumes more significance, when the researcher wishes to conduct a pilot study about a particular research problem. Similarly, in situations, where the sample cannot be generated by probability method, this non-probability method may be unavoidable and it should be regarded as a 'last resort' for designing the sample selection. The following are the important types of Non-probability sampling:

A. Quota sampling
B. Convenience sampling
C. Judgment Sampling
D. Snowball Sampling
E. Volunteer Sampling
F. Panel Sampling
G. Expert Sampling
H. Opportunity Sampling

Comparing the above two methods *viz.*, Probabilistic method and Non-probabilistic method, each method has its own advantages and disadvantages. As mentioned above, the main advantages of non-probability sampling are simple, convenience and cost-effective. However, in this method of sampling, the researcher cannot make probability statements about the units to be included in the sample. Hence, it is not possible to compute the sampling error and a confidence interval for an estimate or a region of acceptance for a hypothesis test. However, reverse was the case for probability sampling. That is, this method of sampling indicate the confidence interval for an estimate or a region of acceptance for a hypothesis test. So, the researcher can indicate, whether the estimated statistic can be generalized to the population parameter or not *i.e.*, he can measure to what extent, the statistic differs from a population parameter. The following factors usually influence the choice of the sampling method:

- ☆ Nature and size of the sampling frame
- ☆ Accuracy and precision at which the results (sample estimates) are expected
- ☆ Analytical techniques to be employed for the data
- ☆ Duration of the research project
- ☆ Budgetary allocation available to the research project
- ☆ Costs to be incurred

Differences between Probability and Non-Probability Sampling Methods

Whatever type of sampling method that is followed, the aim of the sampling process is to draw a part of the target population and to conduct in-depth study on it to make estimates (statistics) of some unknown characteristics of the population in question (Figure 8.8A). Table 8.1 highlights the differences between Probability sampling and Non-probability sampling methods:

Now let us focus on these two methods of sampling in-detail.

Table 8.1: Probability Sampling vis-à-vis Non-probability sampling.

Probability (Random) Sampling	*Non-Probability (Non-Random) Sampling*
The selection of the units into a sample from a population is done depending on the probabilities that are known.	This method of sampling does not consider probability approach for selecting the sample and the selection is due to various criteria like characteristics of these units, easy accessibility of the units, presence of those units at the time of survey, responsiveness of units etc.
Probability sampling methods must meet two criteria: ☆ Every unit in the population must have an equal opportunity of being chosen for the sample (*equality*). ☆ The selection of one unit is not affected by the selection of previous units (*independence*).	Non-random sampling methods lack one or both of these criteria.
Allows use of statistics, tests hypotheses.	Exploratory research, generates hypotheses.
Probability samples are selected in such a way as to be representative of the population. Sample statistics are unbiased estimates of population parameters.	Population parameters are not of interest.
It is possible to compute the sampling error and a confidence interval for an estimate or a region of acceptance for a hypothesis test.	It is not possible to compute the sampling error and a confidence interval for an estimate or a region of acceptance for a hypothesis test.
Eliminates bias.	Adequacy of the sample can't be known.
The advantage of probability sampling is that sampling error can be calculated. (Sampling error is the degree to which a sample statistic might differ from the population parameter). So, when inferring to the population, sample estimates are reported plus or minus the sampling error.	In non-probability sampling, the degree to which the sample differs from the population remains unknown.
Must have random selection of units.	Do not have random selection of units.
Complex, time consuming and not cost-effective compared to Non-Probability (Non-Random) Sampling.	Cheaper, easier, quicker to carry out.
In survey research, most sampling designs are based on Probability approach.	Comparatively, non-probability sampling methods are less followed in survey research and they are especially used in special cases like conducting pilot (case) study.
Probability sampling is more complex, more time-consuming and usually more costly than non-probability sampling. However, because units from the population are randomly selected and each unit's probability of inclusion can be calculated, reliable estimates can be produced along with estimates of the sampling error, and inferences can be made about the population.	Comparatively, non-probability sampling is less time consuming and cost-effective, but the sample estimates are not representative or generalized to the population.

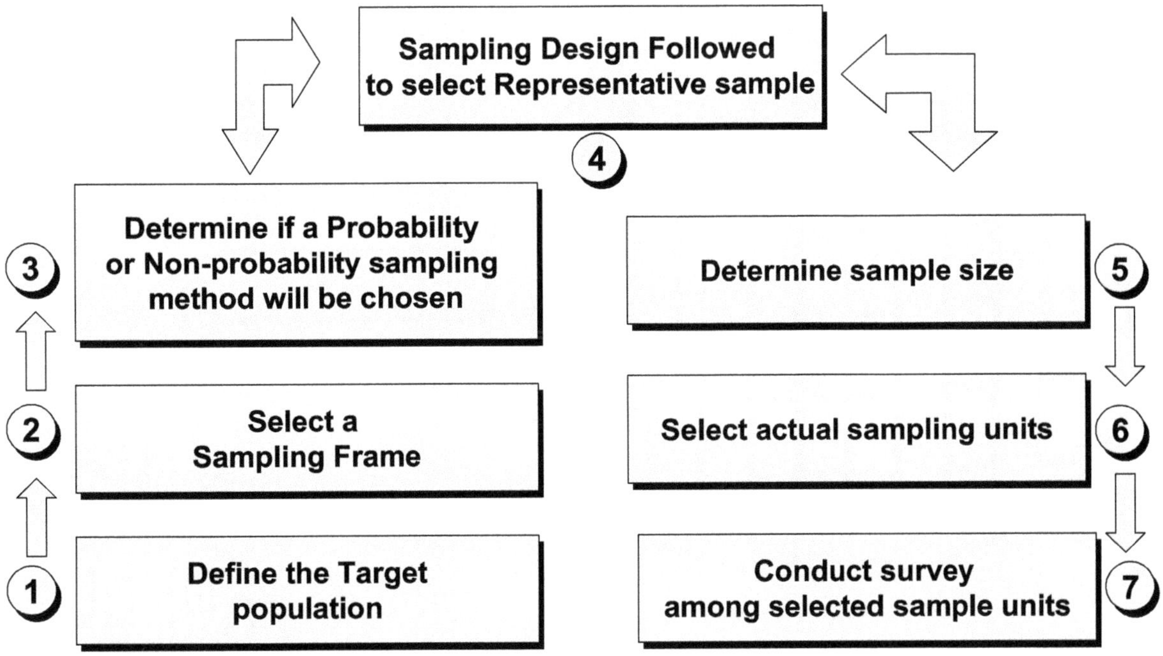

Figure 8.8A: The Sampling Process.

I. Probability (or Random) Sampling

a. *SRS*

This is the simplest and most common technique of drawing a sample by using probability from a population. It is the most recognized probability sampling procedure. Moore and McCabe define a simple random sample as follows: *'A simple random sample of size 'n' consists of 'n' individuals from the population chosen in such a way that every set of n individuals has an equal chance to be the sample actually selected'*. The term 'random' has a very precise meaning. Each unit in the population of interest has an equal chance or likelihood or probability of selection. This technique helps to select 'n' units (from a population of size N) such that, every unit of the NCn possible samples has an equal chance of being chosen. That is, the sample is drawn in such a way that, every unit or element in the population has an equal chance or probability of being included in the sample at any stage during the sampling process (Figure 8.8B). Further, the selection of each unit into the sample is independent of the selection of every other unit *i.e.*, the selection of one unit does not affect the chances of any other unit. This method resembles a lottery method, wherein a chits of names are placed in a box, the box is shuffled, and the names of the winners are then drawn out in an unbiased manner. For example, 'N' farmers are cultivating paddy in Kurnool district, but the researcher requires only 'n' number of units in a sample (*i.e.*, $n<N$). So, every farmer is given a number from 1 to N (or 0 to N-1) and random numbers are generated, either electronically or from a table of random numbers. So, SRS is always an Equal Probability of Selection (EPS) design. But, note that, not all EPS designs represent SRS. So, each unit being selected into the sample is free from personal bias. The sample, thus, obtained is referred as the Random Sample. This type of sampling is highly suitable, if the population is small, if the list of the units or elements in the population can be made *i.e.*, sampling frame can be prepared and if all the units of a sampling frame are homogeneous. For simplicity sake, assume there are 11 units in the population

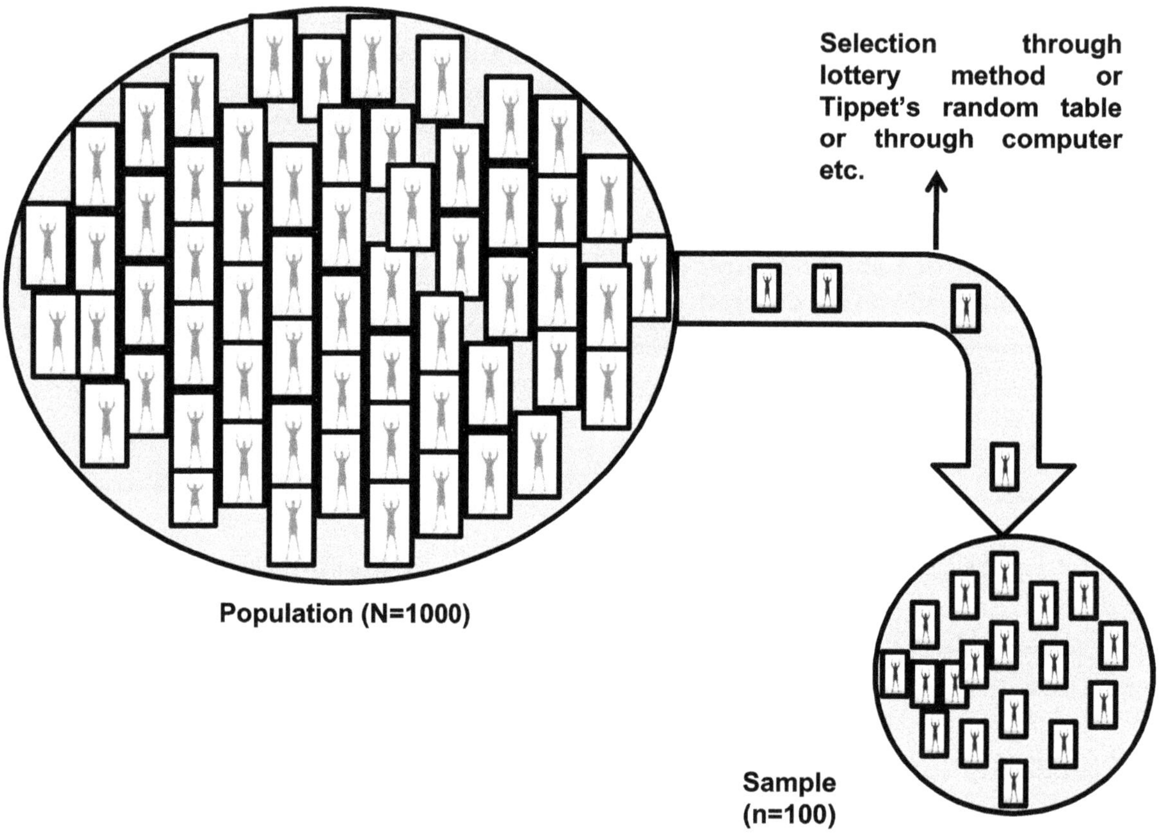

Figure 8.8B: Selection of Random Sample from Population by SRS.

and hence, these units are numbered from 1 to 11. The researcher wishes to select a sample of four units from this population. To do so, four random numbers need to be selected from a random number table, as found in most statistics text books and the procedure is shown through Figure 8.9. The random numbers table shown consists of six columns of two-digit non-repeatable numbers listed in random order. The intention is to sample four numbers between 1 to 11 *i.e.*, the total number of units (11) in the population. Starting at the top of column A and reading down, two numbers are selected, 2 and 5. In column B there are no numbers between 1 and 11. In column C the first random number is 8. In the fourth column, the appropriate number is 10. Thus, in this example, the randomly selected numbers are 2, 5, 8 and 10 and the same are shown as selected sample in Figure 8.10. In this example, the list of population units numbered refers to Sampling frame. This list should be numbered in sequential order from one to the total number of units in the population. Since the random numbers are mutually exclusive (*i.e.*, there are no duplicates), each unit or element in the population is only sampled once. This is called as SRS without replacement (SRSWOR) and this is discussed here under. Whenever a unit is selected for the sample, the other units of the population are equally likely to be selected. It must be noted that, the probability of selecting the first unit or element is not to be compared with the probability of selecting the second unit. When the first unit is to be selected, all the units of the population have the equal chance of selection which is (1/N). When the second unit is to be selected, all the existing N-1 units of the population have 1/(N-1) chance of selection.

A	B	C	D	E	F
73	86	**8**	50	88	37
16	63	91	72	71	81
76	75	77	92	84	23
2	35	22	70	62	90
78	33	64	55	36	46
56	53	95	97	43	67
17	26	15	14	16	18
31	98	29	47	98	52
51	87	30	**10**	13	19
58	39	83	96	94	74
5	29	21	66	24	38
28	61	82	99	79	59
57	27	12	44	80	39

Figure 8.9: Selection of Simple Random Sample from a Random Number Table.

SRSWOR and SRS with Replacement (SRSWR)

SRS has two distinct flavours *viz.*, SRSWOR and SRSWR.

- ☆ *SRSWOR:* The above example illustrates the concept of SRSWOR. That is, in SRSWOR, once an element or unit is sampled, the same is not placed back in the population for re-sampling. So, there is no question of repetitive occurrence of the same unit of population in the sample. Thus, SRSWOR removes previously selected units from the population before subsequent selections and hence, avoids the possibility of the same unit appearing in the sample more than once. That is, in SRSWOR, after every drawing of the selected sample unit, the same is not returned back to the population. The probability of selection of an element or unit into the sample gradually decreases after each draw, as the selected sample units are distinct. The probability of selection of first unit is given by 1/N, for second unit 1/(N-1), for third unit 1/(N-2) and so on. Strictly speaking, the sample selected by without replacement is called simple random sample.
- ☆ *SRSWR:* In SRSWR, once an element or unit is sampled, the same is placed back in the population for re-sampling. So, there is a chance of repetitive occurrence of the same unit of population in the sample again. Thus, SRSWR does not remove previously selected units from the population before subsequent selections and hence, the unit already selected earlier can be re-selected again and the same unit may appear in the sample more than once. That

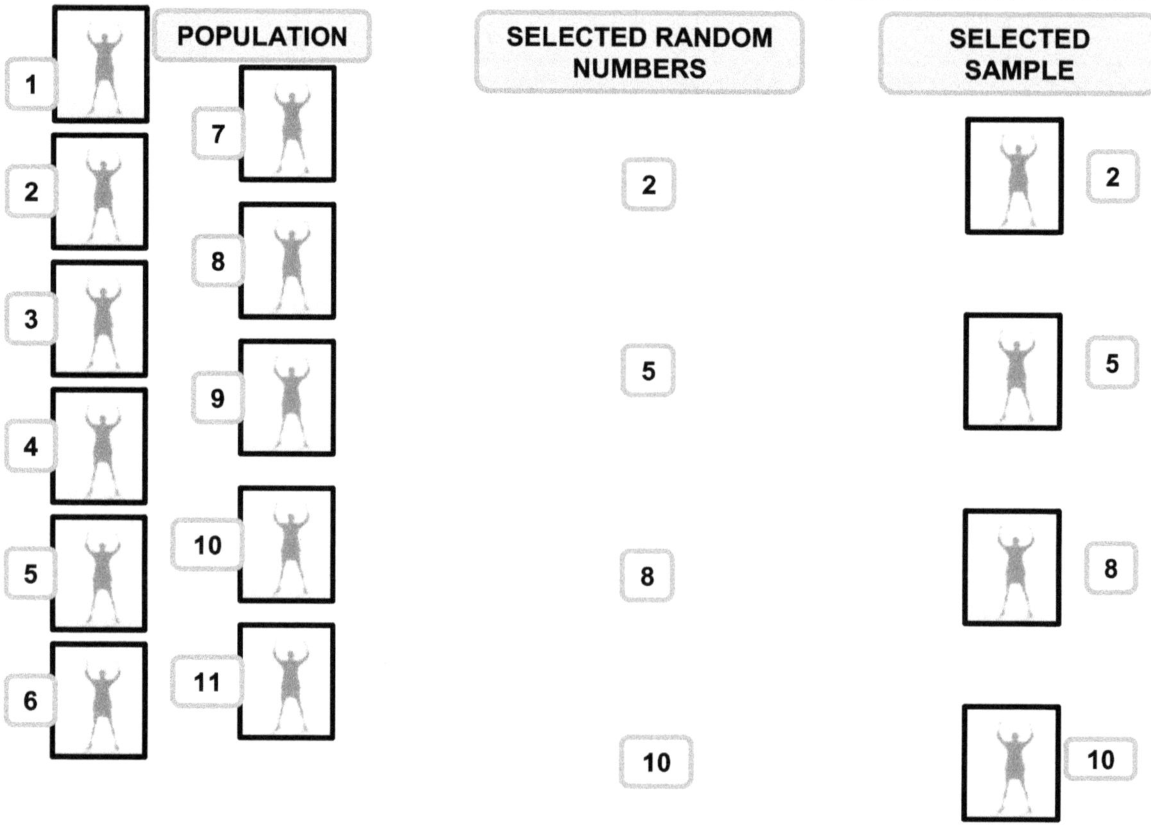

Figure 8.10: Selection of Four Random Sample Units from a Population of 11 Units.

is, in SRSWR, after every drawing of the selected sample unit, the same is returned back to the population before the next draw to be made. So, the probability of selection of an element or unit into the sample remains unchanged after each draw and this is given by 1/N. Although SRSWR can be conducted instead, this is less common compared to SRSWOR. When a small sample is drawn from a large population, SRSWOR is approximately the same as SRSWR, since the odds of choosing the same individual twice is low. That is, though these two methods are different, when the size of the population is infinite, or it is so big that we can consider that it is infinite, both methods will lead us to similar conclusions. Nevertheless, if the sampling fraction 'n/N' is greater than 0.1 (we sample more than 10 per cent of the population), the difference between the conclusions we get may be important.

The examples of these procedures are shown through Figures 8.11 and 8.12 for the selection of four units of sample from a population of 11 units. Since there are four units in the sample, the selection procedure has four steps and they are listed here under:

Step 1: It is the selection of the first sample unit.

Step 2: It is the selection of the second sample unit.

Step 3: It is the selection of the third sample unit.

Step 4: It is the selection of the fourth sample unit.

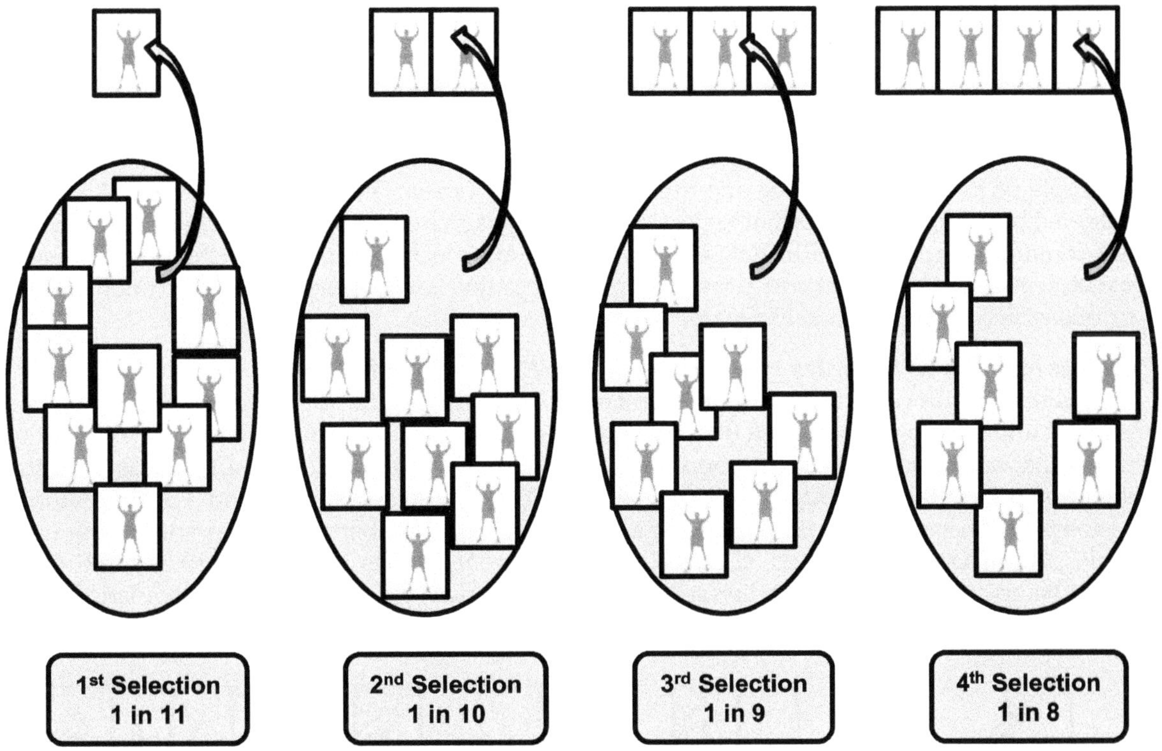

Figure 8.11: Selection of Simple Random Sample without Replacement.

In SRSWOR (Figure 8.11), the selection process involves four steps, as four sample units have to be selected from a population of 11 units. The steps include:

Step 1: It is the selection of first sample unit. The first unit gets selected at a probability of 1 in 11, and each unit in this step have an equal chance or probability of 1 in 11 (or 1/11) to enter into the sample. However, in the subsequent steps, the probability changes, as the first unit once selected is selected *i.e.*, it is not replaced back into the population.

Step 2: So, at this step, the situation changes. Once the first unit is chosen, it is not placed back in the population. Thus, at step two, the second unit to be sampled comes from the remaining 10 units in the population and hence, all of them have the same probability of being selected (*i.e.*, 1 in 10).

Step 3: At this step, the selection is derived from a population of nine units, with each unit having a probability of 1 in 9 of being selected.

Step 4: At this step, the selection is derived from a population of eight units, with each unit having a probability of 1 in 8 of being selected.

Once the steps are completed, the sample contains four different units. Thus, in SRSWOR, the probability of each unit being selected into the sample gradually decreases (whereas, it remains same in SRSWR). Unfortunately, the reduced selection probability from the first to the fourth step is at odds with statistical theory for deriving the variance of the sample mean. Such theory assumes the sample was selected with replacement. However, in practice, most simple random samples are drawn without replacement, since the researcher want to avoid the strange assumption of one unit (say, farmer) being surveyed again and again.

In SRSWR (Figure 8.12), all 11 units have the same probability of being selected (*i.e.*, 1 in 11) at steps one, two, three and four, since the selected unit at each step is again placed back into the population. With this form of sampling, the same unit could be sampled multiple times. In the extreme, the sample of four units could be one unit selected four times. The only reason, why SRSWR is considered because, it is a somewhat easier procedure to analyze compared to SRSWOR. But, SRSWR is less common compared to SRSWOR. Usually, the SRS approach is conducted without replacement because, it is more convenient and gives more precise results. For the purpose of descriptions, when we discuss SRS, we will refer to SRSWOR.

Possible Number of Samples can be Drawn by SRSWOR Technique

As discussed above, by SRS, many different number of samples (*i.e.*, many different combinations of sample units) can be drawn from the same population. Accordingly, in SRSWOR, the number of samples that can be drawn from the population is given by: NCn or [(N!)/n!(N-n)!], where 'N' is the number of units in the total population, 'n' is the number of units in a sample and '!' is the factorial notation for the sequential multiplication of a number times a number minus one, continuing until reaching one. That is, *N!* (termed '*N* factorial') is *N* times *N-1* times *N-2* and the like with the last number being one. For example, when selecting four units (say, farmers) from the population of 11 units, the number of samples could have been is given by: [(11!)/4!(11-4)!] = [(11x10x9x8x7!)/

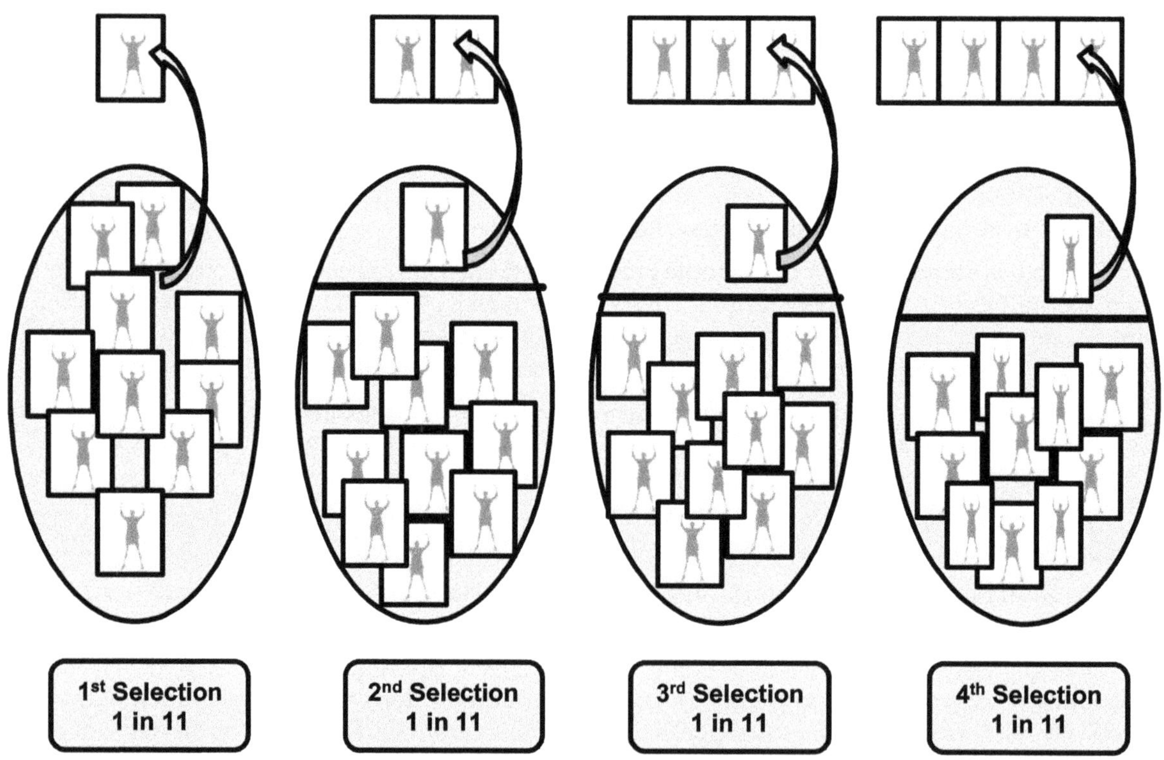

Figure 8.12: Selection of Simple Random Sample with Replacement.

(4x3x2x1(7!)]=[7920/24] = 330. The probability of selecting each sample is given by 1/NCn =1/ [(N!)/ n!(N-n)!] = (n!(N-n)!)/N! = 1/330 = 0.003. If samples are made with permutations, each sample has a probability of selection equal to 1/NPn.

Possible Number of Samples can be Drawn by SRSWR Technique

In SRSWR, the number of samples that can be drawn from the population is given by: N^n, where 'N' is the number of units in the total population and 'n' is the number of units in a sample. For example, when selecting four units (say, farmers) from the population of 11 units (A, B, C, D, E, F, G, H, I, J and K), the sample could have been ACEK or BEIK or ADGK or any of many other combinations (Figure 8.13). To be exact, in SRSWR, the total number of possible samples (each sample of four units) that can be drawn from a population of 11 units is given by 11^4 = 14641 samples. The probability of selecting each sample is given by 1/ N^n = 1/14641.

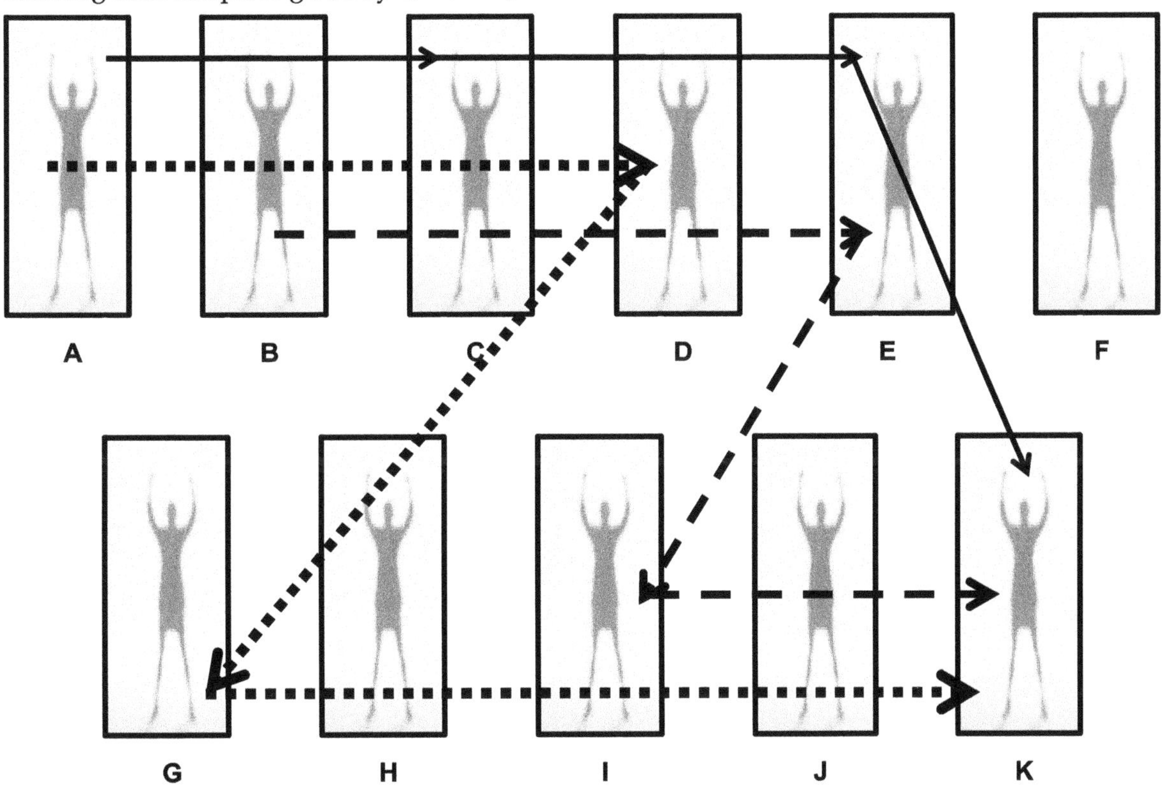

Figure 8.13: Number of Samples Collected through SRSWR Technique.

Methods of Selecting Simple Random Sample

Random sampling can be performed with the help of certain specific methods and some of these methods are discussed here under.

a. Lottery Method

This is an old classical method but powerful and modern method of selection are very close to this method. In this method, *a* lottery is drawn by writing a numbers or the names of various units of the sampling frame on paper chits or metallic balls and then put them in a container. Note that, the paper chits or the metallic balls should be of the same size, otherwise, the selected sample will not be truly random. The chits or balls are completely mixed and then certain numbers are picked up from the container. Those picked chits or balls are considered as sample units. In this manner, the number of slips drawn represents the sample size 'n'. But, this procedure would be tedious and the quality of the sample would depend on how thoroughly the researcher mixed the chits in the container. Perhaps, a better procedure would be to use the kind of a machine to mix the chits thoroughly in the container. This method of selection is commonly used, when size of the population is small. For a large population, there will be a big heap of paper chits or metallic balls and it is difficult to mix them properly.

b. Tippet's Numbers Method

This method was developed by L.H.C. Tippet. The random numbers table so developed comprises of list of numbers written at random on each page (Panel A of Figure 8.14). From these numbers, sample units are drawn at random. This is a very reliable and a dependable method. That is, the best way to choose a simple random sample is to use a random numbers table. Random numbers table is found in every statistics book. The procedure to draw simple random sample is explained through the following steps:

- ☆ ***Step 1:*** Assume there are 500 farmers in a village cultivating paddy. The researcher has to collect 50 farmers from that village to include them in a sample. So, the sampling fraction is f = n/N = 50/500 = 0.10 or 10 per cent.
- ☆ ***Step 2:*** The researcher has to assign numbers to these 500 farmers from 1 to 500.

Panel A : Random numbers table

92630	78240	19267	95457	53497	23894	37708	79862	76471	66418
79445	78735	71549	44843	26104	67318	00701	34986	66751	99723
59654	71966	27386	50004	05358	94031	29281	18544	52429	06080
31524	49587	76612	39789	13537	48086	59483	60680	84675	53014
06348	76938	90379	51392	55887	71015	09209	79157	24440	30244
28703	51709	94456	48396	73780	06436	86641	69239	57662	80181
68108	89266	94730	95761	75023	48464	65544	96583	18911	16391
99938	[illegible]	03621	[illegible]	[illegible]	[illegible]	[illegible]	[illegible]	[illegible]	[illegible]

Panel B : Selection of random sample units

92630	78240	19267	95457	53497	23894	37708	79862	76471	66418
~~79445~~	~~78735~~	~~71549~~	~~44843~~ →	26104	67318	00701	34986	66751	99723
59654	71966	27386	50004	05358	94031	29281	18544	52429	06080
31524	49587	76612	39789	13537	48086	59483	60680	84675	53014
06348	76938	90379	51392	55887	71015	09209	79157	24440	30244
28703	51709	94456	48396	73780	06436	86641	69239	57662	80181
68108	89266	94730	95761	75023	48464	65544	96583	18911	16391
99938	[illegible]	03621	[illegible]	[illegible]	[illegible]	[illegible]	[illegible]	[illegible]	[illegible]

Figure 8.14: Selection of Random Sample Units from Random Numbers Table.

- ☆ ***Step 3:*** Choose a starting place at random on a table of random numbers. Let us suppose, the researcher has chosen 5th column in 2nd row at digit '26104' for selecting the random numbers (Panel B of Figure 8.14).
- ☆ ***Step 4:*** Since the total number of units in a population (sampling frame) are 500, the researcher has to consider three digits in given random numbers. Read three digit numbers in horizontal direction and select the numbers as: 261, 046, 731, 800, 701, 349, 866, 675, 199, 723, 596, 547, 196 .... etc. Since, the numbers 731, 800, 701, 866, 675, 723, 596, 547 ... etc., are greater than population size of 500, ignore them. That is, the numbers outside the range of population units are ignored. So, the researcher considers 261, 046, 349, 199, 196 .... etc., numbers for selecting the sample. That is, members of the population (that are numbered earlier) that correspond to these numbers become members of the sample. Note that, the researcher may begin reading from anywhere in the random number table. However, once started, the researcher should continue to read across the row or down a column.

c. Using a Computer

Instead of using the above methods, recently with the development of inexpensive computers there is a much easier way to assign random numbers and to select the sample from the population. Microsoft Excel has a function to produce random numbers. This procedure involves the following steps:

- ☆ ***Step 1:*** List out the population (or sampling frame) units *i.e.,* their names (say, N = 500) and assume that, the researcher has to select 50 sample units (n) from N.
- ☆ ***Step 2:*** Since, many computer programmes can generate a series of random numbers, copy the list of clients (farmers) names into a column (say Column A) in an EXCEL spreadsheet.
- ☆ ***Step 3:*** In the second column (Column B), type the function '=RAND( )'. Enter this function into a cell and it will produce a random number in that cell. Drag the formula vertically (or even horizontally) and it will produce random numbers. The researcher can modify the formula to obtain whatever range he wish. For example, if he wants to have random numbers from one to 250, he could enter the following formula: =INT(250*RAND( ))+1, where INT eliminates the digits after the decimal, 250* creates the range to be covered, and +1 sets the lowest number in the range. Paired numbers could also be obtained using: =INT(8000*RAND())+1000.
- ☆ ***Step 4:*** Then, sort both columns: the list of names (Column A) and the random numbers (Column B), by the random numbers. This rearranges the list in random order from the lowest to the highest random number. Then, the researcher have to select the desired sample (n=50) from the first fifty names in this sorted list.

This method is quite simple, easily feasible and the researcher can complete the sampling process in the shortest time possible.

Steps in Creating a Simple Random Sample

The above explanation infers that, selection of a simple random sample involves six important steps as listed below.

- ☆ Define the target population
- ☆ Choose the sample size to be selected

- List the population units (or sampling frame)
- Assign numbers to the units of sampling frame
- Draw the random numbers (either through lottery method or Tippets random numbers of through computer)
- Select the sample of desired size that is representative of the population.

Merits of SRS

The advantages of SRS are explained here under. Many of these are similar to other types of probability sampling technique, but with some exceptions.

- Conceptually, SRS is the simplest of all the probability sampling techniques.
- The selected sample is free from bias.
- It minimizes sampling error (provided the sample is large in size and representative of the population) and it requires minimum advance knowledge of the population other than the frame.
- It is best suited for the situations, where not much information is available about the population and data collection can be efficiently conducted on randomly selected units.
- The cost of sampling is small compared to other types of probability sampling.
- It provides the sample that is highly representative of the population being studied.
- Since the units selected for inclusion in the sample are chosen using probabilistic methods, SRS makes the researcher to generalize the statistical inferences from the sample to the population. This is a major advantage because, such generalizations are more likely to be considered to have external validity.
- SRS is a basic type of sampling, since it can be a component of other more complex (probability) sampling methods.
- Unlike other probability methods like stratified random sampling, advanced auxiliary information on the units or elements in the population is not required.
- Each unit of population has an independent equi-probability of selection in SRS, unlike systematic sampling.
- Demands less complicated statistical procedures to analyze the data compared to those required for other probability sampling procedures.
- Since, SRS is a simple method and the theory behind it is well established, standard formulas exist to determine the sample size, the estimates and so on, and these formulas are easy to use.

Demerits of SRS

They include:

- Due to its very randomness, 'freak' results can sometimes be obtained in analyzing the data and these results are not representative of the population parameters. In addition, these freak results may be difficult to spot. So, the researcher has to increase the sample size to overcome such freak results.
- A simple random sample can be obtained only, if the list of the population (or sampling frame) is completely available. Other probability measures like cluster sampling do not

require a sampling frame of the elements of the target population. A complete list of sampling frame is difficult to obtain due to the following reasons:

- ❖ A number of lists of population units may be available from different sources and all these lists may not be alike. This creates lot of confusion among the researcher in selecting the sample units. Further, it consumes more time on the part of the researcher to compile the information from several lists and to prepare a genuine list. Master's level and Doctoral level dissertation students may not have sufficient time to do this.
- ❖ Sometimes, the population list may not be accessible, as it is protected by privacy policies or require a lengthy process to attain permissions.
- ❖ It may be more expensive on the part of the researcher to obtain population list.
- ❖ Though population list is available, the units may be difficult to contact especially, if they are geographically scattered.

☆ SRS technique requires that the population be up to date and complete and such details might be quite hard to get for large populations.

☆ This technique of sampling may generate a representative sample, when the population units are homogeneous in characteristics that the researcher is interested in the study. If the population units are heterogeneous, the sample units may not be representative of the population.

☆ In practice, SRS is rarely seen and especially it is considered difficult, when large samples are needed.

☆ SRS should be used with a homogeneous population. However, if the population units are heterogeneous, say farmers (small farmers, medium farmers and large farmers depending upon the size of their land holdings), simple random sample may not adequately represent the population (characteristics), unless the sample is representative in each category. That is, if the researcher is interested in choosing a sample to be representative of a population on the basis of the distribution in the population with reference to gender, age, land holding, economic status etc., a simple random sample will need to be very large to ensure all these distributions are equivalent to (or representative of) the population. (To obtain a representative sample across multiple population attributes, it is better to use the technique of stratified random sampling and this is discussed in the ensuing pages).

☆ No doubt, an unbiased random sample is important to represent the population. However, through SRS, there is no guarantee that, a particular sample is a perfect representation of the population. So, SRS can also be vulnerable to sampling error. This can be understood from the following example: A researcher wants to select 200 farmers from a village having 2000 farmers to study the cost of cultivation of paddy crop (*i.e.*, 10 per cent of total population). First, the researcher has listed the sampling frame and assigned the numbers from 1 to 2000. From Tippet's random numbers table, he has selected the random numbers (please see row-wise in Panel A of Figure 8.14). Since the population has only 2000 farmers, he is interested to consider only the last four digits of each number. So, he will select the sample units with numbers (starting with 3rd column and 2nd row randomly): 1549, 0701, 1966, 0004 etc. Applying this procedure, the researcher has selected the remaining numbers to be included in the sample till the sample of 200 farmers have been selected. But, there is no guarantee that, this sample will be representative of the population. This is because, say, the population consists of 50 per cent marginal farmers (1000), 30 per cent small farmers (600) and 20 per cent large farmers (400). But, the selected sample may not represent the population, as out of 200 farmers of sample units, 20 farmers may represent marginal farmers, 60 farmers may represent small farmers and 120 farmers may represent large farmers. If the researcher now believes to select the sample in accordance with the size of land holdings of farmers, he would not want to leave representation of these three sub-groups to chance. Stratified random sampling technique, discussed below, attempt to overcome this problem by using information about the (heterogeneous) population to choose a more representative sample. So, SRS may also be cumbersome and tedious, when sample is to be selected from an unusually large target population and that too, if the population units had different characteristics.

- SRS tends to have larger sampling errors and less precision compared to stratified random samples of the same sample size.
- Sample respondents may be widely dispersed and hence, data collection costs might be higher than those for other probability sample designs such as cluster sampling.

b. Stratified Random Sampling

This is also called as *Proportional* or *Quota* random sampling. This sampling offers significant improvement to SRS. As discussed above, SRS is followed when the population units are homogeneous in the characteristics the researcher is interested to study. However, when the population units are heterogeneous in their characteristics, SRS is not suitable to generate a representative sample and in such case, stratified random sampling is the best option. This technique is one of the most important sampling techniques followed by the researcher in social sciences research. We know, a heterogeneous population is composed of unlike elements or units in terms of characteristics like size of land holdings, income levels, gender, age group, irrigation facilities with the farmers etc. Based on one of these characteristics (say, size of land holding), the population units (say, farmers) are categorized into different groups called 'Strata' (singular 'Stratum') such as Marginal farmers, Small farmers and Large farmers. Note that, the basis for grouping must be known before sampling. This process of splitting the population into different strata or sub-populations is called Stratification. (Note that, stratification is sometimes introduced by the researcher after the sampling phase and this is called as 'Post-stratification'. However, this post-stratification is done rarely by the researcher, when he does not have adequate knowledge about the population elements or units or when he could not identify an appropriate stratifying variable etc). The variable or characteristic based on which the population is divided into different strata is called Stratification variable. Sometimes, more than one variable is used for stratification purpose. Say, based on both irrigation facility and size of land holding, the population units can be categorized into different strata such as: marginal farmers with irrigation facility, marginal farmers without irrigation facility, small farmers with irrigation facility, small farmers without irrigation facility, large farmers with irrigation facility and large farmers without irrigation facility. Similarly, in a school, the entire population (students) can be divided into relatively homogeneous groups. That is, all the students of a school may be divided into groups (strata) of boy and girls. Once this is done, random sample from each of such group is drawn independently. So, this technique is suitable, when there are identifiable sub-groups exist within the total population that differ significantly in respect of characteristic under study. Each stratum is a subset of the population that share at least one common characteristic. The researcher first identifies the relevant stratums based on the stratification variable(s) and their actual representation to the target population. SRS is then used to select a 'sufficient' number of units or elements from each stratum. 'Sufficient' refers to a sample size that is large enough to be representative of each stratum and, in turn, each stratum represents the population with reference to the identified stratification variable(s). The number of strata to be formed for the research is left to the discretion of the researcher, but the optimum number should be six, as it convenient for the researcher. The number of strata to be considered is also determined by the characteristic based on which the researcher is intended to classify the population. After dividing the population into different strata, from each stratum, a representative sample is drawn by SRS and hence, the name Stratified random sampling technique (Figure 8.15). Note that, different sampling techniques can be used to select sample from each stratum, but SRS technique produce the representative sample with minimum sampling error. The samples selected from the various strata are then combined into a single sample. The strata are mutually exclusive segments such that, there is heterogeneity between the strata and homogeneity within each stratum. Say, if the population units are N and this is divided into K strata such as N_1, N_2, N_3 ...., N_K, etc., then, $N_1 + N_2 + N_3 + ... + N_K = N$. From each stratum, a random sample is selected in proper proportion to the size of each stratum say, n_1 from N_1, n_2 from N_2, n_3 from N_3 ...., n_K from N_K. Each unit in the stratum has an equal chance (or non-zero probability) of being included in the sample. That is, if one stratum is proportionally larger than another, its sample size should also be proportionally larger and *vice versa*. Thus, the stratified random sampling technique aims to create homogeneous subgroups from a large heterogeneous group and to select a representative sample from each stratum so that, the total sample (*i.e.*, sum of samples of each stratum) is representative of the total population. Say, if a stratum comprises of 40 per cent of total population, then 40 per cent of the sample is selected from that stratum. Thus, stratified sample is a mini-reproduction of the population. This stratified random sampling is characterized by (or it can be distinguished from the SRS and Systematic sampling in the sense that),

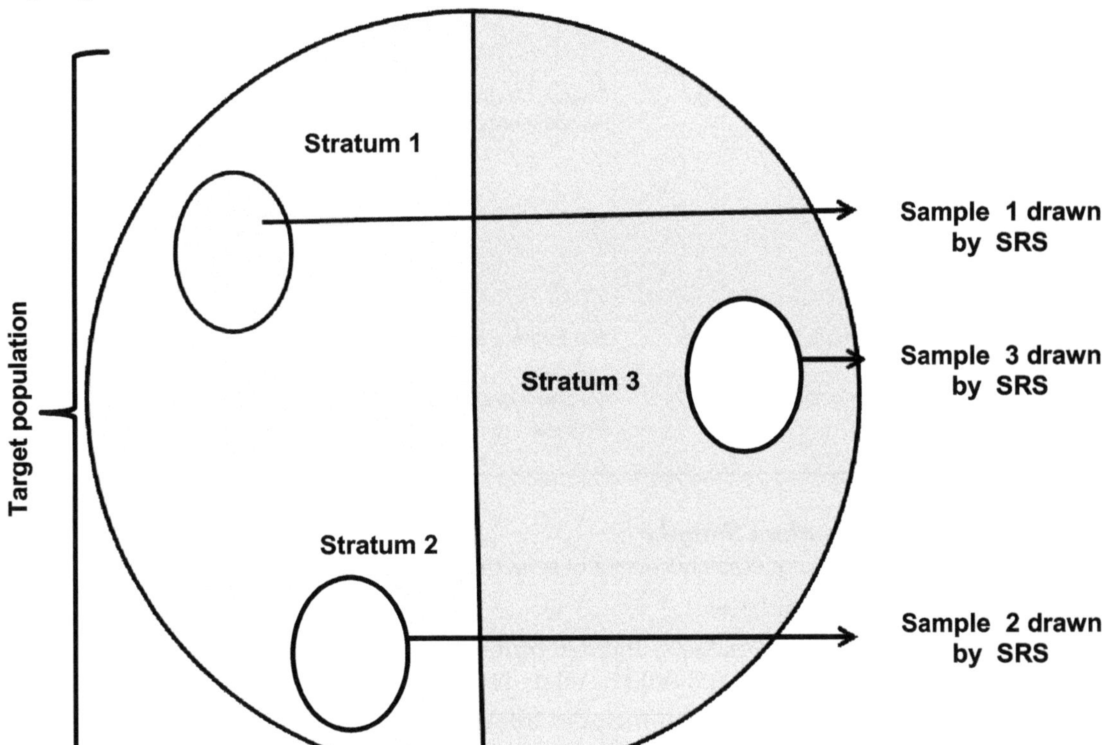

Figure 8.15: Selection of Representative Sample through Stratified Random Sampling Technique.

- It requires division of the parent population into mutually exclusively strata.
- The sample units are chosen independently and randomly from each stratum.
- Each unit in the population will be assigned to one and only stratum and no population units or elements should be omitted.

Further, stratified random sampling differs from Quota sampling in that, the sample elements are selected probabilistically rather than based on convenience or on judgemental basis.

Examples of Stratified Random Sample

The following Tables 8.2 and 8.3 shows some examples of ways to obtain a stratified sample:

Table 8.2: Selection of Stratified Random Sample from an Heterogeneous Population.

Item	*Example 1*	*Example 2*	*Example 3*
Research topic	Economic analysis of paddy cultivation in Kurnool district of Andhra Pradesh	Performance of Regulated Markets in Andhra Pradesh	Economic analysis of different water saving paddy production technologies in selected irrigation commands of Andhra Pradesh
Population	All farmers cultivating paddy in Kurnool district	All farmers selling major agricultural produce in the regulated markets of Andhra Pradesh.	All farmers cultivating paddy under different irrigation commands of Andhra Pradesh
Stratification variable	Size of holding	Regions of Andhra Pradesh	Irrigation command/Source and Size of land holding
Strata	Marginal, small and large farmers	Coastal Andhra, Rayalaseema and Telangana.	***i. Type of irrigation command:*** KWD canal command in Guntur district, Tank irrigation command in Chittoor district and Bore well irrigation command in Karimnagar district of Andhra Pradesh ***ii. Size of land holding of farmers:*** Marginal, small and Large farmers
Obtain a Simple Random Sample	100 farmers from each stratum	200 farmers from each region	50 farmers from each category
Sample	3 x 100 = 300 selected farmers	3 x 200 = 600 selected farmers	150 farmers (50 marginal, 50 small and 50 large) from each irrigation command.

Selecting a Stratified Random Sample

The following are the major steps involved in selecting a stratified random sample:

- Define the target population.
- Identify stratification variable(s) and determine the number of strata to be used. Note that, the stratification variable(s) should be related to the objectives of the investigation.
- Define the sampling frame that includes information on the stratification variable(s) for each element.
- Divide the sampling frame into different strata based on the stratification variable(s) identified earlier. Note that, the strata should not be overlapping, should be independent and mutually exclusive subsets of the population. If different strata overlap, it will give some units a

Table 8.3: Stratification Variables Chosen to Classify Target Population.

Stratification Variable	*Strata Identified*
Spatial	Coastal Andhra Region
	Rayalaseema region
	Telangana region
Time	Peak marketing season
	Off-marketing season
Enterprises	Co-operatives
	Private enterprises
	State enterprises
Size of land holding	Marginal farmers
	Small farmers
	Large farmers
People or households	Farmers
	Rural artisans
	Agricultural labour

higher chance of being selected as subjects in the sample. If this happened, it would not be a probability sample. One element of the sampling frame must be in one and only one stratum. This sampling technique is most effective when the following three conditions are met:

- ❖ Variability within each stratum is minimized.
- ❖ Variability between strata are maximized.
- ❖ The variables upon which the population is stratified are strongly correlated with the desired DV.

- ☆ Assign a unique number to each element in the sampling frame of each stratum.
- ☆ Determine the sample size for each stratum. Note that, the researcher has to identify the sub-types of stratified random sampling technique in fetching the sample size in each stratum *viz.*, proportionate stratified sampling or one of the various types of disproportionate stratified sampling (discussed in the ensuing pages).
- ☆ Randomly select the sample elements or units from each stratum. The number of units drawn by random sampling from each stratum depends on the homogeneity of the units. A smaller sample can be drawn from each stratum, if they have highly homogeneous units, whereas sample can be drawn in much higher proportion from another stratum where the units are known to differ. This is because, in the former case, the information from the smaller number of respondents can be enumerated to the whole sample stratum. However, in the latter case, with much variability among the units, the higher sample units will keep the errors to minimum value.
- ☆ The sum of sample elements across the strata gives the total sample size and this should be representative to the total population.

Types of Stratified Random Sampling

Stratified random sampling technique again fall into one of two categories, each of which has strengths and weaknesses as discussed below.

a. Proportionate Stratified Random Sampling

In this type of stratified random sampling, the size of the sample drawn from each stratum is proportional to the relative size of that stratum in the target population. The researcher employs the same sampling fraction (n/N) from each stratum in collecting the sample. So, in this method, the researcher stratifies the population according to known characteristics of the population (say, size of land holding) and subsequently, randomly draws the individuals in a similar proportion from each stratum of the population. For example, there are 1000 farmers in a village, comprising 500 marginal farmers, 400 small farmers and 100 large farmers. Now the researcher wants to draw a sample of 150 farmers from the three strata in similar proportion as they appear in the population. So, the sample to be selected from each stratum is given by ni = [(n.Ni)/N] = [(n/N)Ni] where, ni = size of the sample selected from ith stratum, n=total sample size to be collected from all strata, Ni = Size of ith stratum, N = Total population units. This is done as shown through the Table 8.4.

Table 8.4: Selection of Stratified Random Sample through Proportionate Method.

Stratum	*Size of each Stratum (Ni)*	*Proportion of Sample to be Selected (per cent) (Ni/N)*100*	*Size of Sample in each Stratum ni = (n/N)*Ni*
Marginal	500	50	(150/1000) x 500 = 75
Small	400	40	(150/1000) x 400 = 60
Large	100	10	(150/1000) x 100 = 15
Total	**1000 (N)**	**100**	**150 (n)**

As shown through Table 8.4, proportionate stratified random sampling increases the representativeness of the sample drawn from each stratum. That is, more sample is drawn from the stratum in which the population (Ni) is higher and *vice versa*. So, even from the small sized stratum, there is a representativeness into the final sample. By this, the sampling error can be minimized. If the researcher goes for SRS for getting this 150 units of sample, he might not get (or have a very meager chance) any sample unit from large farmers category, as their population is very low compared to marginal and small farmers categories or strata. So, to make sure to get representative sample, the researcher can divide the farmers into these three strata, and then select the same proportion of farmers (*i.e.*, n/N, as given in Column 4 of Table 8.4) from each stratum using a SRS method. Hence, this method is called as Proportionate stratified random sampling. Thus, in this method of stratified random sampling, each stratum has the same sampling fraction (n/N) and hence, the sample size of each stratum is proportionate to the population size of each stratum. Since, same sampling fraction is applied to each stratum, it gives every element in the population an equal chance to be selected. So, the resulting sample is a self-weighting sample. This can be better understood from the following example (Table 8.5). Assume that, a researcher wants to select a proportionate stratified random sample from four districts such that, the proportion of size of each stratum (district) in total population in all the four districts is equal to the proportion of size of sample selected from each stratum in total sample to be selected from all the strata. Note that, the sampling fraction here is : 0.11.

Table 8.5: Selection of Proportionate Stratified Random Sample from a Population.

Districts	*Population Units*	*(Proportion of Size of Stratum in Total Population) Per cent Share*	*Proportionate Stratified Random Sample*	*(Proportion of Size of Sample of Each Stratum in Total Sample) Per cent Share*
District 1	7600	33.04	826	33.04
District 2	900	3.91	98	3.91
District 3	6000	26.09	652	26.09
District 4	8500	36.96	924	36.96
Total	23000 (N)	100.00	2500 (n)	100.00

Note that, in stratified random sampling, the researcher is not only interested to estimate the statistics that are representative to the population parameters, but also to compare the estimates of different strata. In such case, proportionate stratified random sampling may not yield sufficient and reliable data findings, as shown through above Table 8.5, where in District (Stratum) 2, the selected sample size is highly inadequate (only 3.91 per cent) compared to other strata. So, comparing District 2 elements to the elements in the other districts or strata may not give realistic conclusions. So, proportionate stratified random sampling is a poor sampling choice for carrying out such analyses. Hence, disproportionate stratified sampling may be a better choice and this is discussed here under. However, proportionate stratified random sampling provides equal or better precision than a simple random sample of the same size. The gains in precision are greatest, when units within strata are homogeneous.

b. Disproportionate Stratified Random Sampling

As discussed above, in proportionate stratified sampling, the number of elements allocated to the various strata is proportional to the representation of the strata in the target population. This condition is not satisfied in disproportionate stratified sampling. However, in disproportionate stratified random sampling, the size of the sample drawn from each stratum is not proportional to the relative size of that stratum in the target population. So, the sampling fraction (n/N) employed for each stratum in collecting the sample is not the same across the strata *i.e.*, the sampling fraction may vary from one stratum to the next. That means, in disproportionate stratified sampling, the number of elements or units sampled from each stratum is not proportional to their representation in the total population. This is because, population elements are not given an equal chance to be included in the sample. So, in this method, the researcher stratifies the population according to known characteristics of the population (say, size of land holding) and subsequently, randomly draws the individuals in a dissimilar proportion from each stratum of the population. For example, there are 1000 farmers in a village, comprising 500 marginal farmers, 400 small farmers and 100 large farmers. Now the researcher wants to draw a sample of 150 farmers from the three strata in such a way that, 50 units are selected from each stratum. So, the sampling fraction (n/N) varies across the strata (in case of marginal farmers, the sampling fraction is 50/500, small farmers: 50/400 and large farmers: 50/100). This is done as shown through the Table 8.6.

Table 8.6: Selection of Stratified Random Sample through Disproportionate Method.

Stratum	*Size of each Stratum (Ni)*	*Sample Units Selected (ni)*	*Sampling Fraction in each Stratum = (ni/Ni)*
Marginal	500	50	(50/500) = 0.10
Small	400	50	(50/400) = 0.125
Large	100	50	(50/100) = 0.50
Total	1000 (N)	150 (n)	—

Thus, in this type of sampling, equal number of the sample units are drawn from each stratum, not depending on the size of the strata. Since, the same sampling fraction is not used across the strata, the size of sample drawn from each stratum is not necessarily distributed according to the size of the stratum and hence, some of the strata of the population may be over represented and some are under-represented. As given in the above example, out of 500 marginal farmers, only 50 farmers were selected as sample from that stratum and the same sample size (50 units) was selected from large farmers stratum though its size is only 100 units. Thus, marginal farmers are under-represented and large farmers are over-represented. Since, the researcher put equal weight to each stratum in selecting the sample it over represents one stratum (large farmers) and under represents the other (marginal farmers). This method is comparatively less time consuming than proportionate stratified random sampling. But, due to over-representation of some strata of the population, some bias in the sample may be introduced and this results in skewed results. Such type of sample may not be truly representative. The precision of using disproportionate stratified random sampling is highly dependent on the sampling fractions chosen by the researcher. If variances differ across strata, disproportionate stratified random sampling can provide better precision than proportionate stratified random sampling, when sample points are correctly allocated to strata. However, if costs and variances are about equal across strata, choose proportionate stratification over disproportionate stratification. If the variances or costs differ across strata, consider disproportionate stratification.

Sub-types of Disproportionate Stratified Random Sampling

There are three important sub-types of disproportionate stratified random sampling technique *viz.*, Disproportionate allocation for within strata, Disproportionate allocation for between-strata and Disproportionate stratification with optimum allocation.

- ☆ ***Disproportionate allocation within strata:*** Through this method, the researcher mainly aims at conducting the detailed study for each stratum in the total sample. This method is so chosen because, the proportionate stratification may not yield large size sample for a stratum (or few strata) and thus, the researcher may not conduct a detailed study of a particular stratum. For example, in Table 8.5, the sample pertaining to stratum: District 2 is not sufficiently large enough to conduct sample study or to compute reliable sample estimates for that stratum. In such cases, one option is to oversample that small stratum. Such oversampling would create a disproportional distribution of sample units among the strata in the sample when compared to the population. The above Table 8.5 can be re-written now, considering Disproportionate allocation within strata and this is shown through Table 8.7.

Table 8.7: Stratified Sampling with Disproportionate Allocation within Strata.

Districts	*Population Units*	*(Proportion of Size of Stratum in Total Population)* *Per cent Share*	*Disproportionate Stratified Random Sample*	*(Proportion of Size of Sample of Each Stratum in Total Sample)* *Per cent Share*
District 1	7600	33.04	697	27.86
District 2	900	3.91	518	20.72
District 3	6000	26.09	534	21.37
District 4	8500	36.96	751	30.05
Total	23000	100.00	2500	100.00

Considering the Table 8.7, now the researcher can compute sample estimates for stratum: District 2. Instead of using Table 8.5, now the researcher can employ Table 8.7, if he wishes to conduct a detailed analysis of District 2 by oversampling the elements in that district. So, instead of sampling only 98 elements from a sample of 2500 elements, now by Disproportionate allocation within strata, he will conduct a meaningful detailed analysis for District 2, as its sample size was increased to 518 elements.

☆ ***Disproportionate allocation for between-strata:*** Through this method, the researcher mainly aims at drawing comparisons across each strata in the selected sample. So, this makes the researcher to select sufficient number of elements from each stratum. So, he mainly aims at maximizing the sample size of each stratum to the extent possible and hence, he tries to select equal number of sample units from each stratum. The same is presented through Table 8.8 and please note the comparisons with Table 8.5 and Table 8.7. Note that, in Table 8.8, the sample elements included under each stratum are same (625 each) and this facilitate the researcher to draw comparative analysis between the strata. Compared to the proportionate sample distribution in Table 8.5, the sampling allocations presented in Table 8.8 provide a maximum possible sample elements for each stratum and thus, make the researcher to have a more balanced comparative analysis across the districts (Strata). Moreover, a relatively large number of elements are sampled from District 2 thereby, permitting detailed analysis within that stratum.

Table 8.8: Stratified Sampling with Disproportionate Allocation between Strata.

Districts	*Population Units*	*(Proportion of Size of Stratum in Total Population)* *Per cent Share*	*Disproportionate Stratified Random Sample*	*(Proportion of Size of Sample of Each Stratum in Total Sample)* *Per cent Share*
District 1	7600	33.04	625	25.00
District 2	900	3.91	625	25.00
District 3	6000	26.09	625	25.00
District 4	8500	36.96	625	25.00
Total	23000	100.00	2500	100.00

☆ ***Disproportionate stratification with optimum allocation***: This type of allocation was first proposed by Jerzy Neyman (1934), and is often referred to as the Neyman allocation. From the above discussion, we inferred that, proportionate stratified random sampling may yield smaller margins of error than SRS in estimating population parameters. However, still the margins of error can be minimized through this method of disproportionate stratification with optimum allocation and this techniques aims at two important aspects *viz.*, minimizing the costs and increasing the precision. We know, the sampling fraction generally varies across the strata considering the costs and variability of the variables of researcher's interest. Further, it is a known fact that, homogeneous strata with a smaller sample size can have the same level of precision as heterogeneous strata with a larger sample size. So, by applying this principle, it may be useful to select the number of elements from each stratum considering the standard deviation of the variable of interest in each stratum. The greater the variability of the variable in a stratum, the higher the sample size of the stratum should be. That is, variability of the variable in each stratum and the number of elements to be selected in each stratum are directly related. However, the number of sample elements to be selected in each stratum varies inversely with the data collection costs. Thus, this techniques assumes these two factors *viz.*, variability of variable of researcher's interest and data collection costs in selecting the representative sample and hence, this is more appropriate over proportionate stratified random sampling. Note that, this Optimum allocation technique may be applied focusing on cost only, precision only, or both cost and precision jointly.

Stratified Random Sampling vis-à-vis SRS

The following are the differences between Stratified random sampling and SRS:

- In SRS, the sample estimates simply represents overall population. But, in Stratified Random Sampling, the sample drawn (from each strata and overall sample) represent not only the overall population, but also the respective strata population. Thus, stratified random sampling makes inferences within a stratum and compares across the strata.
- Stratified random sampling will generally have more statistical precision than SRS. Of course, this will only be true, if each stratum is purely homogeneous. The precision is higher, when there is minimum within-stratum variances and maximum between-stratum variances of the study variables.
- The researcher selects random samples from within each stratum, whereas in SRS he selects the random sample from the overall target population.
- Stratified random sampling makes inferences for each stratum and compares the same across the strata. But, SRS provides estimates for the overall sample that is representative of the total population and do not divide sample into different strata.
- When the population does contain important differences between groups, stratified random samples yield smaller random sampling errors than those obtained with a SRS of the same sample size, especially if optimum allocation is used.
- Compared to SRS, stratified random sample tend to be more representative of total population because, the sample drawn from each stratum is representative to its corresponding sub-population and hence, the overall sample tend to be more representative to the total population.
- Data collection costs of stratified random sampling are lower than that of SRS, as the former breaks up the whole population into homogeneous strata such as geographical area, size of land holding, irrigation facility, income etc.
- Stratified random sampling permits the researcher to use different sampling procedures within different strata.
- Stratified random sampling provides better results than the SRS, when the strata are more heterogeneous among them and more homogeneous internally.
- In using stratified random sampling, the researcher gains more knowledge about the population units compared to SRS, as in the former, he has to classify the population units into different strata based on stratifying variable(s).
- Stratified random sampling guarantees better coverage of the population than the SRS. The researcher has control over the sub-groups that are included in the sample, whereas SRS does not guarantee than any one type of unit or element will be included in the final sample.
- Researchers, who are interested in rare extremes of a population often use stratified random sampling because, he can representatively sample even the smallest and most inaccessible sub-groups of the population. SRS does not allow this.
- Stratified sample will always achieve greater precision than a simple random sample, provided that, greater heterogeneity prevails between the strata and greater homogeneity prevails within each stratum. The greater the differences between the strata, the greater the gain in precision.

- Administratively, it is often more convenient to stratify a sample than to select a simple random sample. For example, interviewers can be trained on how to best deal with a particular age group of respondents, while others are trained on the best way to deal with a different age group. This way, the interviewers can concentrate on and refine a small set of skills and it is less timely and costly for the researcher.
- Compared to SRS, in using stratified random sampling, the researcher needs prior information about the population units and stratification variables, so as to classify them into different mutually exclusive data strata. Acquiring such information may consume significant time on the part of the researcher and it is also costly.
- Stratified random sampling consumes more time compared to SRS regarding the preparation of sampling design, using the technique to draw the sample and analyzing the data collected. Further, the analysis of data collected is more complex than the analysis of data collected *via* SRS. If there is misclassification of units or elements into strata, it may increase variability. This is not the case with SRS.

Merits of Stratified Random Sampling

The above discussion also highlights the strengths of stratified random sampling. The following are the other important merits of this technique of sampling:

- It combines the use of SRS with potential gains in precision.
- It gives the estimates of the population parameters from each sub-population or stratum.
- Stratification generally produces more precise estimates of the population than estimates that would be found from a simple random sample.
- Each element in the population is assigned to a particular stratum only. Then, random sampling is applied within each stratum independently. This often improves the representativeness of the sample by reducing the sampling error.
- A stratified sample can guard against an 'unrepresentative' sample (*e.g.*, selection of men farmers only from a mixed-gender population).
- Researcher can highlight a specific stratum within the population.
- Researchers also use stratified random sampling when they want to observe and compare relationships between two or more strata.
- Stratified random sampling enjoys the following advantages over other sampling methods:
  - Focuses on important subpopulations and ignores irrelevant ones.
  - Allows use of different sampling techniques for different subpopulations.
  - Improves the accuracy/efficiency of estimation.
  - Permits greater balancing of statistical power of tests of differences between strata by sampling equal numbers from strata varying widely in size.
- Stratified random samples generally require smaller sample sizes, which in turn, can save a lot of time, money and effort on the part of the researchers. This is because, this type of sampling technique enjoy higher statistical precision compared to SRS due to the fact that, the variability within the subgroups is lower compared to the variation of dealing with an entire population. If the researcher wishes to go for SRS in the whole population without stratification, the sample would need to be larger than the total of all strata samples to get an

estimate with the same level of precision. Stratified sampling ensures an adequate sample size for sub-groups in the population of interest. When a population is stratified, each stratum becomes an independent population and the researcher will need to decide the sample size for each stratum.

Demerits of Stratified Random Sampling

However, this technique suffers from the following limitations:

- It is quite difficult to correctly estimate the size of sample in each stratum such that, samples drawn from all strata should or be representative of total population.
- In some cases, it may be difficult to identify reliable stratification variables.
- This technique is more expensive than SRS, on a per-individual-sampled basis, since data must be collected and tracked separately for each stratum.
- When multiple stratification variables are involved, it increases both the cost and complexity of sample design.
- This technique demands a larger sample than would other methods.
- This technique is not useful, when there is no heterogeneity of elements or units in the population.
- It may require more administrative effort than in SRS.
- Stratified sampling is more complicated than SRS.

Considering the above merits and since these outweigh the demerits associated with stratified random sampling, it is more preferred over SRS.

c. Systematic Sampling

Compared to other techniques of probability sampling, systematic sampling is the easiest one to use. This method involves the selection of a sample in a very systematic manner. This technique is also known as Interval random sampling. In this sampling technique, a random selection is made regarding the selection of the first sample element or unit from the elevation factor ('i') or sampling ratio or skip interval (*i.e.*, i = N/n) and then subsequent units are selected using a fixed or systematic interval of calculated elevation factor, until the desired sample size is reached. That is, the first unit is selected at random and not automatic and the subsequent sample units are selected automatically in a systematic manner. For example, there are 100 farmers cultivating paddy in a village and the researcher wishes to select 10 sample farmers by systematic sampling. So, the elevation factor would be 100/10 = 10. Now the researcher number the farmers from 1 to 100. He then choose a random number from 1 to 10 (say, '5' from Tippet's Random Numbers Table) and this would be the first farmer (*i.e.*, the farmer numbered 5th in the frame) selected in the sample. Then, he selects subsequent sample units with numbers as 5+(1*10) (=15th unit) as second sample unit, 5+(2*10) (=25th unit) as third sample unit, 5+(3*10) (=35th unit) as fourth sample unit and so on (Figure 8.16). That is, to generalize, if 'Xth' unit is selected as first sample unit randomly from the elevation factor, the subsequent sample units will be in a systematic fashion like [X+(1*Elevation Factor)] as second sample unit, [X+(2*Elevation Factor)] as third sample unit, [X+(3*Elevation Factor)] as fourth sample unit and so on. In every case, the researcher has to determine the elevation factor first and proceed thereafter systematically. That is, he has to select the first sample unit randomly in the population of the elevation factor from 1 through 'i' and thereafter, follow the elevation factor or skip interval. This process will ensure that, there is no over representation or under representation of sample units (say, large farmers or small farmers), but rather

Target population (N) = 100

Sample size (n) = 10

Elevation factor = (N/n) = 10

Select a random number from 1 to 10. Say it is '5'.

Start with #5 and take every 10th unit

1	26	51	76
2	27	52	77
3	28	53	78
4	29	54	79
5	30	**55**	80
6	31	56	81
7	32	57	82
8	33	58	83
9	34	59	84
10	**35**	60	**85**
11	36	61	86
12	37	62	87
13	38	63	88
14	39	64	89
15	40	**65**	90
16	41	66	91
17	42	67	92
18	43	68	93
19	44	69	94
20	**45**	70	**95**
21	46	71	96
22	47	72	97
23	48	73	98
24	49	74	99
25	50	**75**	100

Figure 8.16: Selection of a Representative Sample through Systematic Sampling Technique.

that all types of sample units (say, farmers of all sizes) are generally uniformly represented, as it is in the sampling frame. To overcome human bias in this method, the researcher will select the first unit at random. In other words, the sample is representative of the population, at least on the basis of the sorting criterion. Strictly speaking, systematic sampling is not purely a SRS because, all the sample units are not equally probable. Hence, systematic sampling is technically called as 'Systematic sampling with a random start'. Systematic sampling resembles SRS only up to the selection of first sample unit and from the selection of second sample unit, its procedure altogether differs from SRS.

Steps in Selecting a Systematic Sample

Following are the steps involved in selecting a systematic sample from the population:

1. Define the target population and identify the sampling frame (N). The researcher should arrange all the units in the frame in a serial manner. Note that, the population units should be finite in nature and also should be defined very clearly.
2. Assign numbers to each unit of the sampling frame.
3. Determine the desired sample size (n).
4. Compute the elevation factor ('i') as (N/n) *i.e.*, dividing the number of elements in the sampling frame (N) by the targeted sample size (n). The elevation factor indicates the sampling interval.

It is to be noted that, if the elevation factor is not a whole number, then it is rounded off to the nearest whole number.

5. From the elevation factor, select a random number 'r', from '1' through 'i'. This random number 'r' represent the first unit of sample selected.
6. Select the subsequent sample units systematically as, r, (r + i), (r +2i), (r +3i), and so on, until the desired sample size that is representative of the population is reached or the frame is exhausted.

Thus, systematic sampling is strictly not a true SRS and it is often regarded as 'pseudo random sampling', or 'pseudo SRS', or 'quasi-random sampling'. This is so because, as mentioned above, only for selecting the first unit of sample, it resembles SRS in its probability selection. That too, the probability is restricted to the units of elevation factor only and not related to the entire sampling frame. Once the first unit is selected, the selection of other units will have a zero probability, as they are selected systematically and not randomly. As discussed earlier, in SRSWOR, after every drawing of the selected sample unit, the same is not returned back to the population. So, the probability of selection of each successive unit into the sample gradually decreases after each draw, as the selected sample units are distinct. The probability of selection of first unit is given by 1/N, for second unit 1/(N-1), for third unit 1/(N-2) and so on. However, in SRSWR, once an element or unit is sampled, the same is placed back in the population for re-sampling. So, SRSWR does not remove previously-selected units from the population before subsequent selections and hence, the unit already selected earlier can be re-selected again and the same unit may appear in the sample more than once. Hence, the probability of selection of each unit into the sample remains unchanged after each draw and this is given 1/N. However, in systematic sampling, the probability of selection of first sample unit is given by 1/Elevation factor or [1/(N/n)]. But, the subsequent sample units are selected systematically without any probability. So, in systematic sampling, once after calculating the elevation factor, first sample unit is selected randomly from the units of elevation factor and the subsequent sample units are selected through some combinations of elevation factor and other units are completely discarded from the selection.

Sub-Types of Systematic Sampling

There are three sub-types of systematic sampling *viz.*, Linear systematic sampling, Circular systematic sampling and Repeated (or Replicated) systematic sampling. The procedure discussed above in drawing sample infers about Linear systematic sampling. This is the most frequently used form of systematic sampling. However, the steps in selecting a sample through Circular systematic sampling is same as above, but it differs with respect to Step 5. In case of Linear systematic sampling, the random sample unit 'r' is selected from '1' through 'i'. In Circular systematic sampling, instead of select-ing a random number between '1' and 'i', a random number is selected between '1' and '*N*'. When one gets to the end of the list in selecting the sample, he would continue from the beginning of the list. This creates a circular pattern in selecting the sample. So, this type of sampling can be viewed as a sub-type of Linear systematic sampling. Both Linear systematic sampling and Circular systematic sampling create a single sample. However, Repeated systematic sampling involves the selection of multiple samples from the target population and then combining them into a single sample. That is, instead of only one random start, several smaller systematic samples are selected using multiple random starts. This makes the process more time-consuming than linear systematic sampling.

Examples for Systematic Sampling

This systematic sampling is generally employed to select systematically every 'i^{th}' unit in a rubber tree plantation, in a book library, in a housing colony after drawing a random unit from elevation factor for drawing the sample. This saves lot of time, cost and energy on the part of the researcher. Say, for example, a researcher has to conduct a survey among the households regarding the acceptance of a processed food product. He wish to select sample of 100 households from a long street that starts in a poor area (House No. 1) and ends in an expensive area (House No. 1000) *i.e.*, N=1000. If the researcher draws a simple random sample (n=100) from the population of 1000 houses, he may end up with more number of rich households and less number of poor households (or *vice versa*) thereby, leading to an unrepresentative sample. However, if the researcher goes for systematic sampling and if he selects the first sample unit on random basis (say, 5th unit) and every 10th unit as subsequent sample units systematically, it ensures that the sample is spread evenly representing the population. However, systematic sampling may also result in drawing an unrepresentative sample of the overall population. In the above example, consider a residential colony, where the odd-numbered houses are all on the north side of the road (high income group) and the even-numbered houses are all on the south side of the road (low income group). Under the sampling scheme given above, it is impossible to get a representative sample, as either the houses sampled (n=100) will *all* be from the odd-numbered (north side – high income group), or they will *all* be from the even-numbered (south side – low income group).

This sampling technique can also be used in food processing industries, where the firm manager checks the product items, whether they are produced as per the desired quality specifications or not? The manager can select every i^{th} item to ensure consistent quality or for detection of defects. Therefore, he requires the first item to be selected at the random as the starting point and subsequently he can choose every i^{th} item for evaluation against specifications. It also finds its applicability, while questioning farmers or traders in a sample survey, where the researcher may catch hold of every 10th person entering a regulated market. The classical example is that, say, a supermarket wants to study the buying habits of their customers. They use this systematic sampling technique and can choose every 5th or 10th or 15th or 20th customer entering the supermarket and conduct the study on this sample. This sampling technique is so convenient because, the supermarket can conduct this study even if it do not have a list of customers arriving on that day and it do not even know, how many customers will arrive on that day to the supermarket. So, the main advantage of this technique is its simplicity. Similarly, this technique is often used in industry (machinery say, tractor), where an item is selected for testing from a production line to ensure that machines are of a standard quality. For example, a tester in a manufacturing plant might perform a quality check on every 15th product in an assembly line. The tester might choose a random start between the numbers 1 and 15. This will determine the first product to be tested; every 15th product will be tested thereafter.

Merits of Systematic Sampling

The following are the merits of using systematic sampling technique:

- ☆ This technique yields a more representative sample than the random sample (especially, if the sample size is small).
- ☆ This technique is very simple and convenient in use, as it saves a lot of time.
- ☆ It seeks to eliminate sources of bias. For example, a food inspector checking sweets on a conveyor belt might unconsciously favour red colour sweets.
- ☆ This technique ensures the selected sample is evenly distributed over the entire population.
- ☆ This technique is more economical and less time consuming than SRS.
- ☆ This technique is useful for selecting large samples. So, it is less cumbersome than SRS, as the researcher need not use either a table of random numbers or a lottery method.

- Systematic sampling might be more precise than SRS. In effect, systematic sampling like stratified random sampling also stratifies the population into different strata like 1st elevation factor or skip interval units, 2nd elevation factor or skip interval units and so on. Thus, we might expect that, the systematic sample to be as precise as a stratified random sample. But, the difference between systematic sampling and stratified random sampling is that, in systematic sampling, right from second sample unit, subsequent sample units are selected at the same relative position in each stratum (*i.e.*, in accordance with the computed elevation factor), whereas in stratified random sampling, the concept of SRS is used in each stratum to draw representative sample from each stratum. Further, in systematic sampling, only one unit is selected from each stratum randomly, whereas in stratified random sampling a representative sample (a group of sample units) is selected from each stratum randomly.
- This technique can be adopted as long as the sampling frame does not contain any hidden order and in such case, this technique is as good as the SRS. Even, when the researcher is not aware of the sampling frame clearly, still he can use this method. For example, as mentioned earlier, the supermarket manager can interview every 5th customer entering the supermarket to study about their perceptions towards the quality of the products they purchased in the super market.
- If the ordering of the units in the sampling frame is randomized, systematic sampling may yield results similar to SRS.
- Systematic sampling eliminates the possibility of autocorrelation, but this may occur in SRS. For example, in using a voter's list for the selection of a sample for a study of voter preferences, it is possible that, members of the same family may get selected using SRS, but this is not possible using systematic sampling.

Demerits of Systematic Sampling

This sampling technique suffers from the following limitations:

- Since systematic sampling is not a true random sampling, the resulting sample will not necessarily be representative of the population from which it was drawn. So, this will affect generalizing the results back to the population. Hence, if a researcher does not have adequate experience in conducting systematic sampling technique and have full knowledge of the population to be sampled, it is better to avoid using this method.
- This technique can also introduce bias. For example, if a food inspector wants to inspect the quality of sweets produced from a firm, the period chosen for inspection might coincide with the period of faulty machine, thus yielding an unrepresentative number of faulty sweets.
- Estimating variances is more complex than that for SRS.
- The biggest drawback of the systematic sampling method is that, if there is some cycle in the way the population is arranged on a list and if that cycle coincides in some way with the sampling interval, the possible samples may not be representative of the population. This can be understood from the following example: Suppose, the researcher wishes to select a sample of professors from an Agricultural College. The college is having the following 10 departments: Agricultural Economics, Agricultural Extension, Agronomy, Genetics and Plant Breeding, Soil Science, Plant Physiology, Plant Pathology, Entomology, Biochemistry and Horticulture. Each department has 10 employees, including a Head of the Department. Thus, there are 100 employees in total in the college. The sampling frame is ordered by the

department, with the Head of the department listed first and then, the other employees by descending order of their seniority. If the researcher wishes to survey these employees about their views on their working environment, he might choose a small sample to answer the questions. If the researcher use a systematic sampling approach and the sampling interval is 10, then he could end up selecting only Heads of the departments or the newest employees in each section. So, this type of sampling would not give a complete or appropriate picture of employees' thoughts.

d. Cluster Sampling

It is also called as Area random sampling. Sometimes, in selecting the individual sample units, the researcher may be confronted with some practical problems like, it is difficult for him to get the sampling frame of a target population and even if the sampling frame is available, but the target population may be widely spread over a largely geographical area thus, making data collection costs relatively high. In such case, it is quite difficult for the researcher to select individual sample units from the target population to conduct surveys. So, it is ideal for the researcher to go for Cluster sampling rather than above discussed three techniques of sampling. This cluster sampling is also a probability method of sampling, where instead of individual sample units, the researcher takes a group of units on random basis called 'Clusters' as sample units from a wider geographical area. That is, in cluster sampling, 'cluster', *i.e.*, a group of units, constitutes the sampling unit, instead of a single unit of the population. So, the term 'cluster' refers to a group of individual sample units and such clusters are drawn randomly by the researcher on random basis from a wider geographical area. That is, in SRS, stratified random sampling and systematic sampling, the researcher randomly selects individual population units as sample units, but cluster sampling involves the selection of population elements not individually, but in aggregates. These clusters may be space-based, such as naturally occurring geographical or physical units (*e.g.*, mandals, states, countries etc); organization-based (*e.g.*, schools, colleges etc.) or telephone-based (*e.g.*, area codes or exchanges of telephone numbers).

In the above discussed three random sampling methods, problem arises when the target population is distributed over a large geographical area. For example, if the researcher wishes to conduct a research study on *'Prioritizing the marketing problems of farmers in transacting paddy in Andhra Pradesh'*, and if he resorts to SRS, he may get the sample of respondents from all-over the State. In such case, he has to go for extensive travelling throughout the State in getting the information from the sample respondents. So, to precisely address this problem the cluster sampling was invented. This sampling technique basically consists of the following steps:

- ☆ Define the target population.
- ☆ Determine the desired sample size of the study.
- ☆ Divide the population into different clusters.
- ☆ Determine the number of clusters to be selected. This may be done by dividing the sample size by estimated average number of population units in each cluster. Note that, as cluster number increases, precision increases. On the other hand, as differences between clusters increases, precision decreases.
- ☆ Randomly select sample clusters.
- ☆ Randomly select the units or measure all units within sampled clusters.
- ☆ The clusters should be designed so that, they collectively include all members of the target population.
- ☆ Each member must appear in one and only one cluster.

It is to be noted that, for a good cluster sample design, heterogeneity within the cluster should be high and heterogeneity between clusters should be low. That is, the clusters selected should be alike. But, the units or elements within each cluster should be as heterogeneous as the target population. Further, the clusters would be small, but not so small as to be homogeneous. This is where, cluster sampling differs from stratified random sampling. In the former, homogeneous group is divided into heterogeneous units, while in the latter heterogeneous group is divided or stratified into homogeneous units.

Sub-types of Cluster Sampling

Cluster sampling can be classified into different sub-types considering two important dimensions *viz.*, based on the number of stages in the sample design and based on the proportional representation of the clusters in the total sample.

a. Sub-types Based on Number of Stages of Drawing Sample

Based on this dimension, there are three subtypes of cluster sampling as discussed below.

- ✰ ***Single-stage cluster sampling:*** As the name indicates, there is only stage in drawing the sample *i.e.,* sampling is done only once. As an example of single-stage cluster sampling, consider a researcher is interested in studying the 'Marketing problems involved in transacting paddy in Rayalaseema region of Andhra Pradesh'. Let us suppose, the researcher wishes to interview about 800 farmers. There is no compiled list of all farmers selling paddy in Rayalaseema region, but they are served by 25 regulated markets scattered over a large region. He know that, each market on an average serve about 200 farmers. Therefore, he randomly sample four regulated markets from the total of 25 regulated markets, and then study *all* farmers transacting paddy in those four regulated markets, which would provide the desired sample of 800 farmers (Figure 8.17). Thus, in Single-stage cluster sampling, the researcher has to list out all the clusters (markets) in the population, and from that list he will select four clusters usually by SRS technique. All units in the sampled clusters are selected for the survey.
- ✰ ***Two-stage cluster sampling:*** This sub-type is similar to the above one-stage cluster sampling, except at the last step. That is, instead of including all the farmers, a random sample say, either a simple random sample or systematic sample is taken from the farmers in the selected regulated markets. This sort of sampling beyond the first stage is sometimes referred to as Sub-sampling. So, in the above example, the researcher might select eight regulated markets (clusters) randomly instead of four. Then, he might draw a simple random sample of 100 farmers from each market (cluster). This provides a sample of 800 farmers, just as in the above case, but they come from a larger number of markets (clusters – Figure 8.17). This would be somewhat more expensive than collecting data from just four regulated markets, but it is likely to provide less biased estimates of the population parameters. Generally, unless the clusters are homogeneous, a two-stage cluster sample design is better than a one-stage cluster sample design. In this sampling, the initial clusters are called Primary sampling units (Regulated markets) and clusters within the primary sampling units are called Secondary sampling units (farmers). Thus, in two-stage cluster sampling, the researcher will first select the clusters by SRS. The units or elements in the selected clusters of the first-stage are then sampled in the second-stage, usually by SRS or often by systematic sampling.
- ✰ ***Multi-stage cluster sampling:*** If the cluster sampling proceeds beyond the two stages as discussed above, it implies multi-stage cluster sampling (Figure 8.17). Thus, it involves the

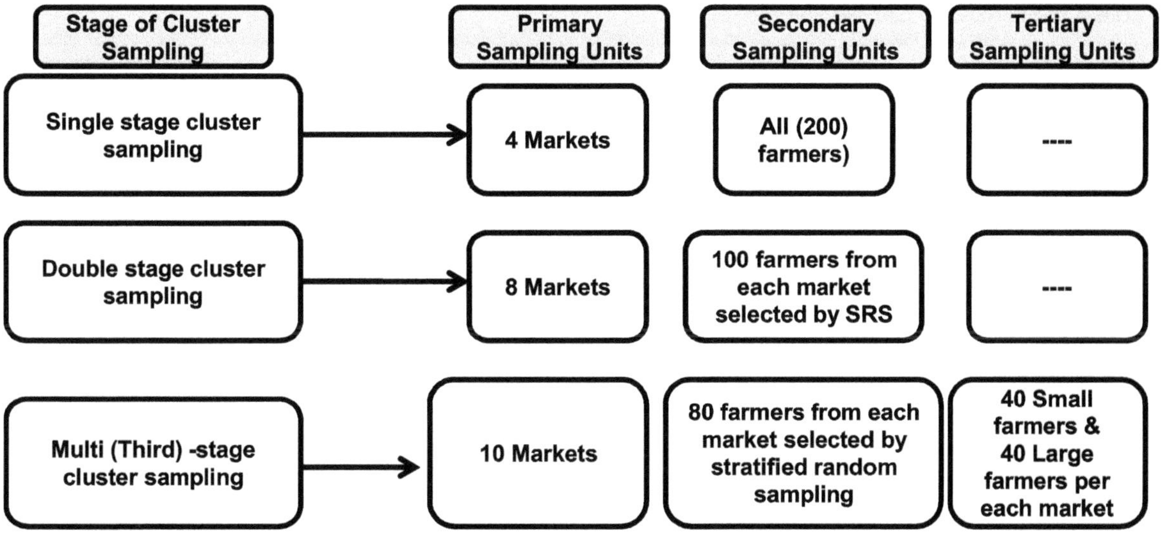

Figure 8.17: Selection of Cluster Sample from different Sub-types of Cluster Sampling.

repetition of two basic steps: listing and sampling. Typically, at each stage, the clusters get progressively smaller in size; and at the last stage, unit or element sampling is used. Sampling techniques (SRS or Stratified random sampling or Systematic sampling) at each stage may differ. It is not necessary that, the sampling procedures at each stage be the same. The number of stages that are used is often determined by the availability of sampling frames at different stages. In the above example, at the last stage, stratified random sampling may be employed to classify the farmers based on their size of land holding as small and large farmers. That is, the researcher may stratify sample elements (farmers or here called, tertiary sampling units) as small farmers (n=40) and large farmers (n=40). So, the sampling units comprise of primary sampling units, secondary sampling units (farmers), tertiary sampling units (small and large farmers) etc., until one gets to the 'final' or 'ultimate' sampling unit and desired sample size. Typically, as the sampling process moves from the selection of primary sampling units to the other sampling stages, the sampling units become smaller and smaller and more and more homogeneous. It is evident that, with increase in number of stages of cluster sampling, the size of cluster decreases and the sample size increases. The large clusters tend to be more heterogeneous than small clusters. Because of the greater heterogeneity of the primary sampling units (high income and low income markets), sampling error is minimized if the selected sample has more primary sampling units than secondary sampling units, more secondary sampling units than tertiary sampling units, and so forth. Thus, in multi-stage cluster sampling, sampling is done in more than one stage. However, most commonly, it begins with random cluster sampling and then applies SRS or stratified random sampling.

The above discussion infers that, single-stage cluster sampling requires a sampling frame of the clusters only; and two-stage cluster sampling and multi-stage cluster sampling require a sampling frame of the elements of the population only for the clusters sampled at the last stage of the process. The Tables 8.9 to 8.11 highlight the selection of cluster samples.

Table 8.9: Selection of Cluster Sample.

Type of Survey	*Target Population*	*Cluster Definition*	*Sampling Frame (Clusters)*	*Cluster Sample Selection*	*Sampling Frame (Sample Elements)*	*Sample Element Selection*
Agricultural Marketing	Farmers	Regulated Markets	List of markets where jaggery is transacted	Use systematic sampling to select markets, from which the farmers will be selected	List of all the farmers transacting jaggery in each market	Use SRS or systematic sampling to choose sample farmers from sampling frame of farmers in each cluster.

Table 8.10: Cluster and Elementary Units Selected Under Cluster Sampling.

Cluster	*Listing Unit*	*Elementary Unit*
Mandal	Household	Person
District	Regulated market	Farmer
State	Hospital	Patient
School	Classroom	Student
Page of book	Sentence	Word

Table 8.11: Sample Elements and Possible Clusters.

Sample Elements	*Possible Clusters*
Rural youth	Students of Agricultural College
	Students of Agricultural Polytechnic
	Students of Home Science College
	Students of Agricultural Engineering
Professors	Professors in Agricultural College
	Professors in Agricultural Engineering College
	Professors in College of Home Science
	Professors in College of Veterinary Science
Input suppliers	Pesticide dealers
	Fertilizer dealers
	Seed dealers
	Private money lenders
Extension service providers	Scientists from District Agricultural Advisory and Transfer of Technology Centres (DAATTC)
	Scientists from Regional Agricultural Research Station (RARS)
	NGOs
	Private Input Dealers
	KVKs

Another example is to take a state-wide sample to analyze child immunization programme compliance. The first stage would be a sample of districts within the State. The second stage would be to sample of mandals within each district. The third stage would be to sample of villages within each mandal. The fourth stage is to sample schools within each village. The fifth stage is to sample classrooms within selected schools. Finally, the researcher take all the students within the classroom. In this example, students (children) are the elementary units and classrooms are the listing units. Thus, there are five stages of sampling (*i.e.*, districts, mandals, villages, schools and classrooms) with four levels of clusters: districts, mandals, villages and schools. In sampling designs that involve more than one cluster, the clusters at the first stage are called Primary Sampling Units. Note that, these clusters can be selected by SRS or by systematic sampling. If these clusters can be grouped into strata, then we can take a stratified random sample. It is also clear that, in cluster sampling, more than one sampling frame might be involved. In this example, the sampling frame at the first stage is a list of all districts. At the second stage, the sampling frame is a list of all mandals within each selected district. At stage three, the sampling frame is the list of villages within each mandal. At stage four, the sampling frame is the list of schools within each village. Finally, the sampling frame at stage five is the list of classrooms in the selected schools. Thus, in cluster sampling, after the first stage of sampling, the sampling frame is compiled from only those clusters selected into the sample. When clusters are selected by SRS, it is generally referred to as Simple Cluster Sampling. The plan is called Simple first-stage cluster sampling, when the sampling is done in one stage and clusters are selected by SRS. Simple two-stage cluster sampling is a plan in which the clusters selected at the first stage are by SRS, the listing units in the second stage are selected independently within each cluster by SRS and the fraction of listing units sampled within each cluster is the same for each cluster.

b. Sub-types Based on the Proportional Representation of Clusters in Total Sample

According to this criterion, clusters may be selected in two ways *viz.*, based on equi-probability or based on PPS.

- ☆ ***Equi-probability sampling:*** In case of equi-probability selection method, each cluster will have the same size *i.e.*, each element in the population had an equal chance of being selected into the cluster.
- ☆ ***PPS:*** In this, the probability of selecting a cluster is dependent on the proportional distribution of its elements in the target population. So, the probability of selecting any cluster varies with the size of the cluster. Large clusters will have a greater probability of selection and *vice versa*. This concept can be well understood from the following example:

  Assume a researcher want to sample some cities in a country X by Multi-stage cluster sampling to obtain a representative sample of 500 residents to conduct a market survey for the acceptance of a RTE processed food product. There are 200 cities in the country. Now the researcher will be in dilemma regarding:

  - ❖ Should he select many cities (clusters) and few people per cluster? (say, Should he specifically (randomly) select 50 cities (clusters) and follow up by interviewing 10 randomly chosen individuals per selected cluster? or
  - ❖ Should he select few cities (clusters) and many people in each one? (Should he specifically (randomly) select 10 cities (clusters) and follow up by interviewing 50 randomly chosen individuals per selected cluster?

It is essential to note that, '*the general guideline for cluster design, then, is to maximize the number of clusters selected while decreasing the number of elements within each cluster*'. Hence, it is appropriate for the researcher to select specifically (randomly) 50 cities (clusters) and follow up by interviewing 10 randomly chosen individuals per selected cluster (Panel B of Figure 8.18).

However, when the cities (clusters) are of greatly different sizes, it is not the correct approach to draw equal size of sample from each cluster. In the above statement, the researcher has randomly selected 50 cities from the full set of 200 cities, and then randomly selected 10 people from each selected city (cluster). But, among the total 200 cities, some have very large populations, such as city 'A (1.7 lakhs), city 'B' (1.4 lakhs), city 'C' (1.3 lakhs) and city 'D' (1.1 lakhs), whereas some cities harbour less number of people say, city 'E' (5000), city 'F' (4780) etc. So, considering different number of people living in different cities, it is not appropriate to assign equi-probability for them (clusters) based on SRS to get included in the sample and that too equi-probability based on SRS again may also not be done to select the people from the selected cities. For example, if each city had an equal chance of being selected in the first (cluster or big-unit) stage, even larger cities may very well be excluded, as each city would have a 50/200 chance (about 1-in-4) of making the cut. Because of their size and prominence, it could be argued that, the large cities should be given a larger chance based on their population size. Further, after sampling cities, the researcher would survey 10 residents per selected city (cluster). Thus, if city 'A' was included in the sample at the first stage of sampling, each city 'A' resident's chance (10/170000 = 0.00006) would be less than that of a resident of, say, city 'E' (10/5000 = 0.002),

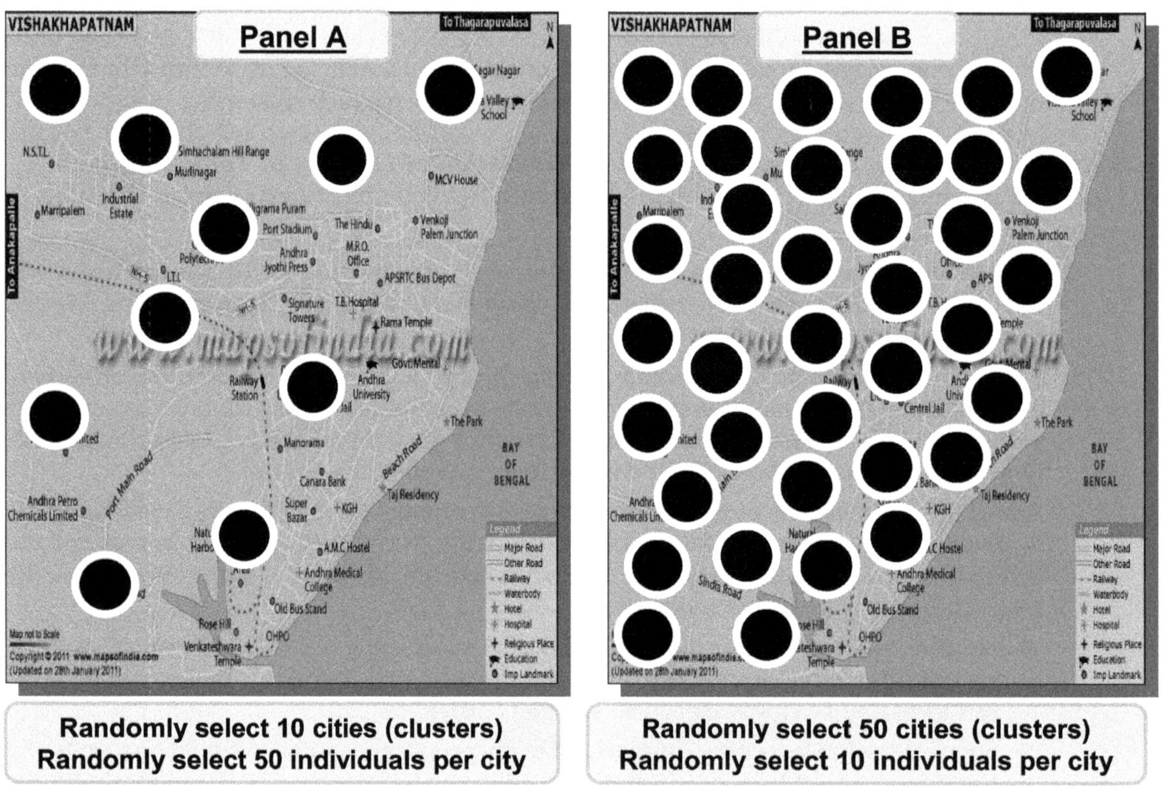

Figure 8.18: Cluster Sampling – Selection of Clusters and Sample Units from each Cluster.

if it is selected in the sample. Thus, every individual of city 'A' had a very low chance and every individual of city 'E' had a very high chance of being included in the sample of 50 residents from those respective cities (clusters). Thus, this concept of selecting the clusters based on PPS highlights the following aspects:

- ✰ ***First stage:*** PPS sampling - larger clusters have bigger probability of being sampled and *vice versa.*
- ✰ ***Second stage:*** Sampling exactly the same number of individuals per cluster such that, individuals in large clusters have smaller probability of being sampled and *vice versa.*
- ✰ ***Overall:*** Second stage compensates the first stage so that, each individual in the population have the same probability of being sampled

Thus, in the long run, any individual resident of city 'A' and any individual resident of city 'E' have an equal chance of being selected. City 'A' with its large population, has an increased chance of being selected as a sample in the first stage compared to smaller cities. However, city 'A' residents, as individuals, have a lower chance of selection in the second stage than do residents of smaller cities like 'E', which balances everything out.

So, when SRS and systematic sampling are used, both large and small cities will have the same probability of being included in the sample, which is incorrect. So, the correct way of compensating for these differences is to choose clusters from the sampling frame with PPS. The above example clearly indicates that, if PPS is applied, each resident in the sample will have an equal chance of being selected. The sample is then said to be *self-weighting*, which will simplify the analysis. The procedure for PPS sampling is described in the following section. The example given below deals with selecting respondents (farmers) from a district, but the procedure is exactly the same whatever the sampling units you choose.

Numerical Example to Derive Sample from different Clusters Based on PPS

Let us suppose, the researcher has to select 300 sample units (farmers) from a population of 6700 farmers in a district to elicit their views about the performance of a new pesticide product in their cotton fields. The selection of respondents based on PPS have proceeded in the following sequence:

- ✰ Divide the entire population of 6700 farmers into 10 clusters (villages)
- ✰ Estimate the population size (or the number of farmers) in each cluster and list them say, by following a geographical pattern.
- ✰ Prepare the Table 8.12 with three columns A, B and C. First column indicates the number of clusters in total (10), 2nd Column indicates the units of population in each cluster and 3rd Column indicates the cumulative total of population units across the clusters.
- ✰ Assume the sample size as: number of clusters = 3 and farmers per cluster = 100.
- ✰ Calculate the sampling interval using the following formula:

  Sampling interval = Cumulated total population(b) / Number of clusters required

  In this example, the Sampling interval = 6700 ÷ 3 = 2233.
- ✰ Select a random number, which is equal to or less than the sampling interval. In this example, choose a random number between 1 and 2233 by using a Tippet's random numbers table or a computer etc. Let us suppose that, the chosen number is 1814.

- ☆ As shown in the Table 8.12, locate the first cluster by finding the farmers size whose cumulative population just exceeds than this random number selected (1814). In this example, the first cluster would be located in village 4, where the cumulative population is 1,900.
- ☆ Now, add the sampling interval to the selected random number *i.e.*, 2,233 + 1,814 = 4,047.
- ☆ Again choose the cluster (village), whose cumulative population just exceeds this number. In this example, the second cluster will be located in village 6.
- ☆ Identify the location of each subsequent cluster by adding the sampling interval to the number which located the previous cluster. So, to locate the third cluster, add 2233 to 4047 and it is equal to 6280. So, the third cluster will be located in the village 10. Stop when you have located as many clusters as you need.
- ☆ From the above calculations, compute Prob 1 *i.e.*, probability of selection of each sampled cluster (5th Column) and this is given by: Prob 1= [(a x d)/b]*100, where a = Cluster population, b= Total Population, d = Number of Clusters.
- ☆ Now, calculate Prob. 2 *i.e.*, probability of selection for each individual in each of the sampled clusters (7th column) and this is given by: Prob 2= (c/a)*100, where a=Cluster population, c=Number of individuals to be sampled in each cluster.
- ☆ Calculate the overall basic weight of an individual being sampled in the population. This basic weight is the inverse of the probability of selection and it is given by: =[1/(Prob 1*Prob 2)] = 23.31. This overall basic weight is same across the selected clusters and this implies, each individual in the population has the same probability of being sampled.

Table 8.12: Multi-stage Cluster Sampling Based on PPS.

(A) Cluster (Village)	*(B) Size (a)*	*(C) Cumulative Sum*	*Clusters Sampled (d)*	*Prob 1 (%) = [(a*d)/b]*100*	*Individuals per Cluster (c)*	*Prob 2 (%) = (c/a)*100*	*Overall Weight (%) = [1/(Prob 1*Prob2)]*
1	1000	1000					
2	400	1400					
3	200	1600					
4	300	1900	1814	13	100	33.3	23.31
5	1200	3100					
6	1000	4100	4047	45	100	10.0	23.31
7	1600	5700					
8	200	5900					
9	350	6250					
10	450	6700 (b)	6280	20	100	22.2	23.31

From the Table 8.12, the following are the important conclusions:

- ☆ ***First stage:*** In PPS sampling, larger clusters have bigger probability of being sampled and *vice versa*. That is, Cluster 6 enjoys more probability (45 per cent) of being selected compared to Clusters 10 and 4 (5th column).
- ☆ ***Second stage:*** By sampling exactly the same number of farmers per cluster, farmers in large clusters have smaller probability of being sampled and *vice versa*. That is, in Cluster 4,

farmers enjoy more probability (33.3 per cent) of being selected compared to Clusters 10 and 6 (7th column).

- ***Overall:*** Second stage compensates the first stage so that, each individual in the population has the same probability of being sampled and this indicated by 23.31 per cent (8th column).

Differences between Cluster Sampling and Stratified Random Sampling

Cluster sampling and stratified random sampling are two different sampling techniques and though they bear some superficial similarities, they are substantially different in the following aspects:

- In cluster sampling, homogeneous group is divided into heterogeneous units, while in the stratified random sampling, heterogeneous group is divided or stratified into homogeneous units. So, in cluster sampling, the members are grouped into relatively heterogeneous groups. But, in stratified random sampling, the members are grouped into relatively homogeneous groups.
- In cluster sampling, the entire cluster is treated as the sampling unit and hence, analysis is done on a population of clusters (at least in the first stage). In stratified sampling, the analysis is done on units or elements within strata. The units are drawn randomly from each stratum, whereas in cluster sampling only the selected clusters are studied. That is, in stratified random sampling, a random sample is drawn from all the strata, where in cluster sampling only the selected clusters are studied, either in single stage or multi stage.
- The main objective of cluster sampling is to reduce costs by increasing sampling efficiency. This contrasts with stratified sampling, where the main objective is to increase precision.
- Cluster sampling can be opted, if the group consists of homogeneous members whereas, stratified random sampling is a good option for heterogeneous members.
- In stratified random sampling, the variables used for stratification should be related to the research problem. However, in cluster sampling, the variables used for clustering should not be related to the research problem.
- In stratified random sampling, in order to minimize sampling error, intra-strata differences should be minimized and inter-strata differences should be maximized. Reverse is the case for cluster sampling.
- In stratified random sampling, all the strata of the population need to be sampled. In cluster sampling, the researcher proceeds by first selecting a number of clusters at random and then sampling each cluster or conduct a census of each cluster. But, usually not all clusters would be included. So, all strata are represented in the sample in case of stratified random sampling; but only a subset of clusters are represented in the sample in case of cluster sampling. Therefore, only a part of the total number of clusters are sampled and all the other clusters are left unrepresented.
- Stratified random sampling requires more prior information about the population than cluster sampling.
- Stratified random sampling takes a longer period of time to accomplish, while cluster sampling is time efficient.
- Stratified random sampling requires a larger number of samples, since the population is divided into several strata, while cluster sampling does not.

- Cluster sampling is very cost efficient, since samples are already specified, while stratified random sampling can be expensive.
- Stratified sampling allows researchers to use different approaches for each stratum and see which approach works best, while cluster sampling does not.
- Through stratified random sampling, the best survey results will occur, as the units within each stratum are internally homogeneous. However, with cluster sampling, the best results occur when elements within clusters are internally heterogeneous.
- In stratified random sampling, a sampling frame is needed for the entire target population. However, in cluster sampling, say in single-state cluster sampling, a sampling frame is needed only for the clusters. In case of two-stage and multi-stage cluster sampling, a sampling frame of individual elements is needed only for the elements in the clusters selected at the final stage.
- Common stratification variables include: age, gender, income, size of land holding etc. Common classification variables in cluster sampling include: geographical area, school, hospital etc.
- In cluster sampling, the cluster is treated as the sampling unit and hence, analysis is done on the entire population within selected clusters. However, in stratified random sampling, the analysis is done on randomly selected sample units from each stratum.
- Cluster sampling is appropriate when it is very time consuming or expensive to choose the individuals one at a time compared to stratified random sampling.
- Cluster sampling divides the population into groups or clusters. A number of clusters are selected randomly to represent the total population, and then all units within selected clusters are included in the sample. No units from non-selected clusters are included in the sample and the sample units are represented by those from selected clusters. This differs from stratified random sampling, where sample units are drawn randomly from each stratum.

Merits of Cluster Sampling

They include:

- This sampling technique is the most economical form of sampling. That is, it is the most cost-effective probability sampling procedure. Two major sources of cost are almost always reduced with cluster sampling: listing costs and travel cost. For instance, compiling research data about each farmer in the district can be very hard. However, compiling data about several blocks in the district is going to be simpler. Here, listing and also travelling costs are lessened significantly.
- The researcher can increase the sample size at low cost.
- Less time for listing the clusters and implementation of sampling design than SRS.
- Generating sampling frame for clusters is often easier and is even readily available at cluster level.
- It ensures easy field supervision and survey administration. This is important, since better supervision of interviewers will result in better data quality.
- Cluster sampling is used to estimate high mortalities in cases such as famines, natural disasters etc.

- No doubt, SRS has been regarded as the method-of-choice by many researchers, since standard errors are minimized when it is used. However, conducting face-to-face interviews to a sample selected by SRS will be prohibitively expensive (in terms of man power and other costs), as the selected sample will most likely be geographically dispersed all over the country. SRS is also ineffective when there is a need to report findings within a short time frame. In such cases, cluster sampling is more preferred over SRS.
- This sampling technique is mainly used when doing marketing research.
- Cluster sampling provides practicability, when survey research has to address several number of respondents in short span of time.
- For the same level of costs, cluster sampling with a higher sample size may yield less sampling error than that resulting from SRS with a smaller sample size.
- Cluster sampling permits subsequent sampling because, the sampled clusters are aggregates of elements.
- Unlike SRS, cluster sampling permits the estimation of characteristics of subsets (clusters) as well as the target population.
- This technique can show and explain 'regional' variations.
- This technique involves dividing the target population into 'clusters' (blocks in a city, schools, hospitals mandals in a district etc.), randomly choosing some of the clusters, and then sampling individuals from each chosen cluster. If carefully done, this can generate a representative sample with less expenditure, less time and less efforts than is involved in using any other probabilistic method.
- Cost reduction is a reason for using cluster sampling. It creates 'pockets' of sampled units instead of spreading the sample over the whole territory. Another reason is that, sometimes a list of all units in the population (a requirement when conducting SRS, Systematic sample or Sampling with PPS) is not available, while a list of all clusters is either available or easy to create.
- Sometimes, it is too expensive to spread a sample across the population as a whole. Travel costs can become expensive if interviewers have to survey people from one end of the country to the other. To reduce costs, researcher prefers to choose a cluster sampling technique.
- Cluster sampling involves selecting population units that are 'close' together and does not require all the population units to be listed. Cluster sampling has two advantages:
    - it eliminates the need for a complete list of all units in the population; and
    - it ensures that selected population units will be closer together, thus enumeration costs for personal interviews will be reduced, and field work will be simplified.

Demerits of Cluster Sampling

Despite all the above merits of cluster sampling, there are certain demerits associated with this technique and they are discussed here under.

- Other probabilistic sampling techniques provide lesser errors as compared to cluster sampling.
- Compared to other probabilistic sampling techniques, this technique offers sample that is the least representative of the population. So, with a cluster sample, there is a chance that the researcher can have an overrepresented or underrepresented sample, which can skew the results of the study.

- ☆ This technique often leads to high sampling error (for samples of comparable size than other probability samples), especially when limited clusters are included in the sample leaving off a significant proportion of the population unsampled. This is because, in this technique, each respondent is not chosen independently of the other respondents and after all, they are all neighbours and this may increase the sampling error. But, this error can be minimized by increasing the number of clusters. Further, if the clusters are similar to each other, this error can also be minimized. However, note this has the effect of increasing data collection costs.
- ☆ Elements in the same cluster may share similar characteristics.
- ☆ Provides less information per observation than a SRS of the same sample size.
- ☆ Standard errors of the estimates are high, compared to other sampling designs with same sample size. The reason for this is that, the listing units within the same cluster tend to be homogeneous with respect to many characteristics. Because of the homogeneity, the information collected from within a cluster tends to be redundant. In statistical terms, there is a high positive correlation between attributes within the respondents in the same cluster.
- ☆ The sampled clusters may not be as representative of the population. Hence, they likely to yield somewhat less accurate estimates of the population parameters than SRS or stratified random samples of the same size.
- ☆ Cluster sampling involves more complexity in analyzing the data.
- ☆ The more stages there are in a cluster sample design, the greater overall sampling error.
- ☆ If clusters are not similar to each other, the fewer the number of clusters, the greater the sampling error.
- ☆ Since elements within a cluster tend to be alike, we receive less new information about the population, when we select another element from that cluster rather than from another cluster. This lack of new information makes a cluster sample less precise than a simple random sample.
- ☆ Cluster sampling involves loss of efficiency when compared with SRS. It is usually better to survey a large number of small clusters instead of a small number of large clusters. This is because neighbouring units tend to be more alike, resulting in a sample that does not represent the whole spectrum of opinions or situations present in the overall population.

The above discussion infers that, if feasibility and economy are the only two criteria to choose a sampling technique, cluster sampling is the only technique to be used. On the other hand, if the researcher has to choose a sampling plan based solely on the precision of the estimates (low standard errors), then cluster sampling would never be used. So, cluster sampling generally provides less precision than either SRS or stratified random sampling. This is the main disadvantage of cluster sampling. But, still cluster sampling is preferred, as the cost per sample point is less compared for other sampling methods. As mentioned earlier, given a fixed budget, the researcher may be able to use a bigger sample size with cluster sampling than with the other techniques. When the increased sample size is sufficient to offset the loss in precision, cluster sampling may be the best choice.

When to Use Cluster Sampling

As discussed above, cluster sampling is the best choice, when the researcher aims at minimizing the cost for conducting a survey and these reduced costs can offset the losses in precision of the research study. In the following situations, cluster sampling is the best choice for conducting survey research:

- ☆ If it is difficult for the researcher to get the sampling frame of a target population and even if the sampling frame is available, but the target population may be widely spread over a largely geographical area, thus making data collection costs relatively high.
- ☆ If it is essential on the part of the researcher to reduce survey costs.
- ☆ If the researcher aims to check loss in precision of the estimates through increasing the sample size.
- ☆ If the population is concentrated in 'natural' clusters (say, city blocks, schools, hospitals etc.). For example, to conduct personal interviews of school teachers, it might make sense to randomly select a sample of schools (stage 1 of cluster sampling) and then interview all of the teachers at that selected school. Using cluster sampling, the interviewer could conduct many interviews in a single day at a single school. However, if SRS technique is employed, it might require the researcher to spend all day traveling to conduct a single interview at one single school and another interview at another school.

So, considering the above situations, cluster sampling is considered as the most cost-effective probabilistic approach, as it offsets the loss in precision of the estimates through increasing sample size at the same fixed costs.

Choice between SRS, Stratified Random Sampling and Cluster Sampling

The Table 8.13 and Figure 8.19 summarizes the choice making in using these three discussed probabilistic sampling techniques.

e. Sampling with PPS

As discussed earlier, there are two kinds of probability sampling through which the representative sample units can be drawn from the population *viz.*, equi-probability sampling and PPS. In the former, each unit or element in the population had an equal chance of being selected into the sample. The same is studied in case of SRS technique. For example, a researcher wishes to conduct a study on *'Economic Analysis of paddy production in Nandyal region of Kurnool district in Andhra Pradesh'*. Assume, there are 1000 farmers in a village. If the researcher wishes to select 300 farmers as a sample, each unit of the population has an equi-probability of entering into the sample (*i.e.*, 300/1000 = 0.3). However, the selected sample may not be true representative of the population. This is because, say, out of 1000 farmers, there are 500 marginal farmers, 400 small farmers and 100 large farmers. Since there is equi-probability for each unit of the population to enter into the sample, there is a chance of getting all 300 marginal farmers into the sample or 250 marginal farmers and 50 small farmers or 250 marginal farmers, 40 small farmers and 10 large farmers or all 300 small farmers or 250 small farmers and 50 large farmers or 250 small farmers, 40 marginal farmers and 10 large farmers etc. Since, the economics of crop cultivation varies across the categories of farmers, the above selected sample through equi-probability (*i.e.*, through SRS technique) may not be true representative of the population. Hence, to overcome this, the technique of PPS sampling is usually followed. That is, when the target population has different size categories and the number of elements in each sub-population category are different or not of equal size, then the sample is selected based on PPS. That is, the probability of selecting a number of sample units depends upon the size of the population units, when the population (farmers) has different size-categories. That is, when the population sizes is in the order marginal farmers>small farmers>large farmers, then sample sizes are also selected in the same order: marginal farmers>small farmers>large farmers. The same is discussed through Table 8.4.

Table 8.13: Choice between SRS, Stratified Random Sampling and Cluster Sampling.

Item	*SRS*	*Stratified Random Sampling*	*Cluster Sampling*
Sample size required to achieve reliable estimates	Medium	Low	High
Cost involved per each respondent sampled	Medium	High	Low
Circumstance under which the technique used	When the population has no natural divisions (*i.e.*, homogeneous)	When the population units are heterogeneous based on a stratification variable(s) identified. There is heterogeneity between the strata and homogeneity of units with in each stratum (Figure 8.20)	When the population divides itself naturally into geographic 'clusters', and there is homogeneity between the clusters and heterogeneity within each cluster.
Procedure	Sample units are drawn randomly from population just like a lottery method	After dividing the population into different heterogeneous strata, sample is drawn from each stratum through SRS. Note that, the SRS adopted in each stratum may be proportionate or disproportionate.	The entire target population is divided into different homogenous clusters, then some clusters are selected randomly. From each cluster, sample units are selected either through SRS or systematic or stratified random sampling. PPS is also adopted to select both clusters and sample units.

Thus, the technique of PPS is adopted when the target population can be divided into different sub-groups and these sub-groups vary in their sizes. So, if representative samples are taken from each sub-group, then the overall sample can be representative to the total population. In such a situation PPS is best technique. This is so because, it is a known fact that, when sample units are drawn with equal probability from different sized sub-groups, the chances of selecting a unit from a larger sub-group are less than selecting a unit from a smaller sub-group. In the above example, for selecting each marginal farmer the probability is 1/500, for selecting each small farmer the probability is 1/400 and for selecting each large farmer, the probability is 1/100. So, the probability of selecting each unit from different sub-groups vary inversely with the size of the sub-group. Hence, it is appropriate to go for PPS technique, where large samples are drawn from large sized sub-groups and *vice versa* such that, the overall sample drawn will be representative of the total population. Thus, PPS takes varying sample sizes from different sized sub-groups into account and this helps to avoid under-representing or over-reporting one subgroup in a study and yields more accurate results. PPS can be combined with other techniques for selecting samples. For example, it can be combined with a systematic sampling and it is exemplified here. Suppose, there are six hospitals with populations of 130, 195, 225, 185, 240 and 525 patients respectively (total 1500 patients), and we want to use these patients population as the basis for a PPS sample of size three hospitals. To do this, we could allocate the first hospital numbers 1 to 130, the second hospital 131 to 325 (= 130 + 195), the third hospital 326 to 550 (325 + 225), the fourth hospital 551 to 735 (550 + 185), the fifth hospital 736 to 975 (735 + 240) and sixth hospital

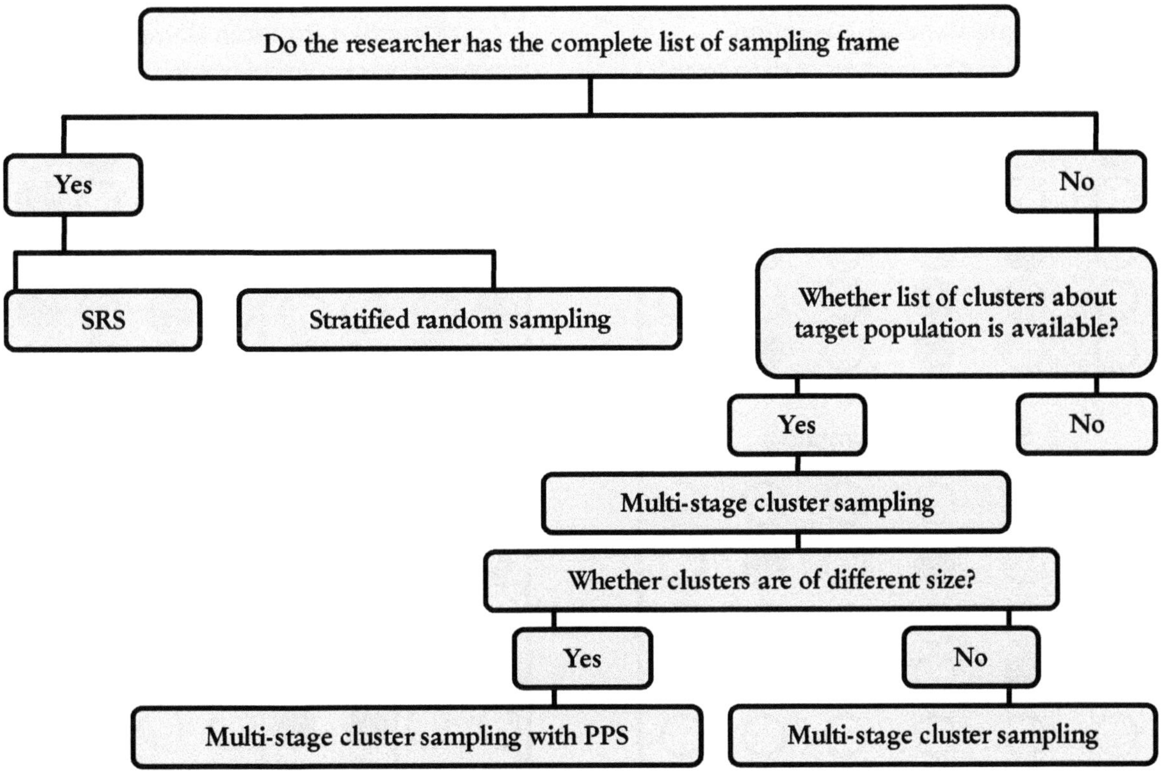

Figure 8.19: Choice between SRS, Stratified Random Sampling and Cluster Sampling.

976 to 1500 (975 + 525). The sampling interval is given by 1500/3 = 500. Let, the random number selected between 1 to 500 is 140 and then count through the hospital populations by multiples of 500. So, the subsequent numbers will be 640 and 1140. So, the researcher will select second, fourth and sixth hospitals for the investigation. Similarly, PPS can be combined with cluster sampling and the same is already discussed earlier with reference to the Table 8.12.

This PPS sampling approach can improve accuracy for a given sample size by concentrating sample on large elements that have the greatest impact on population estimates. Note that, probability sampling requires that, each member of the survey population have a chance of being included in the sample, but it does not require that, this chance be the same for everyone. If there is information available on the sampling frame about the size of each unit (*e.g.*, number of farmers in each category *i.e.*, marginal, small and large) and if those units vary in size, this PPS can be used in the sampling selection in order to increase the efficiency. So, with this method, the bigger the size of the sub-group, the higher the chance it has of being included in the sample and the success of this method (to bring increased efficiency) depends upon the accurate measurement of the sizes of the sub-groups in the target population.

Merits of PPS Sampling

They include:

- ☆ This approach can improve accuracy for a given sample size by concentrating sample on large elements that have the greatest impact on population estimates (Figure 8.20).

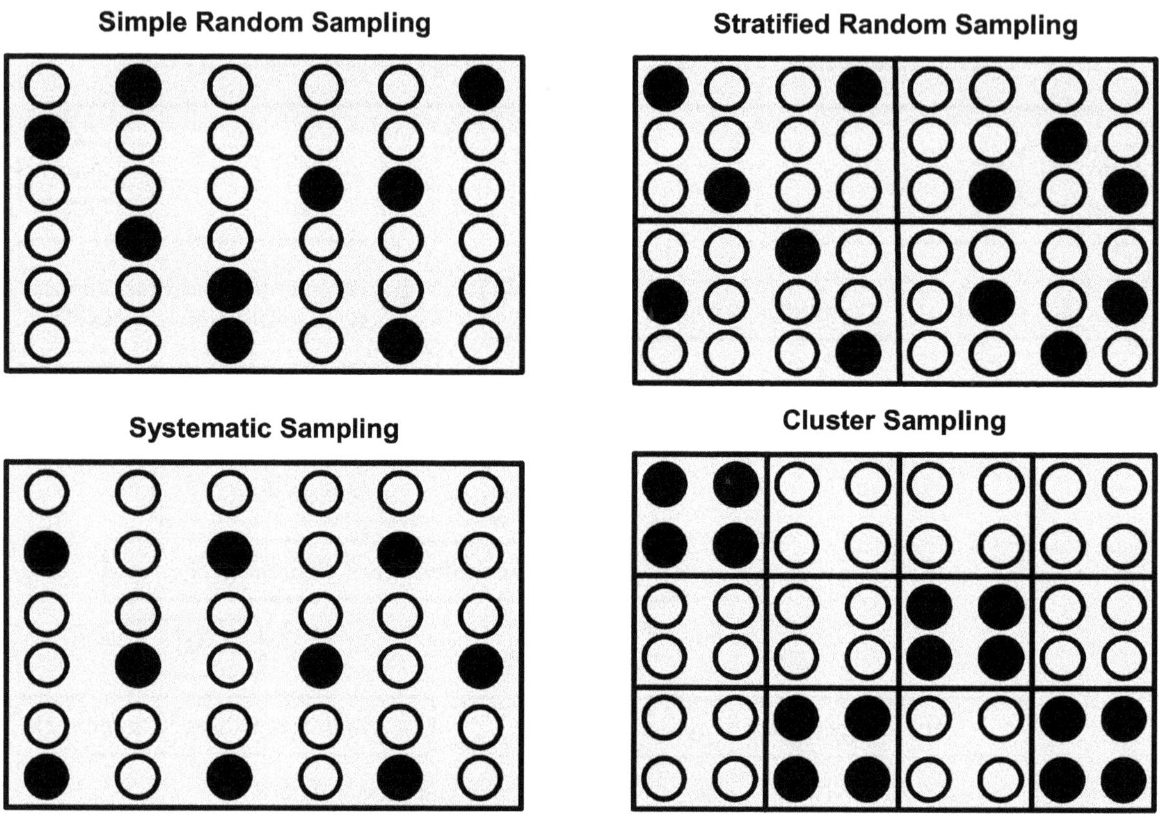

Figure 8.20: Selection of Simple Random Sample from different Probabilistic Sampling Techniques.

- ☆ This technique can improve the statistical efficiency of the sampling strategy by using auxiliary (variable) information. This can result in a dramatic reduction in the sampling variance compared with SRS or even stratified sampling.

Demerits of PPS Sampling

The disadvantages of PPS sampling are:

- ☆ It requires a sampling frame that contains good quality, up-to-date auxiliary information for all units on the frame that can be used as size measures.
- ☆ It is inappropriate, if the size measures are not accurate. In such circumstances, it is better to create strata and perform stratified sampling.
- ☆ Estimation of the sampling variance of an estimate is more complex.
- ☆ Frame creation is more costly and complex than SRS or Systematic sampling, since the size of each unit in the population needs to be measured.

f. Multi-Stage Sampling

Thus far, the discussion has centered around one stage sample designs. But, in multi-stage sampling, the process of selecting a sample is done in two or more successive stages. This technique is a complex form of cluster sampling and is used under the following situations:

- ☆ When listing of all the units in the target population does not exist, impossible and costly.
- ☆ When conducting studies involve a very large population. This is because, when the target population is very large, the SRS becomes unfeasible.
- ☆ Considering all the elements in all the selected clusters may be prohibitively expensive or not necessary. So, instead of enumerating all the sampling units in the selected clusters, the researcher can resort to sub-sampling within the clusters and randomly selects representative units from each cluster till the desired final sample is reached.

So, under the above circumstances, researcher must use other forms of sampling. One such form is multi-stage sampling. The simple form of multi-stage sampling is a two-stage sampling, where the naturally occurring clusters being termed as 'primary units' and the units randomly selected within these clusters as 'secondary units'. As the name 'multi-stage sampling' suggests, it refers to a sampling technique, which is carried out in various stages. Here, the target population is regarded as made of a number of primary units, each of which is further composed of a number of secondary stage units and so on, till the researcher ultimately reach the desired representative sample size in which he is interested. This technique is called Sub-sampling, since the sample is taken in more than one stage. The firstly selected sample units are called the 'primary units', and then select a sample of sub-units called as 'secondary units' from these primary units. First, the entire population is divided into naturally-occurring clusters. Now the researcher has to decide, what elements within each cluster are to be considered. In some cases, several levels of cluster selection may be applied before the final sample units are reached. That is, after forming the naturally occurring clusters, the researcher again forms several sub-clusters, from which the researcher randomly selects the final sample. For example, a researcher want to conduct a market survey of sale of a particular brand of (new) pesticide among the farmers all over the country. So, the researcher has to select a representative sample in the country and hence, he randomly select States (First stage), randomly select districts in each selected State (Second stage), randomly select mandals in each selected district (Third stage), randomly select villages in each selected mandal (Fourth stage) and randomly select farmers in each selected village (Fifth stage). This technique means that, it is not necessary to create a list of every district in all the States, but only for the selected States. Similar is the case for the selection of mandals, villages and farmers. So, the sampling frame is to be prepared only for the selected units. In remote areas, if possible, an additional stage of clustering is used, in order to reduce travel requirements.

Note that, multi-stage sampling involves selecting a sample in at least two stages. At the first stage, large groups or clusters of population units are selected. These clusters are designed to contain more units than are required for a final sample. At the second stage, units are sampled from the selected clusters to derive the final sample. If more than two stages are used, the process of selecting 'sub-clusters' within clusters continues until the final sample is achieved. The same practical considerations apply to multi-stage sampling as to the cluster sampling.

Merits of Multi-Stage Sampling

The use of this sampling technique ensures the following advantages:

- ☆ This technique is generally used when it is costly or impossible to form a list of all the units in the target population. No doubt, a multi-stage sample gives less precise estimates than a simple random sample of the same size. However, a multi-stage sample is often more precise than a simple random sample of the same cost, and it is for this reason, this method is employed.

- The advantages of multi-stage sampling are similar to those for cluster sampling. However, to compensate for the lower accuracy, either the number of clusters selected in the first stage should be relatively large (but this increases the cost of the survey) or the sampling fraction for later stages should be high (*i.e.*, a large percentage of each cluster should be selected).
- Ensures more convenience of finding the survey sample.
- Normally more accurate than cluster sampling for the same size sample.
- Ensures more administrative convenience on the part of the researcher in handling the desired sample size. That is, it avoid the problems of randomly sampling from a population that is larger than the researcher's resources can handle.
- The multi-stage form of sampling is flexible in many senses. First, it allows researchers to employ random sampling or cluster sampling after the determination of groups. Second, researchers can employ multi-stage sampling indefinitely to break down groups and subgroups into smaller groups until the researcher reaches the desired type or size of groups.
- There are no restrictions on how a researcher divide the population into sub-groups.
- This technique reduces considerably the cost of operations of the survey and helps to reduce the non-sampling errors by efficient supervision.
- This technique is indispensable, especially in forest surveys, where it often happens that detailed information may be easily available only for groups of sampling units, but not for individual unit.
- The researcher may employ any type of probabilistic sampling technique selection at different stages in selecting the sub-groups. For example, the first stage may use random sampling, the second stage may use purposive sampling, and the third stage may use stratified sampling.

Demerits of Multi-Stage Sampling

Though this technique offers several advantages, it is not free from certain demerits and they are listed here under.

- Since, no restrictions are involved in selecting the sub-groups, this technique has a level of subjectivity. Thus, there will always be questions as to whether the chosen groups were optimal. So, the researcher must find a way to justify their choices when presenting the study's findings.
- Since the target population is sub-divided at each and every stage, the estimates of the selected sample can never be 100 per cent representative of the population.
- It is not as accurate as SRS, if the sample is of the same size.

Cluster sampling is similar to multi-stage sampling, in that, it involves two or more stages in the sampling process. Similar to multi-stage sampling, cluster sampling is often used, when larger population or geographical area are to be considered for collecting the sample. But, the essential difference between generalized multi-stage sampling and cluster sampling is that, in the latter, at the final point of selection, clusters are small enough for everybody in that cluster to be interviewed by the researcher.

g. Multi-Phase Sampling

In this technique of sampling, some general information is collected from the whole sample and additional or detailed information is collected at the same time or later, from sub-samples taken from the whole sample. That is, sub-sample or only some units of whole sample provide detailed information than others. This multi-phase sampling is generally followed, when the variable of interest is difficult or expensive to measure. Hence, from the whole sample, a small sub-sample will be selected to draw more information about the interested variable. From this sub-sample, a relationship is then derived, which can be used to predict values of the required variable for the whole sample. Thus, although multi-phase sampling also involves taking two or more samples, all samples are drawn from the same frame and the units have the same structure at each phase. A multi-phase sample collects basic information from a large sample of units and then, from a subsample of these units, collects more detailed information. The most common form of multi-phase sampling is Two-phase sampling (or Double sampling or Post-stratification sampling), but three or more phases are also possible. This technique is also used, when the sampling frame lacks auxiliary information that could be used to stratify the population or to screen out part of the whole population. This technique can be understood from the following examples:

- ☆ ***Example 1:*** Suppose, the Department of Animal Husbandry in Kurnool district wishes to collect information on cattle farms in the entire district. But, the Department has the sampling frame listing various farms like agricultural farms, piggery units, sheep units, dairy units, poultry units etc. So, the best way the Department personnel can collect relevant information is by posing a question through a telephone: 'Is part or all of your farm devoted to cattle farming?' randomly across the units. This makes the Department to draw a large sample from the sampling frame and this can be termed as 'whole population' with reference to cattle farms in the district. Once the first sample (whole population) has been drawn, a second, smaller sample can be extracted from among the cattle farmers and more detailed survey can be conducted among them to elicit information on the interested variables. It is to be noted that, in drawing the first sample, a low expensive method (telephonic enquiry) is used and while studying the sub-sample for drawing information on interested variables, a more expensive method (personal interviews) is used. So, by using this method, the Department avoids the expenses of surveying units that are not relevant to the scope of the study (*i.e.*, non-cattle farmers).
- ☆ ***Example 2:*** A researcher want to collect information from the people in Kurnool district regarding their health consciousness through posing some basic questions about their diet, smoking habits, exercise routines and alcohol consumption. Further, the respondents are to be tested through blood tests for analyzing lipid profile and blood sugar levels (pre and post). Conducting random survey and interviewing the persons directly is less expensive, but conducting blood tests for analyzing lipid profile and blood sugar levels is highly expensive. So, the best way to conduct this survey would be to use a two-phase sample approach. In the first phase, the interviews are performed on an appropriately sized sample. From this sample, a smaller sub-sample is drawn. This sub-sample will take part in the blood tests.
- ☆ ***Example 3:*** This technique of sampling can be employed in selecting the requisite sample from a larger population through aerial surveys. Say, a researcher wishes to take field measurements on teak wood trees in a forest. In the first phase, he conducts an aerial survey to classify 'forest area' and 'non-forest area'. After locating the forest area, in the second phase, he takes detailed photographs about teak wood plantations. In the third phase, he further selects a sub-sample of teak wood plots for conducting field measurements.

The above examples also imply that, multi-phase sampling can be employed, when there is insufficient budget to collect information from the whole population and when lot of expenditure is to be involved to measure the interested variables from the whole population.

Differences between Multi-Stage Sampling and Multi-Phase Sampling

Though there is a similarity in their names, multi-phase sampling is quite different from multi-stage sampling and they are listed here under.

- ☆ In multi-stage sampling, the sampling units for the different stages are different. On the other hand, in multi-phase sampling, the same sampling unit is sampled multiple times.
- ☆ In multi-stage sampling, there is a single purpose throughout the sampling, but in multi-phase sampling, the purpose may change at each stage.
- ☆ In multi-stage sampling, at each successive stage, the sample unit is different, whereas in multi-phase sampling, the unit of sample remains same for getting overall information and additional information is collected from only a part from the original sample.

Merits of Multi-Phase Sampling

Multi-phase sampling is similar to multi-stage sampling except that, the primary and secondary units are the same type. The merits of multi-phase sampling include:

- ☆ Multi-phase sampling is carried out to increase precision, reduce costs, and reduce non-response.
- ☆ In case of stratified random sampling, the researcher must have the complete list of sampling frame (also SRS) and he should be aware of stratification variables(s) for conducting in depth study. However, multi-phase sampling is feasible, even when the researcher do not have complete or relevant sampling frame.
- ☆ Multi-phase sampling is employed to reduce data collection costs, if it takes more time and effort to collect data on interested variables. In Phase 1, the easily accessible data may be collected from the whole sample. In Phase 2 and other subsequent phases, detailed investigation about interested variables is conducted. Since, the sample size is reduced in sub-samples, the data collection costs are minimized.

Demerits of Multi-Phase Sampling

They include:

- ☆ Its planning and administration are complicated.
- ☆ It is more complicated to implement this technique and may increase sampling error for estimates based on second set of variables, if strong relationship with variables in first set does not exist.

II. Non-Probability Sampling

The following are the reasons for using non-probability sampling:

- ☆ No sampling frame is available.
- ☆ Bias resulting from non-probability sampling is negligible.
- ☆ The sample size is too small.
- ☆ To demonstrate that a particular trait exists in the population.

- ☆ It can also be used when the researcher aims to do a qualitative, pilot or exploratory study.
- ☆ It can be used when randomization is impossible like when the population is almost limitless.
- ☆ It can be used when the research does not aim to generate results that will be used to create generalizations pertaining to the entire population.
- ☆ It is also useful when the researcher has limited budget, time and workforce.
- ☆ This technique can also be used as an initial study, which will be carried out again using a randomized or probability sampling.

a. Quota Sampling

This is one of the most common forms of non-probability sampling and is similar to stratified random sampling, but with some major differences. Both these techniques involve the division of target population into different categories and then selection of a representative sample from each category. Thus, both these techniques facilitates sub-group analyses. However, there are important differences. Stratified random sampling involves random selection of sample units from each stratum, whereas quota sampling utilizes availability sampling (*i.e.*, it is usually left up to the interviewer to decide who is to be sampled). For classifying the target population into different strata and for selecting sample from each stratum, a sampling frame is required for stratified random sampling and it is not needed for quota sam-pling. More importantly, stratified sampling is a probability sampling procedure permitting the estimation of sampling error. This is not possible with quota sampling. In stratified random sampling, the target population is divided into different strata so that, each stratum has a common characteristic, such as size of land holding or gender or income etc. Through using equi-probability (random) sampling, samples are drawn from each stratum, unlike in quota sampling where the samples are drawn at the will or interest of the researcher on non-random basis. Thus, the sample selected under quota sampling is on the basis of pre-specified characteristics so that, the same distribution of characteristics is assumed to exist in the population being studied.

Thus, like stratified random sampling, quota sampling also involves selection of sample units from different categories of target population. The target population is first segmented into mutually exclusive sub-groups, just as in the stratified random sampling. But, from each sub-group, the sample units are not drawn at random as in the case of stratified random sampling, but based on the criteria considered by the researcher such as income, sex, size of land holding etc., and the researcher may also select the sample respondents, who look most helpful in providing the requisite information *i.e.*, it is usually left up to the researcher to decide who is to be sampled. So, the size of the sample drawn from each sub-group does not reflect the population as a whole. The number of respondents (quota) that are to be drawn from each of sub-group is specified in advance and the final selection of respondents is left to the interviewer, who proceeds until the quota for each category is filled. For example, a researcher may select a sample of 50 men farmers and 50 women farmers between the age of 30 to 50. So, it is in the second step, which makes the technique as non-probability sampling, as the selection of the sample is non-random. So, the sample selected is considered as a biased sample, as every unit in the target population had not an equal chance of being entered into the sample. This absence of random element is its greatest weakness compared to stratified random sampling. So, it can be said that, quota sampling is the non probability version of stratified random sampling. Similarly, if a researcher wishes to conduct a national quota sample, he might need to know, what proportion of the population is male and what proportion is female, what proportions of each gender fall into different age categories, educational categories, etc. The researcher would then collect a sample from each sub-group with the same proportions as the national population. Quota sampling is similar to Stratified random sampling in the following aspects:

- ✰ Population is divided into sub-groups and sample elements are selected from each sub-group.
- ✰ Purpose is to select a representative sample from each sub-group and this facilitates sub-group analysis.

However, quota sampling differs from stratified random sampling in the following aspects:

- ✰ In quota sampling, each element from the sub-group is not selected randomly, unlike stratified random sampling. So, a sampling frame is not required in the former.
- ✰ Sampling error is more in quota sampling and this is due to selection bias. However, in stratified random sampling, selection bias is totally absent, as the sample units are selected randomly from each sub-group.

Quota sampling ensure the sample chosen by the researcher is a representative by selecting elements in such a way that, the proportion of the sample elements possessing a certain characteristic is approximately the same as the proportion of the elements with the characteristic in the population (Figure 8.21). For example, if the researcher wishes to select a proportional quota sample of 100 farmers based on gender, first, he should know, what is the proportion of male farmers and female farmers in the target population. If the data indicate that, there are 70 per cent and 30 per cent of male and female farmers in the target population (N=1000), then the researcher has to select 70 men farmers

Category	Population		Sample	
	Mandal 1	Mandal 2	Mandal 1	Mandal 2
Marginal farmers	682 (42.55)	735 (43.88)		
Small farmers	596 (37.18)	623 (37.19)		
Large farmers	325 (20.27)	317 (18.93)		
Total	1603 (100.00)	1675 (100.00)	300	300

Figure 8.21: Selection of Quota Sample in different Categories in Proportion to the Elements in Respective Sub-populations.

and 30 women farmers in the total of 100 sample respondents (farmers). That is, the researcher will sample from each sub-group based on the gender (and note that, not based on random selection) till he gets the desired quota. If the researcher already got 30 female farmers and not men farmers in the process of sampling from each sub-group, he would continue to sample men farmers alone and discard sampling of women farmers because, he has already met the quota of 30 women farmers. So, the important aspect in quota sampling is that, the researcher has to specify the characteristic in advance and based on this, he has to sample each sub-group up to the required quota only that too non-randomly. So, in this technique, the researcher interviews as many people in each sub-group as he can find, until he has filled his quota and note that, the quota or size of sample in each sub-group will be based on the population proportions. This above methodology was well-explained considering another good example as given below.

Example 1

The Academic Council of the University wants to collect the students opinion about the revised syllabi (course curriculum) prescribed to B.Sc. (Agriculture) students in Agricultural College, Mahanandi. The Council has decided to survey 100 out of 1000 students in the college (Table 8.14).

Table 8.14: Year-wise Number of Students Selected as Quota Sample.

Year	*Number of Students*	*Percentage of Students (per cent)= (Number of students in the year/Total number of students) x 100*	*Quota of Students in a Sample of 100 =(Percentage of students) x 100*
I year	170	17	17
II year	250	25	25
III year	350	35	35
IV year	230	23	23
Total	1000	100	100

The Table 8.14 also highlights the difference between stratified random sampling and quota sampling. In the former, the students are selected from each stratum (year) through a probability sampling method (SRS), whereas in quota sampling, no such technique is used. Say, in I year, the 17 students might be selected by choosing the first 17 students, who first entered the class room on a certain day or by choosing 17 students from the first three rows in the classroom. Note that, the researcher simply exclude those students who arrive late or sit at the back of the class from fourth row. He may even assume that, the students from fourth row onwards hold different opinions about the revised course curriculum than those, who arrived earlier or sat in the front three rows. So, considering the proportions of categories (year-wise students) in total population, the researcher has to select the sample across the categories in the same proportion in the total sample to be derived. The researcher cannot choose more than these quotas, even if he comes across eligible students after reaching the quotas.

This quota sampling can be considered preferable to other forms of non-probability sampling (*e.g.*, judgment sampling) because, it forces the inclusion of members of different sub-populations. This quota sampling is generally viewed as two-staged restricted judgmental sampling technique. The first stage consists of developing categories of population elements based on sex or income or size of land holding etc. Then, based on the selected category (say, sex of the individual), its distribution in the target population is determined.

Merits of Quota Sampling

The following points highlights the advantages of quota sampling technique:

- ✫ Quota sampling is generally less expensive than SRS.
- ✫ It is also easy to administer, as the researcher simply aims at selecting the requisite quota of sample from the target population.
- ✫ Quota sampling is an effective sampling method when information is urgently required and can be carried out independent of existing sampling frames.
- ✫ In many cases, where the population has no suitable frame, quota sampling may be the only appropriate sampling method.
- ✫ It is simpler to undertake than a stratified random sample.
- ✫ This method is convenient, less costly and can include the respondents from all the sub-groups of the target population.
- ✫ This technique is used, when the researcher knows about the proportions of particular sub-groups within a population and this ensure him to select the sample units in the same proportion.
- ✫ This technique is useful when time is limited, research budget is very tight or when detailed accuracy is not important.

Demerits of Quota Sampling

The following are the limitations involved with this sampling technique:

- ✫ Since the sample units are selected from each sub-group based on non-random basis, it results in selection bias. The selection bias may result in:
  - ❖ the quota reflecting the population in terms of superficial characteristics.
  - ❖ selecting the respondents based on availability rather than on their suitability to the study.
- ✫ This technique does not meet the basic requirement of randomness. Some units may have no chance of selection or the chance of selection may be unknown. Therefore, the sample may be biased.
- ✫ The selected sample is not as representative of the population as a whole as other sampling methods.
- ✫ Because the sample is drawn on non-random basis, it is impossible to assess the possible sampling error.
- ✫ The quota frame *i.e.*, the list of populations in each sub-group must be available and it must be accurate. This is often difficult because, it can be hard to find up-to-date information and data on certain characteristics of the population.

b. Convenience Sampling

Convenience sampling is, otherwise, referred to as *Accidental sampling* because, the elements or units will enter into the sample by 'accident'. It is also called as 'Haphazard sampling', as the selection of the respondents is left to the discretion of the interviewer. That is, based on his convenience and availability of units, the researcher will select the sample. Hence, the selected sample may not be true representative of the target population. So, the findings of the study are biased in nature. This technique is also known as 'Grab' or 'Opportunity' sampling. The following are the examples for selecting a convenience sample:

- A food processing firm may select few persons in its locality to evaluate the quality of its food product.
- A researcher may interview few farmers in a village near to his research station to study the yield of the paddy variety released by him.
- A cell phone shop owner interview first 100 customers entering into his showroom.
- A television reporters interview first 10 persons walking on the street about how people view an issue.
- A pollution expert may interview few people in the village to determine, whether the tank water meant for crop irrigation is polluted or not.

In all the above cases, the sample is chosen as convenient by the researcher, without use of a survey method. It is a type of non-probability sampling, as the researcher selected the sample from the population, which is close to his hand or at his nearby village. He may collect the information from the sample units either through informal discussions or through telephone or through internet etc. But, the findings generated from the study cannot be generalized for the target population, as the selected sample units are not true representative of the target population. In one of the above examples if the cell phone shop owner interviews the customers entering his showroom in the morning (10.00 AM to 11.30 AM) and in the evening (7.00 PM to 8.30 PM), the customers he could interview would be limited to those given there at that given time, which would not represent the views of other members in that locality. This is a non-probability sample because, the shop owner has systematically excluded all other people, who visit the showroom other than the above working hours and people visiting other showrooms in that locality.

Merits of Convenience Sampling

They include:

- This technique of sampling is most useful for pilot testing, where the objective is to measure validation rather than obtaining generalizable inferences.
- The technique is easy to use.
- It can deliver accurate results, when the target population is homogeneous.
- It is the least expensive and least time consuming of all sampling techniques.
- This technique of sampling is suitable for exploratory studies.
- This technique is useful in pre-testing of the schedule.

Demerits of Convenience Sampling

They include:

- The disadvantage with convenience sampling is that, the researcher would have no way of knowing whether the sample chosen is representative to the target population.
- Since selected sample is not representative of the target population, the results that are received are biased in nature. Hence, the scientific generalizability of research findings will be very limited.
- This technique of sampling is not suitable for survey research.

c. Judgment Sampling

This technique is also known as Purposive sampling. In this technique, the researcher will select the sample based on certain pre-determined criteria based on his experience in conducting the research.

The criteria judged by the researcher for selecting the representative sample is considered to be most appropriate for the purpose of conducting the research. The type of data needed, tools of analysis to be used etc., also guide the researcher in selecting the sample. So, the personal judgement of the researcher guides him in selecting the sample. The distinguishing feature of judgment sampling is that, the sample units are purposively selected. Again, the sample selection is not based on probability and hence, it is not representative to the target population, but because the selected sample contribute to requisite data and information. That is, this technique involves selecting the sample units from the population purposively that specifically meet some prescribed purpose or possess specific attributes that the researcher is interested in conducting the investigation. Hence, all the population units do not get the chance to get selected into the sample. This technique is usually an extension of Convenience sampling. This judgment sampling is most frequently used in marketing research. When using this technique, the researcher must be confident that, the chosen sample is truly representative of the entire population. For example, a pesticide firm will select few districts in a State for test marketing of its new brand of pesticide. This is because, it presumes that, the population of farmers in those few districts is representative of the total farming community in the whole State. The same is the case regarding the exit polls survey conducted in few States of the whole country with the presumption that, the voting pattern in these selected States reflect the same trend throughout the country.

Merits of Judgment Sampling

They include:

- ✰ This technique of sampling is considered to be most appropriate, if the target population to be studied is difficult to locate or if some members are thought to be better (more knowledgeable, more willing to answer the questions framed in the schedule etc) than others to interview. Say, for example, if the researcher wishes to collect information from farmers pertaining to the cost of cultivation of paddy, he might find that, women farmers are relatively few in number in the selected village and some of them could not provide information to the researcher will plan they are busy with their household activities. In such case, the researcher has to collect more data from the men farmers and comparatively less from women farmers.
- ✰ In some cases, it is better for the researcher to rely on few knowledgeable members for accessing the data and information rather than wasting his time and other resources in interacting with a huge representative sample of the target population.
- ✰ This sampling technique also guides the researcher in selecting the nature of sample (say, based on gender). For example, if the researcher wishes to collect information regarding, 'what it takes the women professionals to reach to the top positions in agricultural sector', then the obvious choice for him is to collect information from professional women that too working at top level in agricultural sector (say, in cooperative banks, agro-processing firms, SAUs etc). The researcher expects that, the selected sample will provide quality information by virtue of their knowledge and enough experience in their respective professions.
- ✰ The above discussion also highlight that, judgment sampling technique is used, when a limited number of total population has the requisite information sought by the researcher. In such case, any type of probability sampling as discussed earlier is not useful.
- ✰ Ensures lower cost of sampling
- ✰ Lesser time involved in sampling process

- As the selected sample are well versed with the data and information to be provided to the researcher, there are lesser chances of distortion of data.
- This sampling technique can be used for pre-testing of instruments like schedules, questionnaires etc.
- This technique serves as a popular method for student's research projects.
- This technique is used primarily in *ex-post facto* research, where the researcher is interested in finding a possible cause-and-effect relationship between two variables, one of which has already occurred.

Demerits of Judgment Sampling

They include:

- Since each member in the population does not have an equal chance of being chosen, a judgment sample is also a non-random sampling method. Hence, the selected sample is not a true representative of the target population. So, the sample estimates cannot be generalized to the population.
- Researcher's bias might set in selecting the sample units.
- If the population is very large, it is not possible to judge the reliable sample units to elicit the information.

d. Snowball Sampling

This is another popular non-probability sampling technique widely employed in academic research. This technique is also called as Chain sampling or Chain-referral sampling or Network sampling or Referral sampling or Respondent-driven sampling. In this technique, first the researcher selects a random sample of respondents and elicits information from them. Later these respondents are asked to identify other sample units, who belong to the target population of interest. So, subsequent respondents are selected based on the information provided by the initial selected respondents. So, this technique of sampling involves both probability and non-probability methods *i.e.*, the initial respondents are chosen by a random method (probabilistic method) and the subsequent respondents are chosen by non-probability methods (as suggested by the initial respondents). This process of enquiry for further respondents is to be continued, until the researcher have studied the desired sample size. That is, this sampling technique stops, when the so-called saturation sample is spotted. That is, a saturation sample is a point of sampling, where new respondents are now unable to give new information that generally depart from those given by previous respondents (referrals). A saturation respondent or informant does not give new data-value and this gives the signal that, getting new samples units is not anymore necessary. Thus, the technique of sampling, which involve the selection of additional respondents from initial respondents is known as Snowball sampling. In the words of Arce, *'snowball sampling results from one key informant being interviewed and asked for suggestions on who else might be good informants, and the next informant is chosen on the basis of this suggestion.'* Note that, the characteristics of sample respondents chosen for the study by both probability and non-probability methods must be similar so that, the total sample selected for the investigation is representative to the target population. This sampling technique can be particularly useful, when complete list of sampling frame is not available or when it is difficult to locate the target population or time constraint or cost constraint on the part of the researcher. For example, this technique is used to locate sample units in case of homeless individuals, migrant workers etc. Similarly, a researcher wishes to conduct a study on *'Economic analysis of paddy production through organic farming in Kurnool district of Andhra Pradesh'*.

In the first enquiry, he may select 10 random farmers in his nearby village. After collecting the requisite information, he asks them, if they know any more farmers practicing paddy production through organic farming. They give him several further referrals, who in turn provide additional contacts. In this way, he manages to contact sufficient farmers practicing paddy production through organic farming. Thus, 'snowball' effect occurs, as referrals multiply at each step. In this example, initially researcher surveyed 10 farmers in a village and later the number may multiply as 20, 40, 80 and so on.

Merits of Snowball Sampling

The following are the advantages of using snowball sampling technique:

- ✰ This sampling technique is a useful choice of locating the sample units, when they are hard-to-reach or when sampling frame is not available.
- ✰ This technique of chain referral process helps the researcher to locate the sample units that are otherwise difficult in accessing them through using other sampling methods.
- ✰ The process of sample selection is cheap, simple and cost-efficient.
- ✰ This sampling technique needs little planning and fewer workforce compared to other sampling techniques.
- ✰ The individuals, who are engaged in illicit or illegal activities say, drug users can be easily located through this technique. These drug users are likely to be less willing to identify themselves in the society and do not take part in a piece of research than many other social groups. So, this technique helps the researcher to locate them, as it involves individuals identifying other individuals with similar characteristics to take part in the selected research.
- ✰ Since this sampling technique is hardly representative of the (larger) target population, it is primarily used for exploratory research.

Demerits of Snowball Sampling

The following are the disadvantages associated with this sampling technique:

- ✰ Though this technique is cost-efficient, it leads to bias in the sampling process and hence, the sample estimates are not representative to the target population. To get the referral names from the initial respondents, the researcher has to maintain a good relationship with them and this leads to bias in the sampling process.
- ✰ Sometimes, the referral names obtained from those sampled in the initial stages may be similar to those initially sampled. So, the selected sample may not represent a cross-section of the total population.
- ✰ The sample respondents selected in the initial stage may refuse to disclose the names of those, whom they know.
- ✰ Since the researcher has to depend upon the referral names given by the initial respondents, he has little control over the sampling method.
- ✰ Since this technique does not select total sample units based on random selection, unlike in probability sampling techniques, it is impossible to determine the possible sampling error and make statistical inferences from the sample to the population.
- ✰ This sampling technique provides inaccurate results because, the sample units selected in the initial stage will have a strong impact on the overall sample. Further, this technique is heavily relied on the skill of the researcher in locating and selecting the sample units and in collecting the requisite information.

- There is no way to know the total size of the overall population.
- There is lack of knowledge on the part of the researcher to know, whether the selected sample is accurate enough to read the total population, as first of all, he does not know the total population size.
- Even though the initial respondents provide the referral names of the subsequent respondents, it may consume lot of time in locating the correct sample units, which renders the benefits only slightly outweighing the costs.

However, with care in selection of sample respondents and by avoiding personal bias, snowball sampling can still be a useful method, particularly if it is difficult to locate the target population.

e. Volunteer Sampling

As the term implies, this type of sampling occurs, when people volunteer their services for the study. That is, in this type of sampling, the researcher appeals to people to voluntarily participate in a survey. The resulting sample is called a 'voluntary response sample'. So, a voluntary sample is made up of people, who self-select into the survey. Often, these sample respondents have a strong interest in the main topic of the survey and hence, they themselves become part of the research study. Often, 'voluntary response sample' over sample people who have strong opinions and under sample people who don't care much about the topic of the survey. This introduces biasedness in getting the information for the selected topic of investigation. Thus, inferences drawn from a voluntary response sample are not as trustworthy as conclusions based on a random sample of the target population. For example, TV and Radio media may ask the audience to participate in an on-line poll for rating an agricultural technology dissemination programme. This would be a voluntary sample, as the sample is chosen by the viewers themselves and not by the survey administrator of the media. It should also be noted that, the people who contribute to these surveys might have different views and are much more interested in that topic than others who do not participate in the survey. Similarly, TV and Radio media often use call-in polls to informally query an audience on their views about a particular topic say on 'irrigation policy framed by the Government', 'MSP policy being followed by the Government', 'quality of agricultural education in the State' etc. Oftentimes, the survey administrator will not impose any limit on the frequency or number of calls one respondent can make. But, it may lead to biasedness, as the same person might be able to vote repeatedly. Note that, all types of sample surveys involve some degree of volunteering and there is no fixed line between a volunteer sample and a probability sample. The main difference between a pure volunteer sample and a probability sample of volunteers is that, in the former case, volunteers make all the effort and no sampling frame is used.

Merits of Volunteer Sampling

The merits of volunteer sampling include:

- No sampling frame is required to select the respondents. This saves considerable time and energy on the part of the researcher.
- The researcher need not convince the respondents about the importance of the selected topic.
- It is relatively easy and quick to select the sample. This technique can also reach a wide variety of participants.

Demerits of Volunteer Sampling

They include:

- This technique may introduce strong bias. Often in opinion polling, only the people who care strongly enough about the subject will respond and the silent majority does not typically respond and this results in large selection bias. So, the selected sample may not be true representative of the target population.
- Most individuals volunteer because, they have something to say, more often negative than positive. The responses are then biased. The researcher may not have adequate control over the makeup of the sample.
- If too few of the population volunteer for the survey, then the researcher should investigate what was so special about them. But, there is no way of finding out, how those who volunteered are different from those who did not.
- The respondents may well have volunteered because, they have their own views about the selected topic, but these may clash with the researcher's aims.

f. Panel Sampling

In this technique, the researcher first select the sample through a random sampling method from a target population and then question the same sample at regular intervals during subsequent surveys to observe the trends of opinion. So, the same sample of respondents were asked for the same information again several times over a period of time. Therefore, each respondent is interviewed at two or more time points and each period of data collection is called a 'wave'. Thus, with this sort of sample survey, as opposed to cross sectional (one off) studies, change of observations or data or information over time can be monitored. However, the demerit of this technique is that, due to frequent surveys, the respondents may lose interest in offering the information during the subsequent interactions with them. This method of sampling was developed by sociologist, Paul Lazarsfeld in 1938 as a means of studying political campaigns. Being a type of longitudinal sampling method, it allows estimates of changes in the population. There have been several proposed methods of analyzing panel data, including Multivariate Analysis of Variance (MANOVA), growth curves, and structural equation modeling with lagged effects. A good example of panel sampling will be studying the performance of a group of fertilizer companies or edible oil mills in a specified region. These samples can be re-visited at a later stage and hence, it represent panel sampling.

g. Expert Sampling

In this sampling technique, the respondents are chosen in a non-random manner based on their expertise on the selected topic for investigation. So, the researcher purposively select the respondents, who enjoy special skills or expertise or high level of knowledge in a particular area. For identification of these 'expert' respondents, the researcher may use some criteria like qualifications, experience etc., and their selection will be taken up as a separate pre-study. For example, if the State Government wishes to bring desirable changes in the agricultural education system suiting to the present competitive environment, a survey will be conducted duly selecting the 'experts' working in that sector and the 'experts' may include: Professors working in the SAUs, Scientific community in the research stations, Entrepreneurs of different private input agencies, Banking Officials, Progressive farmers etc. So, the survey administrator will fix some criteria and select only those experts, who pass the identified criteria for further study. So, expert sampling is an example of purposive sampling and is a non-probability method. The important reason that actually guide the researcher to opt for expert sampling is that, it would be the best way to elicit the views of persons (experts), who have specific expertise in the selected field of investigation, as their opinions are more easily respected by other people. The advantage of this technique is that, since experts tend to be more skilled and knowledgeable with the selected topic than non-experts, opinions from a sample of experts are more credible than a sample that includes both experts and non-experts, although the findings are still not generalizable to the overall population at large.

h. Opportunity Sampling

In this sampling technique, the researcher select the sample from people, who are available at the time, when the study is being carried out and fit the criteria he is looking for. For example, a TV reporter ask the opinions of the viewers coming out of the movie-theater after watching the premier show of the movie. This technique of sample collection saves time and money on the part of the researcher. However, it produces a biased sample and the sample estimates are not representative to the target population. In some cases, some of the participants may decline to take part and thereby, the sampling technique may turn into a self selected (volunteer) sample. Both opportunity and volunteer samplings appear to be quite similar, but the selection of respondents differ across these techniques. This is because, the type of person who volunteers for a research project (volunteer sampling) may be different from the type of person who would be willing to take part in a research project, if asked.

Guidelines to Use the Sampling Techniques

Considering the above discussion regarding the features, merits and demerits of different sampling (both probability and non-probability) techniques, the following useful guidelines can be offered regarding the application of a particular sampling technique in a particular situation. These guidelines are presented here under through Table 8.15.

Determination of Representative Sample Size for Conducting Research Study

As discussed earlier in this chapter, it is almost always impossible to study the entire population that the researcher is interested in. For example, if the researcher is interested in conducting the study on *'Economic analysis of paddy production in Kurnool district of Andhra Pradesh'*, it would be nearly impossible to survey every single farmer cultivating paddy in the selected district. If he goes for enumeration study, it would be extremely timely and costly. As a result, researchers use sample analysis as a way to gather data. Now, the question here, what sample size should be considered by the researcher in drawing the conclusions that should be representative to the total population in the district. So, one of the most common questions the researcher often face is regarding fixing the sample size to be considered for in depth investigation. After formulation of the research problem, the researcher must decide, how many respondents he need to be studied in order to answer the framed objectives of the study. If the study is too small and that too in a small location, the researcher may go for enumeration study. On the other hand, if the research study is extensive covering a wider area, then he has to go for a sample study. Here comes the importance of selecting representative sample size for conducting the in depth study. Some experts say, 100 units of sample (*i.e.*, n=100) is sufficient to draw statistical inferences. Note that, when the target population is spread over a wider area, there lies no meaning in increasing the sample size indiscriminately and it is essential to concentrate more on the accuracy of the data collection (by providing training to the data collectors and using effective data collection tools). This is because, if sample size is increased indiscriminately, it may consume lot of time, cost and other resources on the part of the researcher. So, the eventual sample size is usually a compromise between what is desirable and what is feasible. Hence, it is not only important for the researcher to select the representative sample size, but also to handle the collected data and to analyze it. The sample size to be considered by the researcher may vary across quantitative and qualitative researches. In quantitative research, the researcher prefer a large sample to ensure that, the effect of any variables

Table 8.15: Choice among different Types of Sampling Techniques.

Choice of the Sampling Techniques	*Situation Under the Technique Should be Applied*
I. Probability sampling techniques	
a. SRS	☆ When complete list of sampling frame is available ☆ Target population is homogeneous ☆ Target population is spread over a smaller area ☆ To get a representative sample from target population without a bias (Figure 8.8B) ☆ When the target population does not contain auxiliary information that may be used for stratification purposes
b. Stratified random sampling	☆ When complete list of sampling frame is available *i.e.,* target population is heterogeneous ☆ When the target population contain auxiliary information that may be used for stratification purposes ☆ It is possible to divide a population into two or more homogeneous strata and construct a sampling frame for each stratum. ☆ Units of population between strata are heterogeneous and within each stratum are homogeneous (Figure 8.15). ☆ To permit researcher to use different sampling procedures for different strata. ☆ It is likely that answers to the research questions of a study are likely to be different for different strata. ☆ When the findings are to be reported separately for each stratum. ☆ To draw comparative analysis between strata. ☆ To minimize sampling error. ☆ To select a representative sample from heterogeneous population ☆ When the researcher has complete knowledge about the target population ☆ To lower the data collection costs
c. Systematic sampling	☆ When it is difficult to identify items using a SRS method. ☆ When sampling frame is not available or impractical to prepare, but a stream of representative elements of the population is available. ☆ When the sample selection should be based on probability approach, but in less time and with less cost.
d. Cluster sampling	☆ When population units are widely separated and access to all is difficult ☆ When there are substantial fixed costs associated with data collection of each sample unit ☆ To minimize data collection costs ☆ When sampling frame of individual population elements is not available, but a sampling frame of clusters of elements is available. ☆ To save researcher's time and travel costs
e. PPS	☆ When the target population can be divided into different sub-groups and these sub-groups vary in their sizes. ☆ When population units across the sub-groups widely differ in their size ☆ When selection of the sample units from different sub-groups in the same proportion will not yield a sample that is representative of the target population. ☆ When representative samples are taken from all sub-group in proportion to their size, then the overall sample can be representative to the total population. ☆ To draw large samples from large sized sub-groups and small samples from small-sized sub-groups such that, the overall sample drawn will be representative of the total population.

Contd...

Table 8.15–*Contd...*

Choice of the Sampling Techniques	*Situation Under the Technique Should be Applied*
	☆ To avoid under-representing one subgroup (large sized sub-group) and over representing other sub-group (small sized sub-group). ☆ To improve accuracy for a given sample size by concentrating sample on large elements that have the greatest impact on population estimates.
f. Multi-stage sampling	☆ When listing of all the units in the target population is impossible and costly. ☆ When conducting study involve a very large population. This is because, when the target population is very large, the SRS becomes unfeasible. ☆ Considering all the elements in all the selected clusters may be prohibitively expensive or not necessary. ☆ The complete list of all the units of the population does not exist and is inappropriate.
g. Multi-phase sampling	☆ When some general information is to be collected from the whole sample and additional or detailed information is to be collected, at the same time or later, from sub-samples taken from the whole sample. ☆ When detailed information is to be collected from only some units of whole sample ☆ It is generally followed, when the variable of interest is difficult or expensive to measure. ☆ This technique is also used, when the sampling frame lacks auxiliary information that could be used to stratify the population or to screen out part of the whole population. ☆ To increase precision, reduce costs, and reduce non-response. ☆ This is followed even when the researcher do not have complete or relevant sampling frame, unlike stratified random sampling.
II. Non-probability sampling techniques	
a. Quota sampling	☆ When the population has no suitable frame ☆ When the researcher requires information urgently ☆ When the researcher knows about the proportions of particular sub-groups within a population ☆ When the research budget is very tight or when detailed accuracy is not important. ☆ When the sample is to be selected on the basis of pre-specified characteristics so that, the same distribution of characteristics is assumed to exist in the population being studied and not based on random selection.
b. Convenience sampling	☆ When the selection of the respondents is left to the discretion of the interviewer based on his convenience and availability of units ☆ When it is used for pilot testing of a technology, where the objective is to measure validation rather than obtaining generalizable inferences. ☆ When the researcher plan for least expensive and least time consuming sampling technique ☆ For pre-testing of questionnaire or schedule.
c. Judgment sampling	☆ When the researcher will select the sample based on certain pre-determined criteria based on his experience in conducting the research. ☆ When the criteria judged by the researcher for selecting the representative sample is considered to be most appropriate for the purpose(s) of conducting the research. ☆ When sample is to be selected not based on that they are representative to the target population, but because they contribute requisite data and information. ☆ When the target population to be studied is difficult to locate or if some members are thought to be better (more knowledgeable, more willing to answer the questions framed in the schedule or questionnaire etc) than others to interview.

Contd...

Table 8.15–*Contd...*

Choice of the Sampling Techniques	*Situation Under the Technique Should be Applied*
	☆ Instead of interacting with a huge representative sample of the target population, it is better for the researcher to rely on few knowledgeable members. ☆ When a limited number of total population has the requisite information sought by the researcher. ☆ To collect lesser distorted data.
d. Snowball sampling	☆ When some of the sample units are difficult to locate, hard-to-reach and when sampling frame is not available. ☆ When information is to be elicited from the persons engaged in illicit or illegal activities say, drug users etc. ☆ When the data is to be collected from referral sample units guided by the initial sample
e. *Volunteer sampling*	☆ When the researcher appeals to people to voluntarily participate in a survey. ☆ The sample respondents have a strong interest in the main topic of the survey and hence, they themselves become part of the research study. ☆ The sample is chosen by the respondents themselves and not by the survey administrator. ☆ This sampling is applicable, when the people who contribute to the survey might have different views and are much more interested in that topic than others who do not participate in the survey. ☆ No sampling frame is needed.
f. Panel sampling	☆ When the same sample of respondents were asked for the same information again several times over a period of time. ☆ To analyze the trends in the observations made.
g. Expert sampling	☆ When the respondents to be chosen in a non-random manner based on their expertise on the selected topic for investigation. ☆ When the researcher fix some criteria and select only those experts, who pass the identified criteria for the study, as their opinions are more respected by other people.
h. Opportunity Sampling	☆ When the surveyor has to select the sample from people, who are available at the time, when the study is being carried out and fit the criteria he is looking for.

within the sample are not exaggerated. In qualitative research, however, the sample size is likely to be smaller because, the researcher prefer to gather more in-depth data. However, it is not possible to say with any certainty, what size of the sample is appropriate and hence, the researcher should consider the following aspects before deciding the representative sample size:

- ☆ ***Size of the population:*** Whether the population to be studied is very large or large or small? In general, the larger the population size, the larger the sample size to be considered so that, it will be representative to the target population. If the target population is very small and easy to handle, it is better for the researcher to go for enumeration or census study, as it removes the sampling error component. However, if the researcher goes for a sample (representative) study, he should aim at minimizing the random sampling error.
- ☆ ***Nature of the population:*** Whether the population to be studied is homogeneous or heterogeneous? If the population is homogeneous, a small sample size is sufficient and if the population size is heterogeneous, a large sample size is needed.

- ☆ ***Purpose of the research:*** Whether the research is meant for exploratory or descriptive or explanatory purpose etc? Among these, in explanatory research, large sample size is needed.
- ☆ ***Whether the study is quantitative or qualitative:*** In case of quantitative studies, the sample size is generally large compared to qualitative research.
- ☆ ***Available budget with the researcher:*** Whether the budget available with the researcher is adequate enough to conduct the research study? The more is the budget available with the researcher, the more is the sample size he can consider for the investigation. In case, if the budget is not sufficient, it may equally warn the researcher to drop the survey.
- ☆ ***Method of data collection:*** Trade-off generally arises between the data collection method and sample size that can be afforded. In case of personal interviews or data collection through schedules, which costs too much on the part of the researcher, a small size of sample is preferred compared to eliciting the information through mailed questionnaires.
- ☆ ***Level of variability between responses of the respondents:*** If this is high, the researcher has to go for larger sample size and *vice versa*. For example, there exists more variation regarding the cost of cultivation of paddy among the farmers in a district, but the extent of variation regarding their family expenditure pattern is less. Hence, cost of cultivation survey would need a larger sample size to achieve the same accuracy as a survey on the family expenditure pattern.
- ☆ ***Burden on the respondents:*** If the research survey pose more burden on the respondents in consuming most of their time, they are less likely to participate in the survey seriously. So, the sample size should not be larger than necessary to obtain the accuracy needed.
- ☆ ***Comparison of sub-groups within the target population:*** If the researcher classifies the target population into different sub-groups or strata and if he wishes to compare these sub-groups in terms of their sample estimates, he has to opt for an adequate and representative sample size in each sub-group and thereby, the overall sample size will increase.
- ☆ ***Accuracy and precision level of sample estimates required:*** It is essential that, compared to sample size, it is important for the researcher in deciding the accuracy and preciseness at which the sample estimates are representative to the target population.
  - ❖ ***Accuracy of the sample estimates:*** The accuracy of the sample estimates mean, how confident the researcher is, regarding the sample results that can be true for the target population. The standard confidence level is 95 per cent. This means that, out of 100 cases, in 95 cases, the sample results are an accurate estimate of the target population. That is, the statistical estimates of the sample are representative to the target population in 95 per cent cases out of 100 cases and only in five cases, the estimates may go wrong and not representative to the target population. Sometimes, even the stricter level of 99 per cent confidence level is taken. If the researcher is willing to be 90 per cent confident, the sample size will be smaller. If the researcher is willing to be 99 per cent confident, he will still need a larger sample. Though 95 per cent is the most commonly used confidence interval, but the researcher may use 90 per cent or 99 per cent confidence interval depending on the survey. Decreasing the confidence interval below 90 per cent is not recommended.
  - ❖ ***Precision of the sample estimates:*** After deciding the accuracy or confidence level of the sample estimates, the next aspect the researcher considers is, how precise the sample estimates should be. This is analyzed in terms of 'sampling error' or 'margin of error'.

For example, very frequently, we see in the news papers regarding the exit poll survey that, say, 48 per cent of the population favour the ruling party and 52 per cent disfavor the ruling party or favour the opposition party (+/-3 per cent). What this '+/-3 per cent ' means is that, if everyone in the population were asked, the actual proportions would be somewhere between 45 per cent to 51 per cent (*i.e.*, 48 per cent +/- 3 per cent) favouring ruling party and 49 per cent to 55 per cent (*i.e.*, 52 per cent +/- 3 per cent) opposing the ruling party. This '+/- 3 per cent ' is called margin of error, which measures the accuracy of the per cent estimated in the survey. This margin of error of +/- 3 per cent at 95 per cent confidence interval can be interpreted in a different way. It means that, if the surveys are conducted 100 times – asking a different sample of respondents each time — the overall percentage of people who responded the same way would remain within 3 per cent of the original result in at least 95 out of 100 surveys. This 45-51 per cent (in favour of ruling party) and 49-55 per cent (in favour of opposing party) implies confidence intervals. In this way, most evaluations accept a margin of error up to +/- 5 per cent. Note that, even from the confidence interval, we can compute margin of error. The margin of error is equal to the radius of the confidence interval or half the width of the confidence interval. So, margin of error (per cent) = (b-a)/2, where the confidence interval is (a,b). For example, considering 45-51 per cent confidence interval, the margin of error is given by (51-45)/2 = 3 per cent. So, this 3 per cent is added on either side of the calculated number (48 per cent of the population favour the ruling party).

So, the standards the researcher should aim for 95 per cent confidence level and a margin of error of +/- 5 per cent. If in this example, if the researcher had a 7 per cent margin of error, then the true picture of opinions would be between 41 per cent to 55 per cent favouring ruling party and between 45 per cent to 59 per cent opposing ruling party, then it implies, there is more variability (or less precision) in sample estimates with a +/-7 per cent margin of error as compared to a +/- 3 per cent margin of error. So, the larger the margin of error, the less precise the results will be. So, to increase the precision of the estimates, the sample size should be increased, as it reduces the margin of error. Thus, accuracy and precision can be improved by increasing the sample size. In other words, by increasing sample size, accuracy can be increased and margin of error can be reduced or precision can be increased. In both the cases discussed above, note that, the sampling errors overlap. This means that, these results are too close to call. When working with real numbers, such as age or income of the farmers, precision is presented in terms of the confidence interval (note that, this is not to be confused with the concept of confidence level as explained earlier). The concept of confidence interval and the difference between confidence interval and confidence level can be easily understood from the ensuing discussion. The formula used to compute margin of error is given by '$(1/\sqrt{n})$ expressed in terms of percentage'. For example, if the sample size, n = 1000, the margin of error is 3 per cent (*i.e.*, $1/\sqrt{1000} = 0.03 = 3$ per cent). It is to be noted that, margin of error formula depends only on the size of the sample and not on the size of the population. So, irrespective of the population size, margin of error remains same as long as the sample size remains constant. Suppose, a researcher conducted a survey on 1000 sample farmers to elicit the information about the effectiveness of Minimum Support Price (MSP) policy offered by the Government. The survey revealed that, 55 per cent of the farmers opined the MSP policy is effective in safeguarding their interests against distress sales of farm produce. Since the sample size is 1000, the margin of error is 3 per cent. The interpretation of margin of error is as follows: 55 per cent ± 3 per cent (*i.e.*, 52 per cent to 58 per cent) of all the sample farmers favoured the MSP policy being implemented by the Government. That is, the difference between selected sample per cent and the true population per cent will be within 3 per cent in at least 95 per cent of the cases. This range of values (52 per cent to 58 per cent) is called Confidence interval and it is certain at 95 per cent level (confidence level). That is, the researcher is 95 per cent certain (this is the confidence level) that, 52 to 58 per cent (this is the confidence interval) of the target population favoured the MSP policy implemented by the Government. Similarly, assume that, the average income of a sample of 1000 rural poor in Kurnool district is Rs. 15000 per year. So, the margin of error is +/-3 per cent. The confidence interval at 95 per cent confidence level is given by 15000 +/- 3 per cent (*i.e.*, 14550 and 15450). That means, the researcher is 95 per cent certain (this is the confidence level) that, the average income of the rural poor of the target population in Kurnool district also lies in the range between 14550 and 15450 (this is the confidence interval).

The Rule of Thumb to decide the 'percentage near' for fixing the confidence intervals is shown through Table 8.16 A. For example, a researcher is asked to draw a sample of 500 balls of different colours from a drum containing 100000 balls. Assume that, the balls selected by the researcher comprises of 30 per cent red balls, 10 per cent purple balls and 60 per cent are of some other colors. Looking at the matrix (Table 8.16 A), the researcher can report that, 30 per cent of the balls are 'red', +/- 4 per cent. To further elaborate, he can say, with 95 per cent confidence level, red balls make up 30 per cent {+/- 4 per cent or the range of 26-34 per cent } of the total balls in the drum. Likewise, he can report that, purple balls make up 10 per cent {+/- 3 per cent or the range of 7-13 per cent } of the total balls in the drum. It is also evident from the Table 8.16 A that, as the sample size increases, there are diminishing returns in margin of error (percentage error or statistical error). At percentages near 50 per cent, the margin of error drops from 7 to 5 per cent as the sample size is increased from 250 to 500. But, if the sample size is increased from 750 to 1,000, the statistical error drops from 4 to 3 per cent. As the sample size rises above 1,000, the decrease in marginal returns is even more noticeable.

Table 8.16A: Recommended Allowances of 'Percentage Near' for Fixing the Confidence Interval (in Percentage Points) at 95 per cent Confidence Level.

Percentage Near	*Sample Size*				
	1,000	*750*	*500*	*250*	*100*
Percentage near 10	2 per cent	2 per cent	3 per cent	4 per cent	6 per cent
Percentage near 20	3	3	4	5	9
Percentage near 30	3	4	4	6	10
Percentage near 40	3	4	5	7	10
Percentage near 50	3	4	5	7	11
Percentage near 60	3	4	5	7	10
Percentage near 70	3	4	4	6	10
Percentage near 80	3	3	4	5	9
Percentage near 90	2	2	3	4	6

Table 8.16B: Inverse Relation between Sample Size and Margin of Error.

Sample Size (n)	*Margin of Error (per cent)*
50	14.0
200	7.1
400	5.0
700	3.8
1000	3.2
1200	2.9
1500	2.6
2000	2.2
2500	2.0
3000	1.8
4000	1.6
5000	1.4

So, there exists an inverse relationship between sample size and margin of error and this relationship can also be predicted as shown through Table 8.16 B and Figure 8.22. It is evident that, the amount by which the margin of error decreases is substantial between samples sizes of 200 and 1000. For example, a very small sample, such as 50 respondents, has about a 14 per cent margin of error, while a sample of 1,000 has a margin of error of 3.2 per cent. This implies that, the accuracy of the estimate is strongly affected by the size of the sample in this range of sample size. However, the margin of error does not substantially decrease at sample sizes above 1000. That is, there are diminishing returns. For example, by doubling the sample to 2,000, the margin of error decreases from plus or minus 3.2 per cent to plus or minus 2.2 per cent. So, it is not worth for the researcher to spend his additional time and money for the samples that contain more than 2000 people.

Note that, if the researcher want a margin of error of 'm' per cent, the sample size required is given by, $1/m^2$. For example, if the researcher want a margin of error of 2 per cent, then the sample size required is $1/(0.02)^2 = 2{,}500$. This infers that, sample size is a function of size of the target population, the desired confidence level (accuracy) and level of precision. So, we can have a formula to compute the relation between these three to determine the sample size or simply, we can use the Table 8.17 to guide to a minimum sample size. This Table 8.17 shows the sample size needed, when estimating a population percentage (or proportion) at 95 and 99 per cents confidence levels and a +/-5 and +/-1 per cents margin of errors. It is evident that, the smaller the population, the higher proportion of sample size or respondents needed for the investigation and *vice versa*. For example, if the target population is 300, we need a sample size of 169 respondents *i.e.*, just over half of the target population at a confidence level of 95 per cent. Similarly, for a target population size of 800, the required sample size is 260 respondents *i.e.*, only less than one-third of the target population size at 95 per cent confidence level. When the target population size is larger than 250000, only 384 respondents' sample size will be needed *i.e.*, a still smaller fraction. So, the smaller the population, the larger the needed ratio of the sample size to the population size and *vice versa*. Note that, the sample sizes indicated through Table 8.17 are minimum sample sizes and if the researcher feels that, there is lower response rate (*i.e.*, less than 100 per cent), he has to increase the sample size. However, mere increasing the sample size is not the solution for low response rate of the respondents. When there is low response rate from the respondents, it leads to the threat of non-response bias. So, instead of over-sampling, the researcher should put other efforts like multiple follow-ups, offering some incentives etc., to increase the response rate from the respondents. So, all these should be planned to collect reliable information from the sample respondents in tune with the objective of minimizing the data collection costs. No doubt, census survey eliminates the sampling error, but even there is a chance of getting biased data, if there are low response rates.

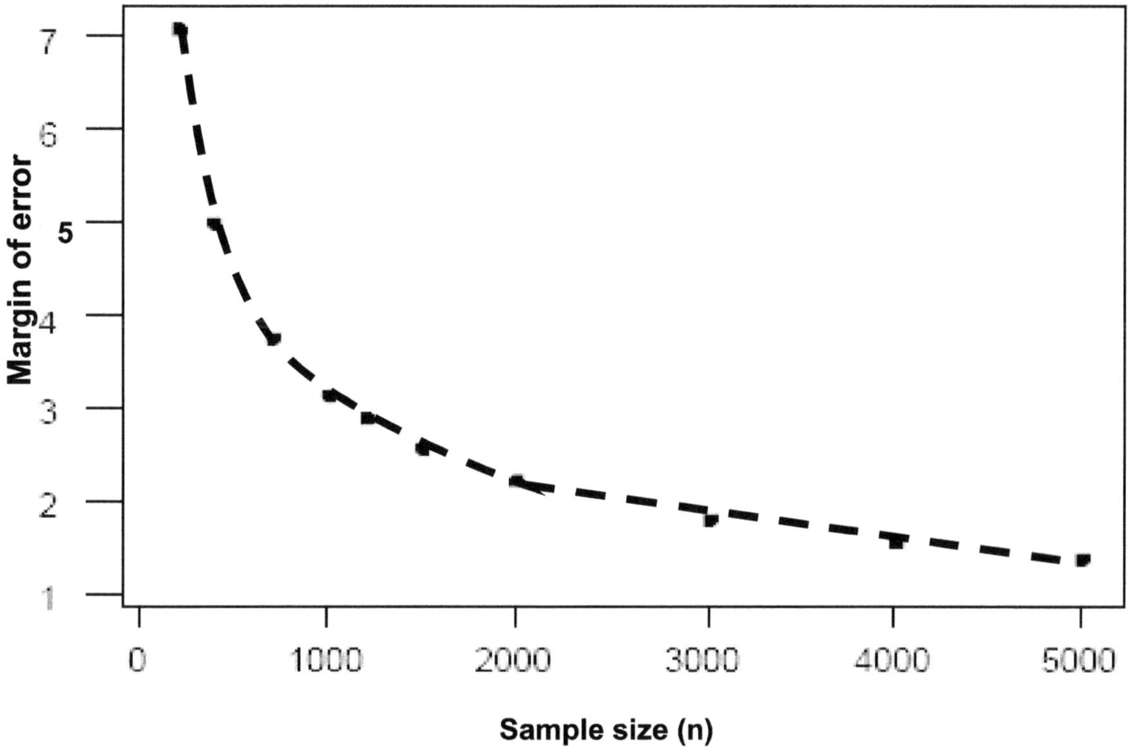

Figure 8.22: Inverse Relation between Sample Size and Margin of Error.

Formulae Used to Determine Sample Size

Statistical tables (Table 8.17) will directly indicates the desired sample size to be selected considering three aspects *viz.*, population size, confidence level and margin of error. Besides this, following are the formulae employed to determined the sample size needed to conduct the research study:

☆ ***Sample Size - Infinite Population (where the population is greater than 50,000):*** The researcher need to determine a few things about the target population and the sample to compute the sample size:

- ***Population size:*** 'How many farmers are cultivating paddy in Kurnool district?'. For example, if the researcher wishes to conduct a survey on '*Economic analysis of paddy production in Kurnool district of Andhra Pradesh*', the population size would be the total number of farmers cultivating paddy in Kurnool district. The researcher need not worry, if he is unsure about this number. It is common for the population to be unknown or approximated.
- ***Margin of error (Confidence Interval):*** It is a known fact that, no sample will be exactly representative of the total population and hence, the researcher should decide, how much error to be allowed. As discussed earlier, the confidence interval determines, how much higher interval or lower interval than the population parameter, the researcher will let the sample estimate to fall.

Table 8.17: Required Sample Size at different Confidence Levels and Margin of Errors.

Population Size	*Required Sample Size**							
	Confidence - 95%				*Confidence 99%*			
	Margin of Error				*Margin of Error*			
	5.0%	*3.5%*	*2.5%*	*1.0%*	*6.0%*	*3.5%*	*2.6%*	*1.0%*
10	10	10	10	10	10	10	10	10
20	19	20	20	20	19	20	20	20
30	28	29	29	30	29	29	30	30
50	44	47	48	50	47	48	49	50
75	63	69	72	74	67	71	73	75
100	80	89	94	99	87	93	96	99
150	108	126	137	148	122	135	142	149
200	132	160	177	196	154	174	186	198
250	152	190	215	244	182	211	229	246
300	169	217	251	291	207	246	270	295
400	196	265	318	384	250	309	348	391
500	217	306	377	475	285	365	421	485
600	234	340	432	565	315	416	490	579
700	248	370	481	653	341	462	554	672
800	260	396	526	739	363	503	615	763
1,000	278	440	606	906	399	575	727	943
1,200	291	474	674	1067	427	636	827	1119
1,500	306	515	759	1297	460	712	959	1376
2,000	322	563	869	1655	498	808	1141	1785
2,500	333	597	952	1984	524	879	1288	2173
3,500	346	641	1068	2565	558	977	1510	2890
5,000	357	678	1176	3388	586	1066	1734	3842
7,500	365	710	1275	4211	610	1147	1960	5165
10,000	370	727	1332	4899	622	1193	2098	6239
25,000	378	760	1448	6939	646	1285	2399	9972
50,000	381	772	1491	8056	655	1318	2520	12455
75,000	382	776	1506	8514	658	1330	2563	13583
100,000	383	778	1513	8762	659	1336	2585	14227
250,000	384	782	1527	9248	662	1347	2626	15555
500,000	384	783	1532	9423	663	1350	2640	16055
1,000,000	384	783	1534	9512	663	1352	2647	16317
2,500,000	384	784	1536	9567	663	1353	2651	16478
10,000,000	384	784	1536	9594	663	1354	2653	16560
100,000,000	384	784	1537	9603	663	1354	2654	16584
300,000,000	384	784	1537	9603	663	1354	2654	16586

**Courtesy*: www.research-advisors.com.

- ❖ ***Confidence Level*** : This implies, how confident the researcher is that the sample estimate fall within the confidence interval? The most common confidence interval considered is 95 per cent and even 90 and 99 per cents are also considered.
- ❖ ***Standard of Deviation:*** This implies, how much variance do the researcher expect from the collected responses. Since the researcher has not yet started the survey, the safe decision is to use 0.5 for large sample size.

Considering the above four aspects, now the researcher can calculate the needed sample size. Assuming the large sample size, the confidence level corresponds to a Z-score. The Z-scores at different confidence levels include: at 90 per cent confidence level, Z score is 1.645; at 95 per cent confidence level, Z score is 1.960; and at 99 per cent confidence level, Z score is 2.326. If the researcher prefer to choose a different confidence level, collect the Z-score from the statistical table. The following is the formula used to determine the sample size needed for the study:

Sample Size (SS) = ((Z-score)^2 * S.D.(1-S.D.)) / (margin of error)^2

Considering, 95 per cent confidence level, 5 per cent margin of error and 0.5 units standard deviation and Z score at 95 per cent confidence level as 1.96, the necessary sample size is calculated as below:

$SS=((1.96)^2 \times 0.5(1-0.5)) / (0.05)^2$

$SS=(3.8416 \times 0.25) / 0.0025$

$SS=0.9604 / 0.0025 = 384.16$

So, 385 sample respondents are needed for the survey.

Note that, if the computed sample size is too large to handle by the researcher, try slightly decreasing the confidence level (*i.e.*, Z score will also decrease) or increasing the margin of error. Note that, by making these changes, the number of sample respondents can be decreased, but it will increase the chance for error in the sampling.

- ✰ ***Sample Size - Finite Population (where the population is less than 50,000):*** When the population size is less than 50000, the following is the formula employed to determine the sample size:

  New Sample Size (NSS) = (SS)/[1+ ((SS-1)/Population size)]

  Assume the population as 5000. Now, calculate the SS using the infinite population formula first. Then use the SS derived from that calculation to calculate a NSS for a finite population. So, the NSS is calculated as below.

  NSS = (385)/[1+ ((385-1)/5000)] = 357

  So, NSS is 357 ($\approx$ 360), when the assumed population is 5000. Please compare this result in the Table 8.17 at 95 per cent confidence level and at margin of error of +/- 5 per cent.
- ✰ ***Sample size, when proportions of population units are available:*** Assume the researcher wishes to conduct a research study on '*Economic analysis of SRI production technology of paddy in Kurnool district of Andhra Pradesh*'. Now the researcher has to collect the requisite information from those farmers, who practice SRI technology of paddy. He came to know from the scientific community working in the Regional Agricultural Research Station and Officials from Line Department of Agriculture that 20 per cent of the total farmers in the district are practicing SRI production technology of paddy. The researcher can select a

representative sample from this proportion of population by using the following formula:

SS = $[PqZ^2]/e^2$

where, P = Proportion of total population, q = 1 – P, Z = Z tab value at 95 per cent confidence level, e = Margin of error (assume it is 5 per cent)

So, SS = $[0.2\,(1 - 0.2)\,(1.96)^2]/(0.05)^2 = 245$

The above methodology implies, sample size depends on the values of P, q, Z and e. Larger the product of P, q and Z and smaller the value of e, larger the required sample.

☆ Yamane(1967) provided the following simplified formula to calculate sample size:

SS (n) = $N/1+N(e)^2$, where 'n' is the sample size, 'N' is the population size, and 'e' is the margin of error. For example, at 95 per cent confidence level and margin of error of 5 per cent, for a given population size of 3000, the sample size is:

SS (n) = $3000/1+3000(0.05)^2 = 353$ farmers.

So, considering the above aspects, it is evident that, it is not an easy task on the part of the researcher to determine the accurate sample size for a research study. Here, it is essential to compare two aspects *viz.*, 'larger sample size' and 'smart sample size'. No doubt, larger sample size minimizes sampling error and ensures higher accuracy or confidence level of the sample estimates, but it is not cost-effective. On the other hand, a clever researcher will try to achieve the accurate estimates within the accepted confidence level, but with lower sample size and that too at lower costs. In general, most sample surveys are conducted using 'smarter' sample sizes, as they ensure accurate estimates that too through cost-effectiveness. However, sometimes, the researcher may face an essential problem in conducting the research study *i.e.*, even the selected 'smart' sample size is not affordable by him in terms of costs or does not fit into the time frame allotted for conducting the survey. In such case, the researcher has left with the following options in continuing with the present research:

☆ He has to accept a slightly lower level of accuracy, particularly for sub-groups of the population.

☆ He has to divert the resources from other components of the research say, reducing the length of the schedule, replacing schedule with a questionnaire and collecting the information through mailing the questionnaire instead of personal interviews etc.

So, the researcher has several ways to choose a desired sample size from the above formulae. Rather he can simply use a rough 'rule of thumb' as given through Table 8.18.

Common Misconception in Selecting the Sample Size

Assume that, there are 10000 farmers in a mandal and the researcher wishes to select a sample of 500 farmers to conduct a survey on marketing pattern of major agricultural commodities by the farmers. Now the question is, what sample the researcher is required for conducting the same study at State level, in order to maintain the same level of accuracy, if the farmers' population in the State is 10,000,000? (*i.e.*, 1000 times the population of the mandal). Generally, the researchers assume it will be 500*1000 = 500000. But, the correct answer is 500.5 ≈ 500. The important aspect here is, Finite Population Correction is used here *i.e.*, FPC = [OSS/(NPS/OPS)] and this added to the old sample size. Here, OSS = Old Sample Size, NPS = New Population Size, OPS = Old Population Size. So, the New Sample Size (NSS) is given by: OSS + [OSS/(NPS/OPS)]. As per this example, NSS = 500 + [500/(10000000/10000)] = 500.5 ≈ 500. Note that, unless the OSS is more than about 5 per cent of the initial population, this FPC makes almost no difference to the newly required sample size. Doubt arises to the researchers in

Table 8.18: Rules of Thumb followed for Selecting Sample Size under different Conditions.

Conditions	*Recommended Sample Size*
If the researcher had no previous experience in conducting surveys and there is no existing survey data.	100 to 200
If the researcher has some previous survey experience or some previous data and he wishes to divide sample into sets of 2 groups (say, small and large farmers).	200 to 400
If the researcher enjoys good previous survey experience and previous data and he wishes to divide sample into sets of up to 4 groups and further want to compare the analysis with previous survey data.	400 to 600

selecting the same sample size (NSS) even when the population increased by 1000 times. This can be better understood from the following example: A researcher want to know the taste of new soft drink that is going to be released into the market. So, he will taste a spoonful of drink from a glass. In this case, the volume of soft drink in the glass represents the population and the quantity intake by spoon represents the sample. So, each spoonful represents the random. Even if the glass was twice the initial size, the researcher will need only spoonful of soft drink to test its taste and he wouldn't need to take two spoonfuls to assess the taste *i.e.*, one spoonful would still be fine. So, the size of glass is irrelevant. This is equally true for human populations.

Advantages of Sampling

As discussed earlier, in the sampling techniques, the researcher study only a selected representative and finite sub-set of the target population. Then, the results will be analyzed and the conclusion about the characteristics of the population will be drawn. If the sample is properly selected, the conclusions will hold good for the whole population. The chief advantages of the sampling technique may be enumerated as under:

- ✰ Manageable (representative) sample will reduce the cost of data collection, time and manpower compared to census survey. Since it is economical than census method, the researcher can employ trained persons or experts for data collection and analysis and thereby, the error may be minimized.
- ✰ Sampling technique ensures administrative convenience, as the researcher can employ limited number of both field and administrative staff.
- ✰ Data collection, compilation and analysis will proceed at a faster pace compared to census survey.
- ✰ Due to limited coverage of representative population, sampling ensures accuracy of research findings.
- ✰ Facilitate the researcher to complete the research in the planned time frame.
- ✰ Better rapport will be established with the sample respondents and this improves validity and reliability of the results.
- ✰ In cases, when the population is very large, then the sampling method is the only practical method for collecting the data.
- ✰ Makes the researcher to easy to handle the data for various statistical analyses.
- ✰ The sampling technique of data collection is very much adaptable to the changes in the circumstances of the population. This means that, the size of the sample can be increased or

decreased according to the size of the population, availability of resources and the degree of accuracy desired. Similarly, the nature of the sampling technique employed also can be changed according to the nature, object and scope of the enquiry. Thus, the sampling technique of data collection is very much flexible and adaptable to the changes in conditions of the enquiry.

- Response rate from the respondents will be high.
- It is easier on the part of the researcher to supervise fewer (sample) respondents than the whole population.
- The results of census method can be checked with sampling techniques. That is, sample data is also used to check the accuracy of the census data.
- If the researcher goes for a census survey, there is no guarantee on the reliability or confidence about the quality of information generated. In such case, sampling technique assume greater advantage. That is, the inference about the population parameters is possible only when the sample data is collected from the selected sample.
- The sampling technique is very much scientific in its approach. Particularly, the techniques of random sampling are based on the Theory of Probability, which is a mathematical concept. This technique is based on certain important laws and principles *viz.*, Law of Statistical Regularity *(Law of Statistical Regularity says that a moderately large number of the items chosen at random from the large group are almost sure on the average to possess the features of the large group)*, Law of Inertia of Large Numbers (*the other things being equal – the larger the size of the sample; the more accurate the results are likely to be)*, Principle of Persistence, Principle of Optimization, Principle of validity etc. This facilitates the researcher to compute the sampling error and degree of reliability of the results by using sampling techniques.
- The analytical results of sampling technique are more dependable than those derived from the census survey. This is because, under the sample technique, it is always possible to determine the extent of sampling errors and the degree of reliability of the results in terms of probability whereas, in census all the units of population may not respond well for the study.
- In some cases, sampling is indispensable. For example, if the researcher want to inspect the cocoons of sericulture firm, to study the extent of infestation in mango fruits etc., the researcher must take sample units, as all the objects cannot be destroyed during the process of inspection and he has to take care that the wastage should be minimum. Thus, sampling is essential when the units under study are to be destroyed.
- Sometimes, the experiments are done on sample basis. The fertilizers, pesticides, seeds etc., are initially tested on samples and if found useful, then they are recommended on a large scale. Most of the research work is done on the samples.

To summarize the advantages of sampling techniques, A.C. Rosander has stated that, *'the sample has many advantages over a census or complete enumeration. If carefully designed, the sample is not only considerably cheaper but may give results which are just accurate and sometimes more accurate than those of a census technique. Hence, a carefully designed sample may actually be better than a poorly planned and executed census.'* In the words of Prof. R.A. Fisher (Father of Statistics), *'the sample technique has four important advantages over census technique of data collection viz., 'Speed, Economy, Adaptability and Scientific approach'.* To ensure the above advantages of sampling, the researcher should be aware of the following considerations:

- Compared to smaller sample sizes, larger sample sizes are more accurate representations of the target population.
- The sample size chosen should be within the available resources *viz.*, time, energy, money, labour, equipment etc., of the researcher.
- The sample selected with the minimum of bias is the most statistically valid.
- A 95 per cent probability or confidence level is usually assumed.

Disadvantages of Sampling

Compared to census survey, sampling technique suffers from the following limitations:

- If sampling is done erroneously, it leads to biasedness and thereby, the conclusions derived from the analysis are not representative of the target population.
- Judging a right sampling technique to draw a representative sample is quite difficult. If a representative sample is not selected, the results will not be accurate.
- The sampling techniques demand thorough knowledge, training and experience on the part of the researcher in selecting the representative sample, collection, compilation, analysis and interpretation of findings. Lack of these qualities on the part of the researcher may lead to serious mistakes.
- In some cases, the sample respondents may not be accessible to the researcher or may be uncooperative. In such a case, they have to be replaced. This introduces a bias in the study.
- In some cases, the target population is too small or too heterogeneous to select a representative sample. In such cases 'census study' is the only alternative for the researcher to collect the data and information.
- The sampling technique may be expensive both in terms of time and cost, especially if the size of the sample is considerably large, procedure adopted is complicated, personnel conducting the survey are not properly trained enough etc.
- The sampling technique demands the services of qualified, skilled, experienced and trained personnel, which entail heavy amount of expenditure.
- In certain cases of research study, where information is needed from each and every unit of the population and that too if the population is too heterogeneous in nature, the sample technique of data collection is not at all suitable. Moreover, if the field of enquiry is very small, this sampling technique of data collection is also not suitable.
- As discussed earlier, each of the various techniques of the sampling suffers from certain inherent defects, which can never be rooted out. Hence, the results obtained through the sample techniques can never be flawless or free from all defects.

B. Identification of Variables Relevant to the Objectives Framed in the Study

The persons from which the data and information are collected by the researcher are called (sample) 'respondents' in social sciences research. Each respondent possess a variety of characteristics. If a characteristic remains same for all the respondents *i.e.*, it does not vary, it is called a Constant. If a characteristic differs across the respondents, it is called a Variable. In research, we do not get excited about constant (since everyone is the same on that characteristic); but we are more interested in variables. So, a variable is any entity that can take on different values. That is, anything that can vary can be considered as a variable. It is defined as a characteristic that may assume more than one set of values to which a numerical measure can be assigned. So, a 'variable' enjoys two defining characteristics *viz.*, it is an attribute that describes a person, place, thing, or idea and the value of the variable can 'vary' from one entity to another. For instance, age of the farmer can be considered as a variable because age can take different values for different farmers or for the same farmer at different times. Similarly, plant height can be considered as a variable because, it can be assigned a value and varies across several plants in a crop stand. This implies, variable is a concept or an abstract idea that can be described in measurable terms. In research, this term refers to the measurable characteristics, qualities, traits, or attributes of a particular individual, object, or situation being studied.

After determining the sample size, now the researcher has to identify and choose the variables that are to be considered and relevant for the study. This is one of the essential steps in conducting research. As mentioned above, 'variable' is a characteristic or factor that has values, that vary and can be measured. So, the variable is a factor that is capable of taking on different values. For example, fertilizers cost, human labour cost etc., are variables, as these values changes with the level of output. If the characteristic takes a same value in all cases, it is said to be a Constant and not a variable. As shown in the Figure 8.23, the number of students joined in Agricultural College, Mahanandi vary year by year, but the list of courses with course numbers of the Department of Agricultural Economics say, AECO:141, AECO:142, AECO:241, AECO:341 and AECO:342 remains same irrespective of the number of students joined in the college. Note that, the numbers assigned to the course numbers are only labels and these numeric codes wouldn't mean anything as numbers. So, some characteristics change easily (variables such as fertilizer cost, human labour cost, temperature, rainfall etc), while other characteristics are almost constant (like course titles, course numbers etc). The quantification of these variables is utmost important in designing all quantitative research projects, where they will be manipulated, measured, described or controlled. Although qualitative researchers do not define variables to the same extent that quantitative researchers do, they still must outline what kinds of phenomena they are studying.

Before going to discuss in-detail about different types of variables employed in the social sciences research in agriculture, it is essential to understand, how to identify the variables that are relevant to the selected research study. As discussed earlier in the Chapter 4, every researcher will identify and take up a research activity based on some pre-determined ideas. So, the prior knowledge of the researcher helps him to take up a particular research problem. For example, a researcher opined that, there is a drastic decline in the profitability of paddy cultivation in Kurnool district since past five years. Note that, when the researcher expressed this 'concept' regarding the drastic decline in the profitability of paddy cultivation, he does on the basis of certain criteria in his mind. His judgement (concept) is based upon certain 'indicators' that lead him to conclude and express that opinion. So, the judgement given by the researcher must require a sound basis and they are called as 'indicators'. So, indicators are a set of criteria reflective of the concept and these can be converted into variables. As mentioned earlier, these variables can be measured and they change their value frequently. So, the correct sequence of measuring the 'concept' is 'Concept → Indicators →Variables → Working definition (measurement)'. This same discussion is exemplified here under through Table 8.19.

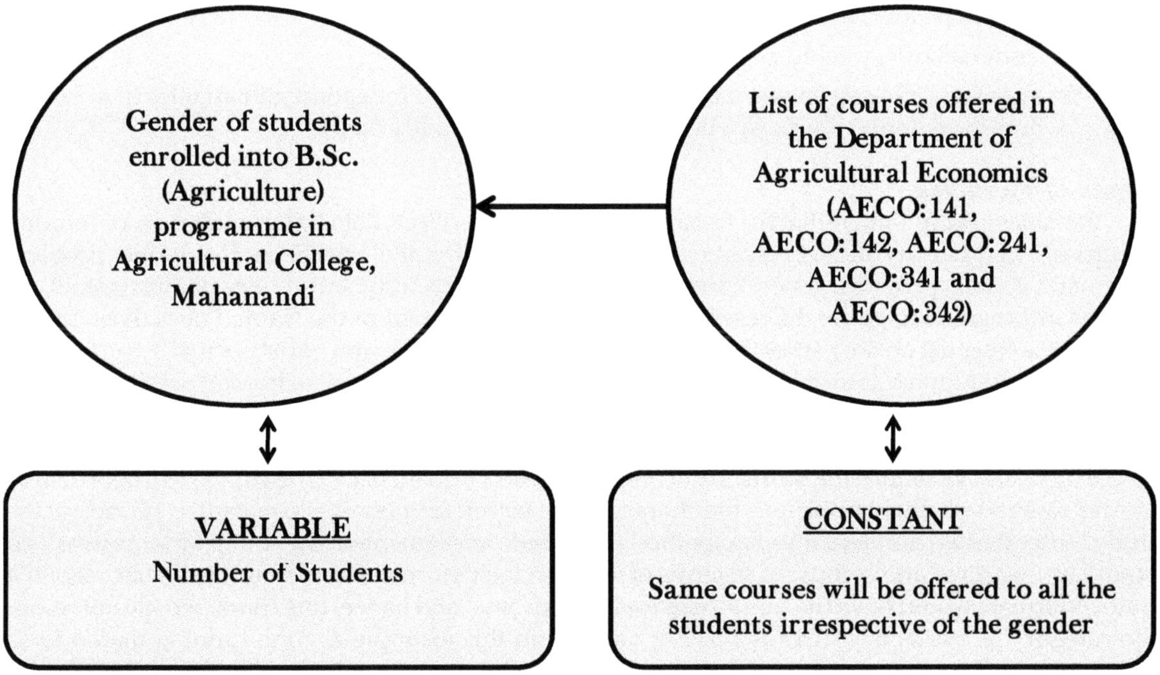

Figure 8.23: Difference between 'Variable' and 'Constant'.

Table 8.19: Concepts, Indicators and Variables in a Selected Research Problem.

Concept	*Indicators*	*Variables*	*Working Definition (measurement)*
Decline in profitability of paddy cultivation since past five years	Gross income	Output of paddy, Market price of paddy	Trends in output and market prices
	Cost of cultivation	Human labour costs, machinery labour costs, seed cost, fertilizers cost, pesticides cost, irrigation cost etc.	Trends in costs of these variables.
Effectiveness of IPM technology training programme on cotton farmers	Total number of farmers trained in the mandal	Number of farmers trained in each season	Differences in net income of the farmers before and after training on IPM.

The above discussion implies that, concepts are mental images or perceptions of the researcher about a research problem. Note that, these perceptions may change from one researcher to another. The major difference between concept and variable is that, the former cannot be directly measured, whereas a variable can be subjected to measurement. That is, measurability is the main difference between a concept and a variable. It is, therefore, important for the concept to be converted into variables.

Characteristics of the Variables

The following are the three important characteristics of the variables:

- ✩ ***Period:*** When it starts and stops (*e.g.*, Rainfall).
- ✩ ***Pattern:*** Daily, weekly, monthly etc (*e.g.*, Market prices).
- ✩ ***Latency:*** Time gap between measuring dependent and independent variables (it may take time to take effect). (*e.g.*, Application of fertilizer to paddy crop).

Types of Variables

Variables are the 'things' that the researcher measures or collects data (information) on, or forming groups on, in order to conduct research study and answer framed objectives. No doubt, 'problem statement' usually provides general direction for the research study, but it does not include all the specific information to guide the researcher towards achievement of the framed objectives. In this context, it is essential on the part of the researcher to identify the relevant variables that are to be dealt in the research study, so as to gain useful understanding for the fulfillment of framed objectives. These variables can be classified in several ways and this can be better understood from the following example. Assume that, the researcher wishes to compare the effects of two different training methods on the increased income of the farmers for controlling bollworm menace on cotton in Kurnool district of Andhra Pradesh. In this example, the characteristic 'cotton farmers' is a 'constant' condition of the study. After the two different training methods have been implemented, the cotton farmers involved would be measured and compared in terms of increased net income. It is very unlikely that, all of the trained farmers would have the same increased 'net income' and hence, this characteristic fall under the category of 'variable'. There is another variable in this example *i.e.*, the training methods. In contrast to the increased net income of all the sampled cotton farmers, which undoubtedly have many possible values, training method is a categorical variable consisting of only two categories *i.e.*, the selected two methods. Thus, we have different kinds of variables and different names or classifications for them. In the above example, net returns derived by the farmers can be considered as a DV, costs incurred by the farmers in applying pesticides represents IV, as it influences the net returns of the selected farmers. Besides IPM costs, skills of the farmers also influences the DV (*i.e.*, net returns) and this represents a qualitative variable. Age of the farmers is an example of continuous variable, but the number of male and female farmers is an example of discontinuous or discrete variable. Thus, there are different types of variables employed in the research study and the important variables are studied here under (Figures 8.24A and 8.24B).

a. IV
b. DV
c. Qualitative (Categorical) Variable
d. Quantitative Variable
e. Extraneous Variable (EV)
    i. Confounded Variable (CV)
    ii. Moderating Variable (MV)
    iii. Controlled Variable
f. Mediating Variable
g. Endogenous Variable
h. Exogenous Variable

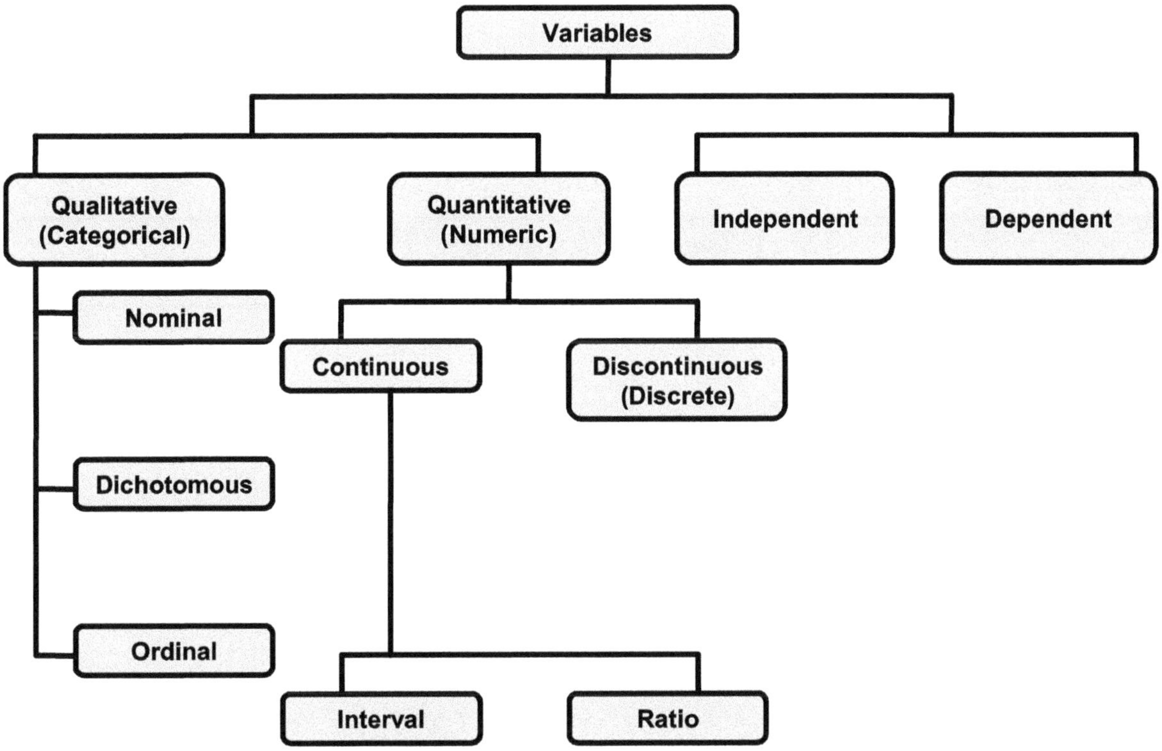

Figure 8.24(A): Classification of Variables.

a. IV

These are the variables that can be manipulated by the researcher in order to study their effects and differences in their effects on those variable (DV) that is being dependent on them. The researcher has some degree of control over the IVs. The variable is independent because, its value can be manipulated by the researcher whatever he wishes it to be. That is, its value may be changed or altered independently of any other variable. The different synonyms for the term 'IV' include: causal variable, determinant, explanatory variable, predictor variable, manipulated variable, stimulus variable, treatment variable, experimental variable, regressor, controlled variable, exposure variable, risk factor (in medical statistics), input variable etc. IV is, otherwise, defined as the variable, whose effect upon the DV, the researcher is trying to understand, explain, and predict. So, if one variable truly causes a second, the cause is the IV and this IV is independent of the outcome itself. That is, in the cause-effect relationship, IV is the 'cause' variable (Table 8.20). For example, in studying the demand function, price of the commodity is the IV (*i.e.*, causal variable). Note that, the influence of the IV on the DV can be either a positive or a negative. That is, with each unit of increase in the IV, there is an increase or decrease in the DV. So, the variance in the DV is accounted for by the IV. This relationship is called 'Cause and Effect'. Thus, in cause and effect relationship, the variable that depends upon the other variable is called DV and the variable that is antecedent to the DV is called IV. IV is often represented by using the letter 'X' and in graphs, it is considered on the X-axis.

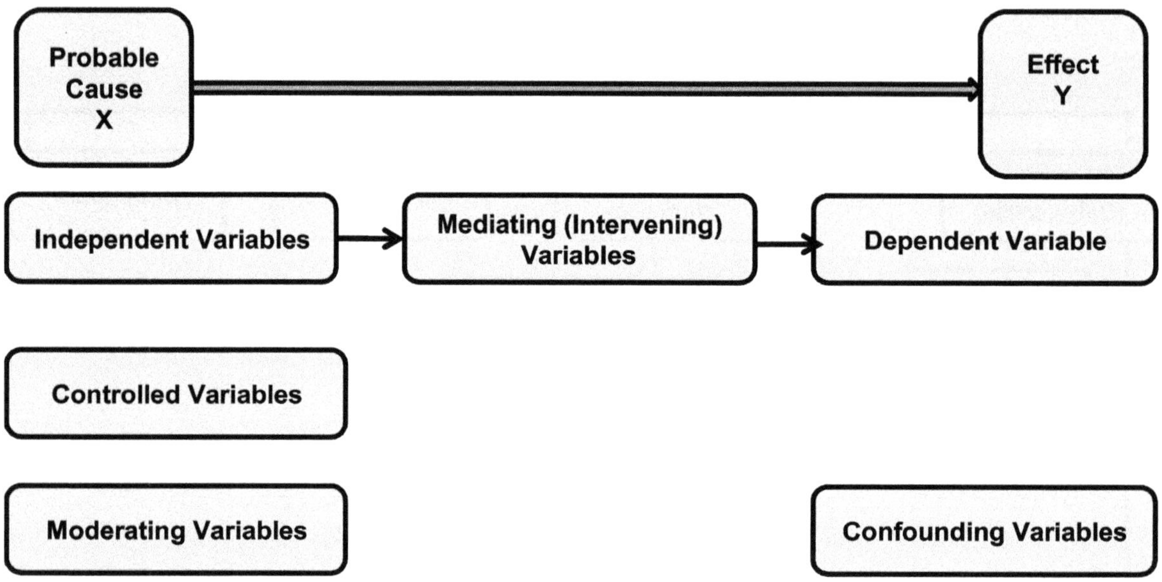

Figure 8.24(B): Family of Variables in Quantitative Research.

Table 8.20: IVs and DV Considered in Addressing the Research Objectives.

Research Objective	*DV*	*IVs*
To analyze the factors responsible for farmers indebtedness in Kurnool district of Andhra Pradesh	Magnitude of debt	Land holding, Cost of cultivation, Family expenditure, Family income
To study the determinants of market arrivals of paddy in Kurnool regulated market of Andhra Pradesh	Quantity of Market arrivals	Production of the commodity, Price of paddy, Price of substitute commodity (wheat)
Data Envelopment Analysis and Commercial Banks Performance in India	The banks' outputs are interest income, non-interest income and total loans.	The banks' inputs are interest expenses, non-interest expenses, transaction deposits and non-transaction deposits.
To analyze the factors responsible for indiscriminate pesticides usage by the farmers on paddy in Kurnool district of Andhra Pradesh	Pesticides cost	Area under the crop, farmers yearly income, Managerial ability of farmer (dummy variable), Irrigation costs

b. DV

The DV is simply that, a variable that is dependent on an IV(s). That is, DV is the response variable, which is observed and measured to determine the effects of the IV(s). So, the DV changes, when the IVs varies and in their levels. The different synonyms for the term 'DV' include: response variable, explained variable, predicted variable, criterion variable, outcome variable, regressand, measured variable, responding variable, experimental variable, output variable etc. In analyzing the cause and effect relationship, DV is the outcome or effect variable of the study (Table 8.20). This measurement of DV helps the researcher in forecasting the trend. DV is, otherwise, defined as the variable, which changes as a result of the IVs being changed in the study and hence, DV is measured as an outcome. So, the outcome of the DV depends upon the influences of the IV(s). That is, in the cause-effect relationship, DV is the 'effect' variable. For example, in studying the demand function, quantity demanded of the commodity is the DV (*i.e.*, predicted variable). So, what the researcher wishes to measure or study the outcome or effect in the research constitutes the DV. This DV is the variable of primary interest to the researcher and the goal of the research is to understand and describe, or to explain its variability, or to predict it. In solving the research problem, the researcher tries to analyze the DV (*i.e.*, finding what variables influence it). For this, the researcher tries to analyze the cause and effect relationship between DV and IV(s) and will be interested in quantifying and forecasting the DV. Note that, it is possible to have more than one DV in a study (in case of Data Envelopment Analysis (DEA)). DV is often represented by using the letter 'Y' and in graphs, it is considered on the Y-axis.

Note that, only causal-comparative, experimental and quasi-experimental research designs have both IVs and DV. Descriptive research simply studies and only describe the current state of a variable. So, there are no presumed cause or effects and hence, there are no IVs and DV. Likewise, correlational research studies only examine relationships between the variables. Since neither variable in a correlational design is manipulated, it is impossible to determine which is the cause and which is the effect. Thus, even in correlational research designs, there is no classification of variables into IVs and DV. In case of DEA, the researcher studies the relationship between multiple outputs and multiple inputs to be estimated. DEA is a linear programming methodology to measure the efficiency of multiple decision-making units (DMUs) when the production process presents a structure of multiple inputs and outputs.

Differences between IV and DV

The following Table 8.21 highlights the differences between IV and DV:

Table 8.21: Differences between IV and DV.

IV	*DV*
The IV is the one that will be changed or manipulated in the research study.	The DV will be the one to be observed to see how it is effected by the changes in the IVs. That is, the DV is the observed result of the IV(s) being manipulated. So, DV's outcome depends on the IV(s).
The IV is usually located on the X-axis in graphical representation.	The DV is usually located on the Y-axis in the graphical representation
An IV variable influences a DV	The DV is influenced by an IV(s)
IVs are changes that occur in a research study that are directly caused by the researcher	DV involve changes that occur due to IVs
The IV has a big effect on the research study and is very important in drawing the conclusions of the study.	The DV is the one being measured by the researcher in the research study. It is the variable that shows how strong the effect of the IV is.
First, IV is changed, then the DV is measured to come up with the accurate conclusions.	The outcome of the DV depends on the IVs
The IV is the factor altered or varied to give various amounts of the stimulus.	The DV is the factor that receives the stimulus.

Contd...

Table 8.21–*Contd...*

IV	*DV*
An IV stands alone and won't be changed by other variables.	A DV is dependent on other variables (IVs) and will change depending on how those IV(s) change.
Presumed cause	Presumed effect
Stimulus	Response
Predicted from.	Predicted to.
Antecedent	Consequence
Manipulated	Measured outcome
Predictor	Criterion

c. *Qualitative Variable or Categorical Variable*

Any variable that is not quantitative is categorical. This variable takes a value that is one of several possible categories. As naturally measured, categorical variables have no numerical meaning. Examples like marital status (single, married, divorced and widowed), occupation (civil servant, farmer, student etc.), hair colour (black, white, brown etc.), gender (male and female) etc., are categorical variables. Since the values of these variables are expressed in categories, we call them as categorical variables. Often categorical variables are disguised as quantitative variables. For example, the researcher might record gender information coded as Male = 0 and Female =1. This is because, data is generally easier to manipulate in an analysis spreadsheet when it is coded quantitatively. Still, the variable is categorical and it is not naturally measured as a number. Other examples of categorical variables include:

- ☆ The blood type of a person: A, B, AB or O.
- ☆ The political party that a voter in India might vote for: Congress, BJP, CPI, CPM etc.
- ☆ The type of a rock: Igneous, Sedimentary or Metamorphic.
- ☆ The employment status of an person: Unemployed, Part-time, Temporary full-time and Permanent full-time.

Note that, categorical variable is one for which each response can be put into a specific category. These categories must be mutually exclusive and exhaustive. Mutually exclusive means that each possible survey response should belong to only one category, whereas, exhaustive requires that the categories should cover the entire set of possibilities. Categorical variables can be further categorized as either nominal, dichotomous and ordinal variables.

- ☆ ***Nominal variables:*** These are the variables that have two or more categories, but they cannot be ranked in an order or no natural order for them. That is nominal variables represent categories with no intrinsic order. The assignment of numbers to these categories is purely arbitrary. Hence, one cannot perform arithmetic (+, -, /, *) or logical operations (>, <, =) on the nominal data. For example, classifying the people living in the State of Andhra Pradesh by region. In this case, there will be three categories *viz.*, Coastal Andhra region, Rayalaseema region and Telangana region. Similarly, marital status of respondents can be categorized as Unmarried, Married, Divorcee and Widower. Other examples could be different crops like wheat, maize, gram, tomato etc., or different irrigation methods: flood, furrow, alternate wetting and drying, sprinkler etc. The important thing is, there is no measure of distance between the categories. A respondent may be either married or not married. The answer is determined, 'Yes' or 'No'. So, there is no question of how far apart in a quantitative sense those categories are. They are just names. Nominal scales name and that is all that they do. Some other examples are sex (male, female), political party (Congress, BJP, CPI, CPM etc), blood type (A, B, AB, O) etc. These examples indicate that, a nominal variable is one that describes a name or category. Contrary to ordinal variables, there is no 'natural ordering' of the set of possible names or categories.

☆ ***Dichotomous variables:*** These are the nominal variables, which have only two categories or levels. That is, this variable takes on one of only two possible values when observed or measured. It is, otherwise, defined as a categorical variable that can take on exactly two values. It is also called as binary variable or dummy variable and is typically treated on its own as a special case. For example, if we were looking at gender, we would most probably categorize somebody as either 'male' or 'female'. Another example might be, if the researcher asked the farmers, if they adopted the IPM technology on cotton crop. Here, the researcher can classify the adoption of IPM technology among the farmers as either 'Yes' or 'No'. Since, dichotomous variables have only two possible outcomes (*i.e.*, 'YES' vs 'NO' or 'SUCCESS' vs 'FAILURE' etc), they are also known as Bernoulli variables. Because of their importance, these variables are often considered a separate category, with a separate distribution (the Bernoulli distribution) and separate regression models (logistic regression, probit regression etc.). In the examples above, with reference to 'YES' vs 'NO' answers, it may be with reference to the presence or absence of a weed or insect species for the different crops, or a 'YES' vs 'NO' answer to a particular irrigation method followed etc. Note that, in compilation of data for analysis purpose, it does not matter whether the variables are numeric or alphabetic. The 'YES' vs 'NO' type variable could be entered as 'YES' or 'NO' or 1 or 2 or 'Y' or 'N' etc. However, it is important to maintain the same case with alphabetic variables. Note that, the values 'N' and 'n' are not the same in the compilation spread sheet. Even a quantitative variable can also be transformed into a categorical variable, called a Dummy variable by recoding the values. For example, the researcher has classified the quantitative variable 'Age' into four categories *viz.*, up to 25 years, 25 to 40 years, 41 to 55 years and 56 to 60 years. Later, values will be assigned to these categories (up to 25 years = 1, 25 to 40 years = 2, 41 to 55 years=3 and 56 to 60 years = 4) and they are called as Dummy variables.

☆ ***Ordinal variables:*** Ordinal implies order and order means ranking. So, the things being measured are in some order. The researcher can have higher and lower amounts. Less than and greater than are meaningful terms with ordinal variables, where they were not with nominal variables. That is, ordinal variables are the variables that have two or more categories just like nominal variables, but these categories can be ordered or ranked. So, an ordinal variable is similar to a nominal variable, but the difference between the two is that in the former there is a clear ordering of the variables. That is, ordinal variables represent categories with some intrinsic order (*e.g.*, low, medium, high; strongly agree, agree, disagree, strongly disagree). Ordinal variables could consist of numeric values that represent distinct categories (*e.g.*, 1=low, 2=medium, 3=high). To best remember this type of variable, think of 'ordinal' containing the word 'order'. So, in ordinal variable, the order matters, but not the difference between values. For example, the researcher might ask the farmers to express the extent of severity of crop disease on a scale of 1 to 10. A score of 7 means more severity of disease than a score of 5, and that is more than a score of 3. But the difference between the 7 and the 5 may not be the same as that between 5 and 3. The values simply express an order. Another example would be movie ratings, from * to *****. Ordinal data has order, but the intervals between scale points may be uneven. Because of lack of equal distances, arithmetic operations are impossible, but logical operations can be performed on the ordinal data. A typical example of an ordinal variable is the socio-economic status of farm families. We know 'upper middle class' is higher than 'middle class', but we cannot say 'how much higher'. Another good example is, if the researcher asked the farmers whether they liked the MSP policy implemented by the Government, they could answer either 'Not good', 'It is not bad' or 'It is very good', then it implies an ordinal variable. This is because, there are three categories, namely 'Not good', 'It is not bad' and 'It is very good'. So, the respondent can rank these three categories from the most positive ('It is very good'), to the middle response ('It is not bad'), to the least positive ('Not good'). However, though these categories can be ranked, it is not possible to assign a 'value' to them *i.e.*, it is not possible to say, 'It is very good' is twice as positive as 'It is not bad' for example. While collecting the information, the respondents were asked to indicate their preferences by selecting one of the following codes: 'Not good' = 1, 'It is not bad' = 2 or 'It is very good' = 3. Another example is, suppose the researcher deals with a variable 'economic status', with three categories (low, medium and high). In addition to being able to classify people into these three categories, you can order the categories as low, medium and high. Now consider a variable like 'educational experience' (with values such as elementary school graduate, high school graduate, degree graduate and post-graduate). These also can be ordered as elementary school graduate, high school, degree graduate and post-graduate. Even though, the researcher can order these from lowest to highest, the spacing between the values may not be the same across the levels of the variables. If these categories were equally spaced, then the variable would be an Interval variable.

d. Quantitative Variable (Numeric Variable)

A quantitative variable is one in which the variates differ in magnitude. For example, farmers income, fertilizers cost, pesticides cost etc. Since the variates can be quantified, the selected sample units can be arranged in a sequence *i.e.*, from lower magnitude to higher magnitude or *vice versa*. So, with the quantitative variables, precise measurements are possible. Quantitative variables can be further divided into two categories *viz.*, Continuous variable and Discontinuous or Discrete variable.

- ☆ ***Continuous variable:*** A continuous variable is one, which is capable of being measured in any arbitrary degree of finesse or exactness. For example, plant height, farmer's income, temperature etc., are continuous variables. That is, these variables can also be measured in fractions or with decimals. For example, plants height may have different values like 3.35m, 3.15m, 2.95m etc. So, a continuous variable can take on any value between its minimum value and its maximum value *i.e.*, it can assume an infinite number of real values. Other examples include measurement of distance, age, temperature etc. Note that, the measurement of a continuous variable is restricted by the methods used, or by the accuracy of the measuring instruments. For example, the height of a plant is a continuous variable and can be expressed as 3.378654 metres tall. However, when the height of a plant is measured, it is usually measured to the nearest centimetre. Thus, this plant's height would be recorded as 3.38 metres. (Note that, measurement of a continuous variable is always a discrete approximation). To make them easier to handle, continuous variables are usually grouped into 'class intervals' and this organization of data facilitates easier analysis and interpretation of findings. Continuous variables can be further categorized as either Interval or Ratio variables.

- ***Interval variables***: These variables are characterized by their measurement such that, they have a continuum and possess a numerical value. These variables are called 'interval' variables because, the intervals between the numbers represent or infer real meaning, unlike ordinal variables. For example, we know, temperature is measured in degrees Celsius or Fahrenheit. These Celsius and Fahrenheit measurements of temperature variable are called interval variables because, the difference between 20 °C and 30°C is the same as 30°C to 40°C. However, note that, temperature measured in degrees Celsius or Fahrenheit is not a ratio variable. So, interval variables have the property that differences in the numbers represent real differences in the variable. Another way to say this is that, equal differences in the numbers on the scale represent equal differences in the underlying variables being measured. For example, the difference between 30 and 40 degrees of temperature compared to the difference between 40 and 50 degrees of temperature on either Fahrenheit or Celsius scales is the same because, the differences in the numbers are the same *i.e.*, the temperature intervals are the same. So, with interval variables, the researcher will not only know, whether one value is higher than another, but that the distances between the intervals on the scale are the same. Note that, interval variables measures the values that lie even below zero. For example, on the Fahrenheit scale, if the temperature is 0°F, then the same on the Celsius scale is -17.78 °C.
- ***Ratio variables***: These variables have all the properties of interval variables plus a real absolute zero. That is, value of zero represents the total absence of the variable being measured. Some examples of ratio variables are height measured in metric system (say, 3.38 metres), time measured in hours, minutes and seconds, blood pressure measured in millmeters of mercury etc. So, the temperature measured in degrees Celsius or Fahrenheit is not a ratio variable because, 0°F or 0°C does not means there is no temperature. However, temperature measured in Kelvin is a ratio variable as '0' Kelvin (often called absolute zero) indicates that, there is no temperature whatsoever. The name 'ratio' reflects the fact that, the researcher can use the ratio of measurements. Say, for example, a distance of ten metres is twice the distance of 5 metres. Similarly, a Kelvin temperature of 100 is twice as hot as a Kelvin temperature of 50 because, it represents twice as much thermal energy (unlike Fahrenheit temperatures of 100 and 50). So, we can construct a meaningful ratio or fraction with a ratio variable. Note that, the measurement of ratio variable cannot go below zero, as zero value itself implies total absence of the variable being measured. For example, a negative plant height is not possible.

☆ ***Discontinuous or Discrete variable:*** As opposed to a continuous variable, a discrete variable can only take a finite number of real values. That is, the discrete variable can be quantified only in whole numbers or integer values and not in terms of decimals. The values of these variables can be determined by counting. Since, these variables are not capable of being measured in any arbitrary degree of fineness or exactness, they contain a clear gap. For example, the number of farmers practicing SRI technology in three villages in a mandal may include: 35, 41 and 10, but fractional values such as 35½, 41½ and 10½ cannot occur. Similarly, age of farmers is a continuous variable, whereas number of farmers is a discrete variable. Like continuous variable, discrete variables may also be grouped and this facilitates easy handling for analysis and interpretation.

The following examples will clarify the difference between continuous and discrete variables.

- ANGRAU has notified in the leading news paper for 300 Assistant Professors, 130 Associate Professors and 37 Professors categories posts. Here, the number of posts would be an example of a discrete variable; since the numbers cannot be in decimals.
- Suppose, a researcher collected data from the farmers regarding the cost of cultivation of paddy in a sample village, He collected the cost data pertaining to various farm operations like transplanting: Rs. 1050.50, human labour costs: Rs. 2450.75, fertilizers costs: Rs. 2138.80 etc. These variables correspond to continuous variables, as the data can be expressed in terms of decimals.

Differences between Qualitative Variable and Quantitative Variable

The following Table 8.22 highlights the differences between qualitative variable and quantitative variable:

Table 8.22: Differences between Qualitative variable and Quantitative Variable.

Qualitative Variable		*Quantitative Variable*	
A qualitative variable is not measurable with numerical instruments, but indicated in terms of adjectives that do not imply ranking or scales (*i.e.*, gender, colours, taste etc.)		A quantitative variable is measurable with numerical instruments and can be ordered in a quantifiable ranking (*i.e.*, height of a plant).	
Qualitative variables take on values that are names or labels. For example, the colour of a fruit (red, green, yellow) or the breed of a dog (*e.g.*, collie, shepherd, terrier).		Variables that are not qualitative are known as quantitative variables. They are numeric *i.e.*, they represent a measurable quantity. For example, when we speak about the population of farming community in a village, it implies, we are talking about the number of farmers in that village. This is a measurable attribute of the village. So, farmers' population would be a quantitative variable.	
A qualitative variable cannot be measured. These variables are only classified on the basis of their distinct attributes.		A quantitative variable is finite and computable. It can be measured on an interval or on a ratio-scale.	
As qualitative data represent individual (mutually exclusive) categories, the descriptive statistics that can be calculated are limited.		As quantitative data are always numeric, they can be ordered, added together, and the frequency of an observation can be counted. Therefore, all descriptive statistics can be calculated using quantitative data.	
Nominal, dichotomous and ordinal variables are three kinds of qualitative variables.		Interval and ratio variables are two kinds of quantitative variables	
For example, the different types of fruits in the refrigerator is a qualitative variable. Other examples include (Assume the researcher is interacting with the farmer):		For example, the number of mangoes in the refrigerator is a quantitative variable. Other examples include: (Assume the researcher is interacting with the farmer):	
In which village were you residing?	Nandipalle	How many children do you have?	Two
In which town is your village located?	Nandyal	How much do you earn?	Rs. 23000/months
What is your occupation?	Farming	How many hours do you work?	7 hours/day
Do you work as Agricultural labour?	No	What is the size of your land holding?	2 hectares

The knowledge of these four types of variables discussed above i.e., nominal, ordinal, interval and ratio variables guides us in understanding about the use of different statistical tools as presented through Table 8.23.

Table 8.23: Applicability of Statistical Tools for different Variables.

Statistical Tools	*Nominal*	*Ordinal*	*Interval*	*Ratio*
Frequency distribution	Yes	Yes	Yes	Yes
Median and percentiles	No	Yes	Yes	Yes
Add or subtract	No	No	Yes	Yes
Mean, standard deviation, standard error of the mean	No	No	Yes	Yes
Coefficient of Variation	No	No	No	Yes

e. EVs

They are also called as Threatening variables. An almost infinite number of EVs exist that might conceivably affect a given relationship. EVs are undesirable variables that affect the relationship between the variables that the researcher is examining. Note that, these EVs influence either IV or DV or both and thus, they change the outcome of expected cause and effect relationship. That is, EVs influence the outcome of a research study, though they are not the variables that are actually of interest. These variables are considered undesirable because, they add an error in the study. Hence, the researcher should aim to decrease or control the influence of EVs as much as possible.

At M.Sc (Ag) and Ph. D. theses level, the researcher will often focus on just two variables: an IV(s) and a DV and that too on a fewer number of IVs. Only in a minority of cases, the student is likely to examine a large number of variables at once. However, just because the student is focusing only on a small number of variables, this does not mean that, these are the only variables that relate to the research he is performing. In this respect, an EV refers to any variables that the researcher is not intentionally studying or cannot study, perhaps because of reasons of cost or difficulty. There can be hundreds or even thousands of these EVs and hence, it is difficult to include all these variables under study. So, the researcher will include only those variables that he is primarily interested in the study and to avoid other (extraneous) variables.

So, EVs are those undesirable variables that primarily are not related to the purpose of the study, but they do influence either the IV or DV or both in a given independent and dependent variables (cause and effect) relationship. So, the EVs must be controlled otherwise, they may influence research outcome. So, unlike IV, these EVs are not manipulated in the study, but still affect the IVs-DV relationship *i.e.*, they intervene the cause and effect relationship of the desired variables. The researcher must note that, whenever he is interested in studying the cause and effect relationship, he must always determine whether there are any EVs that are influencing the IVs and DV. For example, the researcher wishes to study the effectiveness of different teaching methods of Agricultural Economics in different Agricultural Colleges of ANGRAU. Besides different teaching methods (IV), other variables such as teacher's competence, teacher's enthusiasm, age, infrastructural facilities etc., also contribute substantially to the teaching-learning process. These variables cannot be controlled by the researcher and hence, the conclusions lack credibility. Such variables like teacher's competence, teacher's enthusiasm, age, infrastructural facilities etc., are called EVs. Thus, an almost infinite number of EVs exist that might conceivably affect a given IVs-DV relationship. Such variables have to be identified by the researcher before studying cause and effect relationship. So, in order to study the true relationship between the

IVs and DV, the effect of these EVs may have to be controlled. EVs may be categorized as CVs, MVs and Controlled variables. The concept of EVs can be better understood from the following examples:

Example 1

Assume a researcher wishes to study the relationship between background music and task performance amongst employees in a food processing firm. In this firm, the job of employees is to collect items ordered by customers from the warehouse, package them, stick on a label with the customer's address, and put the package on the delivery line. Each time an employee does this, he complete one task. The main objective of the study is to find out, whether background music influences the working efficiency of employees *i.e.*, how many packages (*i.e.*, tasks) the employees process in a given hour? So, the IV considered in the study is the background music and the DV is the task performance. The IV *i.e.*, background music consists of a 'control' and a 'treatment'. The control refers to the normal conditions experienced by employees in the firm godown and it means that, the employees selected under this category are not being provided with background music (*i.e.*, employees work without background music). The treatment category comprises of employees, who are exposed to background music, but other working conditions are similar as in control group. In other words, the treatment is providing the selected employees with background music. It is this IV (*i.e.*, background music) that the researcher is manipulating to examine its effect on the DV (*i.e.*, task performance). The word 'manipulating' is used because, the selected IV is being changed across treatment and control categories. Assume 100 respondents (employees) have been selected randomly from both treatment and control categories. At a given day and time, the experiment is conducted in such a way that, treatment group perform all the tasks listening to the music and the control group do the same tasks without listening to the music. This experiment continues for about 7 hours shift and task performance is measured for every each hour. After the experiment is over, now the researcher compute the working efficiency both for treatment group and control group. This analysis indicate that, whether there are any differences in the number of tasks performed between the control group and treatment group and here the researcher considered only two variables *viz.*, IV and DV. Based on the findings, the researcher may conclude that, background music has positive influence on the task performance of the employees. Notice that, in concluding the above relationship, the researcher has not taken EVs into consideration. That is, there are some other variables not considered in this experiment, but may influence either the IV or DV or both. They include:

EVs that could Influence IV

- ✰ Type of background music (Western music/Classical music/Folk songs/Movie songs etc.)
- ✰ Loudness of background music (low, medium, high volumes etc.)
- ✰ Time of day when the background music was played (morning, afternoon, night etc.)

EVs that could Influence DV

- ✰ Employee tiredness
- ✰ Employee motivation
- ✰ Job satisfaction

Example 2

Suppose, a researcher wishes to measure the effects of alcohol consumption (IV) on working efficiency of the skilled workers (DV) in running a processing machinery in a sugar factory. However, some EVs listed below may affect the outcome of the research study:

- ☆ ***Familiarity in working with the processing machinery:*** Some workers may work better even after alcoholic consumption because, they are more familiar with the same machinery since many number of years.
- ☆ ***Familiarity with the test:*** Some workers may do better than others because, they know what researcher is expecting from the test.
- ☆ ***Used to drinking***: The effects of alcohol on some workers may be less than on others because, they are habituated to drinking.
- ☆ ***Full stomach***: The effect of alcohol on some workers may be less than on others because, they have just had a big meal.

So, the above examples indicate that, EVs could interfere in the experiment and may affect the actual outcome of the experiment. Hence, these EVs should be controlled in conducting the experiment and to get the desired results. If an EV really is the reason for an outcome (rather than the selected IV) then, it is likely to be called as a CV, because it has confused or confounded the relationship that he is interested in. That is, if these EVs are not controlled they may become CVs because, they could go on to affect the results of the experiment. There are two types of EVs:

- ☆ ***Situational variables***: These variables are related the environment in which the respondents are working. In the above Example 1, these variables are related to noise, temperature, lighting conditions etc. These situational variables should be controlled so that, they are the same for all the respondents.
- ☆ ***Respondent variables:*** These variables are related to the participants behavior and nature *i.e.*, they include : mood, intelligence, anxiety, dedication towards work, concentration etc. These variables vary across the respondents and they could effects the results. For example, if the respondent had poor eyesight, this could affect his performance and hence, the results of the experiment.

The above discussion also highlights the feature that distinguishes experimental research from social sciences research in agriculture, where in the former, there is tight control of EVs. Thus, EVs are basically all variables other than those the researcher are interested in for conducting the research study. In an experiment research, the EVs are controlled through imposing similar conditions both for the treatment group and control group except the variable (*i.e.*, treatment) that is intended for study on the treatment group. This is essential because, if this is not the case, the relationship between the independent and dependent variables cannot be due to the treatment given, but may be influenced by the EV, popularly called as third-variable problem. As mentioned earlier, EV can be categorized as CVs, MVs and Controlled variables.

i. CV

As mentioned above, EVs can be dangerous, since their presence may damage the outcome of the given cause and effect relationship. This makes it impossible for researchers to know whether the effects in the study were caused by the IVs selected or by some unknown EVs. If an EV is neglected, it may even turn as CV and this is even more dangerous than simple EV because, it could provide an alternative explanation for the research results. So, the EVs must be controlled in conducting a research study, otherwise they do become CVs and still damage the outcome of the study. Note that, not all EVs become CVs. Only, when an EV changes systematically along with the variables that the researcher is interested in the study, then it is called as a CV. So, when an EV systematically varies with the IV and influences the DV, it is called a CV (Figure 8.25). A variable is considered to be confounding because,

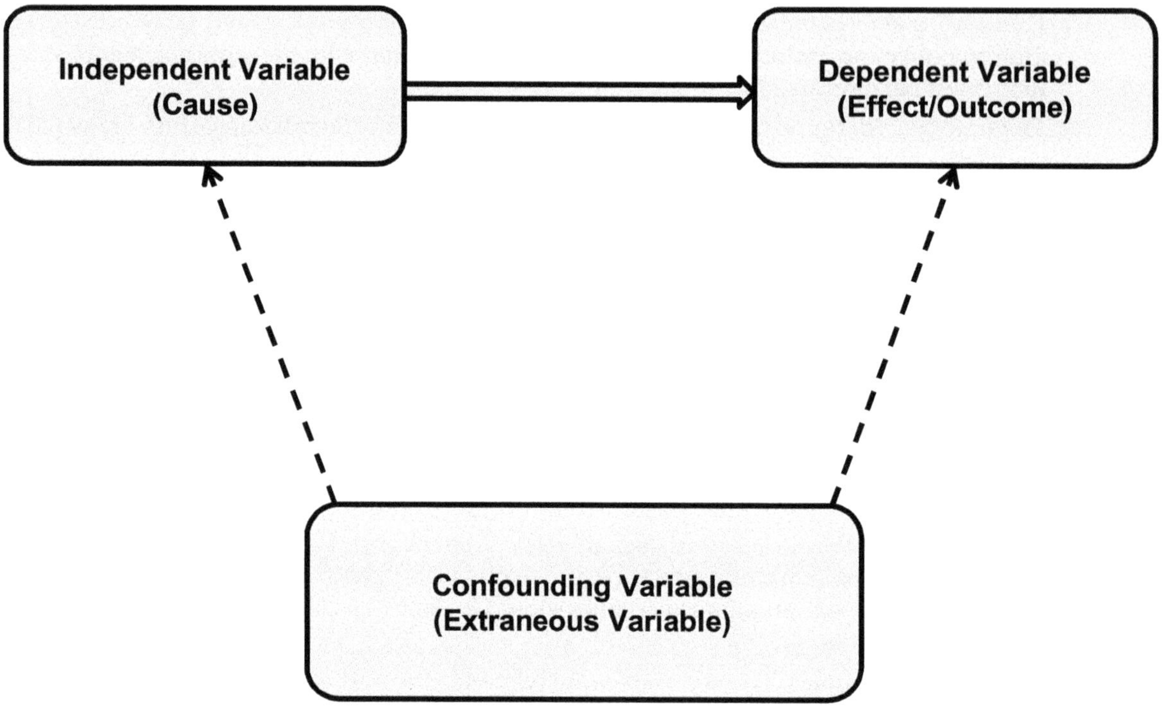

Figure 8.25: Effects of Confounding Variable.

it provides an alternative explanation for the research findings *i.e.*, it gives an alternative explanation for the relationship between the variables actually considered in the study. This threatens the internal validity of research results. The CVs are also called as 'third variables' that the researcher failed to control or eliminate in the research study and thereby, cause the researcher to analyze the results incorrectly. The results may show a false correlation between the IVs and DV thereby, leading to an incorrect rejection of the H_O.

Thus, CV (also known as Confounding factor or Hidden variable or Lurking variable or a Confound, or Confounder) is an EV in a research study that correlates (either positively or negatively) with both IV and DV. So, the perceived relationship between an IV and DV will get affected due to the presence of this CV. So, the outcome between IV and DV in the presence of CV is called Spurious relationship. Thus, CV is a variable that obscures the effects of another variable. To have a better understanding about CV, let us continue with the earlier example studied under EVs *i.e.*, studying the relationship between background music and task performance of employees working in a food processing firm. In the earlier example, we concluded that, there is a positive relationship between the background music and task performance in the food processing firm *i.e.*, with the introduction of background music, the task performance will be improved. However, to analyze the influence of the CV, let us change the way that the original experiment was conducted. Previously, we measured the treatment group and the control group at the same time, once every hour from the beginning of their shift to the end of their shift (*i.e.*, a period of 7 hours). However, let's imagine that, since all the employees in the processing firm work in one giant room, this makes it impossible to provide the treatment group with background music without the control group hearing the music. Since, this would be a clear threat to internal validity, let us change the experimental design. Instead of both groups being measured at once, we turn the music on for the first 3.5 hours of the shift, and then turn it off for the second 3.5 hours of the shift. We record the number of tasks performed correctly from the treatment group during these first 3.5 hours, and then record the number of tasks for the control group during the second 3.5 hours. However, in doing this, we have invited an EV, 'time of shift', into the experiment. Now imagine that, the results from this revised experimental design are the same *i.e.*, the use of background music improved task performance, since the treatment group perform far more tasks in their 3.5 hour shift than the control group did in their 3.5 hour shift. But, the problem is that, unlike before, we do not know whether task performance was improved by the introduction of background music, or the third suspect EV: 'time of shift'. The question arises: Why is this third, suspect EV, time of shift, a confounding variable? The answer is that, the variable, time of shift, changed systematically with the IV that we were measuring (*i.e.*, background music). It changed systematically because, as the time of shift of providing music changed (a) to the first 3.5 hours of the shift (b) from the second 3.5 hours of the shift, so the IV changed from (a) the introduction of music (*i.e.*, the treatment group) to (b) the removal of music (*i.e.*, the control group). So, may be now, the relationship that we are observing is not between background music and task performance, but between time of shift and task performance. This third variable *i.e.*, time of shift of providing music has entered in between the original relationship between background music and task performance and this variable is called as CV. Furthermore, we would expect task performance would drop during the second 3.5 hour shift, when the control group (*i.e.*, no music) were being measured, compared with the first 3.5 hours when the treatment group (*i.e.*, with music) were being measured because, 'tiredness' of the employees in the control group during second shift adversely influenced their task performance. Did the control group simply perform worse in terms of the number of tasks performed because, they were more tired than those employees in the treatment group? The answer is: We don't know. This is the problem, and it threatens the internal validity of the experiment *i.e.*, it provides an alternative explanation for the relationship between the IV (*i.e.*, background music) and the DV (*i.e.*, task performance). So, the discussion infers that, a CV is an EV that is statistically related to (or correlated with) the IV. This means that, as the IV changes, the confounding variable changes along with it. The following are the other examples of CVs:

Example 1

Assume there is a positive correlation between ice cream consumption and number of drowning deaths for a given period. A researcher might attempt to explain this correlation by inferring a causal relationship between the two variables (either that ice cream causes drowning, or that drowning causes ice cream consumption). However, a more likely explanation is that, the relationship between ice cream consumption and drowning is spurious and that a third, CV (the season) influences both variables: during the summer, warmer temperatures led to increased ice cream consumption as well as swimming by more number of people and thus, cause more number of drowning deaths.

Example 2

For example, a researcher wants to know, whether Pesticide X has an effect on controlling a pest. The researcher must take care to design the experiment so that, he can be very sure that, the pest on the crop is controlled because of the influence of this pesticide X and that the control was not caused by other factors (say, biological control). Those other factors would be CVs.

Example 3

Assume a researcher want to know, whether tall people are better basketball players than short people. So, he selects a sample of people of various heights and asks them to shoot baskets from the foul line. The number of baskets they make out of ten shots will be the measure of their basketball ability. The IV variable in this example is height and the DV is the number of baskets. The findings of the study revealed that, small people did make better basketball players than tall people. This outcome is influenced by a CV *i.e.*, 'number of hours of practice'. So, the reason is that, since short people in the sample had more practice with foul shots than the tall people, there is inverse correlation between the selected IV and DV. So, 'number of hours of practice' is a CV.

Example 4

A researcher wishes to examine the relationship between the force he apply to a ball and the distance the ball travels. Naturally, he presume that, '*the more force you apply, the further the ball will travel*'. He conducted the experiment with two conditions. In Condition 1, he applied a 10 lb force to the ball and measured the distance traveled. In Condition 2, he applied a 5 lb force to the ball and measured the distance traveled. On comparing the findings, he concluded that, the ball travelled farther in Condition 2 rather than in Condition 1. That is, he concluded that, '*the less force you apply, the farther the ball travels*'. Should he conclude that, Isaac Newton's law was wrong? No. This is because, this spurious relation was due to the presence of third CV *i.e.*, angle of throw. So, given the presence of this confound variable, the researcher now is in confusion, which variable *i.e.*, force or angle is responsible for the change in the distance the ball travels and the presence of this CV (angle of throw) has led to spurious correlation between original IV (force of throw) and DV (distance the ball travelled).

Example 5

A researcher wishes to conduct a study in two cities to investigate the relationship between income levels of the consumers and quantity of ice creams sold. Without controlling for other variables, the study finds that, these two variables are directly related in the selected cities. This is a flawed conclusion because, there may be any other lurking CV that may influence this supposedly clear-cut relationship. For example, due to rise in temperatures in the city (being a summer season), the sales of ice creams might be increased and this rise in temperature coincided with rise in incomes of the consumers in that financial year. So, without controlling for the CV of temperature, the relationship between income levels and quantity of ice creams sold in the selected city cannot be assumed.

Example 6

If one college, Professor used a Microeconomics textbook 'A' in his class and another Professor used a Microeconomics textbook 'B' in his class, and students in the two classes were given achievement tests to see how well they write the exam. Then the IV (text books) seems to be confounded with third variable *i.e.*, Professor's teaching effectiveness. Hence, there is no way to determine, if differences in test scores of the students between the two classes were caused by either selected IV or by the CV or by both.

The above examples illustrates, why it's so important to always look for the presence of CVs, as their presence lead to spurious relationship between the selected IVs and DV. These confounding (or intervening) variables should be studied because, they may influence or 'mix' the effect of the IVs. Thus, CV is a variable that is associated with the cause and effect relationship and with a possible cause of the relationship. This type of variable may either strengthen or weaken the apparent relationship between the cause and effect. So, unlike control variables, CVs cannot be controlled in the study because, they are associated with the cause and effect relationship and with a possible cause of the relationship. So, the presence of CVs makes the researcher to conclude with wrong answers. Hence, in terms of a more technical language 'CV' refers to 'Bad'. This is so because, this CV is an EV that must change systematically with at least one of the other variables the researcher is measuring (usually the IV). This third variable (CV) changing systematically because, it must behave in a way that is similar to the variable(s) that the researcher is intentionally studying. This makes it difficult to know whether the change in the DV is the result of the IV that the researcher is intentionally measuring, or the third, suspect EV (*i.e.*, CV). The effect of a CV on the DV cannot be separated from the effect of the IV on the DV. These CVs are a bigger problem in observational studies.

ii. Moderating Variable (MV)

MV is also called as Interacting variable. This is a secondary or special type of IV chosen by the researcher, which is hypothesized to change or modify the existing relationship between the IV and DV. In general terms, a moderator variable affects the direction and/or strength of the relation between an IV and DV. This is where, it differs from the Mediating Variable. That is, a MV is the one that influences the strength of a relationship between two other variables (IV and DV), whereas a mediating (or mediator) variable is one that explains the relationship between the two other variables (IV and DV). Thus, the MV shows a strong contingent effect on the IV and DV relationship. This presence of a third variable (MV) may also modify the original relationship between the IVs and DV. According to Baron and Kenny, *'MV is the variable that is related to the direction or strength of the relationship between the independent and dependent variables'*. Thus, the presence of MV modifies the original relationship between the IV and DV. Note that, a MV may be qualitative (such as farmer gender, type of institution) or quantitative (*e.g.*, number of visits made by the local Agricultural Officer). Say, if the researcher wishes to study the effect of the IV 'A' on the DV 'B', but he suspects that a third variable 'C' changes or modifies the existing relationship between 'A' and 'B', then 'C' is considered as the MV. Assume a linear causal relationship in which the variable X is presumed to cause the variable 'Y'. MV is a variable that alters the strength of this causal relationship. Say, for instance, psychotherapy may reduce depression more for men than for women, and so we would say that, gender (MV) moderates the causal effect of psychotherapy (X) on depression (Y). The following are the other interesting examples of MVs.

Example 1

We know, pesticides cost (IV) influences the farmers income (DV). However, the managerial experience of the farmer in terms of selecting right type of pesticide, right dosage of pesticide, right time of application etc. (Figure 8.26) also counts much in influencing the relation between pesticides cost (IV) and farmers income (DV). So, managerial expertise of the farmer can be considered as a MV.

Example 2

Assume there is a strong positive relationship between the quality of library facilities (X) in Agricultural College, Mahanandi and the performance of the students (Y). Although this relationship is supposed to be true generally, it is nevertheless contingent on the interest and inclination of the students. It means that, only those students, who have the interest and inclination to use the library will show improved performance in their studies. In this relationship between X and Y, interest and inclination is MV *i.e.*, which moderates the strength of the association between X and Y variables.

Example 3

A researcher wishes to study the influence of socio-economic background on widow's remarriage. So, in this case, socio-economic background is the IV and widow's remarriage is the DV. But, if widows have children, it may also influence the above causal relationship between socio-economic background and widow's remarriage. So, the third factor *i.e.*, 'widows with children' is the MV in this causal relationship.

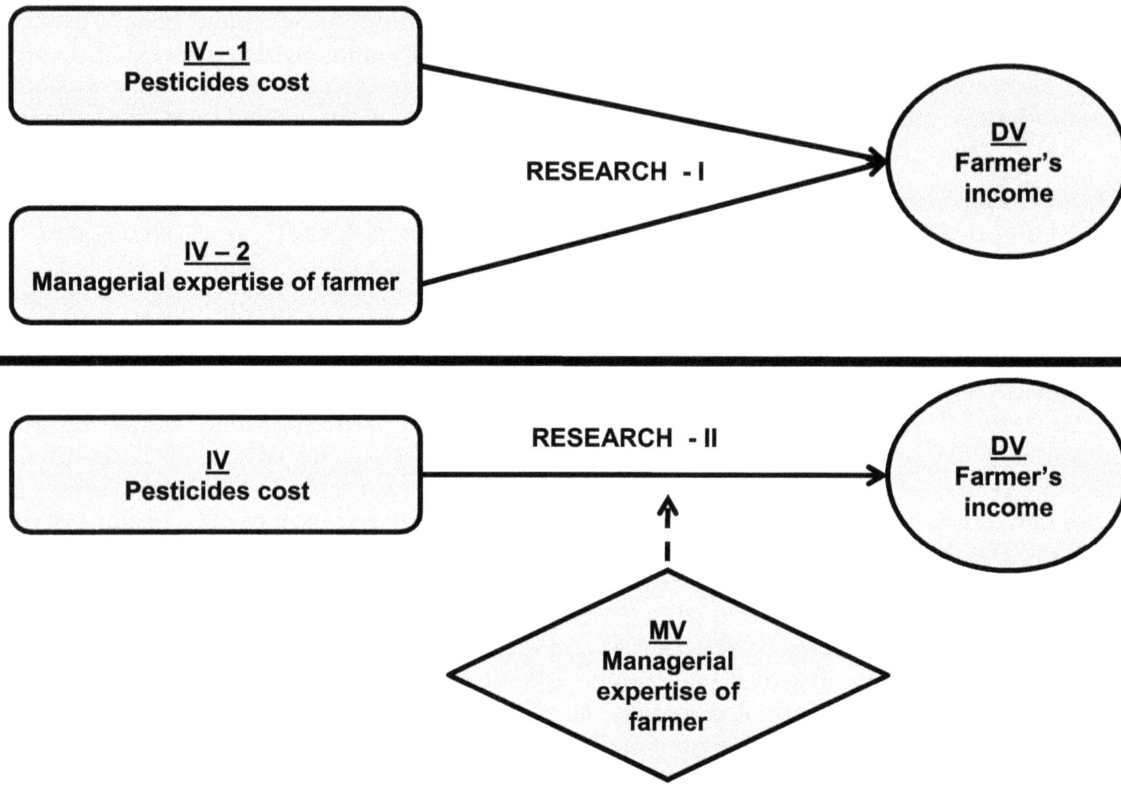

Figure 8.26: Moderating Variable (MV) vis-à-vis Independent Variable.

The decision whether to label them IV or MV depends on how they affect one another. An IV of one research, might become a MV in another different research (Figure 8.26). This infers that, MV is a special type of IV, which the researcher may select for the study in order to investigate, whether it modifies the relationship between the IV and DV. This MV is manipulated by the researcher to discover, whether it produces an interaction effect in the existing IV and DV relationship. The essential difference between IV and MV lies in, how the researcher views each in the study. For IVs, the concern is with their direct relationship to the DV, whereas for MV, the concern is with their effect on that relationship.

CV is an EV that affects an IV and also affects a DV at the same time. That is, when an EV changes systematically along with the variables that the researcher is interested in the study, then it is called a CV. A moderator is a variable (MV) that alters the direction or strength of the relationship between a IV and DV. MV address 'when' or 'for whom' X causes Y. This is where, MV also differs from the Mediating Variable, as the latter, address 'how' or 'why' X causes Y.

iii. Controlled Variables

As discussed earlier, it is virtually impossible to include all the potential variables in the research because of time and cost constraints. As a result, the researcher must attempt to control or neutralize all other EVs that are likely to have an effect on the relationship between the IV and DV relationship. Thus, control variables are those variables that the researcher has chosen to keep them constant, neutralize, or otherwise cancel out or eliminate so that, they will not exert differential or moderating effects on the relationship between the IV and DV relationship. This implies that, control variable is an EV that the researcher does not wish to examine in a study. This control variable is also called as Covariate. The difference between a control variable and a MV is that, the effects of the control variable are minimized or eliminated or held constant, while the effects of the MV are studied. Since both moderator and control variables are IVs, it is up to the researcher to determine the independent, moderator and controlled variables. For example, in Boyle's law, absolute pressure and volume of a given mass of confined gas are inversely proportional, if the temperature remains unchanged within a closed system. Here, temperature is a controlled variable. Similarly, the law of demand states that, quantity demanded of a commodity varies inversely with the price of the same commodity, if the income of the consumer, his tastes and preferences, prices of substitute goods etc., remains constant. Here, income of the consumer, his tastes and preferences, prices of substitute goods etc., are controlled variables.

f. Mediating Variables

Mediating variables may explain the causal relationship between IV and DV, but are not directly observable themselves. It is, otherwise, called as Intervening variable or Intermediary variable. This variable is often confused with MV. As discussed above, MV is one that influences the strength of a relationship between two other variables (IV and DV), whereas a mediator variable is one that explains the relationship between the two other variables (IV and DV). That is, Mediating Variable 'stands between' the dependent and independent variables and shows the link or mechanism between them. Advances in knowledge depend not only on studying cause and effect relationship between the dependent and independent variables, but also in analyzing the mechanisms that account for this causal relation. In this sense, the mediating variable acts as a DV with respect to IV and acts as an IV towards the DV. For example, the statistical association between income and longevity needs to be explained because, just by having more money does not make one live longer. Other variables intervene between money and long life. People with high incomes tend to have better medical care than those with low incomes. So, medical care is a mediating variable, as it mediates the relation between income and longevity. The following are the other interesting examples:

Example 1

A researcher wishes to study a causal relationship between size of land holding (IV) and farmers' net income (DV). The study revealed that, there is a positive relationship between these two variables. But, besides size of land holding, farmers knowledge regarding fertilizer use, quality of seed etc., also influence the net income. So, farmer's knowledge is considered as a mediating variable.

Example 2

We know, parents transmit their social status to their children directly, but they also do so indirectly, through imparting education: *viz.*, Parent's status → Child's education → Child's status.

Example 3

A theory of suicide states that, married people are less likely to commit suicide than unmarried (single) people. The assumption is that, married people have greater social integration *i.e.*, a feeling of belonging to a particular family, unlike unmarried people. Hence, a major cause of committing suicide was that, unmarried people lack a sense of belonging to a particulat family. Thus, this theory can be restated as a three-variable relationship: marital status (IV) causes the degree of social integration (mediating variable), which affects suicide (DV). This specification of the chain of causality between the variables makes the theory clearer and helps a researcher to understand and test complex relationships.

Example 4

An important research finding is that, five-day work week results in higher productivity. The mediating variable here is 'job satisfaction'. Thus, the inference is 'the introduction of five-day work week (IV) will increase job satisfaction (mediating variable), which will lead to higher productivity (DV).

Thus, mediating variable is the one that surfaces between the time the IV start operating to influence the DV and the time its impact is felt on DV. There is thus, a time dimension to the mediating variable. This variable surfaces as a function of the IV opening in any situation and helps to conceptualize and explain the influence of the IV on the DV. A variable functions as a 'mediator' when it meets the following conditions:

- ✰ Variations in levels of the IV significantly account for the variations in the presumed mediating variable.
- ✰ Variations in the mediating variable significantly account for variations in the DV, and
- ✰ When both IV and MV appear in the model, a previously significant relation between the IV and DV is no longer significant.

The following points and examples (Figure 8.27) clearly infer what mediating variable is and how it differs from MV:

- ✰ Mediating variable influence the DV, while MV increase or decrease the influence.
- ✰ MV pin point the conditions under which an IV exerts its effects on a DV, whereas mediating variable simply influences the IV-DV relationship.
- ✰ Generally, MVs are variables that exist prior to data collection, as opposed to mediating variables that are assumed to occur during the phenomena being studied.

g. Endogenous Variable

A variable that is an inherent part of the model being studied and that is determined or caused by one or more variables in the model. Note that, an endogenous variable may also cause another endogenous variable in the model.

h. Exogenous Variable

The term 'Exogenous' is derived from two Greek words '*exo*' and '*gignomi*', meaning 'outside' and 'to come to be' respectively. So, exogenous variable is a variable whose value is determined completely outside the model system and whose behavior is used to describe that of the Endogenous Variable. It is, otherwise, defined as a variable entering from and determined from outside of the model being studied. A causal system says nothing about its exogenous variables. So, an exogenous variable is not caused by another variable in the model. Usually this variable causes one or more variables in the model.

The difference between exogenous and endogenous variables can be better understood from the Figure 8.28. We know equilibrium price of the commodity is determined by two market forces *viz.*, demand and supply. It is clear from Panel B of the Figure 8.28 that, demand curve (DD) is influenced by quantity demanded and price of the commodity and supply curve (SS) is influenced by quantity supplied and price of the commodity. So, quantity of commodity is taken on X-axis and price of the commodity is taken on the Y-axis. These two variables *i.e.*, quantity and price are called as endogenous

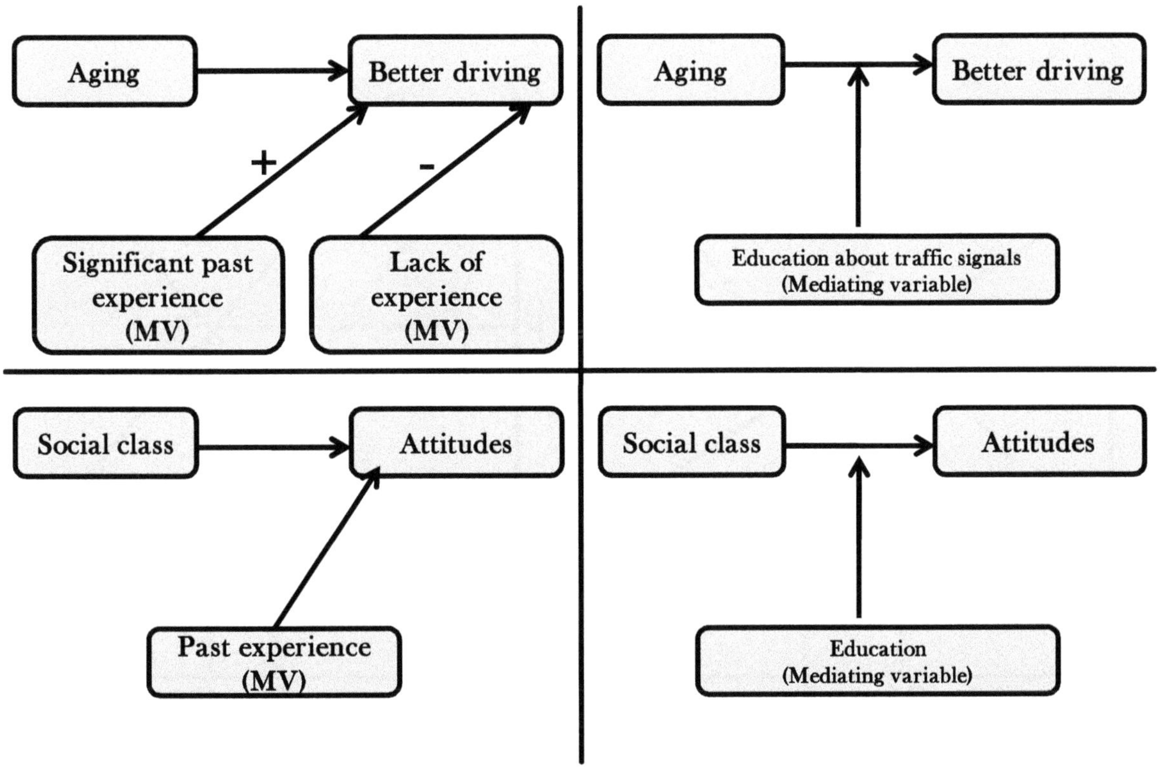

Figure 8.27: Moderating Variable (MV) vis-à-vis Mediating Variable.

variables, as they are controlled within the model of price determination. Every other factor that is not controlled or the variables that lie outside the model (say, income of the consumer, increase/decrease in number of family members, prices of substitute goods etc) would, therefore, be exogenous (and they are held constant). If an exogenous variable say, income of the consumer decreases, demand for the same commodity decreases and the new demand curve D1D1 shifts towards left side or towards the origin. So, the causal factor (decrease in income of the consumer) that lead to shift in the demand curve towards left side in the given model *i.e.*, demand function is called exogenous variable. The same explanation is given in the form of a mathematical equation here under. The demand function, which shows quantity demanded as a function of endogenous variable (price) is given by:

$$Q=a-bP$$

where, 'Q' is quantity of commodity demanded, 'a' is the intercept, 'b' is the slope coefficient for 'P' price. So, when we consider the endogenous variables in this model, we have only one such variable on the right hand side, which has its own coefficient, *i.e.*, price. Quantity is on the left hand side and it is considered as the DV (but also endogenous). The exogenous variables are all considered within the 'a', which gives us the intercept. The same concept is explained considering another example as given below.

$$Q=a+bI$$

In the above model, everything is the same except for the 'I' variable, which represents income of the consumer. Since, income (I) is on the right hand side with a coefficient 'b', (which determines the

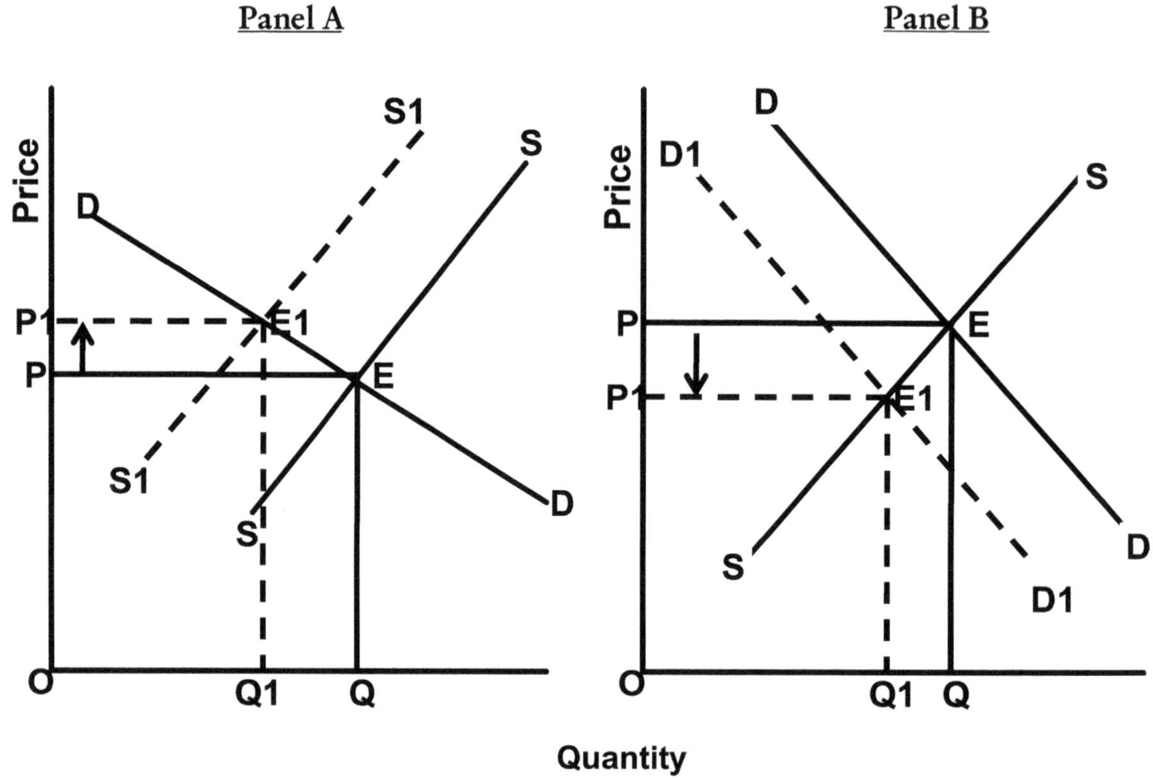

Figure 8.28: Exogenous and Endogenous Variables in Demand-Price Relationship.

slope), it is now the endogenous variable in this model. Now, the price variable is exogenous and is considered in the 'a' which determines the intercept. The positive sign of the slope 'b' implies, the commodity is a normal (if it is negative, it implies the commodity is inferior or giffen good). In Panel A of the Figure 8.28, the same explanation holds good with reference to endogenous variable and exogenous variable (here, the endogenous variables include: quantity demanded and supplied and price) and exogenous variable include rise in prices of raw materials that led to decline in the supply of the product in the market.

Table 8.24 summarizes and draws the comparative picture across different types of variables. So, once the variables relevant to the topic of research have been identified, then the researcher is interested in analyzing the relationship among them. A statement containing the variable is called a Proposition. It may contain one or more than one variable. The proposition having one variable in it may be called as Univariate proposition, those with two variables as Bivariate proposition and then of course Multivariate proposition containing three or more variables. Prior to the formulation of a proposition, the researcher has to develop strong logical arguments, which could help in establishing the relationship. For example, age at marriage and education are the two variables that could lead to a proposition: the higher the education, the higher the age at marriage. The researcher should be capable of explaining the logic to reach this conclusion. Thus, all relationships have to be explained with strong logical arguments. If the relationship refers to an observable reality, then the proposition can be put to test, and any testable proposition is hypothesis.

Table 8.24: Comparative Picture across different Types of Variables.

Type of Variable	*Definition*	*Other related terms used*
IV	These are the variables that can be manipulated by the researcher in order to study their effects and differences in their effects on those variable (Dependent variable) that is being dependent on them. IV is, otherwise, defined as the variable, whose effect upon the DV, the researcher is trying to understand, explain, and predict.	Causal variable, determinant, explanatory variable, predictor variable, manipulated variable, stimulus variable, treatment variable, experimental variable, regressor, controlled variable, exposure variable, risk factor (in medical statistics), input variable.
DV	The DV is a variable that is dependent on an IV(s). That is, DV is the response variable, which is observed and measured to determine the effects of the IV(s).	Response variable, explained variable, predicted variable, criterion variable, outcome variable, regressand, measured variable, responding variable, experimental variable, output variable.
Qualitative variable	This variable have no numerical meaning.	Categorical variable.
Quantitative variable	A quantitative variable is one in which the variates differ in magnitude.	Numeric variable.
EV	An EV is related to the IV or DV that is not part of the research study. So, EVs are undesirable variables that influence the relationship between the variables that the researcher is examining. Note that, these EVs influence either IV or DV or both and thus, they change the outcome of expected cause and effect relationship.	Threatening variable.
CV	CV is an EV in a research study that correlates (either positively or negatively) with both the IV and DV. So, the perceived relationship between an IV and DV will get affected due to the presence of this CV.	Confounding factor or Hidden variable or Lurking variable or a Confound, or Confounder.
MV	A variable that is related to the IV or DV and has an impact on the DV. In general terms, a moderator variable affects the direction and/or strength of the relation between an IV and DV.	Interacting variable.
Controlled variable	A variable that is related to the DV variable, the influence of which needs to be removed.	Restricting variable.
Mediating variable	A mediator variable is one that explains the relationship between the two other variables (IV and DV).	Intervening variable or Intermediary variable.
Endogenous variable	A variable that is an inherent part of the model being studied and that is determined or caused by one or more variables in the model.	—
Exogenous variable	Exogenous variable is a variable whose value is determined completely outside the model system and whose behavior is used to describe that of the endogenous variable. It is, otherwise, defined as a variable entering from and determined from outside of the model being studied.	—

C. Types of Data to be Collected

Data can be defined as the quantitative or qualitative values of a variable. Data is plural of Datum, which literally means *to give* or *something given.* Data is thought to be the lowest unit of information from which other measurements and analysis can be done. Data can be numbers, images, words, figures, facts or ideas. Collection of mere data is not sufficient to draw meaningful conclusions for the framed objectives. In fact, the raw data collected should be analyzed with appropriate statistical tools and techniques and the analytical findings should be interpreted to draw meaningful conclusions objective-wise for the research study. There are different types of data employed in social sciences research in agriculture and each type has its own strengths and weaknesses. The following are the different types of data, the researcher employs in pursuit of his research study:

a. Primary data and Secondary data
b. Cross-sectional data and Time series data
c. Qualitative data and Quantitative data
d. Univariate data, Bivariate data and Multivariate data
e. Subjective data and Objective data

a. Primary Data and Secondary Data

- ✩ ***Primary data:*** These are the data collected from the sample respondents directly through informal discussions. Since the researcher himself collects the data from the (sample) respondents informally, he gets the precise data actually needed for the research project. Further, collecting the data and information through informal interactions with the respondents enable the researcher to build up and maintain healthy relations in relevance to the research study. This primary data is the data collected for the first time by the researcher from the respondents. For example, a researcher wishes to collect information pertaining to the cost of cultivation of paddy in Kurnool district of Andhra Pradesh. He then selects a random sample of farmers representative to the population and then collect requisite data from them through surveys (using schedules) and personal interviews. He personally interacts with the sample respondents and will have a chance to understand their problems in cultivating paddy. This information (*i.e.,* primary data) collected will be reliable and relevant for the research study. So, the information or data collected from the responses constitute primary data and this data represents the original data that has been collected specially for the purpose in mind. This primary data has not been published yet and is more reliable, authentic and objective. This data has not been changed or altered by human beings and therefore, its validity is greater than secondary data. To gain first hand information about the research problem, this primary data collection assumes greater significance and the personnel involved both in provision and collection of such data act as a witness for this data.
- ✩ ***Secondary data:*** These are the data and requisite information collected from authentic published sources. That is, data collected from an authentic source that has already been published in any form is called Secondary data or Published data. So, the researcher need not put special efforts like preparation of schedules and collection of data from the sample respondents informally etc. Thus, secondary data represents the data collected from those, which have already been collected and analyzed by someone else. The review of literature

collected for any research study is based on secondary data. So, secondary data involves sourcing and accessing existing data that were not originally published for the purpose of the study. For example, if the researcher wishes to conduct a study on *'Trends in market arrivals of paddy in selected regulated markets in Rayalaseema region of Andhra Pradesh'*, he will collect requisite (secondary) data from the Annual Administration Reports published by the selected APMCs. Similarly, the researcher collects secondary data pertaining to the MSPs of agricultural commodities from the website: www.agricoop.nic.in, area, production and productivity data of agricultural crops cultivated in Andhra Pradesh from the Statistical Abstract published by the Bureau of Economics and Statistics, Hyderabad, Government of Andhra Pradesh. So, secondary data is less expensive to obtain compared to collect primary data. This is because, someone has already spent the necessary time and money to gather it. Much secondary data is, additionally, already analysed by the people, who gathered it. However, secondary data can be out of date or only partially relevant for the research study under consideration. In some sources, only the published raw data are available without useful interpretations and hence, the researcher may not be clearly aware about the reasons for the trends in data. Further, it won't include the opinions of the respondents of the research study. So, if the research study is exclusively based on secondary data, the conclusions and policy implications of the research study may not be specific to the particular client base and may not be even relevant to the locality under consideration. Though secondary data can be less valid compared to primary data, still it is preferred in the following cases:

- In situations, where it is difficult to obtain primary data and getting information from secondary sources is easier and possible.
- In situations, where primary data does not exist and hence, the researcher has to depend only on secondary data.
- In situations, where primary data is present, but the respondents are not willing to reveal the information.
- In situations, where the budget is too limited and equally there is a time constraint on the part of the researcher to collect primary data.

Characteristics of Secondary Data

The following are the characteristics of secondary data:

- ***Reliability of Data:*** Reliability of secondary data can be established from the following aspects:
  - From which sources the data are collected.
  - Who collected the data.
  - Whether compiler and source both are dependable.
  - The purpose for which the data are originally collected.
- ***Suitability of Data:*** Besides reliability, suitability of its usage should be equally important. That is, the data that are available may be reliable, but may not be suitable for the research study. So, the researcher must be careful on the following aspects:
  - What data are needed for what variables?
  - Time frame for which data are to be collected.

☆ ***Adequacy of the data:*** Note that, the secondary data may be reliable and suitable, but may not be adequate enough to conduct the research study. That is, the data may not be available for the requisite time frame.

Differences between Primary Data and Secondary Data

The following Table 8.25 draws the comparative picture between primary and secondary data:

Table 8.25: Primary Data vis-à-vis Secondary Data.

Item	*Primary Data*	*Secondary Data*
Definition	Data that has been collected from first-hand-experience is known as primary data.	Secondary data are pieces of information that have already been collected for a different purpose, but may be relevant to the research problems at hand. So, secondary data are collected from authentic published sources.
Source of collection of data	Schedules, questionnaires etc.	Authenticated publications like Government reports, Internet web sites etc.
Statistical process	Statistical process is not done (*i.e.,* raw data is available)	Statistical process is done
Originality of data	Data are original *i.e.,* first time collected by the user	Data are not original *i.e.,* data collected and published by other agencies are used by the researcher.
Relevance	The collected data is relevant to the objectives framed for the study. Thus, it is specific to researcher's needs.	The available data may or may not be directly relevant to the objectives framed for the study. So, secondary data are comparatively less relevant to researcher's needs and sometimes only partial information is available.
Interaction with the respondents	The researcher has direct interaction with the respondents and can collect the opinions of (sample) respondents about the research problem, its severity and possible solutions as perceived by the respondents themselves.	The researcher has no direct interaction with the respondents in collecting the relevant information, as someone else has already spent the necessary time and money to gather and, possibly, analyze it.
Relationship development	Getting primary data may often, additionally, benefit the researcher to build relationship with sample respondents in relevance to the research study	No such relationship development is feasible. The researcher will collect the data and information only from the published sources and hence, he may not be fully focused on the respondents' needs. So, the generated findings from the data will not represent the opinions of the respondents.
Availability of current information	Current information can be collected from the respondents directly. The questions the researchers ask are tailored made to elicit the data that is actually required for the study.	It might not be current. Further, researcher usually gets a partial information needed for the research and sometimes he gets an elaborate information more than what actually he needs for the study.

Contd...

Table 8.25–*Contd...*

Item	*Primary Data*	*Secondary Data*
Reliability of research	A research can be conducted without secondary data, but a research without primary data is less meaningful.	Research study based exclusively on secondary data is least reliable and may be biased because, this data has already been manipulated by human beings.
Validity of research	Primary data is the first hand information obtained from a sample that is representative of the target population and hence, the findings will be valid for the entire target population. Hence, using primary data in research can improves the validity of research.	Research study based on secondary data alone may not yield relevant policy implications for the problem under consideration and hence, the research study will have less validity.
Cost-effectiveness	It is more expensive than collecting secondary data. Analysing it can also sometimes require mathematical or statistical expertise.	It is a lot less expensive than primary data. This is because, someone has already spent the necessary time and money to gather it. Much secondary data is, additionally, already analysed by the people who gathered it.
Relevance of data to the current period	The data pertaining to the current period can be generated.	It might not be current. It can be out of date or only partially relevant to the selected study. Certainly, it won't include the opinions of sample respondents or clients.
Time taken for data collection	It can be very time consuming to gather and analyze. So, primary data takes a lot of time and the unit cost of such data is relatively high. Thus, primary data are expensive and difficult to acquire, but they are trustworthy.	It takes less time to collect the data, especially with the advent of internet. Further, secondary data are cheap and easy to collect, but must be treated with caution.
Advantages	✰ Research based only on primary data is more reliable, as that data are authentic and objective oriented and cannot be manipulated. ✰ Using primary data in research ensure the validity (trustworthy), authenticity (genuineness) and reliability (certainty) of the research findings. ✰ Primary data is more accurate and gives detailed information according to the requirement. ✰ The explanation of terms, definitions and concepts are incorporated in primary data. ✰ Less prone to errors.	✰ Collection of data is very economical. ✰ Is available quickly. ✰ Saves a lot of time. ✰ Helps in verifying the primary data. ✰ Leads to widening of the data base. ✰ It provides valuable insights and contextual familiarity with the subject matter.
Disadvantages	✰ Less economical. ✰ More time consuming to get the data. ✰ No scope to verify the primary data. ✰ Narrow data base compared to secondary data.	✰ Research based only on secondary data is least reliable and may be biased because, secondary data has already been manipulated by human beings. ✰ The findings of research study exclusively based on secondary are less trustworthy, authenticity and reliability compared to the studies based on primary data.

Contd...

Table 8.25–*Contd...*

Item	*Primary Data*	*Secondary Data*
		✰ Locating appropriate source and finally getting access to the data could be time consuming. ✰ The data available might be too vast and a lot of time may be spent going through it. ✰ It might have been originally collected for some purpose which is specific and not known to the present researcher. To that extent, it might be erroneous to use it. ✰ The accuracy of secondary data as well as its reliability would depend on its source. ✰ It might not be updated and not of much use in a dynamically changing environment. ✰ There may be some times when this data are not able to provide the necessary information. ✰ Data may not be accurate. ✰ The data may not be latest. ✰ Has a limited utility.
Sources of data	Collection of data by schedules or questionnaires or interviews etc., with the sample respondents.	Collection of data from publications or books or from internet website etc.
Examples	Primary data are employed to study different researchable issues like estimation of cost of cultivation of paddy in a selected area, price spread of paddy in a study area, marketing problems faced by the farmers in transacting paddy in a study area etc.	Secondary data are employed to study different researchable issues like estimating market potential of a pesticide, analyzing competitors for a new brand of RTE food product, sales forecasting of a pesticide firm, export trends of major agricultural commodities from India etc.

The above discussion infers that, both primary data and secondary data have their pros and cons. Primary data offers tailored information, but tends to be expensive to conduct and takes a long time to process. On the contrary, secondary data is not tailored made, but is usually inexpensive to obtain and can be analyzed in less time. But, still it is collected and the researcher has to sort out the requisite information from the total information available and this secondary data are used as complementary along with primary data, so as to have a complete information. However, the type of data the researcher choose can depends up on many things including the type of research study, available budget, researcher's skill in handling the data, statistical software available etc. So, based on these factors, he may choose to use primary data or secondary data or both.

b. Cross-Sectional Data and Time Series Data

✰ *Cross-sectional data:* If the data are collected by the researcher from the sample respondents on the selected variables at a point of time or without regard to differences in time, it represent cross-sectional data and this data are a type of one-dimensional data set. Thus, cross-sectional data gives a snapshot information about the variables at one particular point of time (Table 8.26). This can be shown as X_i; where 'i' = 1, 2 .... N. For example, a researcher wishes to collect the cost of cultivation data of paddy from a random sample of 200 farmers in a village. This 200 farmers represent the cross section of the population in that village. After analyzing the data, the researcher concluded that, for 40 per cent of the sample population, the cost of production is higher than MSP of paddy. Note that, through analyzing cross sectional data, the researcher cannot conclude, whether cost of production is increasing or decreasing and he can infer only about the current situation at that point of time. Generally, Pie charts are used to explain the components' share of cross-sectional data. This type of data are limited in the sense that, it cannot describe changes in variables over a period of time or cause-and-effect relationships in which one variable affects another over time. Thus, time is not considered as one of the study variables in a cross-sectional research design. However, it is worth noting that, in a cross-sectional study, all participants do not provide data at one exact moment. Even in one session, the respondent will complete answering the schedule over some duration of time. Nonetheless, cross-sectional data are usually collected from respondents making up the sample within a relatively short time frame (field period). In a cross-sectional study, time is assumed to have random effect that produces only variance, not bias. Note that, the terms cross-sectional design and cross-sectional survey often are used interchangeably.

Table 8.26: Cross-Sectional Data of Selected Sample of Agricultural Labour.

Agricultural Labour	*Age*	*Wage Rate (Rs/day)*	*Education Status*
Agricultural labour 1	45	150	Illiterate
Agricultural labour 2	38	120	10th Class
Agricultural labour 3	55	120	Illiterate
Agricultural labour 4	41	140	Illiterate
Agricultural labour 5	30	175	Illiterate
Agricultural labour 6	29	200	Illiterate

- ***Time series data:*** In this case, the data are collected on the same variables over a period of time either from the sample respondents (primary sources) or from published (authentic) reports (secondary sources). So, this data enable the researcher to study the changes in the variables over a period of time (Table 8.27), unlike cross sectional data. This can be shown as X_t; where 't' = 1, 2. ——————T. For example, the researcher can fit the trend in prices of paddy in a selected regulated market and he can even forecast the price data into the future. Generally, Line charts are used to plot time series data and these enable the reader to analyze the data with ease, and to compare and contrast the differences between different time periods.

  The major difference between cross-section data and time series data is that, the former does not focuses about the time period, whereas the latter focuses on the time period regarding the variables for which the data are to be collected. So, in cross sectional data, the researcher collects the data regarding the selected variable(s) across the sample respondents at a point of time or considering one particular time frame. That is, when the researcher collects cross-sectional data, he is not concerned about the time period. In contrast, in time series data, the researcher collects the data regarding the selected variable(s) across time period. This

Table 8.27: Time-Series Data on Market Arrivals of Major Agricultural Commodities in Kurnool Regulated Market.

Years	*Paddy (Lakh quintals)*	*Cotton (Kapas) (Lakh quintals)*	*Groundnut (Lakh quintals)*
2004-05	13.18	17.65	16.66
2005-06	12.89	18.75	6.78
2006-07	15.63	18.88	13.54
2007-08	14.65	19.32	19.65
2008-09	12.43	18.35	21.64
2009-10	13.64	17.54	17.63

Source: Annual Administration Reports, APMC, Kurnool.

difference can be easily understood from the following example: if the researcher collects the data pertaining to the differences in consumption pattern of a RTE processed food product from the consumers across the regions in Andhra Pradesh considering one particular time frame (say, in the year 2013), it refers to cross sectional data. On the other hand, if the researcher collects the data pertaining to the consumption pattern of a RTE processed food product from the consumers across the regions of Andhra Pradesh since past five years, it infers time series data.

c. Qualitative Data and Quantitative Data

The terms 'qualitative data' and 'quantitative data' refer to kinds of data. To put it in simple way, we call data 'qualitative' if the data are in non-numerical form and 'quantitative' if the data are in numerical form.

☆ ***Qualitative data:*** As mentioned above, qualitative data are non-numerical data. The term qualitative data refers to observations that are descriptive. These data represent categories. Note that, qualitative data could be much more than just words or text. Examples include: managerial ability of the farmer, gender of the farmer, marital status, soil colours, soil texture etc. These data can be observed, but not measured. The qualitative data are concerned with describing meaning, rather than for drawing statistical inferences. No doubt employing qualitative data alone in a research study affects reliability, but gain in terms of validity. Thus, they provide a more in depth and rich description. Witte and Witte (2009) defined qualitative data as *'when, among a set of observations, any single observation is a word, or a sentence, or a description, or a code that represents a category, then the data are qualitative'*. Note that, 'yes-no' responses, attitudes of farmers towards a technology, descriptions of events such as Kisan Mela etc., are all examples of qualitative data. Certainly, numerical codes can be assigned to qualitative responses (for example, 'Yes' could be assigned 1 and 'No' could be assigned 2), but these numbers do not transform qualitative data into quantitative data. The use of qualitative data in a research study ensure the following advantages:

- They provide an in-depth and detailed picture by studying the attitudes, feelings and behaviours of the respondents.
- The researcher may extend the data collection to newer topics by encouraging the respondents, which are not initially considered.
- If used along with quantitative data, a further detailed picture can be built up about the problem under consideration.

However, the use of qualitative data alone in the research study may pose certain drawbacks like: the findings can be less generalized as only small sample can be analyzed from in depth study, collection of qualitative data is generally more time consuming than quantitative data, difficult to make comparisons as the responses are highly subjective etc.

☆ ***Quantitative data:*** This data deals with numbers and they can be measured. For example, fertilizers costs, irrigation costs, pesticides costs etc., can be quantified and hence, they represent quantitative data. Witte and Witte (2009) defined quantitative data as '*when, among a set of observations, any single observation is a number that represents an amount or a count, then the data are quantitative.*' Note that, counts are also quantitative. So, data showing size of farm family, number of implements the farmer own etc., are quantitative. Quantitative data can be easily analyzed statistically and the findings are fairly reliable, unlike qualitative data. But, this data are criticized for not providing an in depth description. The advantages of collecting quantitative data in a research study include:

- The researcher goes for collecting the data from a large sample or for a lengthier time period so that, he can generalize the research findings.
- Ensures accuracy of the findings.
- The research study can be replicated, and then analyzed and compared with similar studies.
- Personal bias can be avoided by the researcher.

However, the quantitative data analysis may pose a disadvantage that, the researcher may collect a much narrower and sometimes superficial dataset so that, the results simply provide numerical descriptions rather than detailed analytical findings.

d. Univariate Data, Bivariate Data and Multivariate Data

Univariate, bivariate and multivariate data are the various types of data that are based on the number of variables considered in the research study.

☆ ***Univariate data:*** This data explains the variation about a single variable. For example, if the researcher wishes to study the age distribution of farmers in a village and graph the data, it implies there is only one variable *i.e.*, 'the age of the farmers'. This type of data is known as univariate data and it does not deal with relationships, but rather it is used to describe something.

☆ ***Bivariate data:*** Bivariate data is data that involves two different variables, whose values can change. The purpose of bivariate data is to analyze and explain the relationship between the two variables. For example: quantity demanded and price of the commodity.

☆ ***Multivariate data:*** Multivariate data is the data in which analysis is based on more than two variables per observation. Usually multivariate data is used for explanatory purposes. For example: fertilizers costs, irrigation costs, pesticides costs, weeding costs etc., incurred on paddy crop.

There is another way of explaining the terms of Univariate, Bivariate and Multivariate data considering the number of IVs in a statistical model. We know, statistical models are usually expressed in the form of a equation to specify the way in which variables combine with each other to affect DV. This is given by: DV=function of (IV). In case of univariate data, there is only explanatory variable or IV *i.e.*, $Y = a + bX$. (For example, price of a commodity is influenced by the market arrivals). In case of bivariate data, there are two explanatory variables or IVs *i.e.*, $Y = a + b_1X_1 + b_2X_2$. (For example, output of crop is influenced by both area and productivity of that crop). In case of multi-variate data, there are more than two explanatory variables or IVs *i.e.*, $Y = a + b_1X_1 + b_2X_2 + b_3X_3 + b_4X_4$. (For example, gross returns of paddy is influenced by several costs like fertilizers costs, irrigation costs, pesticides costs, weeding costs, human labour costs, machinery labour costs, manure costs etc.).

e. Subjective Data and Objective Data

In general, confusion often arises that, 'qualitative data' is synonymous with 'subjective', and 'quantitative data' is synonymous with 'objective'. But, this is not the case, as explained here under.

- ✰ ***Subjective data:*** This refers to the data that result from an respondent's personal opinion or judgement and not from some external measure.
- ✰ ***Objective data:*** This data does not refer to respondent's internal opinion i.e., it arises 'external to his mind'.

The differences between these data can easily understood from the following example: When the farmer respond to the survey question 'Do you know about the SRI production technology of paddy' and his answer 'Yes' represents qualitative data and it is not subjective. This is because, the knowledge of farmer about the SRI technology is an indisputable fact and is not open to subjectivity. So, this response is both qualitative and objective. If the farmer is asked to give his personal opinion about the cost-effectiveness of SRI technology, and if he responds as 'the SRI technology is not cost effective', then the response can be categorized as both qualitative and subjective. If the farmer is asked to respond about the seed rate to be used in SRI technology and if he replied as '5kg/ha', then this response is both quantitative and objective. If the farmer is asked to rate the SRI production technology on a scale of 1 to 10' and if he replied as 'eight', then this response is both quantitative and subjective (Table 8.28). This example clearly shows that 'qualitative' data is not synonymous with 'subjective', and 'quantitative' data is not synonymous with 'objective'. So, both quantitative and qualitative data can be objective or subjective.

Table 8.28: Farmer's Responses Categorized as Objective Data and Subjective Data in Comparison with Qualitative Data and Quantitative Data.

	Qualitative	*Quantitative*
Subjective	'I think the SRI production technology of paddy is not cost effective'.	'On a scale of 1 to 10, I rate the SRI production technology of paddy as '8'.'
Objective	'Yes, I know about SRI production technology of paddy'.	I used a seed rate of 5 kg/ha while adopting SRI technology of paddy'.

D. Identification of the Sources of Data

There are two ways of obtaining the requisite data for the selected research problem *viz.*,

i. Primary sources, Secondary sources and Tertiary sources.
ii. Census survey and Sample survey.

i. Primary Sources, Secondary Sources and Tertiary Sources

Sources of data or information are generally categorized as primary, secondary or tertiary depending on their originality and their proximity to the source or origin.

- ***Primary sources:*** The following are some of the popular definitions of primary sources:
  - Primary sources are original materials on which others research is based
  - These sources are the first formal appearance of results in the print form or electronic form. That is, the first publication of the results of scientific investigations represent a primary source.
  - These sources present information in its original form, neither interpreted nor condensed nor evaluated by other writers.

  Thus, primary sources include the document or record containing first-hand information or original data or raw material on a selected research topic. These sources may present original thinking or new discoveries or new information collected at the time of an event. For example, the original results of empirical studies that are typically found in research journals or research papers delivered at conferences are considered as primary sources. These primary sources are the works created at the time of an event or by a person, who directly experienced an event. So, primary sources are the first hand evidence left behind by the researchers at the time of events. These sources are created at the time when the events are occurring, but can also include online documents created later. So, primary sources are characterized by their content, regardless of whether they are available in original format or in digital format or in published format. That is, primary source refer to the content and not necessarily the format of the work. Primary sources can include:
  - Interviews, journals, speeches and witness statements
  - Articles containing original research, data or findings never before shared
  - Original hand-written manuscripts
  - Newspaper and magazine clippings

  The following are the examples of primary sources:
  - News paper clippings covering the present agricultural scenario in Andhra Pradesh
  - A research article reporting on the growth rates of regulated agricultural markets State-wise and at All-India level.
  - Bhattacharya H and Robert Innes, 'Income and the Environment in Rural India: Is There a Poverty Trap?, *Am. J. Agr. Econ. (2013) 95(1): 42-69*
  - Video of the speech of Prime Minister of India on the eve of 45th Indian Independence day celebrations.
  - 'The Hindu' newspaper's front page clipping about the pros and cons of Indian budget announced in the year 2011-12.
  - Scientific journal publishing research articles
  - Proceedings of meetings, conferences and symposia
  - Patents
  - Census statistics
  - Autobiographies
  - Interviews, surveys, speeches

☆ ***Secondary sources:*** Secondary sources are less easily defined than primary sources. These are the sources of data comprising the research works based on the works of others. So, the findings or texts that interpret upon data obtained from primary sources constitute secondary sources. For example, textbooks, encyclopedias, dictionaries, handbooks, review articles in news papers or popular magazines, literature reviews, criticisms on works of literature, movie reviews etc., constitute secondary sources. So, any published work that is one step removed from the original source, usually describing, summarizing, analyzing, evaluating, derived from, or based on primary source materials constitute secondary source of data or information. Thus, a secondary source is something written about a primary source or describe, interpret, analyze and evaluate the primary source. They include comments on, interpretations of, or discussions about the original material of a primary source. So, it can be inferred that, secondary sources constitute second-hand information. For example, *'if I tell you something, I am the primary source. If you tell someone else what I told you, you are the secondary source'*. Another good example is that, a newspaper article is a primary source if it reports a event, but it is regarded as a secondary source if it analyses and comments on that event. The following are the examples of secondary sources:

- bibliographies (*may also be tertiary*)
- dictionaries and encyclopedias (*may also be tertiary*)
- dissertations or theses (*more usually primary*)
- handbooks and data compilations (*may also be tertiary*)
- newspaper and popular magazine articles (*may also be primary*)
- review articles and literature reviews
- textbooks (*may also be tertiary*)
- works of criticism and interpretation

☆ ***Tertiary sources:*** Tertiary sources, sometimes viewed as a subset of secondary sources. It is quite difficult to distinguish between secondary sources and tertiary sources. These sources distill primary or secondary materials into publications such as dictionaries, directories, encyclopedias, almanacs, chronologies, handbooks, manuals, guides, library catalogs, online databases (tertiary or secondary) etc. So, tertiary sources include the list of primary and secondary resources in a specific subject area. They show the index or citations and thus guide the researcher how to use, secondary (and sometimes primary) sources. This implies that, they facilitate the researcher to read the list of secondary sources in a convenient and easy-to-read form. The following are the important examples of tertiary sources:

- Bibliographies (*may also be secondary*)
- Dictionaries and encyclopedias (*may also be secondary*)
- Directories
- Manuals
- Handbooks and data compilations (*may also be secondary*)
- Textbooks (*may also be secondary*)

b. Census (Enumeration) Survey and Sample Survey

☆ ***Census survey:*** When the researcher collects data or information from all the units of a population it refers to census survey or complete enumeration. In general, the researcher goes for census survey, when he wishes to collect accurate information about the whole population and for many sub-divisions of the population. Since it is laborious and time consuming, the researcher practices this mode of data collection when the size of population is small. This mode of data collection ensures the following advantages:

- As mentioned above, the researcher gets accurate information about the population.
- The concept of biasedness is totally eliminated in the study.
- In case of small population, satisfactory response rates are achieved.
- The estimates are not subjected to sampling error.
- Provides a true measure of the population (no sampling error).
- Benchmark data may be obtained for future studies.
- Detailed information about small sub-groups within the population is more likely to be available.

However, this mode of data collection suffers from disadvantages like huge cost and consumes more time on the part of the researcher, more laborious, the study may not be completed in the specified time period etc. Further, all the respondents may not provide information to the researcher.

- ***Sample survey:*** Since going for census survey consumes lot of time and budget on the part of the researcher, he generally prefers to go for selecting a representative sample from the target population (Figure 8.29). So, drawing or eliciting the information or data by the researcher from the selected representative sample infers sample survey. Thus, in a sample survey, only part of the total population is approached for information on the topic under study.

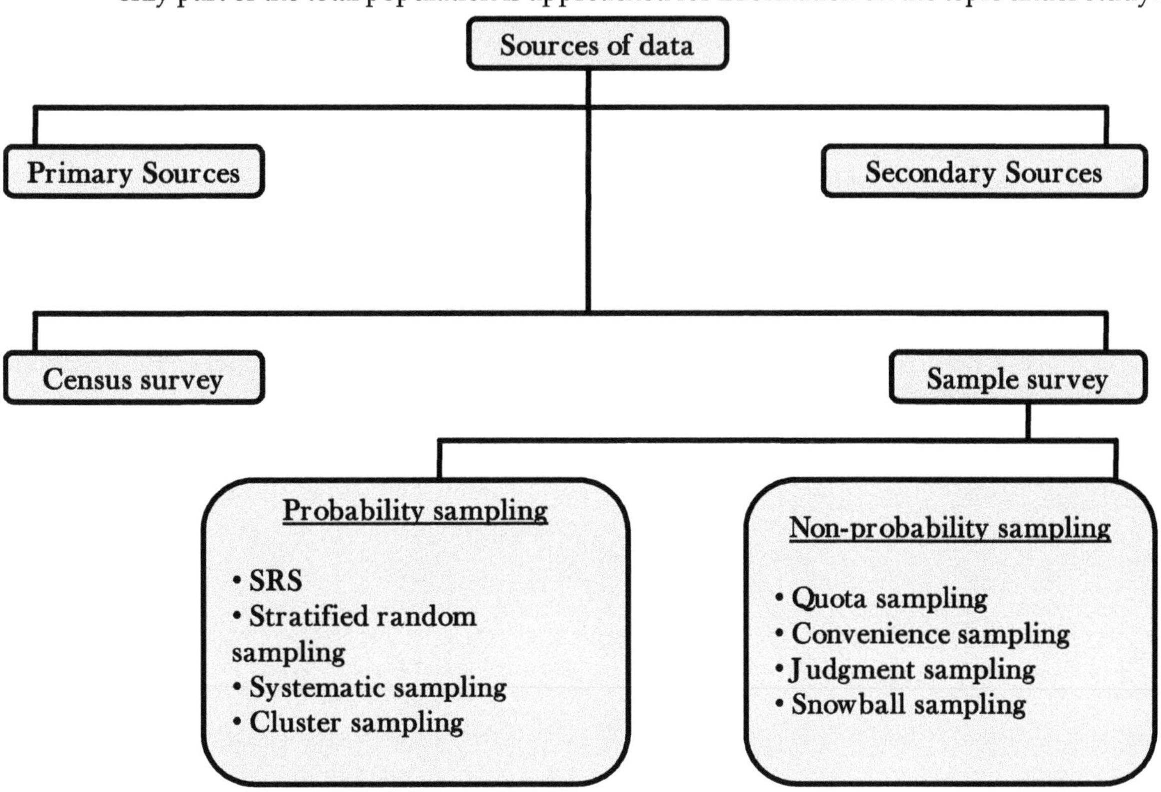

Figure 8.29: Sources of Data for a Research Study.

These data are then 'expanded' or 'weighted' to make inferences about the whole population. The disadvantages of census survey relate to the advantages of sample survey and similarly, the advantages of census survey relate to the disadvantages of sample survey.

The above discussion infers the following differences between census survey and sample survey:

- Census survey considers each and every individual in the population, whereas sample survey considers only representative sample respondents. This means that, census survey is much larger in scope than a sample survey.
- Census survey is much bigger in proportion than sample survey.
- Census survey consumes more time and money compared to sample survey.
- The concept of sampling error arises only in sample survey, unlike census survey. Though this error can be minimized, it cannot be eliminated in a sample survey. Therefore, the results of sample survey always have a margin for error, whereas census survey is always accurate.
- Census survey gives accurate information about the population than sample survey. However, some researchers opine that, sample survey gives more accurate information than census survey, as all the units in population may not provide information to the researcher, whereas the representative sample can be better handled by the researcher.
- Compared to census survey, sample survey needs less time to compile and process the data.
- In sample survey, the researcher can go for detailed or in-depth investigation compared to census survey.
- If the researcher wishes to test the characteristics of population say, to test the infestation in mango fruits, life span of electric bulbs etc., sampling is inevitable and the researcher cannot go for testing all the units in a population.
- If viewed from the respondents' side, in sample survey there is less respondent burden, unlike census survey.
- Since the researcher handles a part of the target population in sample survey, the research findings obtain rather quickly compared to census survey.
- Compared to sample survey, census survey enjoys more data about sub-populations within the target population.

9

Execution of Research Project

After deciding the representative sample size, identifying the variables and data to be collected, now it is high time for the researcher to go for data collection from the selected (sample) respondents or secondary sources or both. Before proceeding for data collection especially from the primary sources, the researcher has to prepare the format comprising of sequence of questions to be posed on the selected sample respondents. The format so prepared by the researcher to conduct agro-economic survey is called Schedule. A similar type of format is used in Agricultural Extension studies and it is called as Questionnaire. These two formats *viz.*, schedule and questionnaire differ widely, but the main difference is that, in the former the data is collected informally (face to face situation) by the researcher and he himself fill the format, whereas in the latter, the questions are filled by the selected sample respondent on his own. The other differences between these two formats were listed out in the ensuing pages. Besides these two formats, there are different ways of collecting data from the respondents like personal interviews, group discussions, case study method etc. The two important tools of data collection employed in social scieces *viz.*, schedule and questionnaire are discussed in-detail in this chapter

9.1. Schedule or Interview Schedule

A schedule is a formalized written set of questions for obtaining requisite information from the respondents. Schedule is an efficient data collection tool, when the researcher knows exactly what is required and how to measure variables of interest. Schedule carry a set of questions on a given research topic, which are asked by the researcher or interviewer in face-to-face interaction with the respondent. It contains direct questions as well as questions in tabular form. If the respondent faces difficulty in answering the questions, the researcher can explain the same and their importance and can collect relevant information. Schedule includes different types of questions *viz.*,

A. Open-ended questions and Closed-ended questions
B. Direct questions and Indirect questions
C. Primary questions and Secondary questions
D. Nominal, Ordinal and Interval questions and
E. Contingency questions

Many schedules include all these types of questions. However, the researcher selects the type of questions according to his need of the study. These different types of questions are discussed here under.

A. Open-Ended Questions and Closed-Ended Questions

☆ ***Open-ended questions or Inventive type questions*** them. That is, the respondent is free to give responses *or Open format questions:* These questions allow the respondent considerable freedom in answering in his own words. These questions are answered in-detail. For these questions, the respondents provide their own answers without any previously provided options. These questions also include fill-in-the-blank responses. That is, these questions are not followed by any ready-made answers. So, the respondent or informant has to think about the answer himself and he is free (not bound by rules) to answer as he likes. Thus, open-end responses are free and spontaneous expressions on the part of the respondent, who is not limited in his replies to a particular question posed to him by the researcher. This type of question is also called as 'inventive type question', as the respondent has to think or invent the answer for himself. The merit of these 'open-ended questions' is that, it gives a wide chance to the respondent to express freely his views on the questions. To record the answers for these open ended questions, the researcher should provide sufficient space in the schedule, as the respondent has to provide answers in-detail for these questions. The following are the examples of an open ended question:

- What are the problems faced by the farmer in marketing paddy?
- Suggest your views to improve paddy marketing in Nandyal region of Kurnool district
- What are the causes for soil salinity in your paddy fields?

So, open-ended questions do not provide any structure for the respondent's reply. The questions and their orders are pre–determined in the nature. These questions are often used during schedule development and pre-testing, as they facilitate in generating more information compared to closed-ended questions. Sometimes, open question: 'Do you have any additional comments?' is often used at the end of sections of questions or at the end of the schedule. This ensures, that the respondent has the opportunity to express his views relevant to the issues being addressed by the schedule. This is important because, some respondents may want to provide additional information to clarify an answer. It is essential to plan for this type of information on the schedule. In some cases, the researcher may formulate open-ended questions that require write-in responses. The following are the examples:

- ❖ What services you expect from the tractor selling firm?
- ❖ What product-mix you expect from Hindustan Unilever Limited firm?
- ❖ What do you think you will be cultivating in your land after two years from now?
- ❖ What policies pertaining to improvement in agricultural marketing you expect from the newly elected Government?

Similarly, some open-ended questions are also framed to elicit numeric responses from the respondents. The following are the examples:

- ❖ How much experience do you have in practicing vermi composting.
- ❖ Since how many years you are purchasing irrigation water from your neighbouring farmer?
- ❖ In a typical month, how frequently you will be in contact with local Agricultural Officer?

☆ ***Closed-ended or Restricted form questions:*** These questions has to be answered by the respondent by choosing an answer from the set of answers given under a question. The researcher has to tick the answer after receiving the response from the respondent. The possible responses listed under each question are called response categories. So, Closed-ended questions are those, where response categories are listed with the question. Compared to open-ended questions, the responses of closed-ended questions may be easily tabulated and statistical measures can be easily applied because, the number of possible answers to each question is fixed. Note that, responses for close-ended questions must be exhaustive (*i.e.,* include all possible answers) and also mutually exclusive (selection of single option from the given multiple options). There are different types of closed-ended questions as given below.

- ❖ ***Multiple-Choice Questions (Multichotomous questions or Cafeteria questions):*** In this case, respondents are asked to select one of the alternatives or from a range of responses given below the question. That is, the multiple choice question asks the respondent to select one response from a list of choices. This multiple choice question is often confused with the Checklist question. In the latter (*i.e.,* in the checklist question), the respondent is asked to pick one or more responses from the list. Generally, it may not be evident to the respondent, whether the question is multiple choice or checklist. So, instructions should be included. The following are the examples for multiple choice and checklist questions:

Multiple Choice Questions

How frequently you are applying pesticide on cotton crop to control boll worm menace?

i. Every week ii. Every fortnight

iii. Once in a month iv. Once in two months

Which of the following is the most important problem faced by the farmer in transacting groundnut in Kurnool regulated market? (Please check one only)

i. Transportation ii. Storage

iii. Processing iv. Market extension services

v. Grading vi. Price fluctuations

vii. Any other please indicate

Checklist Question

In the ensuing kharif season, what crops you might prefer to cultivate? (Check as many as apply)

i. Paddy	ii. Maize
iii. Cotton	iv. Chillies
v. Sunflower	vi. Red gram
vii. Green gram	

As mentioned above, the response categories for multiple choice and checklist questions require careful drafting. The list of response categories should be mutually exclusive and exhaustive. Consider the following example:

Since how many years you are selling paddy in Kurnool regulated market yard?

i. 5 years	ii. 5-10 years
iii. 10-15 years	iv. 15 years or more

In the above example, say, if the farmer is selling the produce in Kurnool regulated market yard since past 10 years, here he will be in confusion to respond to either 'ii' or 'iii' response categories. That is, in this example, the response categories are not mutually exclusive and they overlap. Similarly, use exhaustive response categories in the closed-ended questions. Categories are exhaustive, when there is a category available to all potential responses. The above response categories are not exhaustive. Say, if the farmer is selling paddy since past two years only, how he could answer this question? So, a better choice of response categories would be as given below:

Since how many years you are selling paddy in Kurnool regulated market yard?

i. Less than 5 years	ii. 6-10 years
iii. 11-15 years	iv. 16 years or more

From the above example, it is clear that, response categories are both mutually exclusive (there is no overlap) and exhaustive. Thus, the response alternatives should include the set of all possible choices. The general guideline is to list all alternatives that may be of importance and the end of given choices include an alternative labeled 'Any Other (Please Specify)'

- ☆ *Dichotomous Questions:* This is the simplest version of a closed-ended question. This question usually split the respondents into two distinct groups. That is, the respondents have two choose only one answer out of the two given choices or alternatives for each question. The two choices may include: Yes/No, Approve/Disapprove, Agree/Disagree etc. Sometimes, a neutral alternative is supplemented (say, 'Don't know', 'No idea' etc). So, these two-choice questions are also used as screening questions to prevent respondents from being asked a series of questions that do not apply to them. For example:

Do you intend to buy a new tractor in the coming crop season?

i. Yes

ii. No

iii. No idea

Do you think SRI production technology of paddy is cost-effective compared to transplanted technology?

i. Yes

ii. No

iii. No idea

In some cases, an open-ended question is followed after the closed-ended question. The same is exemplified below.

Do you have any suggestions to reclaim waterlogged induced soil salinity in Mutluru canal command area?

i. No

ii. Yes

If the respondent answer is 'Yes,' then a follow-up open-ended question is:

'If yes, what are your suggestions?'

☆ ***Ranking Questions:*** In this case, the respondents are asked to rank or order the response categories. That is, here the respondent will prioritize the given response categories. For example:

The following is a list of some of the ways that farmers will get the market information. Please rank them in order of priority by placing the number '1' beside the source you think would be most helpful, number '2' beside the source you think would be second most helpful, and so on.

__________Co-farmers

__________Personal visit to nearby regulated market

__________Commission agent functioning in regulated market

__________Radio

__________Television

__________Internet

__________Village traders

__________ Other (please specify)________________________________

However, these ranking questions may not give a clear picture. This is especially true, if the response categories differ greatly. Say, if there are three response categories *viz.*, commission agent functioning in regulated, personal visit to regulated market and internet, first two options are closely related compared to third option. This complicates data analysis. Another problem is that, the respondent may wish to assign the same rank to two or more options than assigning different ranks. In the above example, the respondent might rank both 'commission agent functioning in regulated market' and 'personal visit to regulated market' with same rank say, 1. In some cases, the respondent may not be able to rank all the choices given in the list, as some options (say, internet) are not available for him. In such cases, the respondent may be asked to rank at least five options in the list in the order of priority.

☆ ***Rating Questions:*** As the sub-head indicates, here the respondent is asked to rate their answer. For example:

How satisfied are you with the grading facilities available for cotton in Kurnool regulated market?

i. Very satisfied ii. Satisfied
iii. Dissatisfied iv. Very dissatisfied

The number of ratings should be fixed by the researcher depending upon the question, how in depth information it is going to elicit from the respondent. Usually, ratings will be made on 1 to 10 scale. There may be two ratings say, agree or disagree or as many as 10, ranging from 1(say, 'very much dissatisfied') to 10 ('very much satisfied'). In some cases, neutral alternatives like ('Neither satisfied nor dissatisfied') will be introduced in the response categories. The same is exemplified here under.

How satisfied are you with the grading facilities available for cotton in Kurnool regulated market?

i. Very satisfied ii. Satisfied
iii. Dissatisfied iv. Neither satisfied nor dissatisfied
v. Very dissatisfied

In the above example, fourth response category is the neutral alternative introduced in the rating options. In some cases, response categories such as 'Do not know/No opinion' or 'Not applicable' are introduced, especially when the posed rating question may not be relevant to a particular respondent. So, the researcher should ensure, whether the scale under consideration permit a full range of answers? Sometimes, the researcher also limits the respondent's options. But, he should try to expand the possibilities and also use 'Any Other: ____________' whenever possible. In using numerical scales, the researcher should ask himself, what the pros and cons of a longer or shorter scale would be. For example, would a 5-point scale have any advantages over the 4-point scale? One obvious difference is that, the 5-point scale permits respondents to give a neutral answer. This difference is important when you have an agree/disagree scale, as some people will want to fence-sit. Although the researcher may not want neutral answers, he may want to permit them, if respondents are determined to give them. It is also important, whether to give odd or even number of choices in the response scale? In general, it is suggested that, the researcher should use an odd number of choices in order to allow the respondent to express a middling or neutral strength of opinion. This can be problem with some respondents, who refuse to express an opinion and give the middle category for all questions. However, the alternative can be even worse: respondents who have no or neutral opinion being forced to choose negative or positive and doing it randomly.

In some cases, say, for example, if the researcher wishes to collect information pertaining to the marketing of cotton in Kurnool district and if one particular respondent sell away the produce in the village itself immediately after harvest, then the response category 'Not applicable' will be ticked by that respondent in case of the following question:

How satisfied are you with the grading facilities available for cotton in Kurnool regulated market?

i. Very satisfied ii. Satisfied
iii. Dissatisfied iv. Neither satisfied nor dissatisfied
v. Very dissatisfied vi. Not applicable

Open-Ended Questions vis-à-vis Closed-Ended Questions

As discussed above, open-ended questions allow for self-expression, but can be burdensome and time-consuming as well as difficult to analyze. Closed-ended questions are usually less burdensome for the respondent and data collection and capture is cheaper and easier. Following Table 9.1 highlights the advantages and disadvantages of both open-ended and closed-ended questions.

Table 9.1: Differences between Open-Ended Questions and Closed-Ended Questions.

Open-Ended Questions	*Closed-Ended Questions*
Advantages ☆ The respondent has the freedom to express his views and ideas and he is not influenced by the response categories, unlike closed-ended questions. This facilitates the respondent to interpret the question in different ways. So, an open-ended question expands the scope of the question compared to the closed-ended question, where in the latter, clues are given for the respondent to answer and interpret the question. ☆ Can be used in making intensive studies of the limited number of the cases. ☆ More flexibility on the part of the respondent in answering the questions. ☆ Enable the researcher to clear up any misunderstandings faced by the respondent and may even go in depth into the subject. ☆ Also enable the researcher or interviewer to test the limits of the respondent's knowledge. ☆ Good rapport may be built in between the researcher and the respondent. ☆ Enable the researcher to make a truer assessment of what the respondent believes or thinks. ☆ Researcher can know and learn about respondent's own 'natural' wording about the study and this facilitate him to plan data collection easily with clear meaning. ☆ Permit unlimited number of answers. ☆ Can find the unanticipated answers for the questions. ☆ Reveal respondents thinking processes.	**Advantages** ☆ Consume less time both on the part of the researcher and respondent in eliciting the information. ☆ No extended writing and simply the researcher has to tick or underline or circle the correct answer given by the respondent. ☆ Easy to group and process the data. ☆ Less interviewer training. ☆ These questions are quicker and easier to answer since respondents simply choose the appropriate response categories rather than formulate a response and state it in their own words. ☆ The respondent is more likely to answer the question as intended because, the response categories indicate the focus of the questions. ☆ If the same question is being used for several surveys, using the same response categories facilitates comparisons across surveys. ☆ Answers across respondents are easy to compare. ☆ Answers easier to analyze on computer. ☆ Response choices make question clearer. ☆ Easy to replicate study. ☆ They avoid problems of interpreting respondents' replies or answers. ☆ They can provide better information than open-ended questions, particularly where respondents are not highly motivated. ☆ They provide uniformity of questions and responses, so they are potentially easier for evaluating the opinion of the sample group as a whole.
Disadvantages ☆ Demands more training on the part of the researcher or interviewer in drawing information or data from the respondent. ☆ Difficult to generate the data. ☆ Difficult to draw comparisons. ☆ More costly and time consuming. ☆ Coding may be burdensome and unreliable. ☆ Merely an issue may be raised by these questions. ☆ May result in error-prone processing than closed-ended questions. ☆ In the case of numerical data, open-ended questions are more difficult for respondents to answer than closed-ended questions and more difficult and prone to error during data capture. ☆ Answers can be irrelevant.	**Disadvantages** ☆ No spontaneous responses. ☆ Responses are limited to the stated alternatives. So, the respondent often suggest answers that may not be there in his mind. This may defeat the very purpose of the study. That is, the respondent cannot express his true opinion on a particular issue or cannot express his own judgment. ☆ The given response categories in the questions may sometime irritate the respondents. ☆ When drafting the questions, effort is needed to develop the response categories and these must be mutually exclusive and exhaustive. ☆ Many choices can be confusing. ☆ Can't tell, if respondent misinterpreted the question.

B. Direct Questions and Indirect Questions

- ***Direct questions:*** These are the questions that are related to the respondent directly. For example: *Do you interact with local Agricultural Officer frequently?*
- ***Indirect questions:*** These are the questions that are related to the responses of other people. For example: *Do you think other farmers in the village interact with local Agricultural Officer frequently?*

C. Primary Questions and Secondary Questions

- ***Primary questions:*** These are the questions that are directly related to the research topic. For example: *What is the cost of cultivation of paddy?*
- ***Secondary questions:*** These are the questions that are not directly related to the research topic *i.e.*, the information is of secondary importance. For example: *Who takes decision in your family to go and purchase fertilizer for cultivating paddy on the farm?*

D. Nominal, Ordinal and Interval Questions

- ***Nominal questions:*** For these questions, the responses fall in two or more categories and these categories have no specific order. For example: Are you Married/Unmarried; Are you Illiterate/Educated etc.
- ***Ordinal questions:*** For these questions, the responses also fall in two or more categories and these categories can be ranked in a specific order. But, there is no implication that, these categories have equal interval. For example: How frequently you meet the local Agricultural Officer: Regularly/Occasionally/Never.
- ***Interval questions:*** For these questions, the responses also fall in two or more categories and these categories have equal interval. For example: What is your annual income: Less than Rs. 10000/Rs.10001 to Rs.20000/Rs.20001 to Rs.30000/Rs.30001 and more.

E. Contingency Questions

A contingency question is a question whose relevance to the respondent is determined based on the response received from the earlier screening question. For example:

- *Whether you applied chemical fertilizer to paddy crop?*
- *Whether you applied Nitrate form of fertilizer or Ammonical form of fertilizer to paddy crop?*

In the above example, second question is a contingency question. The framing of contingency question arises because, every question need not be relevant to all the respondents. However, if the sample selected is homogeneous, the framing of contingency questions can be minimized.

a. Characteristics of Good Schedule

Developing a thorough and relevant schedule is essential for gathering information from the target population. For this, the researcher should formulate right questions and this directly contribute to learn a great deal about the target population. When the researcher incorporates wrong questions in the schedule, the information gathered may not be very helpful and, in some cases, it might not be valid because, those being interviewed may have been confused by the questions. So, a good schedule should have the following characteristics:

- The questions in the schedule should be clearly worded and be unambiguous. That is, the questions framed in the schedule should be clear without any confusion on the part of the respondent.
- The questions framed in the schedule should be in a logical sequence and create interest among the respondents to answer them. Do not split questions over pages, or ask questions that require answers to be completed on subsequent pages. It is equally important in giving answer choices for the questions framed in the schedule. Thus, these two broad issues *i.e., 'how the question and answer choice order can encourage respondents to complete the survey'* and *'how the order of questions or the order of answer choices could affect the results of the survey'*. Ideally, the early questions in a survey should be easy and pleasant to answer. These kinds of questions encourage respondents to continue in the survey. Grouping together questions on the same topic also makes the schedule easier to answer. Whenever possible, leave difficult or sensitive questions until near the end of the survey. Any rapport that has been built up will make it more likely people will answer these questions. If people quit at that point anyway, at least they will have answered most of the questions in the schedule. So, do *not* leave important questions to the end of schedule. This infers that, the more sensitive a question, the closer it should be to the end of the schedule, for two reasons: firstly, rapport takes time to build up, and secondly, if a respondent does get offended and refuses to go on with the interview, little information will be lost. Therefore, the sensitive questions normally come close to the end of the schedule. Answer choice order can make individual questions easier or more difficult to answer. Whenever there is a logical or natural order to answer choices, use it. Always present agree-disagree choices in that order. Presenting them in disagree-agree order will seem odd. For the same reason, positive to negative and excellent to poor scales should be presented in those orders. When using numeric rating scales, higher numbers should mean a more positive or more agreeing answer. Note that, question order can affect the results in two ways. One is that, mentioning something (an idea, an issue, a brand) in one question can make people think of it while they answer a later question, when they might not have thought of it if it had not been previously mentioned. The other way question order can affect results is habituation. This problem applies to a series of questions that all have the same answer choices. It means that, some people will usually start giving the same answer, without really considering it, after being asked a series of similar questions. Respondents tend to think more, when asked the earlier questions in the series and so give more accurate answers to them. The order in which the answer choices are presented can also affect the answers given. Respondents tend to pick the choices nearest the start of a list when they read the list themselves. They tend to pick the most recent answer when they hear a list of choices read to them. As mentioned previously, sometimes answer choices have a natural order (*e.g.*, Yes, followed by No; or Excellent - Good - Fair - Poor). If so, the researcher should use that order. At other times, questions have answers that are obvious to the person that is answering them (*e.g., 'What brand(s) of pesticides do you use?'*). In these cases, the order in which the answer choices are presented is not likely to affect the answers given. However, there are kinds of questions, particularly questions about preference or recall or questions with relatively long answer choices that express an idea or opinion, in which the answer choice order is more likely to affect which choice is picked.
- Should cover questions or statements relating to all significant aspects of the study.
- Should look into the problem analytically, dissecting every, major and significant components of the problem.

- The questions framed in the schedule should be relevant to the objectives framed in the study.
- The answers obtained for the questions framed in the schedule must be meaningful and can be quantifiable as far as possible to permit tabulation and statistical analysis and to generalize the findings.
- The questions should not generate more than one answer from the respondent *i.e.*, they should be specific.
- Questions must be within the scope and interest of the respondent.
- The schedule should be a pre-tested one so that, it generates all relevant information for the study.
- The questions should not be repetitive.
- All the acronyms and abbreviations should be clearly defined.
- Note that, the researcher should limit the number of questions in the schedule considering the nature and scope of the study, objectives framed for the study, tools of analysis to be employed, sample size, budget available etc., and the questions framed in the schedule should generate interest among the respondents to render answers to the researcher.
- The schedule should be designed to minimize possible response errors.
- The schedule should seek only that data, which cannot be obtained from other sources.
- The size of the schedule should be attractive. It should not be too lengthy, but should be comprehensive.
- The questions should follow good psychological order proceeding from general to more specific aspects.
- The questions given in the schedule should be such that, the respondent is able to understand them in the light in which they are asked.
- The questions framed in the schedule should be inter-linked.
- The sequential arrangement of questions should be such that, it stimulates the respondents to give the correct answers.
- The questions should be relevant and pointed. They should not deal with the subjective evaluation.
- Should be designed with good language. Clear and straight forward language be used.
- Should be reliable such that same results are obtained whenever the schedule is used when everything else remains same.
- Should not be too length nor too short. Should give fair coverage to the topic.
- KISS - Keep It Short and Simple. Before employing the schedule on the respondent, the researcher must ask himself, 'what you will do with the information from each question'. If researcher himself cannot give a satisfactory answer for a question, leave it out. Also avoid the temptation to add a few more questions just before conducting the survey in front of the respondent. If necessary, place the questions into three groups: must know, useful to know and nice to know. Discard the last group, unless the previous two groups are very short.

So, accurate information and accurate response are the two essential conditions of a good schedule. For achieving these two by the researcher, he should stimulate the respondents to answer the questions with interest. Besides this, the physical structure of the schedule should be attractive; the questions asked or information sought should be relevant to the enquiry. Above all, a good schedule should facilitate the researcher to tabulate the data or information and to carry out suitable statistical analysis.

b. Design of Schedule

The schedule with above 'good' characteristics makes the researcher to collect relevant data or information from the sample respondents. Besides these characteristics, the researcher should note various aspects in designing the schedule, as its effectiveness in obtaining requisite information also depends on its construction or formation. Note that, it is not an easy task to prepare a good schedule. Hence, serious attention must be paid to the following aspects in designing a schedule in addition to the above characteristics:

- Before eliciting the information from the respondent, first the researcher should motivate him to communicate the required information.
- The beginning questions should not be open-ended questions.
- Put the more important questions at the beginning.
- Try to keep all questions on one subject together.
- The questions to be posed on the respondent should be in a logical sequence from the more general to the more specific. That is, the order of the questions needs to be planned and this order will directly influence the answers obtained from the respondents. Note that, knowledge questions must not be preceded by relevant (specific) information. For example, the researcher should not ask *'Can you name any fungicide to control rice blast'*, if that question is followed by a general question, *'Mention different fungicides for controlling rice blast'*. So, it is essential to start with broad questions about the subject first, then to narrow down to the specific issues. If the opening questions are interesting, simple and easy to answer, respondent's cooperation and involvement can be maintained throughout the schedule. So, it is important to ask general questions before specific questions in order to obtain the freest opinion of the respondent. This procedure is known as Funnel technique or Funnelling.
- In the beginning, in order to make it easier to gather the data, the questions should be written in multiple choice format. This not only benefits the researcher in gathering the requisite data, but also makes the respondents to understand the questions easily. Further, at the beginning of the interview, opening questions must be interesting and should be properly arranged such that, they can be answered at ease by the respondent. This will facilitate to build good rapport between the researcher and the respondent. Gradually, the researcher should proceed in a logical manner from topic to topic and from basic knowledge to specific questions.
- As far as possible, prefer to get numerical data from the respondent rather than simply gathering theoretical information or vague statements. For example:
  - **In a typical month, how often you meet the local Agricultural Officer?**

    i. Never ii. Occasionally

    iii. Sometimes iv. Often

    v. Regularly

The above question is incorrect, as it gives only a vague information to the researcher. To get more information, the same question is modified as given below.

- **In a typical month, how often you meet the local Agricultural Officer?**

  i. Less than once   ii. 2 times

  iii. 3 times   iv. 4 times

  v. More than 4 times

- Use simple words both in schedule and while explaining the meaning of the questions to the respondents.
- Avoid very long sentences in framing the questions. The number of words in a question should not be more than 20. If longer is needed, break up into several shorter sub-questions.
- Each question should be specific and deal with a single idea. Ambiguous questions (say, *'What is your mother's educational level? Illiterate or Literate.* This question is ambiguous because, what does 'illiterate' mean exactly? Further, this item is subjects to the 'social desirability problem.' Respondents always want to put themselves and their family in a positive light when responding. This is called Social desirability *i.e.,* they want to appear desirable to others. Because of this, participants might not give an accurate answer to avoid appearing undesirable. This is sometimes done on purpose, but other times done unconsciously. Who wants to admit that their parents are illiterate? So, such ambiguous questions should be completely avoided. If the researcher wishes to collect the information on this item, then he has to rephrase the same question to avoid the social desirability problem:

  - ***'What is the last level of schooling that your mother completed?***

    i. None   ii. Primary

    iii. Secondary   iv. Tertiary

  - ***Did you smoke during your pregnancy?***   Yes [ ]   No [ ]

    Note that, mothers tend to answer 'No', even if they smoked during pregnancy.

  If such ambiguous questions creep in, different respondents will understand the question differently and this affects the outcome of reliable information. For example, if the researcher poses a closed-ended question *'Is your work is more laborious because you are weeding manually'?* This question may be asked to all farmers in the survey irrespective of whether they remove weeds manually or using herbicides. So, if the answer by the respondent is 'No', it may have different meanings. Similarly, embarrassing questions such as *'Why you have not secured job so far', 'Whether your father shared land property to you', Have you ever thought about committing suicide?* etc., must be avoided. Such social desirable answers given by the respondents represent 'Faking good' (also known as Social desirability or Obsequiousness). Thus, in answering such questions, respondents may systematically alter schedule responses in the direction they perceive to be desired by the investigator. So, socially undesirable answers tend to be under-reported.

- While structuring the schedule, the researcher should use the language keeping in view of the population being studied. As far as possible, communication with respondents should be carried out in their own language and make sure that, all used terms are known to them. Always consider the language skills of respondents when developing questions. Respondents should be addressed in the second person (*you*) for a friendly, personal tone and basic rules of grammar should be followed. While reading the questions, they should sound natural and conversational, and be easy for the listener to follow. For example, the following question does not use simple and common words: *'Have you ever applied 'dimethyl phosphate of 3-*

hydroxy-N-methyl-cis-crotonamide' to your field?' But, it is likely that, a significant number of respondents may not know about the generic name of 'Monocrotophos'. So, many of the respondents may not provide an accurate response. So, better wording would be: *'Have you ever applied Monocrotophos on your crop?'* Further, do not use too long sentences and complex language, as they make it difficult for the respondents to understand and answer.

- The researcher should go for pre-testing of schedule before proceeding for actual data collection from the sample respondents. Pre-testing refers to the trial and error procedure followed by the researcher to find out, whether the designed schedule is capable of eliciting desired responses from the respondents. That is, it involves few test interactions with the respondents to know, how far the formatted schedule is fit enough to collect requisite data from the actual (selected) sample respondents. Pre-testing the schedule is an essential step before its final employment among the selected respondents. The purpose of the pre-test is to check question wording, and to obtain information on open-ended questions if any, with a view to designing a multiple choice format in the final schedule. Note that, it is extremely difficult to work out the survey directly on the sample respondents, without having the knowledge regarding the practicability of the planned questions in the schedule. That is, before going for main survey with the sample respondents, first the researcher has to test the schedule to know, whether he can get all requisite data necessary for his study and whether that data can be (easily) provided by the sample respondents. This pre-testing also benefit the researcher on the following grounds:
  - It enables the researcher to know, how the sample respondents will react to the questions and even the answers they are likely to give.
  - It enable him to know, how much time it will take to complete the survey and how much budget he has to allocate.
  - It makes the researcher to know, whether the framed questions in the schedule are in logical sequence and whether they are meaningful even to the average respondent as to the researcher?
  - Researcher may get the idea regarding the heterogeneity among the population to be surveyed and this guide him in determining an efficient sample design.
  - The non-response rate can be expected and accordingly the questions can be modified or re-arranged.
  - Helps the researcher to determine, whether the research questions or hypotheses are appropriate.
  - Helps to determine, whether the levels of measurement are appropriate for the selected variables.
  - Provides information on the feasibility and the appropriateness of the sampling method.
  - Helps, determine the length of the schedule.
  - Enable the researcher to plan the method of data collection *i.e.*, whether by survey method or by interviews or by group discussions etc., based on various aspects like response rate, relative cost, accuracy of data obtained etc.
  - Enable the researcher to analyze the adequacy of the size of the schedule in eliciting all requisite information in accordance with the framed objectives for the study.
  - Enable the researcher to test, whether any of the framed questions are likely to confuse the respondents, highlight key words in the question etc.

- Allow enough time for the researcher to incorporate any revisions.
- Enable the researcher to know, whether each question is measuring what it is intended to measure?, are all the words understood?, are questions interpreted similarly by all respondents? does each closed-ended question have an answer that applies to each respondent?, does the schedule create a positive impression and motivates the respondents to answer it?, does any part of the schedule suggest bias on the part of the researcher etc.
- This benefits the researcher to know, whether any of the framed questions are confusing or misleading. In general, in designing the questions in a schedule, there will be one main question and one or more dependent or sub-questions. For example, if the researcher want to know whether the respondent cultivate paddy, he will frame a closed-ended question. If paddy farmers are identified, he frame different questions pertaining to the research topic say, related to the cost of cultivation of paddy. In this case, only paddy farmers will answer the dependent questions. So, the researcher must vigorously examine every question and include those questions that are relevant to the study. Here comes the importance of pre-testing of schedule, as it makes the researcher to assess the practicability of the questions in the study area. By asking the respondents only relevant questions, it reduces the length of interviews, cuts down the participation of respondent's time in the survey and lowers the cost of the survey. For example, the question *'What is your present occupation?'* would not apply to all respondents and is applicable only to those who are employed. So, this question should follow a screening question: 'Are you employed?' So, if the answer to this question is 'Yes', then the former question will arise. Though this question *i.e., 'What is your present occupation?'* seems harmless, it could prove annoying to those, who are not employed.

The following are the essential steps involved in pre-testing of schedule:

- Select a sample similar in socio-economic background and geographic location to the one that will be used in the main study. Note that, this sample will not be included in the final survey.
- Note down the data and information provided by the selected respondents and identify gaps, if any. That is, whether the questions framed in the schedule are sufficient enough to collect all relevant information or not.
- Note the wording mentioned in the schedule and if necessary, incorporate the popular words.
- After conducting the survey, the researcher must question himself on the following aspects:
  - Are the questions meaningful to the respondents?
  - Are respondents easily able to understand the terms?
  - Did he feel that, he is receiving valid information?
  - Was the questions order logical and did the interview flow smoothly?
  - Did some parts of the schedule arouse suspicion?
  - Did any part of the schedule seem repetitive or boring?
  - Was the researcher capable of explaining technical terms to the respondents?
  - Was the schedule is too long?

- ❖ Make necessary corrections in the schedule before employing it on the selected sample respondents in the main study.

☆ It is not good asking respondents about data or information of past *i.e.*, too long ago for them to remember accurately or data that they are unlikely to have accurate information or about personal information that are not relevant to the study.

☆ Avoid Double-Barrelled Questions (also known as two questions in one) in the schedule. *A question that actually asks two questions in one is called a Double-barrelled question.* Since two questions are involved in one sentence, it simultaneously deals with two concepts at a time. Such questions make it difficult for the respondent to know, which part of the question to answer and for the researcher to know, which part of the question the respondent actually answered. For example, if the researcher framed closed-ended question: *'Do you plan to leave your implements in farm shed and hire a machinery for ploughing the land from the ensuing Rabi season?* This question may be very difficult for some respondents to answer, since their personal situation may not fit into a simple 'Yes' or 'No'. For example, a respondent may plan to:

- ❖ Use implements in one season and hire a machinery in another season;
- ❖ Stop the cultivation in the ensuing *Rabi* due to lack of adequate power supply;
- ❖ Use implements even in the *Rabi* season.

If the answer is 'No' to what aspect it refers to? That is, whether 'No' refers to l*eaving the implements in farm shed or with reference to hiring a machinery for ploughing the land in the ensuing Rabi season? Similarly, 'Do you agree that, rice blast can be transmitted through water or soil?* A 'No' answer may mean not by water, or not through soil, or both.

A close observation also reveals that, the above question really asks two questions: *'Do you plan to leave your implements in the farm shed during the coming Rabi season?'* and *'Do you plan to hire a machinery during the coming Rabi season?* So, the best solution is to split the above question into these two different questions.

☆ Avoid leading questions in the schedule. A leading question is one that suggests or leads the respondent to a certain answer. That is, in case of leading question, the way the question is phrased will have an influence on the response to be given by the respondent. So, these leading questions will lead to biased answers and affect survey results. For example, the question is as follows: *Please tell me whether you agree with, disagree with, or have no opinion on the following statement: 'Application of organic manures to paddy improves yields and therefore, should be promoted'.* In this statement, the question itself directly indicates the positive influence of organic manures on the yields of paddy and this influences the response of the respondent. So, the better way to ask the same (neutral) question is: *Please tell me whether you agree with, disagree with, or have no opinion on the following statement: 'whether application of organic manures on paddy should be promoted.'?*

Thus, in 'Leading question', different wording can guide or direct the respondents toward an answer. The other examples include: *'Do you do physical exercise, such as cycling?* (This is a leading question because, it will likely lead the respondent to focus only on cycling.), *Don't you agree that. ?* (This negatively worded question leads respondents to answer 'No'). So, the preferred phrasing is, 'Do you agree or disagree that. '?

☆ Avoid loaded questions in the schedule. A loaded question is one, which contains emotional words that automatically lead to approval or disapproval by the respondents. That is, this question suggests a socially desirable answer or is emotionally charged. For example, '*Should the Government take the responsibility in safeguarding the interests of farmers*'? Most of the respondents are going to agree with this question simply because, this is highly socially desirable. Similarly, a question which may hit the emotions of the people will be answered in a socially desired manner. For example, if the researcher asks the respondent (husband): '*Have you ever been scolded by your wife when you consumed liquor*'? Straight away this question may be considered to be a challenge to the masculinity of the person. Hence, it may be embarrassing for the person to admit such an experience. Therefore, even if he was scolded up by his wife, he might give a socially desirable answer. Similarly, if the researcher pose the question, '*State Agricultural Minister advocated the farmers to go for cultivating pulse crops in rabi season in view of acute power and water scarcities. Do you agree with the statement?*' More respondents will tend to agree with this statement.

- ☆ Complex and lengthy questions should be avoided in a schedule. For example, in a population survey, the researcher posed the following question: '*Has it happened to you that over a long period of time, when you neither practiced abstinence, nor used birth control, you did not conceive?* This question is vague ('a long period of time'), too formal ('has it happened to you that'), and complex, because of its length and use of the neither/nor construction.
- ☆ Avoid burdensome questions that may tax the respondent's memory. If researcher ask questions about past behaviour or events on the respondents, it may make serious demand on the respondent's memory. For example, '*How did you feel about your brother when you were 10 years old*'? It may very difficult to recall something from the childhood.
- ☆ Use Filter question, if needed. If the researcher asks a question that does not apply to the respondent or that the respondent is not qualified to answer, it may be irritating or may cause a biased response. So, by including filter question, it minimizes the chance of answering questions that are inapplicable on the part of the respondent. So, Filter question is that question, which screens out respondents not qualified to answer a second question. For example, if the researcher pose the question, '*How much time do you spend playing games with your children?*', then what if the respondent is unmarried? and even if the respondent is married, but does not have the child. In both these cases, the above question is inapplicable to the respondent. Before this question, the researcher may put a filter question: '*How many children do you have*'?
- ☆ In schedule, there will be a number of multiple choice questions, which involve a number of alternatives from which the respondent has to consider only one correct answer. But in ranking item questions, there may be several preferences arranged arbitrarily and the respondent may indicate or rank them. Note that, such ranking questions should, as far as possible, be avoided from being included in a schedule. They are likely to confuse the process of collection of data.
- ☆ As far as possible, hypothetical questions should not be framed in the schedule because, by posing such questions, the researcher does not gain any useful data or information from them. Similarly subjective and qualitative questions should be avoided as far as possible.
- ☆ Don't assume that the respondent is an expert on themselves. For example, the researcher want to test the idea that farmers give better evaluations to trainers, who tell a lot of jokes in the training classes. The wrong way to investigate this is to ask '*Do you rate a trainer higher, if he tells many jokes?*' because, this assumes that the farmer is completely conscious of everything they do and why. The right way is to ask the farmer two separate questions: '*How*

would you rate the following trainer?' and *'How many jokes does the trainer tell in class?'* Then statistically correlate the answers, to see the relation between rating given by the farmers for teachers and number of jokes cracked in the training classes.

- ☆ Avoid false premises. Asking *'What is the most important thing we should do to restructure the deteriorating economy in Andhra Pradesh'?* assumes that, the economy is deteriorating in the State, which the respondent may not agree with. This puts the respondent in a tough spot. It would be better to rephrase as *'What is the most important thing a Government can do to restructure its economy'.*
- ☆ Avoid asking about future intentions of the respondents. That is, hypothetical questions like *'If you were a Chief Minister of Andhra Pradesh, what strategies you plan to overcome farmers' suicides in Andhra Pradesh'* is notoriously unrelated to actual future behaviour.
- ☆ The response categories given under closed-ended questions should ensure that, all possible alternatives are clear to the respondent. If not, it can result in a leading question that adversely affects the survey results. For example:

'Do you think pesticides application is the best method to control boll worm menace in cotton?'

i. Yes ii. No

iii. No opinion

In the above question, the respondent is left with only one response alternative. That is, whether he think pesticide can effectively control boll worm menace in cotton? So, when only one alternative is presented, there is a tendency for the respondent to agree with that alternative. So, in this example, 70 per cent of the respondents opined that, pesticides application is the best method to control boll worm menace in cotton.

Results

i 70 per cent Yes, pesticides application is the best method

ii. 21 per cent No

iii. 9 per cent No opinion

An alternate wording for the same question is as follows:

'Do you think pesticides application is the best method to control boll worm menace in cotton compared to IPM?'

i. Yes, pesticides application is the best method

ii. No, IPM is the best method

iii. No opinion

Results

i. 33 per cent Yes, pesticides application is the best method

ii. 39 per cent No, IPM is the best method

iii. 28 per cent No opinion

The above revised question comprises of two clear alternatives: 'pesticides application is the best method' or 'IPM is the best method'. When this question was asked, the results were more evenly distributed between Yes, No and No opinion than for the previous question. Thus, by providing clear response categories, the respondent can give a more reliable answer.

- Avoid using double negatives in the same question framed in the schedule, as the respondent will not know whether to agree or disagree. For example, *Would you be for or against not allowing the sale of pesticides to farmers without the recommendations of local Agricultural Officer?* In this question, there are double negatives *viz.*, *'for not'* and *'against not'*. The former opinion *i.e., 'for not allowing the sale of pesticides to farmers without the recommendations of local Agricultural Officer'* implies the respondent is *against* allowing it. Similarly, if the respondent opines that, *'against not allowing the sale of pesticides to farmers without the recommendations of local Agricultural Officer'* implies the respondent is *for* it. Understanding this question is quite difficult because, it contains a double negative words *viz.,*: *'against'* and *'not'*. So, it is better to rephrase the above question so that, it contains no more than one negative word and it is as follows: *'Would you be for or against allowing the sale of pesticides to farmers without the recommendations of local Agricultural Officer?*
- Failure to identify a common starting time for getting the response may lead to bias. For example, *'In the last 12 months, have you had a tie up with any processing agency for processing mango produce?* Identification of common starting time is important because, a survey is normally conducted over an extended period. So, the time frame of 'last 12 months' will vary depending on the date of the interview. In this context, the following question is better and will provide a common time frame: *'From January 1 to December 31 of last year, did you have a tie up with any mango processing agency?'*
- It is better to collect accurate, continuous data at source instead of degraded data. Once degraded data have been collected, it is impossible to recover the original continuous data or to change cut-off criteria for categories. The following are the examples for information on age of farmers:

  What is your birth date?

  What is your age in years?

  Which age category do you belong to?

  So, for information on age, the first question is the best because, it can provide accurate continuous data, followed by the second question. The third question is the least desirable because data are degraded.
- When outcome measures make it impossible to detect significant changes or differences, Type II error will occur (discussed in Chapter 12). For example:

  How important is soil health to the crops, on a scale of 1 to 3?

  (Unimportant) 1 - 2 - 3 (Important)

  How important is soil health to the crops, on a scale of 1 to 10?

  (Unimportant) 1 - 2 - 3 - 4 - 5 - 6 - 7 - 8 - 9 - 10 (Important)

  The first question may not have sufficient discriminating power to differentiate the respondents because of the limited categories. The second question may be better.
- Some questions may be framed in such a manner that, respondents choose an inaccurate answer. For example: *'Which technology would you prefer'?*

  [ ] A technology that has a 5 per cent mortality on seed germination.

  [ ] A technology that ensures 90 per cent of the seeds will germinate or survive.

  Respondents may choose the second option, when they see or hear the words '90 percent' and 'survive', but in fact a 90 per cent survival rate (or 10 per cent mortality) is worse than a 5 per cent mortality.

- ☆ The mind-set of the respondent can affect his or her perception of questions and therefore, can affect answers. For example:
  - ❖ *How many times you go to your farm in a week?*
  - ❖ *How many time do you meet your local Agricultural Officer in a week?*
  - ❖ *How many times do you visit nearby regulated market in a month?*

  The change in wording from 'per week' to 'per month' can result in wrong answers for the third question above, because of the possible mind-set of the respondents.
- ☆ If the precise wording of a question changes in different surveys, the results may not be comparable. For example:
  - ❖ *Compared to other persons of your age, would you say your health is excellent, good, fair, or poor?*
  - ❖ *Would you say your health in general is excellent, very good, good, fair, or poor?*

  The first question and the second question use different wording, namely, 'compared to other persons of your age' vs 'in general'. This may guide respondents to evaluate their health in a different context.
- ☆ The researcher should not get confused between multiple-choice questions and multiple-answer questions. A multiple-choice question is one, where the respondent is told the possible answer choices. A multiple-answer question (which need not be multiple-choice) is one that allows more than one answer. Whether a question can have one answer or more than one depends on its meaning. Here is a question with only one possible answer: '*What is the seed rate used in SRI production technology of paddy*?' The following is an example for multiple choice question:

In a typical month, how often you meet the local Agricultural Officer?

i. Less than once　　ii. 2 times

iii. 3 times　　iv. 4 times

v. More than 4 times

The following is an example for multiple-answer question:

'Suggest the measures to improve the functioning of Kurnool regulated market'

i. ______________________________

ii. ______________________________

iii. ______________________________

iv. ______________________________

v. ______________________________

vi. ______________________________

'Which TV channel you see to watch Agriculture related programmes in a typical week?'(Circle all that apply)

i. E TV channel

ii. Saakshi TV channel

iii. Doordarshan channel

iv. TV5 channel

v. Gemini TV channel

vi. ABN channel

In the above example, sometimes respondents may be tempted to give all possible answers to a question. Note that, if every respondent gives every possible answer, this doesn't help much. The researcher can make respondents think a little harder by limiting the number of answers to about three per person. Respondents, who give more than three answers can be asked *'Which are the three most important answers?'*

- ☆ Use the terms or words that are appropriate in space and time. Look at the following example regarding the vernacular terms used for *Kharif* season, *Rabi* season and Summer season for cultivating crops across three regions of Andhra Pradesh:

Season	*Vernacular Language*		
	Coastal Andhra	*Rayalaseema*	*Telangana*
Kharif	Sarwa	Mungari	Abi
Rabi	Dalwa	Hingari	Tabi
Summer	Endakaru	Endakaru	Kathera

If the researcher wishes to study *'Economic analysis of paddy based cropping systems in Andhra Pradesh'*, he has to use the words, *Kharif, Rabi* and Summer seasons in the schedule to elicit correct responses from the farmers across the three regions of Andhra Pradesh.

- ☆ Questions that provide too few categories can force respondents to choose imprecisely among limited options. For example:

  Do you agree? Yes [ ] No [ ]

  Do you agree? Yes [ ] No [ ] Don't Know [ ]

  The first question, which does not have a 'Don't know' category, may produce a bias because, respondents who have no opinion are forced to select an answer that may or may not reflect their true feelings. So, the second question is recommended.

- ☆ Missing intervals in response choices can cause confusion. For example: *'How often does your tractor is repaired'?*

  [ ] Less than once per month

  [ ] Once per month

  [ ] Once per week

  [ ] More than once per week

  In the above example, respondents do not have a place to put 'once in every two weeks'. So, the following response categories are recommended:

  'How often does your tractor is repaired'?

  [ ] Once per month

  [ ] Once in every two weeks

  [ ] Once per week

  [ ] Less than once per month

- ☆ As discussed earlier, overlapping intervals in response choices can cause confusion. For example:

  'How many cigarettes do you smoke per day'?

  [ ] None [ ] 5 or less [ ] 5-25 [ ] 25 or more

  Respondents smoking exactly 5 or 25 cigarettes per day do not know in which category to place themselves. So, the following question is more appropriate:

  'How many cigarettes do you smoke per day'?

  [ ] None [ ] 1 -4 [ ] 5-24 [ ] 25 or more

- ☆ An even or an odd number of categories in the scale for the respondents to choose from, may produce different results.

  Do you agree? (Agree) 1 – 2 – 3 (Disagree)

  Do you agree? (Agree) 1 – 2 – 3 – 4 (Disagree)

  The first question, with an odd number of categories, tends to result in neutral answers (*i.e.,* 2), and the second question, with an even number of categories, tends to force respondents to take sides. The two approaches produce different results, but there is no general consensus as to which one is better.

- ☆ Avoid using acronyms and abbreviations in the questions, as the respondents may not be aware of them, unlike researchers. For example, *'Do you received training from DAATTC of Kurnool district'?* Many respondents will not know that, DAATTC is the acronym for District Agro-Advisory and Transfer of Technology Centre.

- ☆ Avoid using hypothetical questions like 'Would you like to', 'What would you do if', 'What might be your reason' etc. Further, the researcher has to decide carefully, whether he has to personalize or impersonalize the questions to be posed on the respondents.

- ☆ Technical jargon and the profession's technical terms may not be understood by the respondents and hence, their usage should be avoided. For example, *'what is your opinion regarding the efficacy of systemic fungicide, S-benzyl diethyl phosphorothionate'*. The technical term in this question may not be understood by many respondents, so it is preferable to ask the same question in more common terms as *'what is your opinion regarding the efficacy of systemic fungicide, Kitazin?'*

- ☆ Similarly, uncommon and difficult words should be avoided in schedules. Gowers (1954) and Day (1998) have produced lists of words that can be replaced by simpler alternatives. For example:

Uncommon	*Common*	*Uncommon*	*Common*
Assist	Help	Consider	Think
Effectuate	Cause	Elucidate	Explain
Employ	Use	Initiate	Begin/Start
Major	Important/Main	Perform	Do
Quantify	Measure	Require	Want/Need
Reside	Live	State	Say
Sufficient	Enough	Terminate	End
Ultimate	Last	Utilize	Use

Use these common words in schedules, especially schedules targeted for the general population, to avoid misunderstanding.

- ☆ Vague words in vague questions encourage vague answers. For example,

  How often do you meet the local Agricultural Officer?

  [ ] Regularly [ ] Occasionally

  How often do you meet the local Agricultural Officer?

  [ ] twice a week or more often [ ] once a week [ ] less than once a week

  The first question is vague because 'occasionally' and 'regularly' are not defined. The meaning can easily be made more precise, as in the second question.

- ☆ As far as possible, avoid asking sensitive questions on the respondents. In general, sensitive questions, such as age, personal or household incomes, sexual orientation, or marital status, may elicit inaccurate answers and may also affect the interviewer-interviewee relationship so that, all subsequent answers can be affected. If it is inevitable for the study, the researcher has to follow some alternative ways in posing the same on the respondents. The researcher should not ask a sensitive question when others are present. Careful wording can also help reduce the impact of sensitive questions on survey responses. Terms such as '*Many people*' or '*Most people*' can be used to introduce the question. When using this technique, care must be taken not to introduce bias. Another alternative is to use a closed-ended question with a range of response categories. For example,

'Since how many years you are consuming liquor'?

i. Less than one year
ii. 2 to 3 years
iii. 4 to 5 years
iv. More than 5 years

How old are you?

In what year were you born?

In the above example, the first question, which is direct, tends to result in a high percentage of refusals to answer. The second question tends to yield fairly accurate responses.

- ☆ While designing the questions, the researcher should be attentive on the following:
  - ❖ Who?, What?, Where?, and When? for each question should be clearly specified
  - ❖ To 'Whom' the question applies to?
  - ❖ What information should be included in or excluded from the response?
  - ❖ What units the answer should be provided in (*e.g.*, kgs, tonnes, quintals etc.)
  - ❖ What time period the question refers to (When?)

  For example, assume the researcher posed the question: 'W*hat is your income?'*. Note that, this question lacks clarity. In the question, to whom the word *'your'* refers to. Whether it refers to the respondent's personal income or family income or income after paying taxes or income before paying taxes etc. Further, there is no specific time period for which the respondent should provide the answer (*i.e.*, for the year 2011 or 2012 or 2013 etc.). So, while framing each question, care should be taken such that, it will have a specific or clear response from the respondent. The following question gets the clear response: *'What is your personal income after paying direct taxes in the year 2012'*.

- ☆ The leading question framed in the schedule should not restrict the answer to be given by the respondent. For example, the researcher framed as a closed-ended question *'Do you read articles related to agriculture published in Telugu books like Annadata or Rythu vani?'* In this case, the respondent's reply is restricted to only one of two options and if he reads third book (say 'Rythu Nestham'), he could not express the answer.
- ☆ Words carrying dual meaning must be avoided.
- ☆ The schedule must be limited in its length and scope.
- ☆ Questions that are irrelevant to the research topic *i.e.*, not related to the framed objectives must be avoided. That is, care must be taken to avoid ambiguous, too personal and embarrassing questions. The wording should consider the full context of the respondent's situation. In particular, consider the who, what, when, where, why, and how dimensions of the question. For example, the question, *'Which brand of toothpaste do you use?'* might seem clear at first. However, the respondent may consider 'you' to be the family as a whole rather than he or she personally. If the respondent recently changed brands, the 'when' dimension of the question may be relevant. If the respondent uses a different, more compact tube of toothpaste when traveling, the 'where' aspect of the question will matter. So, a better wording of the question might be: *'Which brand of toothpaste have you used personally at home during the past six months? If you have used more than one brand, please list each of them.'*
- ☆ Allow a 'Don't Know' or 'Not Applicable' response to all questions, except to those in which the researcher is certain that, all respondents will have a clear answer. In most cases, these are wasted answers as far as the researcher is concerned, but are necessary alternatives to avoid frustrated respondents. For the same reason, include 'Other' or 'None' whenever either of these is a logically possible answer. When the answer choices are a list of possible opinions, preferences, or behaviours, you should usually allow these answers.
- ☆ The researcher should be well-versed with the subject matter before framing the questions in the schedule. When both the researcher and the respondent are well versed with the subject matter of the research topics, many questions become unnecessary and can be avoided.
- ☆ The questions should be so designed to arouse the interest of the respondent and must proceed from simplicity to complexity.
- ☆ The questions should proceed from one frame of reference to another instead of jumping back and forth.
- ☆ Most schedules include a number of demographic questions. These are questions about the respondent's characteristics and circumstances. Questions about sex, age group, occupation, education, household type, income, religion, are all demographic. These demographic questions normally come close to the end of the schedule. Note that, the interviewer does not need ask some 'questions,' such as the sex of the respondent, and the area where the person lives. The answers to such items are already known, and can simply be written on the schedule.
- ☆ While framing questions in the schedule, the researcher must critically think about how the data will be processed. When the interviewing is complete, the data will require analyzing. In a survey involving hundreds of respondents, it is extremely costly to analyze the free ranging responses to open ended questions. Where there are hundreds of individual responses, they must distilled into a much smaller number of groupings in the search for meaningful patterns of response.

Above all, the survey should be interesting or rewarding for the respondents as well as participants (interviewers). A good topic, easy, quick questions and perhaps recognition given for the respondents (say, acknowledging them in the report) can make the respondents' experience more pleasurable and ensure a reasonably high response rate. However, too many surveys are confusing and lengthy. Cull everything, but the vital questions and make those questions easy and interesting.

c. Types of Schedules

Schedules can be classified into two broad types *viz.*, Structured schedules and Unstructured schedules.

- ☆ ***Structured Schedules:*** Structured schedules are those which pose definite, concrete and prepared questions. It means the questions are prepared in advance and not constructed on the spot during the questioning period. Additional questions may be used only, when needed or inadequate replies by the respondents. As discussed above, the structured schedules may involve two broad types of questions *viz.*, Closed-ended questions and Open-ended questions.
- ☆ ***Unstructured Schedules:*** Unstructured schedules have very little structure at all. The researcher or interviewer goes into the interview with the aim of discussing a limited number of topics, sometimes as few as one or two, and frames the questions on the basis of the interviewee's previous response. Though only one or two topics are discussed, they are covered in great detail. Subsequent questions would depend on how the interviewee responded in the interaction. So, in using unstructured schedules, the interviewer wants to find out about a specific topic, but has no structure or preconceived plan or expectation as to how they will deal with the topic. The main difference between structured and unstructured schedules is that, in the former, the interviewer has a set of broad questions to ask and may also have some prompts to help the respondent or interviewee and this is not the case in the unstructured schedules. Thus, in unstructured schedules, the researcher or interviewer has no control over things in collecting the data and hence, he provides no direction. This is because, he has no predetermined set of questions for asking and thereby, the respondent is free to explain or tell his experience. This form of schedule is used for intensive studies, but generally for a limited number of selected cases. The chief disadvantage of unstructured schedules stems from the danger that, non-additive and non-comparable data will be accumulated, when no structuring is imposed.

There is another classification of schedule depending upon the purpose for which the structured schedule is employed. They are discussed here under:

- ☆ ***Observation Schedule:*** The schedule that is used in observation research is called as Observation schedule. In these schedules, observer records the activities and responses of a worker or a group under specific conditions.
- ☆ ***Rating Schedule:*** In the field of business guidance, psychological research, and social research, the rating schedules are used to assess the attributes, opinions, preferences and other qualitative elements.
- ☆ ***Document Schedule:*** The schedules of this type are used to obtain data regarding written evidence and case histories from autobiography, diary, case histories of the companies or Government records etc.

- *Interview Schedule:* In an interview schedule, the researcher presents the questions of the schedule to the respondent and records their responses.
- *Village or Community schedule:* It is used by census researchers who collect general information on populations, occupations, etc.
- *Family or Household schedule:* It gives full demographic details of farm households, the status of individuals, data on education, age, family relations etc. The details pertaining to the agro-economic aspects like farm holding survey, marketing pattern of agricultural commodities, financial aspects of agri-business etc., can also be studied.
- *Opinion or Attitude schedule:* To schedule the views of population regarding an issue.

d. Steps in Construction of Schedule

As discussed above, the schedule so prepared by the researcher should have a prescribed format in terms of arranging the sequence of questions and use of simple words so that, the respondent is at the ease in providing the requisite information. To ensure this format on scientific note, certain steps (Figure 9.1) must be followed by the researcher in developing the schedule and they are discussed here under.

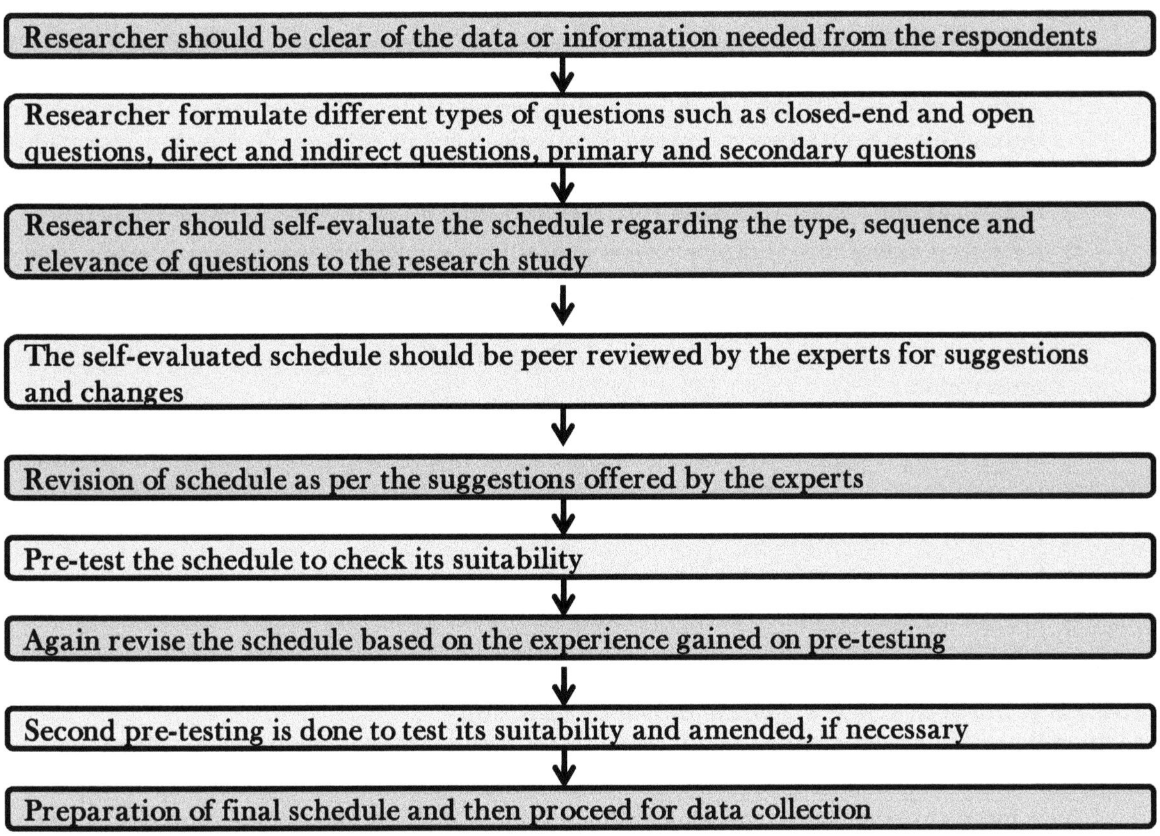

Figure 9.1: Steps in Construction of Schedule.

e. Advantages of Using Schedule in Data Collection

Following are the important advantages of employing schedule in agro-economic research:

- As the researcher or investigator directly interacts with the respondent in eliciting the data or information, the response rate his high. Further, the researcher will clarify doubts if any, confronted by the respondent in answering the questions. This will certainly motivate the respondents in contributing the full information to the researcher.
- As the schedule is filled by the researcher in the presence of respondent, he will incorporate the correct data in the schedule and can cross check the information, whenever necessary.
- Due to personal interaction between the researcher and respondent, a good rapport is built in between them and this further contributes in getting reliable information from the respondent. This good rapport between them also facilitates in creating a healthy atmosphere for different kinds of informal and formal discussions.
- Considering the time element involved in filling each schedule, it takes less time on the part of the researcher to fill the schedule because of two reasons: the researcher himself fill the schedule and that too in the presence of respondent. This makes the researcher to act quickly in accessing the desired information.
- The respondent feels human touch in answering the questions given in the schedule. As the respondent hears the questions posed by the researcher, he feels the importance of the research problem and gives the answers on a responsible note.
- As the researcher fills the schedule through personal interactions with the respondent and through his personal observation and there is also a facility of cross checking the information or data, data collection through schedules greatly minimizes the errors.
- Data collection through schedule can be employed both on literate and illiterate people.
- If the respondent could not answer some questions and if faces any doubts to answer some questions in the schedule, the researcher could try in different ways through personal interactions to obtain the information.
- The collected information through schedule can also be supplemented with personal observation.
- Schedule is a more flexible tool for data collection, as the order of questions can be changed at will by the researcher and in convenience of the respondent for answering the questions.
- Knowledge of past and future is possible.
- In-depth research is possible.
- The researcher can cross-check or verify the data from other sources, if possible.
- The information obtained is very reliable and dependable and helps in establishing cause and effect relationships.

f. Disadvantages of Using Schedule in Data Collection

In spite of the above mentioned advantages, following are the important disadvantages of employing schedule in agro-economic research:

- If the researcher has to collect data from each and every respondent through personal interactions, it consumes more time to complete the total survey and also pose cost burden on him.

- ☆ The personal presence of researcher at the time of interaction with the respondent may lead to bias in answering the questions by the respondent.
- ☆ When the respondents are spread over a larger area, it is difficult for the researcher to move to distant areas and come in contact with each and every respondent.
- ☆ Personal presence of researcher may pose less comfort on the part of the respondents to deliver information especially personal information.
- ☆ As compared with questionnaire, schedule method is generally costlier and more time consuming and this factor becomes a serious limitation when the respondents are physically scattered over a wide area.
- ☆ This method of data collection demands a large number of well trained and experienced field workers for collecting data from the respondents. So, it leads to higher cost on the part of the research project.
- ☆ This method of data collection can be used successfully in a limited region and on limited number of respondents only.

9.2. Questionnaire

Questionnaires are an inexpensive way to gather data from a potentially large number of respondents. This format is employed to collect information from the respondents in Agricultural Extension studies. A questionnaire consists of a number of questions printed or typed in a definite order on a form or a set of forms. It is administered to a respondent either personally or through mail. The respondent answers the questions on his own without being aided by the researcher. Questionnaires are now widely used for collecting data, particularly when data are to be collected from a large number of people, who are scattered over a wide area. They are used both as independent and separate method of collecting data. They are also used as an additional device to check data gathered through observation and personal interview. The following are the popular definitions of questionnaire:

'A questionnaire is a means of gathering information by having the respondents fill in answers to printed questions'.

Wallace and Wallace

'Fundamentally, the questionnaire is a set of stimuli to which literate people are exposed in order to observe their verbal behaviour under these stimuli'.

Lundberg

'Questionnaire is a device for securing answers to questions using a form which the respondent fills in himself'.

Goode and Hatt

Though both schedule and questionnaire are treated as synonyms, technically there are differences between these two. First, let us study the similarities between schedule and a questionnaire.

a. Types of Questionnaires

There are different types of questionnaires as discussed here under:

- ☆ ***Structured questionnaire:*** Structured questionnaires are those, which pose definite, concrete and prepared questions. In these questionnaires, the questions are presented such that, they have fixed response categories. The questions placed in this questionnaire initiate a formal inquiry. They can even supplement and check the data previously accumulated. This questionnaire is widely used in studies of the economics and the social problems, studies of the administrative policies and changes etc.

- ***Un-structured questionnaire:*** Unstructured questionnaire have very little structure at all. Here, every question is not necessarily presented to the respondent in the same wording and does not have fixed responses. Respondents are free to answer the question the way they like. Thus, this type of questionnaire is very flexible in working. It acts as the guide for the interviewer. This questionnaire is used in studies related to the group of families or those relating to the personal experiences, beliefs etc.
- ***Mixed Questionnaire:*** This is a questionnaire which is neither completely structured nor un-structured. It consists of both the types of questions. That is, both close and open ended questions will be used. This questionnaire ids most commonly employed in the field of social research.
- ***Disguised typed questionnaire:*** Here the questions are direct and therefore, respondents may give answer to certain sensitive questions, whose accuracy may be questioned.
- ***Closed and Open Questionnaires:*** The question that call for short check responses are known as restricted or closed form type. For example, they provide for marking a 'Yes' or 'No', a short response or selecting an item from a list of responses. Here the respondent is not free to write on his own, and he has to select the right answer from the given options. But, the researcher can easily tabulate, interpret and summarize the data findings. On the other hand, in case of open ended questionnaire, the respondent is free to response in his own words. Unlike Closed questionnaire, in Open questionnaire, it is difficult for the researcher to tabulate, interpret and summarize the data findings. Many questionnaires include both close and open type questions. The researcher selects the type of questionnaire according to his need of the study.

Open format questions or open-ended questions give the respondents an opportunity to express their opinions in a free-flowing manner. The questions and their orders are pre–determined in the nature. But, these questions don't have pre-determined set of responses and the respondent is free to answer, whatever he/she feels right. By including open format questions in a questionnaire, the researcher can get true, insightful and even unexpected suggestions. Qualitative questions fall under this category. Note that, an ideal questionnaire would include an open-ended question at the end of the questionnaire that seeks feedback and/or suggestions for improvements from respondents. This questionnaire is used making intensive studies of the limited number of the cases, as it merely raises an issue for discussion. But, it do not provide any structure for the respondent's reply. Say for example:

'State your opinion about the quality of products offered by Bayer India Limited'.

____

____

Closed format questions or closed-ended questions involves multiple choice questions (say, YES or NO), where respondents are restricted to choose one among many of the given multiple choices. There is no fixed limit as to how many multiple choices should be given; the number can be even or odd. One of the main advantages of including closed-ended questions in questionnaire design is the ease at performing preliminary analysis. These questions are ideal for calculating statistical data and percentages, as the answers set is known. So, these questions mainly produce quantitative data. Closed ended questions can also be asked to different groups at different intervals to efficiently track their opinion about a product or service or company over time. The demerit of these questions is that the respondent cannot express his own judgement. The following is the example of a closed-ended question:

'What is the seed rate followed under SRI production technology of paddy'

a. 2kg/acre b. 5kg/acre c. 7kg/acre d. 10kg/acre e. 15kg/acre

Four types of response scales for closed-ended questions are distinguished:

- Dichotomous, where the respondent has two options
- Nominal-polytomous, where the respondent has more than two unordered options
- Ordinal-polytomous, where the respondent has more than two ordered options
- (Bounded)Continuous, where the respondent is presented with a continuous scale

- ***Fact and Opinion questionnaires:*** In case of Fact questionnaire, the respondent is expected to give information of facts without any reference to his opinion or attitude about them. But, in case of Opinion questionnaire, the respondent gives the information about the facts with his own opinion and attitude.
- ***Pictorial questionnaire:*** In this questionnaire pictorial representations are used for important items like plough, crow bar, sickle, tractor, sprayer, duster etc., so as to promote interest among the respondents in answering the questions. This questionnaire is used in studies related to the social attitudes and the pre–judices in the children. Compared to other questionnaires, this is used very rarely.
- ***Self-administrated questionnaire:*** This questionnaire can be self-administrated, i.e. administrated to the potential responders by mail, E-mail or Internet. The respondents fill in the questionnaire without the help of the research investigator or a member of the investigation team. So, employing this type of questionnaire is relatively cheap and easy to administer. The respondents' replies to the questions can be kept confidential without the influence of the interviewer. However, the employment of this questionnaire can result in low response rates, as the respondent is less motivated to respond and he may fill the questionnaire at his convenient time. In some cases, the given questions can be misunderstood and they pose wrong answers. However, in recent years, web-based questionnaire have become increasingly popular compared to mailed questionnaire. Compared to mailed questionnaire, the former can be easily set up and the corresponding link can be sent to a large number of persons at low or no expense. Mostly closed-end questions and pictorial questions can be used to make the respondents respond easily. However, this web-based questionnaire is particularly useful in populations with high internet literacy.
- ***Interviewer-administrated questionnaire:*** This questionnaire is used in face to face or telephone interviews. Unlike web-based questionnaire, this questionnaire can be used easily to interview less literate or illiterate people. In case, if the respondents feel difficulty in understanding and answering the questions, they can get clarified from the interviewer. Note that, questionnaires need to be short (up to a maximum of 10-15 min), especially for telephone interviews. So, the response rate is much higher than the self-administrated questionnaire. However, there is a scope for bias, which can be introduced by interviewer's

perceptions and interpretations of the answers (interviewer bias). For studying sensitive issues like sexual behavior, this interviewer-administrated questionnaire is not suitable.

- ***Factual-type questionnaire:*** This questionnaire deals with the questions about public or *observable* information that it would be tedious or inconvenient to get any other way. For example, how many trainings the farmer has undergone on SRI technology of paddy, how quickly did the farmer can complete the winnowing operation?
- ***Opinion-type questionnaire:*** This questionnaire employs the questions to elicit the responses from the respondents, what they think about something or someone. There's no right or wrong answer, but simply the respondent has to indicate his strength of opinion on a particular question say, do you like it or not, or which do we prefer? Will you vote for Mr. A or Mr. B for electing a member in cooperative farming society etc.
- ***Attitude questionnaire:*** This questionnaire employs the attitude questions that focus the respondent's attention to inside themselves, to their internal response to events and situations in their lives. There are a lot of questionnaires consisting of attitude questions about experiences with new plant varieties, new processing machinery, harvesting machinery, purchase a particular brand of fertilizer, use of Information Technology in agriculture etc. This questionnaire is also called as Satisfaction questionnaire.
- ***Exploratory questionnaire*:** If the data to be collected is qualitative or is not to be statistically evaluated, then the researcher employs exploratory questionnaire. This questionnaire harbour broad questions and may include both open ended and closed ended questions. For example, the researcher interview the female head of the household to find out, how decisions are made within the family while purchasing farm inputs and while selling farm output.
- ***Formal standardized questionnaire:*** This questionnaire is employed by the researcher, if he wishes to test and quantify hypotheses and the data is to be analyzed statistically. This type of questionnaire is characterized by:
  - prescribed sequence and wording of questions to ensure that, each respondent receives the same stimuli
  - prescribed definitions for key words used in the questions to ensure that each respondent can easily understand the questions and gives his perception.
  - prescribed response format for rapid completion of filling the questionnaire.

b. Questionnaire Layout or Design of Questionnaire or Characteristics of a Good Questionnaire

It is to be noted that, unlike schedule, in questionnaire, the questions are filled by the respondent himself and not by the researcher or surveyor. Hence, the lay out or order or sequence of questions should be given prime importance, so as to arose interest on the part of the respondents. In addition to sequence of questions, type of questions (*open end questions and closed end questions; Direct questions and Indirect questions; Primary questions and Secondary questions; Nominal, Ordinal and Interval questions; and Contingency questions)*, wording, transitional statements etc., also play an important role in creating a quality questionnaire. In continuation of the guidelines discussed under 'Design of Schedule', the following are some of the additional guidelines that are to be essentially followed, while designing the questionnaire:

- All questionnaires must be supported with some text. This should contain the following features:

- ❖ The purpose of the questionnaire should be communicated clearly to potential respondents.
- ❖ Where deemed appropriate, the confidentiality of responses should be assured.
- ❖ The supporting text should contain simple instructions on how to complete.

☆ ***Sequencing of Questions:*** The order of the questions should be so designed in such that, it makes the respondent to gain and sustain interest in answering all the questions. The sequence of questions should be logical from general to specific *i.e.*, the questions should flow smoothly from one question to the next. The questions' sequence must quickly awaken the respondent's interest and motivate him in answering all the questions fully and completely. Another important aspect is that, the questions on a particular subject should be grouped together.

☆ ***Opening questions:*** These are the beginning questions in the questionnaire. These questions are important in the sense that, they stimulate and encourage the respondents in answering the questions. So, as far as possible, these questions must be related to the purpose of the survey and how well the respondents are related to it. Note that, these questions must be 'easy' to answer. The beginning questions must be closed-end questions. Open end questions which require elaborate answers should be avoided at the beginning of questionnaire, as they make the respondents to feel more burdensome in answering subsequent questions.

☆ ***Sensitive questions:*** After placing the beginning questions, now the researcher should carefully introduce sensitive questions to elicit useful responses relevant to the research project. These questions are considered 'sensitive' in the sense that, they must be posed carefully without affecting the emotions of the respondents. For example, the researcher should not pose the question on a respondent (say, who completed his B.Sc (Ag) degree): *'Since how many years, you remain unemployed'?* If the respondent is unemployed, this question will definitely hurt his feelings. So, the researcher should ask this sensitive question in a different way: *'In which year, you have completed your B.Sc. (Ag.) degree?'* Similarly, careful consideration should be given to place these sensitive questions in the questionnaire. If these sensitive questions are not properly framed and introduced too early, the respondent may lose interest in answering them and even the subsequent questions as well. Therefore, sensitive questions should be introduced at a point, where the respondent is most likely to feel comfortable answering them and where they are most meaningful in the context of other questions. In general, to overcome the respondent's unwillingness to answer the sensitive questions, the following strategies will be followed to elicit the answers:

- ❖ These questions will be placed at the end of the questionnaire.
- ❖ Preface the sensitive question with a statement.
- ❖ Use the third person technique (For example – One farmer in another village mixed insecticides and fungicides for spraying on paddy crop - Is it right?, One person in another State badly needed a job and hence, he used wrong means to get it – Is it right?). This technique enables the respondent to clearly understand the sensitive questions and he attempts to answer them.
- ❖ Categorize the responses in the form of multiple choice questions rather than asking for a specific response.

☆ It is essential to soften the impact of sensitive questions, as these questions leads to 'Social desirability bias'. That is, when the respondents are posed with sensitive or personal questions, they generally respond in such a way that, their answer is in accordance with the social norms and they do not reveal the truth or real belief. For example, the following questions may result in social desirability bias: *'Whether you are alcoholic'?*, *'Whether you work in the farm after consuming liquor'?* These are sensitive questions, as the respondent will not reveal the correct answers, even though he is alcoholic. In such cases, the researcher should follow some techniques like posing the question in such a way that, it is less threatening on the part of the respondent. Another important way is to express the question such that, the sensitive behaviour is more common. So, terms such as *'many people'* or *'most people'* can be used in the question to elicit the correct answer.

☆ ***Transitional Statements:*** When a questionnaire contains different sections of a topic (ie., sub-topics) to be dealt with, these transitional statements are used to separate sections and to introduce sections of related questions. Their use is important, as they indicates to the respondent that a new sub-topic is going to be discussed. For example, following are the four transitional statements in the questionnaire:

- *Part A - We first would like to collect some background information about your farm.*
- *Part B - We would now like to collect information about the production related aspects of crops cultivated by you in the Kharif season.*
- *Part C - We would now like to collect information about the marketing related aspects of produce transacted by you.*
- *Part D - The next questions are about your farm business income statements during the last year.*

Note that, while framing the questions, the researcher should carefully note two important aspects *viz.*, overall sequence of the questions separated topic-wise by transitional statements and intra-sequence of question with in a particular topic area.

☆ ***Instructions:*** It is essential to give some useful instructions to the respondents for answering the questions either at the beginning of the questionnaire or at the beginning of each sub-topic next to the Transitional Statement. In some cases, these instructions will follow the question helping the respondent to answer the question. In the following example, the instructions are included in bold for both first and second questions.

- *For whom did you work? (For example, Seed firm, Pesticide firm, Fertilizer firm etc.)*
- *What kind of service you are rendering in your job? Give a full description.(For example, seed production, seed marketing, pesticides manufacturing, pesticides marketing, fertilizer production, fertilizer marketing etc.)*

Note that, if the instructions are given elsewhere in the questionnaire, it should be clearly indicated to the respondents. In some cases, a reference guide is specially enclosed with the questionnaire and this benefits the respondents in going through all the instructions at a stretch and start answering the questions. For example:

Note: The following questions refer to the adoption of package of practices in SRI technology of paddy production in the Kharif season of the current year. In case, if you did not practice SRI technology in the current year, give similar details for the past year.

In the above example, the directions are given before the questions are asked regarding the adoption of package of practices in SRI technology of paddy production. Similarly, definitions for any technical terms may also be given in the reference guide enclosed to the questionnaire.

- In a questionnaire, there will be a number of multiple choice questions, which involve a number of alternatives from which the respondent has to consider only one correct answer. But in ranking item questions, there may be several preferences arranged arbitrarily and the respondent may indicate or rank them. Note that, such ranking questions should, as far as possible, be avoided from being included in a questionnaire. They are likely to confuse the process of collection of data.
- As far as possible, hypothetical questions should not be framed in the questionnaire because, by posing such questions, the researcher does not gain any useful data or information from them. Similarly subjective and qualitative questions should be avoided as far as possible.
- In a questionnaire, horizontal vs vertical format of the response choices can affect the answers. For example:
  - *'In terms of yield, SRI technology of paddy production is ——'*

    Excellent ... [ ] Good ... [ ]

    Fair ... [ ] Poor ... [ ]
  - *'In terms of yield, SRI technology of paddy production is ——'*

    Excellent .... [ ] Good .... [ ]

    Fair .... [ ] Poor .... [ ]

  The horizontal response format (first example) can cause confusion among the respondents because of poor spacing and may result in the wrong answers being checked or circled. The vertical response format (second example) has been suggested as better for listing response options.
- To rate the responses of respondents between two extremes, 'Bipolar questions' can also be included in the questionnaire. Bipolar questions are the ones having two extreme answers written at the opposite ends of the scale. The respondents are asked to mark their responses between those two. The following is the example:

  'How would rate the services of Mahindra Tractor firm'?

  Very good —— —— —— —— —— —— Very Bad

  Fast —— —— —— —— —— —— Slow

  Reliable —— —— —— —— —— —— Unreliable
- Juxtaposed scales (Hunt *et al*, 1995), a type of self-report response scale that asks respondents to give multiple responses to one item, may elicit different responses than separate scales. For example:

  'Indicate how important and how satisfied you are with each of the following using a scale of 1 to 5':

	(Unimportant) 1 - 2 - 3 - 4 - 5 (Important)	
	(Dissatisfied)	(Satisfied)
a. SRI technology of paddy	____________	____________
b. BPT 5204 variety of paddy	____________	____________
c. Sub-surface drainage technology	____________	____________

The above question is in a juxtaposed scale format. The advantage is that, it can force respondents to think and compare the importance and satisfaction for each item because, they are side by side. However, this questionnaire format has been shown to cause confusion among respondents who are less educated, in which case the following question, with parts A and B in a separate scale format, may be preferred:

A. *Indicate how important each of the following is to you using a scale of 1 to 5:*

(Unimportant) 1 - 2 - 3 - 4 - 5 (Important)

a. SRI technology of paddy ________________
b. BPT 5204 variety of paddy ________________
c. Sub-surface drainage technology ________________

B. *Indicate how satisfied you are with each of the following using a scale of 1 to 5:*

(Dissatisfied) 1 - 2 - 3 - 4 - 5 (Satisfied)

a. SRI technology of paddy ________________
b. BPT 5204 variety of paddy ________________
c. Sub-surface drainage technology ________________

☆ Avoid using skipping questions in the questionnaire, as they may lead to the loss of important information because of logical errors in the flow of questions. For example:

1. *Whether you practice SRI production technology of paddy?*

   [ ] Yes [ ] No (Go to question 7)

2. *Do you interact frequently with the local Agricultural Officer?*

   [ ] Yes [ ] No

3. . . .
4. .....
7. *Do you use internet in accessing marketing information?*

   [ ] Yes [ ] No

The above questions, because of errors in the skipping sequence, will not collect information pertaining to second question, for those who are not practicing SRI production technology of paddy. Pretesting of the survey instrument should prevent such a bias.

☆ Alignment of the response choices to the left or right side of the possible responses can produce a bias (Dillman, 1978). For example:

'In terms of yield, SRI technology of paddy production is ——'

Excellent .... [ ] Good .... [ ]

Fair .... [ ] Poor .... [ ]

'In terms of yield, SRI technology of paddy production is ——'

[ ] Excellent [ ] Good

[ ] Fair [ ] Poor

It has been suggested that, placing the response choices to the right side of (i.e., after) the list of possible responses will result in fewer errors on the part of interviewers in a personal or telephone interview. This facilitates subsequent data input directly from the questionnaire. For mailed and other self-administered questionnaires, placing the response choices to the left of (*i.e.,* before) the possible responses makes it easier for the respondent to circle or check them.

☆ Answering of questions in a questionnaire may include 'No-saying' (also known as nay-saying) and 'Yes-saying' (also known as yea-saying). That is, some respondents tend to answer 'No' to all questions or to answer 'Yes' to all questions. For example:

What are the reasons for adopting SRI production of paddy in your farm?

	Yes	*No*
❖ It yields more output......................	[√]	[ ]
❖ Demands less irrigation water.....	[√]	[ ]
❖ Requires less seed rate...........	[√]	[ ]
❖ More manures application enhances soil fertility......	[√]	[ ]
❖ Neighbouring farmer is adopting............	[√]	[ ]
❖ Local Agricultural Officer suggested me...............	[√]	[ ]
❖ Lack of adequate irrigation facilities...............................	[√]	[ ]
❖ Technology is easy to understand and adopt..............................	[√]	[ ]

In the above example, the respondent chooses yes for all items. One way to reduce the no- or yes-saying bias is to use both positive and negative statements about the same issue in a battery of items to break the pattern), as in the following example:

	Yes	*No*
1. Demands less labour	[ ]	[√]
2. It yields more output	[√]	[ ]

☆ The questions that are to be framed in the questionnaire should consider the respondent's sub-conscious reaction. The following are the few cases:

❖ Respondents usually avoid ends of scales in their answers (called as End aversion). That is, they tend to try to be conservative and wish to be in the middle. For example:

SRI production technology of paddy is cost-effective. Do you agree?

[ ] Strongly agree [ ] Agree [ ] Disagree [ ] Strongly disagree

Respondents are more likely to check 'Agree' or 'Disagree' than 'Strongly Agree' or 'Strongly disagree'.

❖ Respondents tend to give positive responses regarding questions on satisfaction (Positive Skew). So, questions on 'satisfaction' may cause problems.

Example	*Yes*	*No*
1. *Are you satisfied with cooperation from your family members?*	[√]	[ ]
2. *Are you satisfied with your profession?*	[√]	[ ]
3. *Are you satisfied with your marriage life?*	[√]	[ ]

- The questions that are to be framed in the questionnaire should also consider the respondent's conscious reaction. The following are the few cases:
  - Respondents try to appear sick to qualify for support. This is called Faking bad (also known as hello-goodbye effect). For example:

    'Which of the following problems do you have in the production of paddy?

    Respondents tend to provide a wide range of problems than they have.
  - Respondents tend to underreport on socially unacceptable problems or disorders. For example:

    'Do you sexually harass women labour working in your farm?'

    For the above question, even if the respondent sexually harasses women labour working in his farm, he will tend to underreport the answer.
  - For socially unacceptable exposures like smoking, consumption of liquor, drug abuse etc., respondents tend to underreport the answers. For example:

    'Do you consume liquor every day?

    A direct and intruding question like the one above may lead to underreporting by the respondent. To elicit the requisite information, the researcher has to ask the question in a different way say, asking about that behaviour in the past before asking about current practices, because past events are less threatening. For example:

    1. Are you consuming liquor since past five years?
    2. Last year, how much money you spent on consuming liquor?

       In view of the underreporting of answers, the researcher should place such questions toward the end of the questionnaire so that, they will not affect other questions. Also consider using anonymous, mailed questionnaires instead of face-to-face interviews.
- Avoid collecting the data from 'Proxy respondent' (also known as Surrogate data). For deceased cases or surviving cases (e.g., brain tumors), whose ability to recall details is defective, soliciting information from proxies (e.g., spouse, family members) may result in differential data accuracy. In general, it is not advisable to ask someone to answer attitudinal, knowledge, or behavior questions for others. For example:
  1. *What is your wife's occupation?*
  2. *Please tell me how afraid your wife is of getting cancer?*

     Not at all afraid   Extremely afraid

     1   2   3   4   5

  Note that, the first question is appropriate, but the second question is not.
- ***Format Considerations:*** The researcher must note some important considerations for designing the questionnaire, as he should make it attractive to stimulate the respondents in answering the questions. These considerations include:
  - All the questions must be neatly typed without any grammatical or typographical errors.
  - Consistency of font and space between the lines should be maintained for questions, instructions, headings and transitional statements.
  - Titles and section headings are usually in a larger font than the questions and response categories.

- The questions should be numbered serially throughout the entire questionnaire.
- Data entry codes that appear on the questionnaire should not interfere with the questions or the numbering of the questions.
- It is a good idea to provide titles or headings for each section of the questionnaire. For example:

  INFORMATION FOR RESPONDENT

  SECTION 1: General Information about the respondent

  SECTION 2: Statements of production of crops

  SECTION 3: Statements on marketing pattern of agricultural produce

  SECTION 4: Statements on labour employment in the farm
- The front cover of the questionnaire is extremely important, since it must attract the attention of the respondent.
- To make the questionnaire attractive, colour print on white papers is used and even in some cases, colour papers are used. Different sections of the questionnaire are printed on coloured paper, but the response boxes are shown in white or a lighter shade of the same colour. This helps the respondent in finding the correct placement of the answer for each question.
- Graphics can be used to enhance the quality of questionnaire.

☆ The framed questions should be clear and meaningful to the respondents. This is essential because, the researcher can generate reliable data, only if the respondents can easily understand the meaning of the words used in the questions. Respondents will be willing and even provide additional information, if they clearly understand what is being asked of them. Furthermore, it is essential that, respondents must understand the questions in the way intended by the researcher. So, the wording of questionnaire is utmost important and hence, the researcher should note the following aspects:

- Meaningful, simple and everyday words should be given in the questions
- Double meaning words should be avoided
- The language skills of respondents should be considered when developing questions. That is, questions should be framed in the language in which the respondents are familiar.
- Four key principles should be bear in mind, when designing the questionnaire: *viz.*, Clarity (ensure the questions are easily understandable), Brevity (keep the questionnaire short), Neutrality (always phrase questions neutrally) and Simplicity (use simple language):
  - ❑ ***Clarity:*** The researcher must ask himself, whether the intended audience will fully understand the questions. As far as possible, try to keep the questions short.
  - ❑ ***Brevity:*** The questionnaire must be short. Note that, the shorter is the length of the questionnaire, the higher the response rate will be. Further, short questionnaires are also easier to compile and analyze.
  - ❑ ***Neutrality:*** The questions framed in the questionnaire must be neutral *i.e.*, they should not lead the respondent towards any favoured answer. For example: *'SRI production*

technology of paddy is beneficial in the contexts of both cost-effectiveness and higher output. Do you agree with this statement'?

a. Yes b. No

c. Not sure

This phrasing of question may result in a bias towards a positive response, as it firmly states a 'truth' that respondents may be reluctant to deny. A better way of phrasing the above question would be: *'Could you tell us your views on the benefits of SRI technology of paddy in the contexts of cost-effectiveness and higher output'?* Do you believe it to be:

a. Strongly beneficial b. Beneficial

c. Not very beneficial d. Not at all beneficial

e. Not sure

This neutrality in the question is important, as it directs the respondents in giving the correct answer.

- ***Simplicity:*** Phrase the questions in a simple language than be easily understandable by the respondents. Note that, ask questions in a logical sequence and this generates interest among the respondents in answering the specific questions. If the opening questions are interesting, simple and easy to answer, respondent's cooperation and involvement can be maintained throughout the questionnaire. So, it is important to ask general questions before specific questions in order to obtain the freest opinion of the respondent. This procedure is known as Funnel technique.

- The questions framed in the questionnaire must be as specific as possible to ensure that respondents understand exactly what is required from them. For example, look at the following question: *What is your income?* This question is not so easy to answer because, the word *'your'* does not indicate, whether it is the income from the agri-business alone or non-agricultural income or total income of the family or total income of the family before taxes and deductions if any or total income of the family after taxes and deductions if any etc. So, the better way to ask the above question (assuming that 'household' has been defined for the respondent) is: *'What was your household's total income, from all sources after taxes and deductions, for last year?*
- As far as possible, avoid using technical words with which respondents are unfamiliar. If such words are inevitable to be included in the questions, give the meaning (or define) of these words in the reference guide.
- When the respondents are familiar with Abbreviations and Acronyms, these can be used in the questionnaire. For example, 'TV' is the popular acronym used for 'Television' and this can be easily understood by the respondents. However, if the abbreviation is given with reference to an Organization or Institute, which the respondents are not familiar, its complete or full form is to be given. Say, for example, the respondents may not be aware of the expansion of the term 'WALAMTARI'. In such case, give its full form (Water and Land Management Training And Research Institute) when it appears for the first time in the questionnaire or even in the reference guide.
- Ensure that, questions framed in the questionnaire are applicable to all the respondents. This generates common information, reduces the length of compilation of data, cuts

down time the respondents spare for providing the data, lowers the cost of the survey etc. For example, the following question is applicable to all the farmers: *'How much area is cultivated during Kharif season in the current year?* Note the following question is not applicable to all the farmers: *How much profits you realized by taking crops during Kharif season in the current year?* This is because, some farmers may incur losses due to various factors like pests and disease incidences, low market price, low output due to soil salinity etc.

- ❖ For some variables like land area, cost of cultivation, seed rate etc., the units of measurement are extremely important. In such case, the researcher should clearly specify the units of measurement in the questionnaire. For example, *'what is the size of your land holding'?* If this question is posed on the farmers, some farmers may express the size of land holding in acres, some in hectares, some in cents, some in sq. meters, some in sq. yards etc. So, the right way of asking the above question is: *what is the size of your land holding (hectares)'?*
- ❖ As discussed in designing the schedule, even in questionnaire also, avoid using double-barrelled questions, avoid leading questions, avoid using double negatives etc.
- ❖ Try to avoid skip rules in framing the questions. For example: *'If you answered 'Yes' to this Question, go to Question 10' in the next section'*. This leads to confusion on the part of the respondents and they lose interest in completing the filling of questionnaire.
- ❖ The questions should be as brief as possible with shorter sentences and ensure that the respondents can read well. Respondents should be addressed in the second person (*you*) in a friendly or personal tone. Long, boring or repetitive questions may result in a loss of interest in answering or produce inaccurate responses. Similarly, a questionnaire that is too long can lead to respondent fatigue, loss of interest and so poor quality data; too short and it may mean that there is no time to build rapport.
- ❖ When designing the questionnaire, it is necessary to prevent later complications which might arise at the coding stage. Therefore, the researcher should note the following points:
  - ❑ Locate coding boxes neatly on the right hand side of the page.
  - ❑ Allow one coding box for each answer.
- ❖ It is advisable to pre-construct a tabulation sheet, anticipating how the data will be tabulated and interpreted, before the final form of the questions is decided upon in the questionnaire. Working backward from a visualization of the final analysis of data is an important step in avoiding ambiguity in questionnaire form. If mechanical tabulating equipment is to be used, it is important to allow code numbers for all possible responses to permit easy transfer to machine-tabulation cards.
- ❖ Like pre-testing of schedule, pre-testing of questionnaire is equally important before using the designed questionnaire on the selected sample respondents. Pre-testing of questionnaire is done on the units other than selected sample. The main objective of pre-testing of the questionnaire is to judge its suitability and utility. In order to reduce the response error, it is necessary that the questionnaire should be tested before it is actually used as a device on the selected sample. Thus, pre-testing is nothing but testing the validity of questionnaire before it is actually used. The following aspects should be tested in pre-testing of questionnaire:

- Whether the questionnaire meets the objective of research study? It means that, it should help in getting maximum possible data.
- Poor response in pre-testing implies modifications are required in the questionnaire.
- Whether the framed questions reduce the response error?
- To identify difficulties and errors with the questions.
- To identify whether the question order affects the interpretation of questions.
- Whether instructions given in the questionnaire are clear?
- How respondents feel comfortable in answering the questionnaire?
- Whether the questionnaire is respondent friendly in eliciting all the requisite information?
- Time and cost constraints in getting back the filled-in questionnaire from the respondents.

So, pre-testing is a 'dress rehearsal'; as it duplicates the final survey design on a small scale. It makes the researcher confident in imposing the questionnaire now after its pre-testing on the selected sample for eliciting the requisite data and information. Note that, questionnaire design is an iterative process *i.e.*, throughout questionnaire development and testing, changes are continually being made. As and when the researcher decides that, the final stage of the process come, he will not make further changes and impose the pre-tested questionnaire on the selected sample.

❖ Layout of the questionnaire is equally important in eliciting good and prompt responses from the responses. There are two layout issues *viz.*, the overall physical layout of the questionnaire and the format of questions included in the questionnaire.

- Overall physical layout include, both from ethical and practical point of views, the researcher need to provide sufficient background information about the research in the questionnaire. Each questionnaires should have a covering letter or page, on which some information appears about: Type of Organization or Institute, Name of the Research Project, Name of the Investigator, Year of Study, Purpose of the study, Sponsor of the research project, Return address and date, Confidentiality etc. The researcher must also submit that, the data or information provided by the respondents will be utilized for research purpose only and not for any personal usage.
- Regarding the format of the questions, the above discussed guidelines must be followed meticulously. On the first or face sheet of the questionnaire, put the time and date of answering the questions, name of the respondent, identification number of the respondent, telephone number etc. Indicate the instructions and guidelines to be followed by the respondents in answering the questionnaire. These instructions indicate what is expected from the respondents by the researcher. Note that, specific instructions should be given for each question, where the style of questions varies throughout the questionnaire. For example: *'Put a tick mark in the appropriate box'* and *'Put a circle around the relevant number'* etc. Each questionnaire should be numbered and this help the researcher to verify, whether all the filled-in questionnaires are received back from the respondents and also helps to locate the respondent in accordance with the serial number given. Print these instructions and guidelines in a different style from question to distinguish them.

Layout is very important especially for mailed questionnaires because, there is no friendly interviewer to interact with the respondents. Instead the questionnaire's appearance persuades the respondents. Note that, at the end of the questionnaire, attach another letter thanking the respondent for sparing his valuable time and for providing the requisite information. Thus, first the researcher should attach a letter to the (mailed) questionnaire indicating the importance of the survey and this will sets the tone for the whole questionnaire. The letter should include the following information:

- Indicate the title or subject of the survey
- Indicate the sponsor of the survey
- Explain the purpose of the survey
- Request the respondent's co-operation
- Explain the importance of completing the questionnaire
- Explain how the survey data will be used in the report
- Explain how the respondent can access the survey results or findings
- Date for completing the filling of questionnaire
- Address for returning the filled (mailed) questionnaire
- Attach a letter rendering 'Thanks' to the respondent for actively participating in answering the questionnaire

As mentioned earlier, the survey should be interesting or rewarding for the respondents as well as participants (interviewers). A good topic, well-written instructions, easy, quick questions and perhaps recognition given for the respondents (say, acknowledging them in the report) can make the respondents' experience more pleasurable and ensure a reasonably high response rate. Note that, at the end of the questionnaire, the researcher should allow respondent to tell him something that he don't know. No doubt, many questionnaires employed in the survey of social sciences research are well constructed and get useful information, yet fail to reveal a critical fact, because the researcher didn't know to ask about it. For example, if the researcher asks a (open-ended) question on the respondent say, *'if you are asked by the district collector, what would you suggest to improve organic farming'?* This question is designed to get the respondent's imagination going and is likely to generate the sort of creative suggestions that more structured questions cannot. Before going for data collection through employing questionnaire, the researcher should himself the following:

- What information does the questionnaire aim to collect?
- At whom is it aimed?
- Is it for self-completion, or would an interviewer fill it in?
- What types of questions are used?
- How is the questionnaire set out?
- Is it easy to fill in?
- Did the respondent can understand the questions?
- How long did it take the respondent to complete?
- What sort of questions comes first?

- Are the questions in a logical order?
- Were any of the questions sensitive or too personal?

c. Similarities between Schedule and Questionnaire

The following points highlight the similarities between schedule and questionnaire

- Both involve a set of structured questions relating to the research problem.
- The questions in both these tools proceed from general to specific, simple to complex, move smoothly from one topic to another etc.
- In both, certain types of questions have to be eliminated such as vague and ambiguous questions, embarrassing questions, loaded and leading questions, questions eliciting no response etc.
- In both, pilot studies and pre-tests are necessary for formulating the instrument and for bringing them to the final form.
- Both have to go through the same stages of development.

d. Schedule vis-à-vis Questionnaire

Both schedule and questionnaire are employed to collect information through a set of questions from the respondents. Unlike questionnaire, there is personal interaction between the researcher or interviewer and respondent by employing schedule in data collection. Hence, it is often remarked that 'schedule is an informal questionnaire'. The following (Table 9.2) are the major differences between schedule and questionnaire:

Table 9.2: Differences between Schedule and Questionnaire.

Schedule	*(Mailed) Questionnaire*
Schedule is employed for data collection in Agro-Economic research studies.	Questionnaire is employed for data collection in Agricultural Extension research studies.
The format of questions are filled by the researcher or interviewer.	The format of questions are filled by the respondent himself.
There is personal interaction between the researcher and respondent in filling the schedule.	Personal interaction need not be required between the researcher and respondent in filling the questionnaire. Questionnaire can be sent *via* mail or post to the respondent to elicit data or information from the respondent.
If the researcher has to collect data from each and every respondent through personal interactions, it consumes more time to complete the total survey.	Data collection through questionnaire takes less time to complete the total survey, as the researcher can simply mail it and there is no need for personal interactions with the respondents.
Considering the time element involved in filling each schedule, it takes less time on the part of the researcher to fill the schedule because of two reasons: the researcher himself fill the schedule and that too in the presence of respondent. This makes the researcher to act quickly in accessing the desired information.	It takes more time to fill each questionnaire by the respondent, as he will fill the same at his convenient time.

Contd...

Table 9.2–*Contd...*

Schedule	*Questionnaire*
Data collection through schedule is costly, as the researcher has to move everywhere to meet the respondents personally.	Data collection through questionnaire is a cheaper method than schedule.
A schedule has more data collecting ability than a questionnaire.	A questionnaire has less data collecting ability than a schedule.
A schedule is a problem oriented data collecting method.	A questionnaire can cover a very wide field of data.
Schedule is used for collecting data from relatively smaller areas and smaller sample.	Larger areas and larger sample can be covered easily for data collection through questionnaire.
A schedule has to be explained by the researcher or investigator to elicit data or information from the respondent.	A questionnaire take for itself and is self-explanatory.
In case of schedule, researcher should ensure the filling of all the questions.	Questionnaire can be returned without answering all the questions.
Data collection through schedule can be employed both on literate and illiterate people, as the schedule will be filled by the researcher through personal interaction with the respondent.	Data collection through questionnaire can be employed only on literate people.
There is no risk of incomplete and wrong information in schedule.	Risk of incomplete and wrong information is more in questionnaire.
Physical appearance of schedule is not necessarily be attractive, as it is filled by the researcher.	Physical appearance of questionnaire has to be attractive.
Success of schedule depends on honesty and competency of enumerator.	Success of questionnaire depends on its design.
In schedule, there is no chance of non-response.	In questionnaire, if the questions are complex, there are chances of non-response.
The return rate of schedule is short compared to schedule, as the researcher himself informally interacts with the respondents.	The return rate of questionnaire is long compared to schedule.
The researcher can directly motivate the respondent in filling the responses, as there is personal interaction between them.	Appeal letter should be used along with questionnaire, so as to motivate the respondents to fill it.
Schedule is comparatively more difficult to plan, construct and administer than questionnaire.	Questionnaire is comparatively easier to plan, construct and administer than schedule.
Uniformity cannot be ensured in case of schedule.	Questionnaire can be used as a standardized device and hence, uniformity can be ensured.
There is the advantage of cross checking of information in data collection through schedule.	The researcher does not get the opportunity to cross check the information provided by the respondents through questionnaire.

Contd...

Table 9.2–*Contd...*

Schedule	*Questionnaire*
Schedule comparatively puts more pressure on the respondents in providing the requisite information in time.	Questionnaire puts less pressure on the respondents in providing the requisite information.
Response rate is high (or Response error is low).	Response rate is comparatively low (or Response error is high). The return rate of questionnaire is only 30-40 per cent.
There is personal interaction between researcher and respondent and hence, no sample is missed out from the survey.	If the mailing address is wrong, some eligible sample is missed from the survey.
Due to personal interaction between researcher and respondent, the misunderstandings of the respondents regarding the questions in the schedule can be corrected.	Different respondents interpret the questionnaire in a different way and this may affect the quality of survey. Further, these misunderstandings cannot be corrected.
Proper rapport is built between researcher and respondent, when data collection is planned through schedule.	No such rapport is built here.
Difficulties and problems encountered by the respondents in answering the schedule can be clarified on spot by the researcher.	There is no or very less chance for clarifying difficulties and problems encountered by the respondents in answering the questions given in questionnaire by the researcher.
There is an opportunity of getting additional information from the schedule through personal interaction between the researcher and respondent.	There is no possibility of getting additional information from the questionnaire.
Researcher interacts personally with the respondent and hence, he is sure about getting the information as provided by him.	Researcher is no sure whether the respondent actually filled the mailed questionnaire.
The respondent will answer all the questions due to personal interaction between the researcher and respondent.	Many questions may remain unanswered.
The reliability of respondent's background information can be verified.	The reliability of respondent's background information cannot be verified.
Deeper probe is possible through data collection by schedule.	This is comparatively less in case of data collection through questionnaire.
There is possibility of getting an in depth answer for a question.	There is no possibility of getting an in depth answer for a question.
Human element finds recognition.	Human element finds missing.
If the respondent is hesitant in providing certain information, he can be convinced if schedule is employed for data collection.	If the respondent is hesitant in providing certain information, he cannot be convinced, if mailed questionnaire is employed for data collection.
There is scope of interviewer's bias in data collection.	There is no such bias in data collection.
There is hesitancy in answering embarrassing questions because there is direct face to face interaction between the researcher and the respondent.	Even embarrassing questions may also be answered because, the assurance of keeping the confidentiality of information is there.

Contd...

Table 9.2–*Contd...*

Schedule	*Questionnaire*
In case of data collection through schedule (interview method), the schedule remains in the hands of the interviewer, and the respondent does not know what question is likely to follow. Therefore, in this method of data collection, there is a control in the question order.	In a mailed questionnaire, the respondent usually reads the whole of the questionnaire prior to answering the questions. The latter questions may influence the answers to the earlier questions thereby, it is likely to bias the data.
Schedule is based on the direct method.	Questionnaire method is an indirect method of collecting the data.
Information collected through schedule is more reliable.	Information collected through questionnaire is less reliable compared to schedule.
In general, questions that are included in the schedule are short and pointed.	In general, questions that are included in the questionnaire are wide and detailed.
In schedule method, observation is not the sole criterion, but the exchange of thoughts between the researcher or interviewer and the respondent (interviewee) counts much in getting the data or information.	In questionnaire method, data provided by the respondent (written by the respondent himself) is the sole criterion in getting the requisite data or information.
Schedule method can be easily and safely used in sampling method of research. Since there is personal interaction between the researcher and respondent, the former can furnish correct information from the latter.	Questionnaire method is not useful for sampling method. Since the investigator is not present, the respondent who has to furnish answers is sometimes not able to answer all the questions correctly and as required. Due to this limitation, the data collected is not reliable.
In schedule, questions are framed keeping in view of the skills and difficulties of interviewers.	In questionnaire, questions are framed keeping in view of the knowledge or education standards of the respondents.

Note that, in its form and content, a questionnaire is similar to the schedule. Like the schedule, a questionnaire can be structured and unstructured. As in schedule, in questionnaire also, the pattern to be followed for designing the questions and wording of the questions is the same. But, the questionnaire is to be made more attractive and appealing compared to schedule, as it has to stimulate the respondent to fill the questions himself.

e. Advantages of Questionnaire

Unlike Schedule, questionnaire method is an indirect method of data collection. Though indirect, it offers the following advantages as compared to other methods:

- ☆ Questionnaires are economical in terms of materials, money and time in relation to the generation of vast research data and information.
- ☆ When compared with other methods of data collection (Schedule, Case study, Observation), the mailed questionnaire method is the cheapest and quickest method. The cost in this method is only that of getting the questionnaire prepared and its postage to the respondent.
- ☆ It permits wide coverage of sample in short span of time, unlike schedule.
- ☆ It helps in conducting in depth study.
- ☆ Demands less skill on the part of the researcher in administering the data collection compared to other methods of data collection like Schedule, Observation, Case study etc.

- This method of data collection is indispensable, when the respondents are scattered over a wide geographical area.
- Compared to other methods of data collection like Schedule, Observation, Case study etc., employment of questionnaire method saves lot of time on the part of the researcher. This method also ensures approaching a large sample at a point of time.
- Since the questionnaire is prepared by following systematic guidelines, all the questions are self-explanatory to the respondents and this helps the researcher to elicit requisite information. Moreover, the respondents are free to give information as and when they want during their leisure time. Because of this freedom, the information that is provided is more dependable and reliable.
- Since there is no personal interaction between the researcher or interviewer and the respondent, the latter is free from external influences. Hence, the respondent will provide reliable and unbiased meaningful information based on his knowledge, views and attitudes.
- Special responses related to confidential issues like sex, caste, religion, drinking habits, marital relations etc., can easily be obtained by keeping the name of respondents anonymous.
- There are less chances of error in the data and information, as the questionnaire is filled by the respondent himself. So, the information generated in this survey method is more original.
- As discussed earlier, in the questionnaire, there are several sections and these are divided in accordance with the objectives framed for the study. So, the obtained information from the respondents is also according to the framed objectives.
- Each respondent receives the same set of questions phrased in exactly the same way. Questionnaires may, therefore, yield data more comparable than information obtained through a schedule.
- If the questions are highly structured and the conditions under which they are answered are controlled, then the questionnaire could become standardized. That is, a questionnaire can be easily standardized. For example, if every respondent is asked the same question in the same way, the researcher, therefore, can be sure that, everyone in the sample answers exactly the same questions, which makes this a very reliable method of research.
- Questionnaires can be filled out whenever the respondent has time. Thus, there is a better chance that respondents will take time to think about their response. Many hard-to-reach respondents place high value on responding to surveys at their own convenient time.
- Absence of researcher or interviewer can induce respondents to reveal sensitive or socially undesirable information.

f. Disadvantages of Questionnaire

This method of data collection has the following disadvantages:

- If the respondents do not show interest in filling the questionnaire, it will affect the data availability for the research project. Sometimes, due to low interest, the respondents were reluctant in providing accurate information for the questions framed in the questionnaire.
- Some respondents give incomplete answers and this makes the researcher difficult to compile the information. That is, mailed questionnaire has very low completion rate either because the respondents did not understand the questions or they shied away.

- Some respondents fill the questionnaire so badly that, even they themselves find it difficult to read their own handwriting.
- Some respondents do not even send back the questionnaire and this affects the completion of research study. That is, mailed questionnaire has very low rate of return of the filled questionnaires.
- If the researcher requires cross-checking of information with the respondents, it is difficult by this method of data collection. That is, this method is very rigid and no alteration can be introduced.
- Prejudices and bias of the researcher influences the questions framed by the researcher in the questionnaire.
- In some cases, the questionnaire itself is incomplete. That is, all the aspects related to the framed objectives in the study are not considered for framing the questions. In such case, the data collected are incomplete for the study.
- Since there is no face to face contact between the researcher or interviewer and the respondent, it may result in manipulation of data and information by the respondents.
- There is no scope for checking the truthfulness of the answers.
- This method of data collection cannot be employed over illiterate farmers and small children.
- The researcher keeps on waiting for the return of the filled questionnaires. When enough response is not there, then the reminders are sent, and again there is a waiting time. All this adds to the cost of the study.
- Respondents may have different interpretations of the same question. Due to the absence of personal interaction with the interviewer or researcher, the respondents are unable to get any help for needed clarifications.
- In a mailed questionnaire, the respondent usually reads the whole of the questionnaire prior to answering the questions. The latter questions may influence the answers to the earlier questions thereby, it is likely to bias the data. In case of data collection through schedule (interview method), the schedule remains in the hands of the interviewer, and the respondent does not know what question is likely to follow. Therefore, in data collection through schedule, there is a control in the question order.
- In general, lengthy mailed questionnaires lead to low response rate. (A general rule of thumb is that, the length of questionnaire should not exceed six pages).
- The researcher does not know about who filled the questionnaire.

Having understood the advantages and disadvantages of using questionnaire in data collection, it is essential to check its limitations through proper construction of questionnaire. Considering the above advantages and disadvantages, questionnaire is said to be the most 'used and abused' method of data collection by the lazy man. This calls for developing the questionnaire in a scientific manner through considering the following aspects:

- The language and the wordings of the questions should be kept interesting enough for the respondents to give replies. Since, in questionnaire method, the respondent gives reply from a distance, the researcher has to consider the psychology of respondent in framing the questions in a logical sequence. This framing of questions in a questionnaire in a proper manner refers to Construction of questionnaire.

- The questionnaire prepared must ensure reliability. The reliability of a questionnaire is the ability of the questionnaire to give the same results when filled out by like-minded people in similar circumstances. Reliability is usually expressed on a numerical scale from zero (very unreliable) to one (extremely reliable).
- The questionnaire prepared must ensure validity. The validity of a questionnaire is the degree to which the questionnaire is actually measuring or collecting data about what researcher think it should be measuring or collecting data about.
- While designing the questionnaire, the intellectual level of the respondent has to be kept in view.
    - The questions framed in the questionnaire should cover all the aspects of the research problem.
    - The researcher should use his past experience in the formulation of questionnaire. The learnings from past experience enable the researcher to check mistakes or shortcomings of similar questionnaire. If he is framing the questionnaire for the first time, he should contact other researchers in finalizing the questionnaire.
- The following steps are to be followed meticulously in drawing quality information through questionnaire method:
    - Framing the questions in questionnaire scientifically (logically) to meet the data requirements.
    - Pre-testing of the questionnaire to judge its suitability and utility.
    - Compile the names and addresses of the respondents and despatch the questionnaire to all the respondents.
    - A code or serial number is given to each respondent.
    - Receipt of questionnaire from the respondents should be recorded date wise.
    - Proper follow up for receiving of adequate response. If proper response is not coming even after third reminder, change the names of respondents and follow the above procedure.

g. Physical Appearance of the Questionnaire

The physical appearance of a questionnaire can have a significant effect upon both the quantity and quality of data or information to be obtained from the respondents. Note that, the quantity of data to be collected from the respondents is a function of the response rate. Even data quality can also be affected by the physical appearance of the questionnaire. An attractive questionnaire stimulates the respondents to answers the questions in the questionnaire and even influence the completion of task accurately. The following are the important guidelines that are to be followed in designing the physical appearance of the questionnaire:

- Instead of loose papers, the researcher should design the questionnaire as a booklet, as it makes easier for interviewer or respondent to progress through the document.
- The questions framed in the questionnaire must be clear without any ambiguity. This help to improve the ease with which respondents are able to complete a questionnaire.
- With the intention of lessening the number of pages of questionnaire sometimes, the researcher will place too much information on one page by reducing the size of the font. But, this is

counter-productive since it gives the impression that the questionnaire is being complicated. So, questions should be framed with well-adjusted space so that, it makes the respondent easier to fill the questionnaire.

- Colour coding can help in the administration of questionnaires. If the respondents deals with two types of respondents, say men farmers and women farmers, it makes both the researcher and the respondent easier in handling the questionnaire.
- The instructions for answering the questions should be placed alongside the questions to which they pertain.

h. Response Error

As discussed under Chapter 8, one of the sources of non-sampling error was 'measurement error', which is the *difference between the recorded response to a question and the 'true' value*. In the questionnaire design literature, this error is more commonly referred to as Response error. Since in Agricultural Extension studies, the data are mainly collected through employing questionnaires, it can be a major source of response errors. So, designing the questionnaire (discussed above) is extremely important in minimizing this response error.

Sources of Response Error

If the above discussed guidelines are not followed in designing the questionnaire, it will lead to Response errors. So, response errors can occur anywhere in the question-answer-recording process. The type of questionnaire employed also influences the number of response errors. In general, in the absence of interaction between the respondent and the researcher or interviewer (say, in mailed questionnaires), the response errors are highest. The following are the different types of response errors:

- ***Memory error:*** In general, the respondents answer the questions in the questionnaire based on their recall memory. But, it quite difficult to remember all the past events by the respondents. The response error so resulted is called as Memory error or Recall error. So, this error results in under reporting of events by the respondents in the questionnaire.
- ***Telescoping error:*** In contrast to the above error, sometimes the respondents may report activities that occurred outside the reference period as having occurred during the reference period. This source of error is called Telescoping error. This error results in the over-reporting of behaviours or events. This is again of two types *viz.*, Forward Telescoping and Backward Telescoping.
  - ***Forward Telescoping:*** In this error, the respondents tend to report behaviours or events as they have occurred more recently than they actually did.
  - ***Backward Telescoping:*** This is the reverse of Forward Telescoping. In this case, the respondents may report that behaviours or events occurred longer ago than they actually did.

In general, the greater the time gap between the occurrence of an event and responding time by the respondent (*i.e.*, reference period), the greater is the recall loss or response error. However, this is not always true. This is so because, human mind is very peculiar, as sometimes, it will remember child hood events more clearly than the recent past events. Similarly, shorter reference periods tend to increase telescoping errors. For example: *'In last week, how many agricultural related TV programmes have been watched by you?'* Even note that, the interviewer can also be a source of response error. If he asks the same question repeatedly, and if the meaning of the question changes in every instance, it will increase the response error.

Techniques to Reduce Response Errors or Increasing Response Rate

Since response errors are inevitable in questionnaire mode of survey and they are very serious in case of mailed questionnaire, it is essential to impose certain techniques for checking these errors. Response rate is the number of questionnaires returned or completed, divided by the total number of eligible people, who were contacted or asked to participate in the survey. The following are some important guidelines for reducing response errors or increasing the response rate:

- ☆ Questions should be specific, with simple words and having short length so that, the respondent can easily understand them. Some researchers are of the opinion that, lengthy questions call for lengthy answers and this makes the respondent to talk more period of time and this additional talk, in turn, may aid the respondent's recall of past information.
- ☆ The interviewers should be sufficiently trained to elicit good response from the respondents.
- ☆ As discussed earlier under Questionnaire Layout, all the guidelines must be followed to minimize the response error.
- ☆ The reference period can be shortened, if it is determined that the respondent has difficulty remembering all the events that occurred within the reference period.
- ☆ The interviewers must ask the respondents to refer the farm records, if any in providing the requisite information. He may even ask the farmers to maintain a diary for noting down all the major events in the farm business in each year and even important data base.
- ☆ The interviewer can guide the respondents in using a calendar or milestones like festivals or holidays and this can also help to minimize memory error.
- ☆ The cover letter that accompanies the questionnaire or is printed on the first page of the questionnaire is an important means of inducing a reader to complete and return the questionnaire. In the covering letter, the researcher must convey, why the study is important, how the respondent was selected, assuring the confidentiality of information provided by the respondent, motivating the respondent to respond who are the ultimate beneficiaries, impact of the research project on the Nation's economy etc. A personalized letter addressed to a specific individual shows the respondent that he or she is an important part of the research project. Including an individually typed letter on letter head is an important element in increasing the response rate in mailed questionnaire survey.
- ☆ In some cases, the respondent's motivation for returning a questionnaire may be increased by offering monetary incentives or premiums. Such incentive works for all income categories. But, monetary incentives are given generally to respondents, who are economically weak or likely to be influenced by money considerations. This money is given in advance or after receiving the filled questionnaire. Non-monetary inducements may be in the form of offering pens, books, T-shirts and a variety of premiums, as they attracts the attention and creates a sense of obligation. It may also be a letter of appreciation or mentioning of the name in the report of the study or so on. Letters of appeal or appreciation are considered to be very successful method of non-monetary inducement.
- ☆ The questions that are relevant for the objectives framed in the study cannot be manipulated. However, certain interesting questions can be added to the questionnaire, perhaps in the beginning, so as to stimulate the respondents' interest and to induce cooperation.

☆ When response is poor from the respondents, the researcher has to send the reminder. Sometimes, one reminder is sufficient and sometimes more than one. It is proved that, follow-up letters or reminders are of greater value to the respondents. Usually, the reminders may be sent in the form of a post-card. These reminders should be sent at regular intervals and when the researcher does not receive any response from the respondent even after sending three reminders, better drop that respondent from the survey. Sometimes, reminders may be telephonic or through a messenger or an intermediary. Note that, the time of sending a questionnaire and even a reminder should coincide with the leisure period of the respondent and not during busy cropping season.

10

Data Collection

After finalizing the schedule or questionnaire, now the researcher will plan to go for data collection from the selected sample respondents. Note that, the data or information collected by the researcher is not just meant for performing statistical analysis, but will enable the researcher to use the same at various phases of research right from judging the importance of the study to till drawing the policy implications for the study. So, this phase of research process is very crucial, as it forms of the crux of the research investigation. In social sciences research, data is collected from various sources (discussed earlier in the Chapter 8) and a variety of methods or techniques are used in its collection (Chapter 9). Each method has pros and cons. When population is large, it is neither necessary nor advisable to cover each and every member, but a sample survey would suffice. It is quicker, cheaper and, if well-designed, it can provide precise information about characteristics of the population. On the other hand, the researcher uses other important method, 'Case study'. The choice of the method depends upon time, money and objectives of the study. These two methods of data collection in social sciences research are discussed in-detail here under.

10.1. Survey Method

This method of data collection can be classified as Census survey and Sample survey depending upon the size of the population considered for survey. In the former, all the units of population are considered for survey and in the latter a part (representative) of the population (called 'Sample') is considered for survey. The relative advantages and disadvantages of these two methods are already discussed in Chapter 8. The survey method can be administered in a couple of different ways. In one way (known as a structured way), the researcher employs the 'schedule' and informally interacts with the sample respondents to elicit the requisite data or information through posing various questions in relevance to the framed objectives for the study. In the other way, known as a questionnaire, the respondent himself fills out the survey questions on his or her own. These surveys are generally standardized to ensure that they have reliability and validity. Standardization is also important so that, the results can be generalized back to the target population. Whatever may be the tool (*i.e.*, either schedule or questionnaire), it involves the sampling of respondents with a view towards making statistical inferences about the target population. Since survey method is mostly based on a sample of the target population, the success of this method is mainly dependent on the representativeness of the sample for the concerned population.

In agro-economic surveys, this survey method is widely employed by selecting a representative sample from the target population. The researcher often employs an instrument called 'Schedule' for drawing requisite information or data from the sample respondents. As discussed earlier, Schedule is the name usually applied to a proforma containing a set of questions, which are asked and filled by the researcher or an interviewer in a face to face interaction with a respondent. It is a standardized device or tool of observation to collect the data in an objective manner. So, the researcher interacts personally with the respondent to elicit the requisite information or data. This survey method through employing schedule is widely used under the following circumstances:

- ☆ When the researcher wishes to collect data quickly, in depth and at low cost.
- ☆ When the respondents are either literate or illiterate.
- ☆ When the respondents does not maintain farm records.
- ☆ When trained interviewers are available.
- ☆ When statistical analysis is to be employed to generalize the results.
- ☆ When a researcher wants to collect data on a phenomenon that cannot be directly observed.
- ☆ When the respondents are not spread over a wider area.

In Agricultural Extension studies, the researcher usually employs mailed questionnaire as a tool for collecting requisite data or information from the respondents without any face-to-face interactions with them.

A. Techniques of Survey Method

There are mainly four techniques which the researcher can elicit data through the Survey Method

a. Telephone survey
b. Personal survey
c. Mail survey
d. Electronic survey

a. Telephone Survey

As the name itself indicates, here the researcher interacts with the respondent through telephone for obtaining the requisite data. This mode of data collection is generally followed, when the researcher wants to get the information rather quickly and sometimes to clarify doubts or for rechecking the information provided by the respondent through any other survey technique.

Merits

The following are the advantages of Telephonic survey:

- ☆ It is one of the fastest methods of data collection.
- ☆ Even if the respondent is illiterate, the researcher can elicit the requisite information.
- ☆ It permits the researcher to clarify the doubts if any, on cost-effective basis.
- ☆ Response rate of telephone interviewing is better than mail interviews or mailed questionnaires.
- ☆ A good rapport may be built between the researcher and respondent and this facilitates in getting the quality information.
- ☆ This method is more accessible to all the respondents.

Demerits

The following are the disadvantages of Telephonic survey:

- ☆ It is difficult for the researcher to elicit data or information through this technique, especially if the questions are not properly structured.
- ☆ If the researcher requires long descriptive answers for the questions posed by him, this technique is not cost-effective.
- ☆ In some cases, respondents are reluctant to provide the information, as they do not trust the researcher.
- ☆ This survey may interrupt the personal time of the respondents.
- ☆ This survey is hard-to-reach respondents, as many people use call screening to accept only calls that they are expecting.
- ☆ Especially in market research, it is more ideal to conduct a personal survey rather than a telephone survey because, better responses can be elicited when the participants could see, feel or taste the product.

b. Personal Survey

This is the most common or versatile of the all survey techniques. In this technique, the researcher informally interacts with the respondent in eliciting the requisite data or information. The researcher employs pre-tested schedule on the selected respondents to collect the requisite information. The advantages and disadvantages of personal interviewing through schedule were already discussed in-detail in Chapter 9.

c. Mail Survey

This is also a common technique like personal interviewing, but here the researcher employs mailed questionnaire for collecting data or information from the respondents. So, unlike personal interviewing, in this technique, the researcher will not interact informally with the respondent, but post a questionnaire to him so that, the respondent himself will fill and return the same to the respondent. The advantages and disadvantages of mailed questionnaire were already discussed in comparison with schedule survey through Table 9.2 of Chapter 9.

d. Electronic Survey

At present, the world is in the midst of an electronic communications revolution. The rapid evolution of computer hardware and software has provided a catalyst for the researchers in collecting and analyzing the research information in new ways. This electronic collection of data through computers has made research easier than ever. As the internet is gaining popularity world-wide, this electronic survey for data collection is fast growing. This is cheap, fast and does not rely on the traditional geographical boundaries thereby, allowing international data collection. There are three ways of collecting data using the computers *viz.*, Point of contact; e-mail-based; and Web-based based upon the type of technology relied upon.

- ***Point-of-contact:*** This involves having the respondent to fill out an e-survey on a computer provided by the researcher, either on-site or in a laboratory setting (Synodinos, Papacostas, and Okimoto, 1994), for organization members, who do not use computers in their jobs (Rosenfeld, Booth-Kewley, Edwards, and Thomas, 1996). Point-of-contact surveys have also been popular among researchers wishing to have tight control over the context of the study (*i.e.*, lab based).
- ***E-mail based survey:*** E-mail data collection is fairly self explanatory. The respondent receives the questionnaire either embedded in the email or as an attachment and then sends it back after filling it by hitting reply. Note that, these e-mail questionnaires need to be short and simple. E-mail-based surveys are generally seen as being delivered more cheaply and faster than traditional paper and pencil surveys; however, they still require the researcher to manually code the data into a database after receiving completed surveys.
- ***Web-based survey:*** Web-based surveys are constructed using software, which creates a form for respondents to fill in. There are many different types of software to accomplish this and the researcher needs some degree of knowledge to be able to design and create the online questionnaire. Once the questionnaire is ready, the researcher has to think on the important aspect that, how to send it to the respondents either as a link in an e-mail or as a pop-up window on a web site. Both of these have problems, as people often delete e-mail, as they think that, they are not directly relevant to that topic or issue and modern browsers often block pop-up windows automatically. This web-based survey has a number of important advantages over e-mail collection. The survey form can be made more interesting for the respondent. Skips and routing can be built into the form so that, respondents do not have to follow tedious instructions. In addition, these surveys can use animation, voice and video to enhance the user's experience. The researcher can download the data straight into a data analysis package thereby, saving time in coding and typing in responses.

  Electronically administered surveys offer significant advantages over traditional print and telephonic surveys. They include:

  - Web-based and email electronic surveys cost considerably less than mail or telephone surveys. It will eliminate printing, postage and extra personnel costs.
  - An electronic survey is likely to generate a high response rate, as the respondents frequently find that, using their computer to answer questions is easy, quick and convenient.
  - Recipients generally respond to electronic surveys more rapidly than traditional print surveys, and the agencies or companies or researcher can analyze data as it is collected. The faster the data is analyzed, the faster the agency or company or researcher can act upon the new found knowledge.
  - Unlike a printed survey, an electronic survey can be revised and/or customized for different targets quickly and easily.

- ❖ Able to reach a large number of geographically spread population
- ❖ Able to use visual aids in web surveys

However, there are two major problems with this type of data collection: sampling and response rates. For example, if the researcher is conducting a survey on students in a University, he can get their e-mail addresses of the sampling frame easily. Say, if he is doing the survey in a village, it is very difficult for him to get the e-mail addresses of the sampling frame. Further, though internet use is growing, not all members of the population have access to e-mail and the internet and are therefore, automatically excluded. The response rates to such surveys are notoriously low with the added problem of the researcher not knowing who exactly has responded for the survey and what their motives were.

Whatever the technique the researcher employs, the following aspects are common in conducting the survey method:

- ✰ Selection of sample respondents from the target population.
- ✰ Pre-test the questions formulated in the schedule.
- ✰ Select the mode for posing questions and collecting responses.
- ✰ Train and supervise interviewers (if they are involved).
- ✰ Contact sample respondents.
- ✰ Collect the data and check for its accuracy and internal consistency.

B. Types of Surveys

There are different types of surveys as discussed here under.

a. Cross-Sectional Surveys

These surveys are used to gather data or information on a particular sample at a single point of time. For example, a researcher collecting data pertaining to the cost of cultivation of paddy from a sample of 100 farmers in 2012-13 refers to Cross-sectional survey.

b. Longitudinal Surveys

In this survey, the data are collected over a period of time *i.e.*, the data are collected more than once across different years. This survey includes three sub-types *viz.*, Trend studies, Cohort studies and Panel studies.

- ✰ ***Trend Studies:*** This refers to the longitudinal survey that examines changes in sample (or entire population) across time. Each survey collects data from different individuals. For example, in the year 2000, a random sample of 500 farmers are asked about the marketing problems in transacting paddy in Nandyal region of Kurnool district. In 2012, a different set of random sample of 500 farmers are again asked about the marketing problems in transacting paddy in Nandyal region of Kurnool district. The difference in opinions of the farmers in these two years indicates the trend in marketing problems in transacting paddy in Nandyal region of Kurnool district. Thus, trend studies focus on a particular population, which is sampled and scrutinized repeatedly. While samples are of the same population, they are typically not composed of the same people. Trend studies, since they may be conducted over a long period of time, do not have to be conducted by just one researcher. A researcher may combine data from several studies of the same population in order to show a trend.

- ***Cohort Studies:*** This is also a longitudinal survey that examines the changes in cohort across time. So, in each study, the data are collected from individuals within the same cohort. A cohort is a group of people, who share a common characteristic or experience within a defined period (*e.g.*, people born in the same year, 100 students graduated from the college in the same year, patients are exposed to same drug or vaccine). Thus, Cohort studies also focus on a particular population, sampled and studied more than once. But cohort studies have a different focus. For example, a sample of 100 students of 2007 batch in ANGRAU are questioned by a researcher regarding their job prospects in agricultural and allied sectors. Five years later, the researcher could question another sample of 100 students of 2007 batch students on the same aspect and studies any changes in the job prospects. So, a cohort study would sample the same batch every time. Note that, if the researcher surveyed the students of 2012 batch five years later, it would be a trend study and not a cohort study.
- ***Panel Studies:*** In Panel study, the researcher examines the changes in individuals over time. That is, he collects the data from the same individuals across time. For example, in 2002, a random sample of 500 farmers are interviewed to know the efficacy of a fungicide in controlling the sheath blight of paddy. In the year 2007, the same 500 farmers are interviewed again to know the efficacy of the same fungicide in controlling the sheath blight of paddy. The difference in opinions is the change in efficacy of that fungicide in controlling the sheath blight of paddy. Thus, panel studies allow the researcher to find out, why changes are occurring, since they use the same sample of people every time. Thus, same sample interviewed several times constitute a panel. These panel studies provide the opportunity for more in-depth analysis, but they are very expensive. Further, they are difficult to conduct and tend to take a lot of time and they suffer from high attrition rates. 'Attrition' is what occurs when people drop out of the study.

C. Merits of Survey Method

As discussed above, this method consists of a number of questions that the respondent has to answer in a set format. This method ensures the following advantages:

- This method facilitates the researcher to collect a large amount of data in a relatively short period of time.
- Surveys are less expensive than many other data collection techniques.
- Surveys can be created quickly and administered easily.
- Surveys can be used to collect information on a wide range of topics such as cost of cultivation, production and marketing problems of farmers, functioning of market functionaries in a regulated market, attitudes, values etc.
- Can be cost-effective compared to in depth interviews.
- As the data can be subjected to statistical analysis, the findings can be generalized to the target population.
- Can be administered even in remote areas, if the survey is conducted through Web, mobile devices, mailed questionnaires, e-mail, telephone etc.
- Efficient at collecting information from a large number of respondents in short span of time.
- High response rate can be ensured through employing schedules.
- Since the schedules and questionnaires are pre-tested, they are relatively free from several types of errors.

D. Demerits of Survey (Schedule) Method

This method of data collection suffers from the following limitations:

- ☆ Compared to questionnaire, employing schedule in a survey method takes more time in completing the survey and also poses more cost burden on the part of the researcher. These are especially true, if the respondents are physically scattered over a wide area.
- ☆ This method demands a large number of well-trained field workers and this also involves great cost on the part of the researcher.
- ☆ The personal presence of researcher at the time of interaction with the respondent may lead to bias in answering the questions by the respondent.
- ☆ When the respondents are spread over a larger area, it is difficult for the researcher to move to distant areas and come in contact with each and every respondent.
- ☆ Personal presence of researcher may pose less comfort on the part of the respondents to deliver information especially personal information.
- ☆ The answers provided by the respondent may not be an accurate reflection of how he truly feels.
- ☆ Respondents' motivation, honesty, memory and ability to respond influence the reliability of data collected through this method.
- ☆ Structured surveys, particularly those with closed-ended questions may have low validity.
- ☆ Although the respondents are chosen through random sampling technique, non-response errors may exist.
- ☆ The response categories given below the closed-ended questions such as 'moderately agree', 'occasionally', 'less frequently' etc., may lead to vague data sets because, they may indicate different meaning to different respondents.

E. Different Measures to Promote Response Rate of the Respondents

From the above discussion, it is clear that, the schedule method can be used successfully in a limited region and on limited number of respondents. Though limited number of sample respondents are selected that too representative of the population, there is another major worrisome factor *i.e.*, significant non-response rate from the respondents especially from mailed questionnaires, telephonic survey etc. The following are the different ways to overcome this problem of low response rate from the respondents:

- ☆ The researcher has to give advanced information to the respondent regarding the conductance of survey on him in due course and its importance. In case of mailed questionnaire, an appealing letter is to be enclosed requesting him to provide the requisite information and due acknowledgements for him in the survey report. An advanced letter correspondence or telephonic message is the best option to inform the sampled respondent about the upcoming survey. At the end of the survey, an appreciation letter is to be sent to the respondent recognizing his cooperation in smooth conductance of the survey.
- ☆ The interviewers should be specifically trained on how to respect the respondents for sparing their valuable time, questions to be posed on the respondents, how to compile and tabulate the information on the computers, gap filling the information from the respondents who were not reached earlier, cross checking the information, if any etc.

- ☆ Prior to collect the information, first the researcher has to give the particulars regarding his name, designation, name of the institute for whom he is working, objectives of the institute and research study and a letter from the head of the institute requesting the participants to participate and cooperate in the survey. This gives a lot of confidence on the part of the respondents to provide requisite data or information.
- ☆ The researcher should clearly explain the respondents that, the information or data provided by them in solely used for conducting the research study only and not for any misuse.
- ☆ If any financial incentives are to be provided to the respondents for accessing the information, some part should be paid in advance and remaining part after the completion of the survey.
- ☆ Above all, the schedule should be respondent-friendly without indulging him to provide any personal information and only the questions pertaining to the research study should be incorporated.

10.2. Case Study

A case study is an intensive analysis of an individual unit say, a person, a group or an event within a specific context. In most cases, a case study method selects a small geographical area or a very limited number of individuals as the subjects of study. Case studies may be exploratory or descriptive or explanatory. Although case study method remain a controversial approach to data collection, it is widely recognized in many social science studies especially when in-depth explanations of a social behaviour are desired. In the words of Thomas, *'Case studies are analyses of persons, events, decisions, periods, projects, policies, institutions, or other systems that are studied holistically by one or more methods. The case that is the subject of the inquiry will be an instance of a class of phenomena that provides an analytical frame — an object — within which the study is conducted and which the case illuminates and explicates.'* Robert K. Yin defined the case study method *'as an empirical inquiry that investigates a contemporary phenomenon within its real-life context; when the boundaries between phenomenon and context are not clearly evident; and in which multiple sources of evidence are used.'* Thus, case study involves an in-depth, longitudinal (over a long period of time) examination of a single instance or event called a 'case'. It provides a systematic way of looking at events, collecting data, analyzing data and reporting the results. So, the thorough study of a 'case' enable the researcher to gain a sharpened understanding of why that event or instance has happened as it did, and what might become important to look at more extensively in future research. Case studies lend themselves to both generating and testing hypotheses. The following are the important characteristics of a case study:

- ☆ It deals with a single unit and the unit may be an individual or a family or an institution or an event or a technology etc.
- ☆ It involves an in depth or deeper study of the selected unit. So, this method studies in depth rather than breadth. Thus, it places more emphasis on the full analysis of a single event or even a limited number of events and their inter-relations.
- ☆ This method deals with the questions of 'what' and 'why' of the unit under consideration. Thus, it is both descriptive and explanatory in its approach. It also explains the place and role of a unit in its surrounding social milieu.
- ☆ It employs several methods of data collection like interview, observation (primary data), reports, records, newspapers, magazines, diaries (secondary sources) etc., to prevent errors and distortions.

- ☆ It perceives the respondent as a knowledgeable person, rather than as a mere data supplying respondent.
- ☆ It considers a typical case for the study.

A. Basic Assumptions of a Case Study

They include:

- ☆ This method studies the selected unit in its totality with reference to a particular time or over time.
- ☆ It considers the individual unit as a representative unit.
- ☆ The environment in which the selected unit is functioning is so complex and hence, it needs an in-depth investigation.
- ☆ The studying of the unit is influenced by the time factor.
- ☆ The fundamental responses from human beings will be more or less the same.

B. Steps Involved in Case Study Method

As mentioned earlier, case studies emphasize detailed contextual analysis of a single event or even a limited number of events or conditions and their relationships. Social scientists, in particular, have made wide use of this qualitative research method to examine contemporary real-life situations and provide the basis for the application of ideas and extension of methods. Many well-known case study researchers such as Robert E. Stake, Helen Simons, and Robert K. Yin have written about case study research and suggested techniques for organizing and conducting the research successfully. The execution of case study method involves the following six steps:

Step 1: Determine and define the research questions

Step 2: Determine type of case study (single or multi-model approach)

Step 3: Preparation to collect the data

Step 4: Collect the data

Step 5: Evaluate and analyze the data

Step 6: Prepare the report

Step 1: Determine and Define the Research Questions

First, the researcher has to choose a case study and while choosing the one he should pose himself the following questions:

- ☆ Can I gain access to that case (say, an individual or a group or a firm or a company etc)
- ☆ Is the data available on the selected subject?
- ☆ Is there enough published information on that case?
- ☆ Is too much information already available on that case?
- ☆ Will I be able to conduct this study in the available time frame?

If the researcher answers 'Yes' for himself on the above questions, he then establishes a firm research focus by framing the research questions about the case or problem to be studied and determining the purpose for the study. As mentioned earlier, the 'case' may be a event or a individual or a family or a technology assessment etc. The researcher investigates the selected case in depth using a variety of data gathering methods to produce evidence that leads to understanding of the case and elicit data or information for the research questions. In general, case study addresses the research questions which begin with 'how' or 'why'. To formulate these research questions, researcher conducts a literature review. This review establishes, what research has been previously conducted and how best he can refine it and thus, enable him to formulate insightful questions about the problem.

Step 2: Determine Type of Case Study (Single or Multi-Model Approach)

After framing the research questions, now the researcher has to determine the approach to be used *i.e.*, whether he has to go for selecting a single case or a multiple case, so as to conduct an in depth study. When using multiple cases, each case is treated as a single case. Each case's conclusions can then be used as information contributing to the whole study, but each case remains a single case. This selection of a single case or multiple cases is a key element in framing the research design. For example, a case study may involve study of a 'single industry' and a 'firm' involved in that industry. This type of case study involves two levels of analysis and increases the complexity and amount of data to be gathered and analyzed.

Step 3: Preparation to Collect the Data

The key strength for conducting a case study is to collect the requisite data from the 'case' so selected. So, the researcher has to prepare in advance what type of data or information he has to collect and what analytical techniques he has to use in answering the framed research questions. Before these, he has to select the data collection tools like surveys, interviews, observation etc. Throughout the design phase, researcher must ensure that, the study is well constructed to ensure construct validity[1], internal validity[2], external validity[3], and reliability[4]. So, in case study method, a large amount of data can be generated from multiple sources. So, advance preparation by the researcher facilitates proper handling of large amounts of data in a documented and systematic manner. Above all, the researcher must be a good listener and must note down, what the words being used by the sample case.

Step 4: Collect the Data

The advanced preparation of the researcher facilitates him to collect and store the data systematically in prescribed formats designed for the purpose. The researcher should not collect data randomly. There needs to be a purpose for collecting certain data. Refine research question/statement, if necessary. He should carefully identify the causal factors associated with the observed phenomenon. Note that, case study research is flexible, but when changes are made, they must be documented systematically. Field note should also be maintained to record the feelings, stories and illustrations of

---

1. **Construct validity** means selecting the most appropriate measurement tool for the concepts being studied. Does your tool really measure what you want to assess?
2. **Internal validity** is another term for using different methodological tools to "triangulate" the data. What other methods can you use to check for the same phenomenon?
3. **External validity** refers to how well the data can be applied beyond the circumstances of the case to more general situations. Can you apply your data across the industry and to others?
4. **Reliability** means the extent to which the results can be repeated in ways that yield the same results. That is, the extent to which the results are accurate and stable. Are you confident that your study can be repeated by others and the results will be the same?

the 'case' and it should be kept separate from the data being collected and stored for analysis. These two sources of data or information should be used, while interpreting the research findings.

Step 5: Evaluate and Analyze the Data

The researcher should compile and analyze the data to answer the research questions. Since case study employs multiple sources of information, it will enable the researcher to triangulate data in order to strengthen the research findings and conclusions. He categorize, tabulate, and analyze the data and if necessary, he can conduct cross-checks to verify the data. Focused, short, repeat interviews may be necessary to gather additional data. The cross-checking of information and analysis of data enable the researcher to answer the original 'how' and 'why' research questions.

Step 6: Prepare the Report

Finally, the researcher has to submit the report. The report comprises of different sections and the format can be modified as appropriate. In general,

- ✰ Qualitative data like stories, anecdotes, quotations, interview transcripts, recordings etc., are placed in the Results section along with empirical data.
- ✰ Where possible, empirical evaluation techniques are used.
- ✰ Generally, the researcher refers back to the research question(s) with quotations or other qualitative/quantitative evidence collected.
- ✰ The report also includes evidence from published literature that confirms and disconfirms the data collected. This is generally placed in the Discussion section, and analyzed carefully.
- ✰ The report conclusion makes assertions and suggestions for further research activity so that, another researcher may apply these techniques to another study of similar case and this helps to generalize the findings.

In general the case study report comprises of the following sections:

- ❖ Preliminaries.
- ❖ Introduction (introduction about the case, describe the problem and why the researcher has selected that 'case' for study etc.).
- ❖ Background reading (evaluations about previous research, other examples, if any etc.).
- ❖ Methodology followed for the study (tools and techniques employed for collecting the data or information and analysis and explain the reasons why these techniques are used for the study).
- ❖ Results and Discussion (Present the analytical findings of the study).
- ❖ Summarize the findings of the study. Compare these findings with past studies of the same case, if available.
- ❖ Evaluate findings in light of background reading.
- ❖ Draw conclusions (generalize) from the findings.
- ❖ Suggest recommendations for the future.
- ❖ End matter.

C. Sources of Evidence in Case Studies

Stake (1995), and Yin (1994) identified the following six sources of evidence in case studies:

- ☆ ***Documents:*** They refer to letters, memoranda, agendas, administrative documents, newspaper articles or any document that are essential for the study. These documents are also useful for making inferences about events.
- ☆ ***Archival records:*** They refer to service records, organizational records, list of names, survey data etc. However, the investigator has to be careful in evaluating the accuracy of these records before using them.
- ☆ ***Interviews:*** There are several forms of interviews that are possible *viz.*, Open-ended, Focused and Structured or Survey. In an open-ended interview, key respondents are asked to comment about certain events. The focused interview is used in a situation, where the respondent is interviewed for a short period of time, usually answering set questions. This technique is often used to confirm data collected from another source. The structured interview is similar to a survey, and is used to gather data in cases such as neighbourhood studies. The questions are detailed and developed in advance, much as they are in a survey.
- ☆ ***Direct observation:*** Direct observation occurs when a field visit is conducted by the researcher during the case study.
- ☆ ***Participant-observation:*** Participant observation is a unique mode of observation in which the researcher may actually participate in the events being studied. This technique is useful for providing additional information about the topic being studied. Data obtained through participant observation serve as a check against participants' subjective reporting of what they believe and do.
- ☆ ***Physical artifacts:*** Physical artifacts can be tools, instruments, or some other physical evidence that may be collected during the study as part of a field visit.

It is important to keep in mind that, not all sources are relevant for all case studies (Yin, 1994). However, the researcher should be capable of dealing with all of them. It is to be noted that, no single source has a complete advantage over the others; rather, they might be complementary and could be used in tandem. Table 10.1 indicates the strengths and weaknesses of each type of evidence.

D. Types of Case Study Research

Stake (1995) distinguished different types of case studies as follows:

- ☆ ***Intrinsic case study:*** An intrinsic case study is carried out, when one wants to understand a particular case.
- ☆ ***Instrumental case study:*** When one has a research question and wants to get insight into the question by studying a particular case, instrumental case study is conducted.
- ☆ ***Collective case study:*** A collective case study refers to extension of an instrumental study to several cases.

The following are the other important types of classifications of case study research:

a. Depending on the Type of Research Question

Depending on the type of research question, Yin (1994) suggested three different types of case studies *viz.*, exploratory, explanatory and descriptive case studies. If the research is mainly focused on 'what' questions, it may call for exploratory study. An explanatory case study deals with 'how' or 'why' questions. A descriptive study focuses on covering the background information and accurate description of the case in question.

Table 10.1: Strengths and Weaknesses of different Types of Evidence in Case studies.

Source of Evidence	*Strengths*	*Weaknesses*
Documentation	☆ Stable - repeated review ☆ Unobtrusive - exist prior to case study ☆ Exact - names etc. ☆ Broad coverage - extended time span	☆ Retrievability - difficult ☆ Biased selectivity ☆ Reporting bias - reflects author bias ☆ Access - may be blocked
Archival Records	☆ Same as above ☆ Precise and quantitative	☆ Same as above ☆ Privacy might inhibit access
Interviews	☆ Targeted - focuses on case study topic ☆ Insightful - provides perceived causal inferences	☆ Bias due to poor questions ☆ Response bias ☆ Incomplete recollection ☆ Reflexivity - interviewee expresses what interviewer wants to hear
Direct Observation	☆ Reality - covers events in real time ☆ Contextual - covers event context	☆ Time-consuming ☆ Selectivity - might miss facts ☆ Reflexivity - observer's presence might cause change ☆ Cost - observers need time
Participant Observation	☆ Same as above ☆ Insightful into interpersonal behavior	☆ Same as above ☆ Bias due to investigator's actions
Physical Artifacts	☆ Insightful into cultural features ☆ Insightful into technical operations	☆ Selectivity ☆ Availability

Source: Yin, 1994.

b. Depending on the Disciplinary Orientation

Case studies can also be classified into the following four types according to disciplinary orientation:

- ☆ ***Ethnographic case study:*** This case study is more than an intensive, holistic description and analysis of a social unit or phenomenon. It is a socio-cultural analysis and interpretation of the unit of study. Example: 'An Experiential Approach To Teaching Retail Management'.
- ☆ ***Historical case study:*** This study presents a holistic description and analysis of a specific case from a historical perspective. Example: The War of the Triple Alliance (1864-1870): A Historical Case Study on the Causes of Regional Conflict.
- ☆ ***Psychological case study:*** This study focuses on an individual as a way to investigate some aspects of human behaviour. Example: 'Farmers' Rights in India-A Case Study'.
- ☆ ***Sociological case study:*** This study attends to the constructs of society and socialization in studying educational phenomenon. Example: 'The Sociology of Warabamdi: A Case Study From India'.

c. Depending on the Overall Intent of the Study

Irrespective of disciplinary orientations, case studies can be classified according to the overall intent of the study into four categories, as discussed here under.

- ☆ ***Descriptive case study***: This study gives a detailed theoretical account of the phenomenon under study. Lijphart (1971) described such case studies as *'entirely descriptive and move in a theoretical vacuum'*. They are useful in presenting basic information about the study case, where little research has been done.
- ☆ ***Interpretative case study***: This study involves studying relationships among the variables and in constructing theories. This study is inductive in its approach (Merriam, 1988).
- ☆ ***Evaluative case study:*** This study involves description, explanation and judgment for the event under consideration (Yin, 1984).
- ☆ ***Pilot case study***: This study helps researchers to refine their data collection plans with respect to both the content of the data and the procedures to be followed (Yin, 1994).

d. Depending on the Number of Case Studies Handled by the Researcher

Besides above classifications, there is another classification of case studies into Single case and Multiple case studies depending upon the number of cases or events handled by the researcher. As the name itself indicates, Single case study deals with a single event for the study. A study using more than one case is called as Collective case study or Multi-case study or Multi-site study or Comparative case study or Multiple case study. This type of study involves collecting and analyzing data from several cases and may be distinguished from the single case study that may have sub-units or sub-cases embedded within. By looking at a range of similar and contrasting cases, we can strengthen the precision, the validity and the stability of the findings (Miles and Huberman, 1994).

E. Differences between Survey Method and Case Study

These two methods differ in the following aspects:

- ☆ In contrast to surveys, typically the number of units studied in a case study is less.
- ☆ The extent of (detailed) information generated from case study is greater compared to survey method.
- ☆ In survey method, the researcher selects a representative sample the population and hence, he is aware of the population. This is not the case in case study method *i.e.*, the population is unknown. Further, the researcher selects a single unit for the study and not a representative sample.
- ☆ In survey method, the sample is described or analyzed and an inference is made about the population. However, in case study method, the 'case' is described or analyzed and no inference is made about the population, but inferences can be made about the similar cases.
- ☆ Example for survey method: If the researcher collects the data pertaining to cost of cultivation of paddy from 100 farmers in Nandyal region of Kurnool district in Andhra Pradesh in 2012-13, it implies survey research.
- ☆ Example for case study: If the researcher studies the functioning of ITC's *'e-Choupal'*, it implies a case study method.

Similarly, case study also differs from an experimental research. As compared with an experiment, the case study researcher has much less control over the variables, than if an experiment were used to investigate a situation.

F. Merits of Case Study

This method of data collection is very popular considering the following main advantages:

- Enable the researcher to understand fully about the behavioural pattern of the concerned case or unit.
- Good opportunity for innovation.
- Good method to study rare phenomena.
- Good method to challenge theoretical assumptions.
- Facilitate the researcher to gain in depth knowledge about a particular event.
- In this method, the problem is recognized as a unit and various aspects of the unit are subjected to deep and detailed study.
- This study enables the researcher to draw the characteristics of different firms or events and thus, aid in comparative study.
- When several (similar) cases are extensively studied and analyzed, it will enable the researcher to deduce various generalizations, which may be developed into useful hypothesis.
- Case study helps in framing a schedule for a similar problem and this helps in other ways of data collection.
- This method is flexible with reference to different data collection methods employed.
- If the researcher studies different cases, it will enable him to classify them into different classes and thereby, facilitate stratification.
- This method can be conducted in any type of social setting.
- According to Yin (1989), this method helps in studying the case in a situation, where it has not been studied before.
- In cases, where the problem under study constitutes a process and not one incident (say, cartel formation), then case study is the appropriate method.
- This method allow researcher to fully understand how an intervention worked, or why an intervention had an effect in a particular case.
- Case studies tend to have very strong internal validity. (but are often criticized because of their extremely poor external validity. Because case studies look at only one case at a time, and purposefully choose cases that are atypical, external validity is a real concern).

G. Demerits of Case Study

They include:

- Hard to draw definite cause-effect conclusions
- Hard to generalize from a single case
- Possible biases in data collection and interpretation
- Case studies are often labeled as being too long, difficult to conduct and producing a massive amount of documentation (Yin, 1984). Say, for example, ethnographic case studies can elicit a great deal of data over a period of time and it is quite difficult to manage such huge data, especially when not organized systematically.
- Yin (1994) considered case methodology '*microscopic*' because of the limited sampling cases.

- ✰ This study demand more finance, time and energy on the part of the researcher.
- ✰ This method is based on several assumptions, which may not be very realistic at times. So, the reliability of the findings is questionable.
- ✰ This study is based on the information that is given by an individual or a firm. Say, if the information furnished is wrong, there is no other basis for the researcher to test the reliability of the information.
- ✰ In this method, individual is recognized as a unit and the entire study is concentrated on that unit. Through the study of a particular unit, it is not possible to apply the knowledge on other units.
- ✰ If the records on the basis of which data are collected are defective and not based on objectivity, the generalization becomes defective.
- ✰ It is not possible to quantify the feelings, emotions, reactions, values etc., and as such it lacks scientific temper.

To conclude, case studies are considered useful in research, as they enable researchers to examine data at the micro level. As an alternative to quantitative or qualitative research, case studies can be a practical solution, when a big sample population is difficult to obtain. Further, they present data of real-life situations and provide better insights into the detailed behaviours of the subjects of interest. In the words of Eisenhardt (1989), '*case studies are particularly well suited to new research areas or research areas for which existing theory seems inadequate. This type of work is highly complementary to incremental theory building from normal science research. The former is useful in early stages of research on a topic or when a fresh perspective is needed, whilst the latter is useful in later stages of knowledge*'. In spite of the criticisms stated above against this method, the researchers continue to deploy the case study method particularly in studies of real-life situations governing social issues and problems.

10.3. Other Data Collection Methods Employed in Social Sciences Research

In social sciences (agricultural) research studies, besides above two methods the researcher generally employs another two popular data collection methods *viz.*, Observation, and Participatory Rural Appraisal (PRA).

A. Observation Method

This is the most commonly used method of data collection, especially in studies relating to behavioural sciences. This method involves accurate watching and noting of phenomenon or data as they occur in nature with regard to cause and effect relation or mutual relation. In case of interview method of data collection by employing schedule, the participants (say, farmers) may be asked about how they behave in certain situations to elicit their responses. But, there is no guarantee that, they actually provide the reliable information. So, by observing them in those situations gives more reliable information to the researcher rather than by mere interviewing them. This method of data collection can also serve as a technique for verifying or nullifying information provided in face to face interactions. In some research studies, besides observation of participants, observation of practical environment in which the participants are working also counts much and this enable the researcher to plan and execute a research project. For example, by observing the practical environment of a particular agro-ecological situation, the researcher can formulate an action research project (say, reclamation of water logged induced soil salinity in a village). Thus, observation method of data collection offers a distinct way of collecting data, as it does not rely on what participants say they do or what they say they think.

It is more direct than that. Instead, it draws on the direct evidence of the eye to witness events first hand. It is a more natural way of gathering data. Whenever direct observation is possible, it is the preferable method to use, as it involves the systematic recording of observable phenomena or behaviour in a natural setting.

Characteristics of Observation Method

They include:

- ✰ Data collection is through direct contact (observation) with phenomenon under study.
- ✰ Major source for collection of primary data.
- ✰ Researcher has to conduct in-depth study.
- ✰ Collection follows observation *i.e..,* the researcher first observes the phenomenon and then collects the data.
- ✰ Collection of data is based on the observation of phenomenon and this observation is made with a definite purpose.
- ✰ The collected data enable the researcher to analyze cause and affect relationship between the variables.
- ✰ Facilitate the researcher to collect data directly.
- ✰ Observation and data collection in natural setting.
- ✰ Eye witness was given priority over interaction.
- ✰ It is specific, objective and quantitative.
- ✰ The record of observations should be made immediately.
- ✰ The results can be checked and verified.

Types of Observation

There are different types of observation method of data collection:

a. Unstructured and Structured Observations

- ✰ ***Unstructured observation:*** If the researcher does not define the items to be collected and manner of recording the data in advance, then that method refers to Unstructured observation. Generally, in exploratory research, the researcher employs this technique of data collection. This unstructured observation will not allow the researcher to overlook any crucial areas of the project.
- ✰ ***Structured observation:*** If the researcher defines the items to be collected and manner of recording the data in advance, then that method refers to Structured observation. Generally, in action research, the researcher employs this technique of data collection. At the beginning of research investigation, though the researcher employs unstructured observation, but as the investigation proceeds observation gets restricted and structured.

b. Participant and Non-participant Observation

Depending upon the degree of involvement of the researcher in observing and collecting the data in the investigation, these are the two types of observation and they are discussed here under.

- ✰ ***Participant Observation:*** In this method, the observer (researcher) becomes more or less one of the groups under observation and shares the situation as a visiting stranger, an attentive listener, an eager learner or as a complete participant observer. He personally record and interpret the behaviour of the group. The main advantage of participant observation is that, it helps the researcher to get first hand information about the group under investigation. It also helps to get the information in an unbiased way.

- ☆ ***Non-participant observation:*** In this method, the observer (researcher) keeps his observation as inconspicuous as possible. The purpose of non-participant observation is to observe the behaviour of the group in a natural setting. The observer observes the group through maintaining both physical and psychological isolations from them. This method is used with groups like infants, children or abnormal persons. It permits the use of recording instruments and the gathering of large quantities of data.

  Among these two, participant observation gives a better insight into the investigation, but it is time consuming. In participant observation, as the observer develops relationship with the group, there is a chance of losing his neutrality, objectivity and accuracy to rate things as they are. So, some researchers opine that, it is best for the observer to remain only a partial participant and to maintain his status of scientific observer apart from the group.

Steps for Effective Observation

The following are the sequential steps to be followed for an effective observation in a research investigation:

- ☆ ***Planning:*** Of course this is the pre-requisite for any data collection method. In this case of observation method, the researcher should plan for the following aspects:
  - ❖ Adequate sample size.
  - ❖ Units of behaviour to be observed should be clearly defined.
  - ❖ Methods of recording should be simplified.
  - ❖ Plan the variables for which the data are to be collected and note that, all or many of the variables should not be observed simultaneously.
  - ❖ Sequence of data or information to be collected should be planned.
  - ❖ The sample should be observed more frequently.
- ☆ ***Execution:*** It is essential that, the observer should be fully trained and well equipped. This enables the observer to focus attention on the specific activities or units of behaviour of the group under observation.
- ☆ ***Recording:*** As mentioned earlier, 'collection follows observation'. The collected data must be recorded for enabling the researcher to interpret it. Two common procedures for recording observations are: Simultaneous and Soon after the observation. The choice of method of recording depends upon the nature of the group and the type of behaviour to be observed. The former method of recording (Simultaneous form of recording) may distract the subjects, while the latter (after observation), the observer may fail to record the complete and exact information. Therefore, for a systematic collection of data, the various devices of recording should be used such as checklist, rating scale, score card etc.
- ☆ ***Interpretation:*** After recording the data, the researcher interprets the findings. Interpretation can also be done directly by the observer at the time of his observation. In some studies, when several observers are involved in an investigation, the observers will simply collect the data and the interpretation job will be looked after by an expert.

Merits of Observation Method

This method of data collection enjoys the following benefits:

- This method of data collection is common to all the discipline of research.
- Data collected directly in the natural setting.
- On the spot recording of data ensures systematic approach.
- Substantial amount of data can be collected in a relatively short time span.
- Ready-made data are available and this facilitates speedy analysis.
- Inter observer reliability is high.
- Since observations are based on actual and first hand experience, the data are more realistic than the data collected through other methods.
- This method of data collection facilitates the formulation of reliable (testable) hypothesis. For verification of hypothesis, again we depend upon observation. Therefore, it can be said that, the problem presents itself and resolves itself through observation method.
- The conclusions of observations are more reliable than non-observation conclusions, because they are based on first hand perception by the eyes and can be verified by any one by visual perception. So, the conclusions drawn by this method are more realistic.

On account of the above advantages, the observation method is called Classical Technique of Investigation for research purposes.

Limitations of Observations Method

However, this method of data collection suffers from the following limitations:

- Some events like emotions, likes, dislikes, frankness etc., cannot be noticed even by this method.
- Since we have to depend upon on our eyes for observation, we can never be sure that, what we are observing is the true as it appears to our eyes, as eyes are prone to deception.
- Establishing validity is difficult.
- Subjectivity in the explanation may creep in. For example, if we see a person coming out of a police station, we may immediately conclude that, he might have involved in an anti-social activity, which may not be the case always.
- It is a slow and labourious process.
- It is costly both in terms of time and money.
- The data may be unmanageable.
- There is possibility of biasedness.
- Sometimes, the atmosphere in the study group tends to become artificial and this hampers their naturalness in behaviour and thus the purpose of observation which is to know the behaviour of individuals under normal conditions gets defeated.
- Full answer cannot be obtained by observation method alone and hence, this must be supplemented by other methods of study.

Though there are good number of limitations under observation method, these can be minimized by conducting systematic observation by the researcher, as it provides a framework for observing the variables and data to be collected in the investigation.

B. Participatory Rural Appraisal (PRA)

PRA is an approach and range of techniques that enable stakeholders to analyze their problems and then plan, implement and evaluate agreed-upon solutions. PRS is also known as Participatory Learning and Action (PLA). PRA can also be meant in the following ways:

PRA is an approach (and family of methodologies) for shared learning between local people and outsiders to enable development practitioners, government officials, and local people to plan together appropriate interventions. This technique of data collection aims to incorporate the knowledge and opinions of rural people in the planning and management of development projects and programmes. Note that, PRA involves *'not working FOR the people but WITH the people'*.

Evolution of PRA

PRA is a short cut method of data collection compared to schedule and questionnaire employed. It utilizes a range of complementary techniques and also involves local people and outsiders of different sectors and disciplines. Outsiders facilitate the local people in providing the requisite information, sharing the information and ideas and conditions to plan and act. Thus, PRA technique is a bottom up (demand driven) approach. The formulation of the PRA Method has a background of the long experience from an expert in, Robert Chambers (1983), in the field of community development working in various developing countries. The following Figure 10.1 shows the evolution pattern of PRA:

Principles or Key features of PRA

PRA is a systematic process of learning about conditions of an intervention or an issue related to it in an intensive, cumulative and efficient manner. It is information/data collection and structuring, developing hypotheses about problems, constraints and potentials. It is done on-site. This technique harbors the following features:

a. ***Inter-disciplinary team:*** This technique is executed by the members (called facilitators) of different skills and professional backgrounds. These members of the PRA team should be involved in all aspects of the study, design, data collection, analysis, and interpretation. Wherever possible, the PRA team should always include women and, whenever possible, local people. Note that, PRA is a learning experience in which the participants also learn from each other.

b. ***Diversity of information:*** PRA seeks diversity of opinions from the respondents in a qualitative sense rather than attempting complex and quantitative averages and statistical figures. Though gathering or collecting simplified data, accuracy is achieved through drawing from diversified information sources, through cross-checking of data, and other methods for reliability.

c. ***Habour complementary techniques:*** Appropriate techniques are mixed to tailor to the specific requirements of the study. These techniques should be clear, self-evident and simple, appropriate to local conditions and open for any modification that might be suggested by the local people.

d. ***Group interaction:*** PRA is cumulative learning from, with, and by local people, eliciting requisite information. The PRA team comprises of people from different disciplines and they work in a coherent manner.

e. ***Triangulation:*** This is a way of 'cross-checking' different sources of information. The data or information received from the respondents or participants should be looked at from different

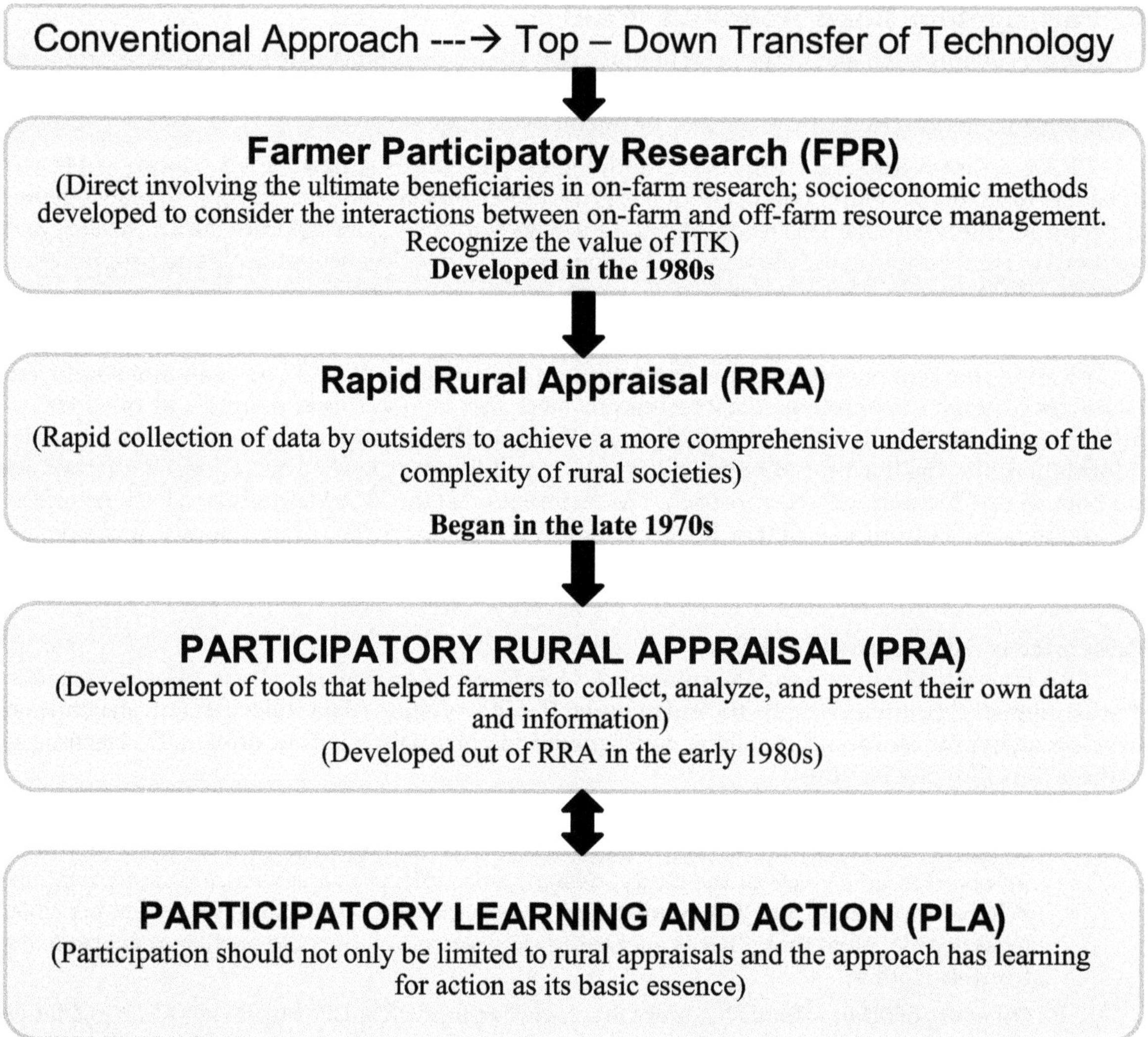

Figure 10.1: Evolution of PRA.

angles *i.e..*, from the point of view of different stakeholders involved in conducting PRA. Thus, triangulation means looking at any problem from as many perspectives as possible, but at least three (Figure 10.2). The different levels of people involved in triangulation may include: women/men, social scientists/technical specialists, insiders/ outsiders, youth/ elders etc.

f. ***Participatory Learning:*** The guiding principle of PRA is learning from, with and by people in the community. The PRA practitioners attempt to see *'with the eyes of the people'* concerned, especially the poor strata of society. That is, local people serve as partners in data collection and analysis.

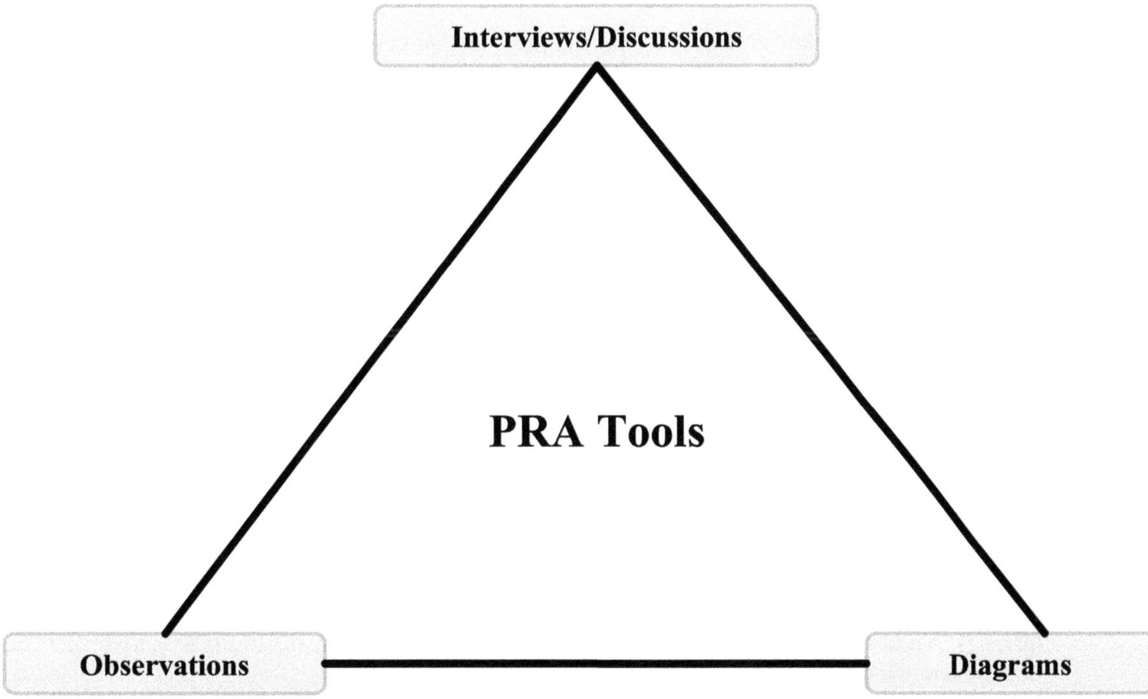

Figure 10.2: Triangulation Employed in PRA.

g. ***Focusing by iteration:*** Step by step analyses of the farming situation focuses more in-depth understanding of the crucial issues.

h. ***On-site Process Work***: The results of the PRA are discussed and reflected by the whole team daily – usually in the evening hours – and the proceedings for the next day are flexibly agreed upon.

i. ***Flexibility:*** PRA employs flexible tools to draw the information and not based on standardized methodology.

j. ***Off-setting biases:*** PRA offsets biases in accessing the information, as it involves more listening less lecturing.

k. ***Optimal Ignorance:*** In taking up the activities, facilitators aim at leaving out unessential details. That is, PRA technique avoids collection of unnecessary data. It optimizes trade off between quality, relevance, accuracy and timeliness.

l. ***Using local materials:*** Local materials like dirt, stones, sticks (or paper), will be used to depict the data or information and no computers/electronic devices are used.

m. ***Localization:*** By learning from and with rural people: directly, on the site, and face-to-face, gaining from indigenous physical, technical and social knowledge.

n. ***Learning rapidly and progressively:*** With conscious exploration, flexible use of methods, opportunism, improvisation, iteration, and cross-checking, there is rapid learning about information and data. Note that, no blueprint programme is followed.

Tools Employed in PRA

PRA employs the following tools in eliciting the requisite data or information:

a. ***Transect Walk:*** Transect walk is a walk inside the village. The facilitators are accompanied by the villagers and during this walk observations are made regarding the farming situation. This walk will help the facilitators to observe, to listen, to identify different zones and conditions, and to ask questions to identify problems and possible solutions. This transect walk also helps the researchers to understand the socio-economic- conditions of villagers, access to resources for different stakeholder groups (traditional, modern, industrial etc.), topography, indigenous technology, soils and vegetation, farming practices, problems and opportunities which are cross tallied with the resources mapping and modeling and possible options for responsible management of natural resources. To begin transect walk, first the researcher(s) should decide the purpose. explain the purpose of walk to the villagers, select a logical starting point for the walk, proceed, taking time for brief discussions and at the end, present outcome in a diagram (Figure 10.3).

b. ***Village Mapping:*** A person, who has links with an area can be not just a key informant but also a key researcher. These key informants will be asked to draw village map in the selected village, as they are well-versed with the local conditions, available resources (like fields, waste land, grazing area, irrigation, wells etc) infrastructure (like schools, hospitals, banks) etc. Select a convenient site in the village to draw village map on the ground and ask the farmers to use stick and other locally available materials like sand, stones etc., to depict the features on the map (Figure 10.4).

Land scape Features	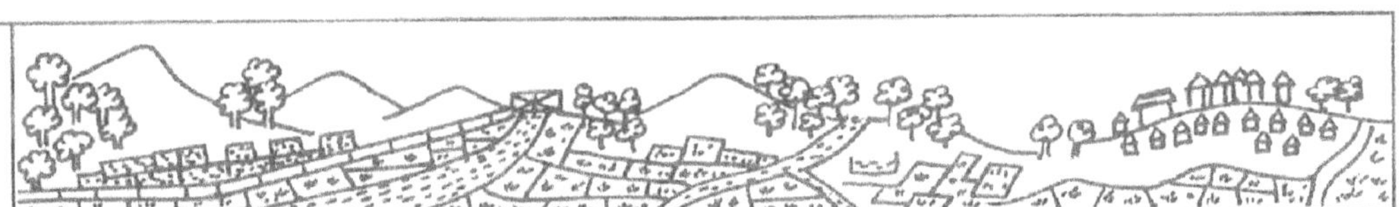					
Area locality	TentelKhunti	Karla Pita	Kara Dungri	Gharla Munda	Latha Kend	Ghantabahal
Land types (local name)	Up land (Aat) Low land (Bahal) Hills (Parbat)	Mostly upland (Aat) Very little low land (Bahal)	Rocky mountain (Parbat) Upland (Aat) Forest (Patra)	Up land (Aat) Low land (Bahal) Plantation forest	Only low land (Bahal)	Mostly low land (Bahali and Bahal)
Water source	Kankara Jor	Dor Jodia	Seasonal stream	Karla Munda Lathakend Munda	Archu Munda	Ningi Munda, wells, tube wells
Species	Mahua, Rengal, Char, Dharua, Bija Sahas, Palsa, Womb, Kendu (wild animal)	No Forest Railway line	Chakunda, Niligiri, Bamboo Bhamar, Neem, Sipol Tental	Nou Munda, Akari Chakunda, Niligiri, Bhamar, Sipo, Mango, Kaju (Patra land)	Mostly Palsa, Mahua	Neem, Banyan, and Mahua
Uses	Less cultivation Low productivity Forest produces	Cultivation of vegetables	Less cultivation Grass land, Plantation	Encroached field plantation, cultivation of more rice	Good cultivation of Paddy and vegetables	High yield, good cultivation of rice, millets, pulses
Ownership	Government and people	Government and people	Government and people	Soil conservation department and people	People	People

Participants: Sundari Rana, Jitendra Jani, Pancha Majhi, and Chate Chatar
Factilitators: Saroj Das and Mahendra

Figure 10.3: Transect Map.

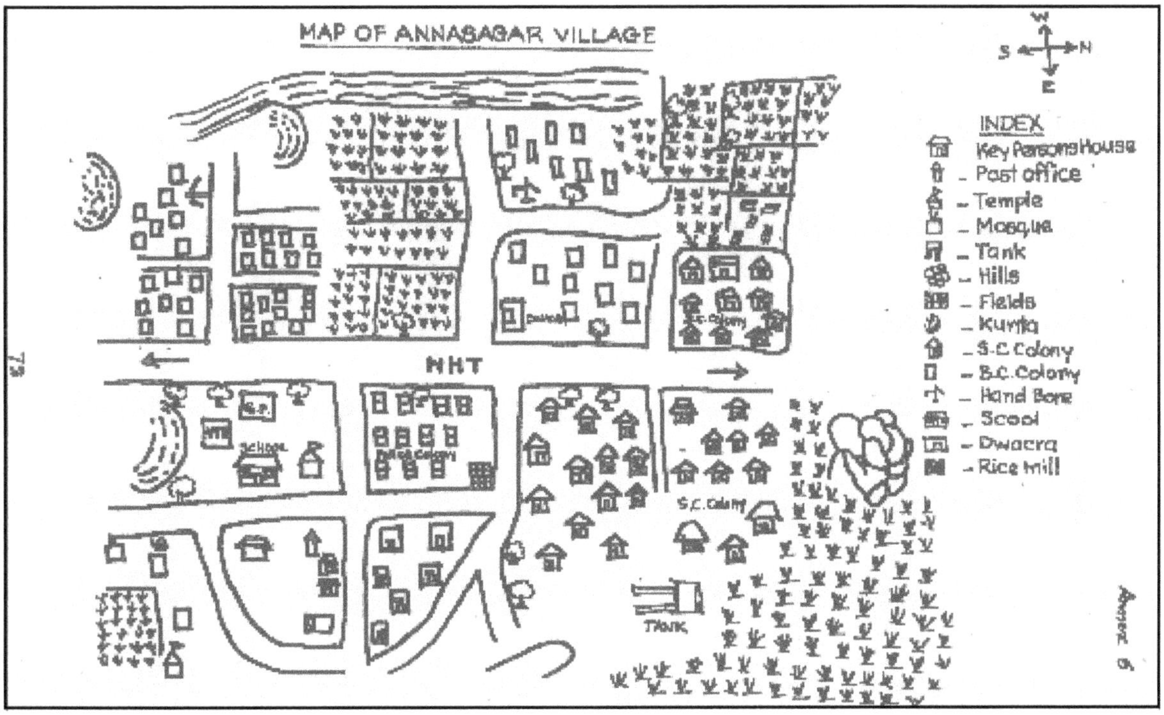

Figure 10.4: Village Map.

c. ***Time lines:*** This is a *time-related* PRA method. It allows people use their concept of time. It involves establishing the chronological sequence and relative importance of events as reported by the local people (Figure 10.5). Major historical community events and changes are dated and listed. This tool is employed to know time and events, history, evolution of a village, agricultural practices, health care practices, etc. (done by constructing a chronology of events that have taken place in consultation with the people). Understanding the cycles of change can help communities focus on future actions and information requirements. Also discuss implication of changes on an aspect over time. It also helps to understand farmers' perceptions on trends and underlying causes. Note that, in this tool, the researcher should select elder persons as resource persons to collect the information.

d. ***Seasonal Calender:*** All the major changes that occur within the rural year are represented in this map (Figure 10.7). This calendar shows the season/month-wise availability of resources, activities, events, climate, cropping patterns, livestock, labour demand, employment, income and expenditure, debt, credit, food and nutrition, disease, fodder, milk production, marketing, etc. (Done with the use of stones, sticks and different coloured seeds to represent months, quantities or rainfall, number of days of employment, income, etc.). So, this tool provides insight into seasonal variation of various issues. Also helps to identify the events that happen often. This map also provides the room to include the types of problems and constraints and point out opportunities to develop the village.

e. ***Venn Diagram:*** This map or diagram is used to depict the strength of association between different local groups, people, villages, institutions, organizations etc. (Figure 10.8).

Figure 10.5: Timeline Map.

Natural resources / Period	 Water	 Tree	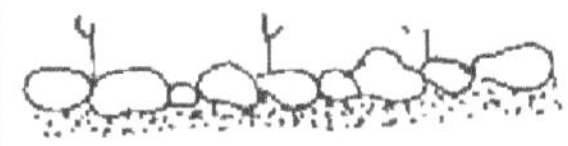 Farmland
1960 (Independence)	• Rains daily during June • No wells • Water always in the river	○ ○ ○ ○ ○ ○ ○ ○ ○ ○	○ ○ ○ ○ ○ ○ ○ ○ ○ ○
1970 (End of civil war)	• Rains at least once in 2-3 days in June	○ ○ ○ ○ ○ ○ ○ ○	○ ○ ○ ○ ○ ○ ○ ○ ○
1979-80 (Shagari)	• Rains at least once in 4 days in June	○ ○ ○ ○ ○	○ ○ ○ ○ ○ ○
1999 Today	• Rains sometimes once in 8-10 days in June • River and well dry in dry season	○ ○ ○	○ ○ ○

Participants: Adama, Elizabeth Sule, Omojo Peter, Ochesija Nichlas and Oglinulu Okpachu
Facilitators: Paul C

Figure 10.6: Trend Analysis.

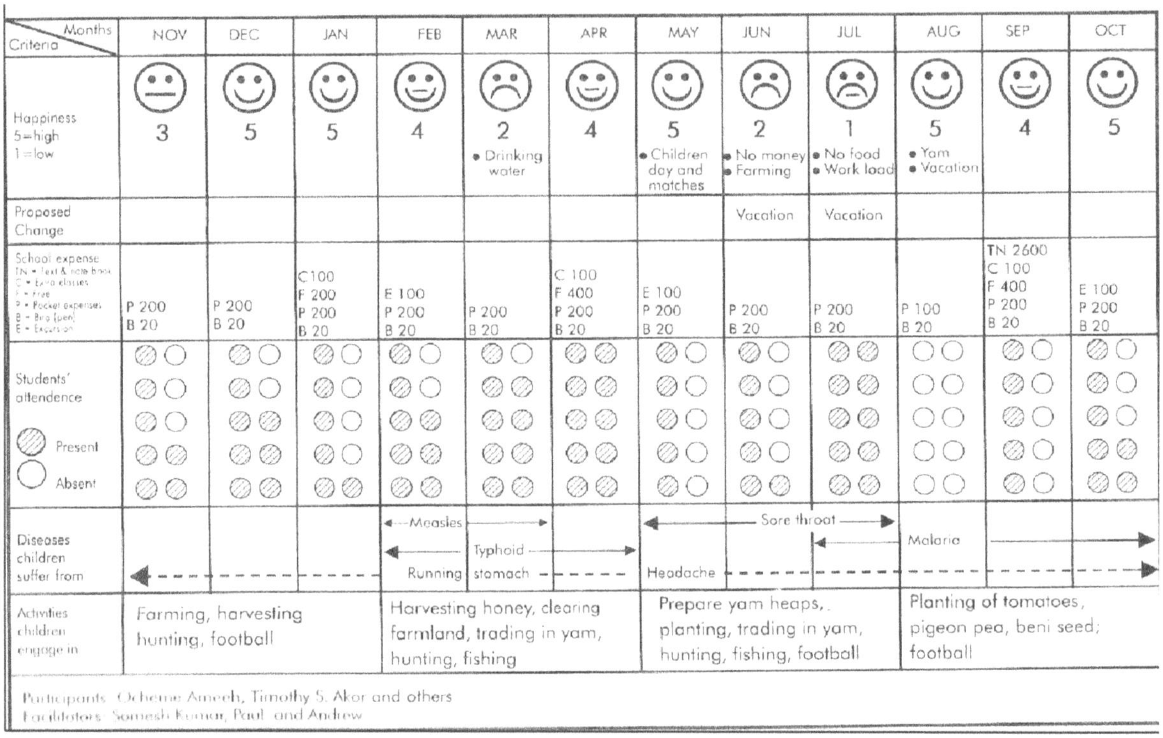

Figure 10.7: Seasonal Analysis.

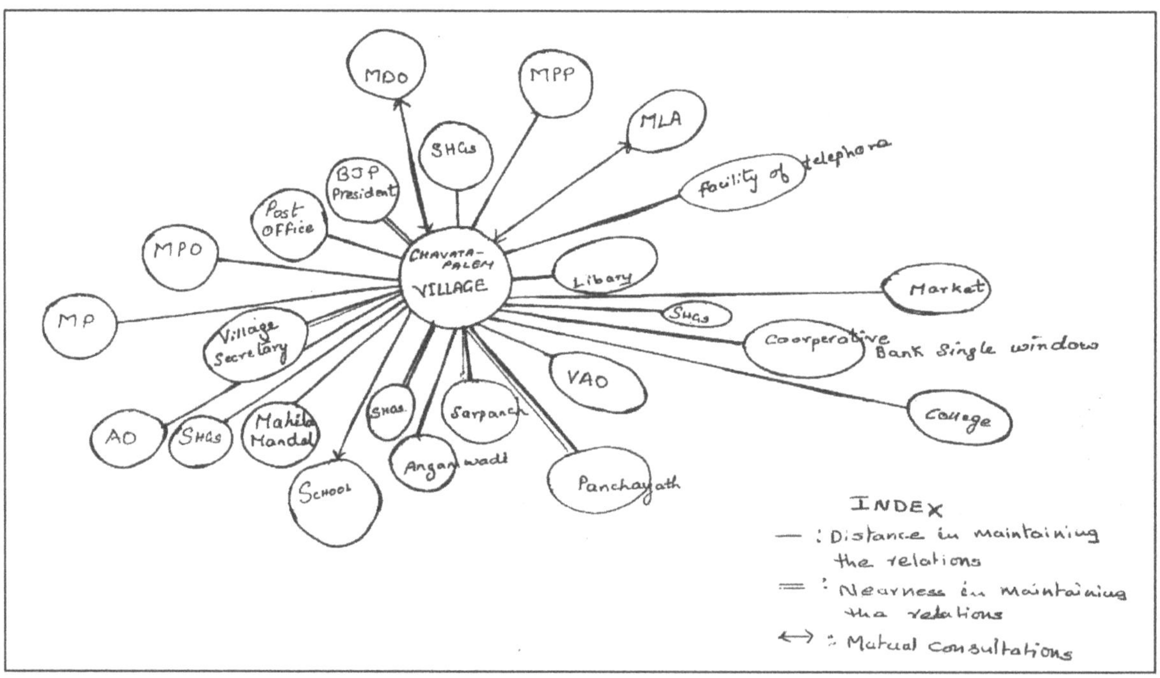

Figure 10.8: Venn Diagram.

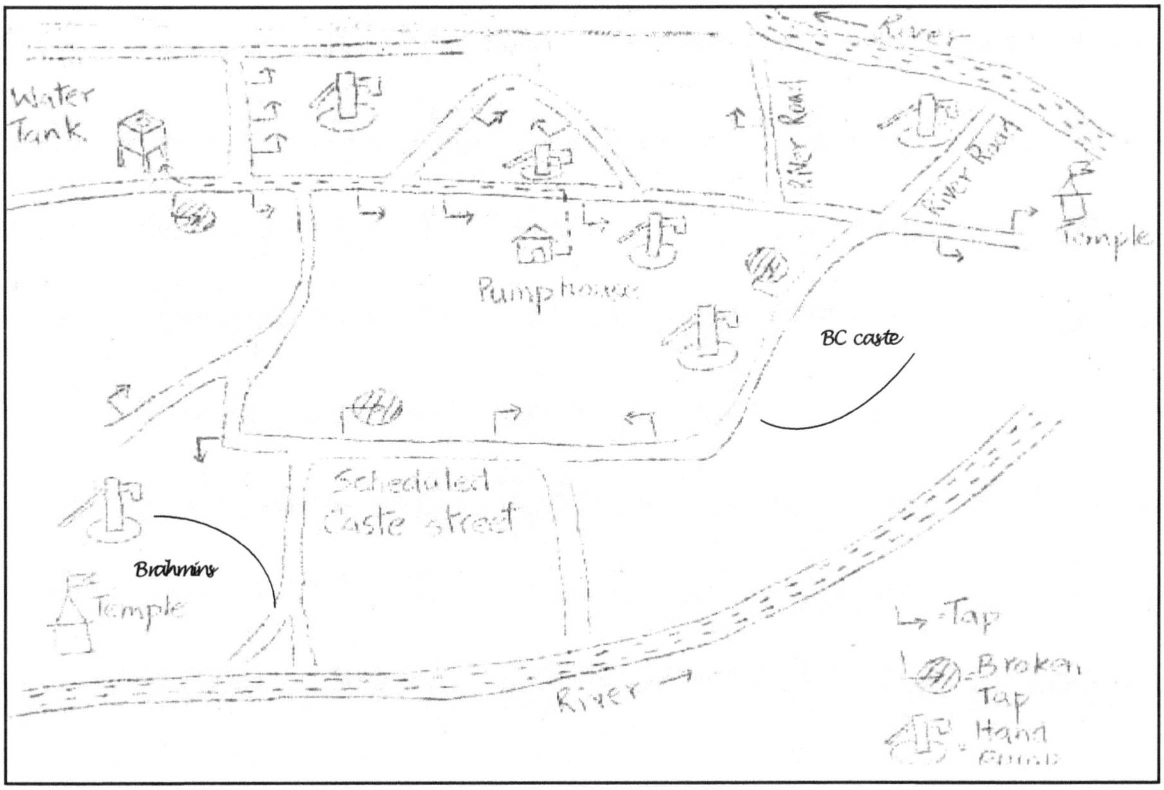

Figure 10.9: Social Map.

Overlapping circles are used to represent the degree of interrelationship between the selected items and their relative importance. Venn diagrams are useful in understanding the usage of the resource environment by multiple users, and the interactions between different user groups. This facilitates the optimization, and a rational exploitation of the resources for different uses. Also used for mapping processes, causes, effects, linkages etc.

f. ***Social Mapping:*** In this tool, farmers are asked to draw a simple map or diagram about the geographical distribution of different households indicating their caste, income group, religion etc. (Figure 10.9).

g. ***Resource Mapping:*** In this tool, farmers are asked to draw a map highlighting physical features, resources available, infrastructure etc., in the village (Figure 10.10).

h. ***Matrix Ranking:*** This tool enables the researcher to identify wealth criteria and establish the relative position of households. Also employed for ranking items such as, crops, varieties, types and breeds of livestock, trees, fodders, supplementary income generating activities etc. (Done by asking farmers to list different items etc.; species of trees and vegetables and different criteria for evaluating them. Each class or category is then given a rank or score by the villagers. This is done by means of quantification with pebbles or seeds). This ranking is useful in discovering, weighting and prioritizing problems as presented by groups during the early stages of identifying and setting the research agenda. It also help the researchers in

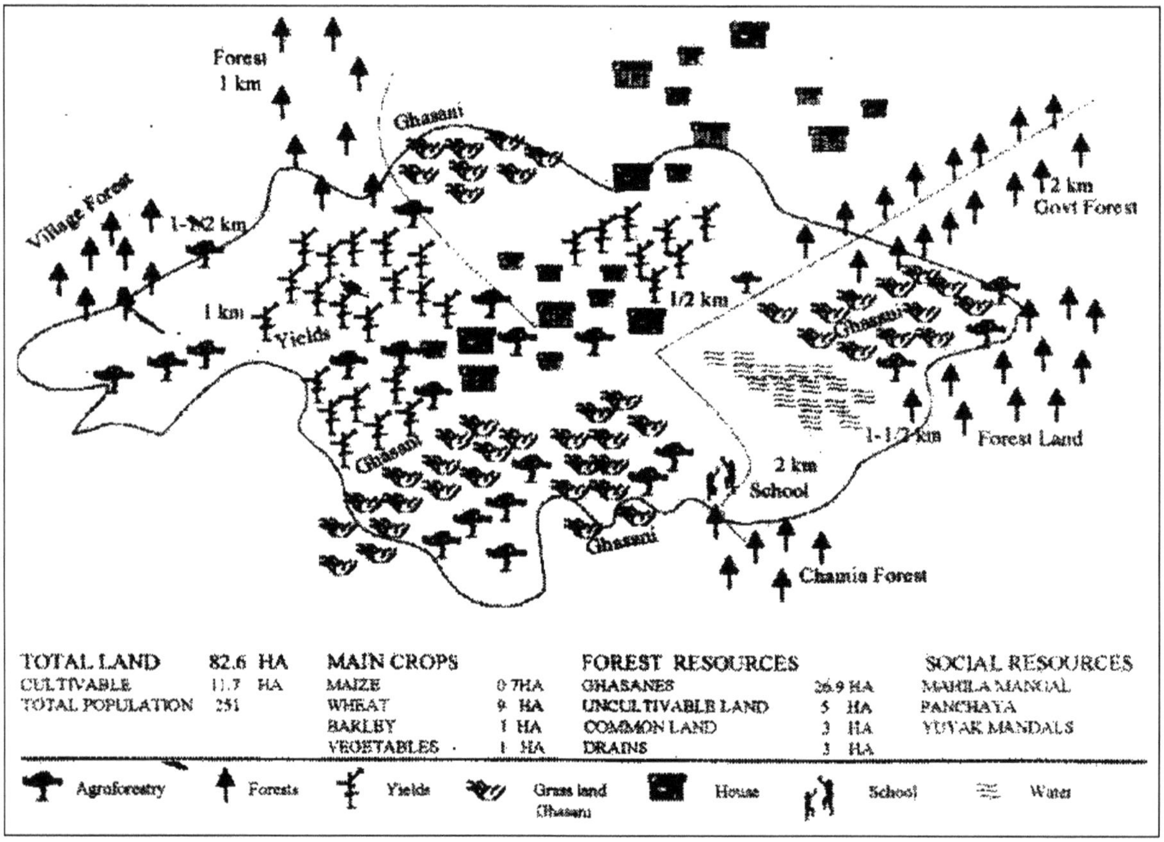

Figure 10.10: Resource Map.

understanding the relative access of different resources to different sections of the society say, in terms of technology and reach (Figure 10.11).

The other tools involved in PRA include: Participatory analysis of secondary sources- aerial photographs, Participatory mapping and modeling, Oral histories and ethno-biographies, Seasonal Calendars, Daily time use analysis, Livelihood analysis, Participatory linkage diagramming, Well being and wealth grouping and ranking, Analysis of difference, Estimates and quantification, Stories, portraits and case studies, Team contracts and Interactions, Presentation and analysis, Participatory planning, budgeting, monitoring, Group discussions etc.

Successful Conductance of PRA

The following hints will help the researcher for Successful conductance of PRA:

- ☆ Begin with a concrete activity that requires group input and will almost certainly to lead to a concrete output. A mapping exercise is a good starter.
- ☆ Use direct observations and ask about what you see.
- ☆ Try to get opinions from all groups, especially women, the poorer, the more remote, those who do not use services, who are liable to be left out.

Well-being Groups (characteristics)	Quality of land	Availability of water for irrigation	Livestock ownership	Land holding	Quality of food	Quality of house (walls, floor, roof etc)	Dependence on labour	Dependence on forest or wood	Dependence on stone cutting	Quality of clothes
Manages well (Grain from own land lasts for 8–10 months) (11)	●●●	●		●●●●●●	●●●●●	●●●●●●●●●●				●●●●●●●
Somehow manages (Grain from own land lasts for 6–8 months) (21)	●●●●		●●	●●	●●●	●●●●●●	●●	●●●●●●		●●●●●
Manages with difficulty (Grain from own land lasts for 3–4 months) (39)			●	●	●●	●●●●	●●●●●●●●●	●●●●●●●●●●	●●●	●●●
Dependent on labour and forest (42)					●	●●	●●●●●●●●●●	●●●●●●●●●●	●●●●●	●

Participants: Anand Chandra Pradhan, Murli Pradhan, Chandrasekhar Pradhan, Arjun Pradhan, Khirodra Pradhan, Sanjay Biswal, Pradeep Behra, Dhoba Behra, and Narayan Pradhan

Facilitators: Sabyasachi Rath, Vijaya Mishra, and Anindo Banerjee

Figure 10.11: Matrix Ranking.

- ☆ For every exercise decide, who is the reporter/observer. This person not only notes the discussion but also the process that is taking place. The reporter/observer monitors input from team members and can thereby correct an exercise if he/she sees it does not work as planned. This specific role is very important in interdisciplinary work while trying to use new research methods.
- ☆ All responses from interviewees must be recorded as they are given. This has to avoid researchers' own judgements dominating the report to be produced.
    - ❖ Do not collect data that you will not use later on.
    - ❖ Do not lecture or dominate.
    - ❖ Do not interrupt or interfere. Once a task has been initiated let people get on with it.
    - ❖ In discussion give people time to think or discuss among themselves.
    - ❖ Embrace error. We all make mistakes. Do not hide it but share it.
    - ❖ Relax, do not rush. Practice is the best way to learn and you do not need to learn everything in the first 30 minutes.
    - ❖ Be around in the evening, at night and in the early morning. Stay the night in villages if you can.
    - ❖ Allow unplanned time, walk and wander around.
    - ❖ Probe and use the six helpers – who, what, where, when, why, and how.
- ❖ Show interest and enthusiasm in learning from people.

- Have second and third meetings and interviews with the same people and discuss your findings.
- Allow time for team interaction and for changing the agenda.
- Revise the checklist at the end of each day.
- The researcher should also look into the following issues **in conducting PRA:**
  - Be an active learner
  - Show interest and enthusiasm
  - Respect villagers knowledge
  - Appreciate their contribution
  - Be sensitive to their moods
  - Follow protocol as required by the situation
  - Create an atmosphere of confidence
  - Involve the silent ones
  - Record names of informants

Layout of PRA

The layout of a typical PRA programme is generally as follows :

Day I:

INTRODUCTORY

- History of the village
- Village Layout
- Village Infrastructure, etc.

Day II

EXPLORATORY(Simple)

- Study of resources
- Livelihoods
- Trends
- Preferences etc.

Day III

EXPLORATORY(Complex)

- Seasonality
- Identifying resources
- Wealth ranking
- Class and caste stratification, conflict, etc.
- Causes and effects etc.

Day IV

CONVERGENCE

- Identifying opportunities

- ☆ Listing priorities
- ☆ Identifying roles and responsibilities of different stakeholders of village development

Day V

CONCLUDING

- ☆ Operational Plan
- ☆ Documentation, etc.

Areas of Application of PRA

PRA is widely employed in different areas as given below:

- ☆ Natural resource management
- ☆ Agriculture
- ☆ Poverty alleviation/women in development programmes
- ☆ Health and nutrition
- ☆ Preliminary and primary education
- ☆ Village and district-level planning
- ☆ Institutional and policy analysis.

Dangers and Drawbacks of PRA

The technique of PRA can cause the following difficulties, if the facilitators does not focus on DON'T'S of the following:

- ☆ Don't feel superior to the villagers
- ☆ Don't feel that there is nothing to learn
- ☆ Don't hesitate to ask villagers for advice
- ☆ Don't hesitate to clear your doubts
- ☆ Don't interrupt
- ☆ Don't forget to say 'Thank-you'
- ☆ Don't take villagers for granted
- ☆ Don't monopolize the interview

So, the facilitator should show the right attitude in terms of Openness, Humility, Curiosity, Sensitivity and Acceptance. He should also have right behavior in terms of showing respect, being friendly, sharing, being patient and interacting, not lecturing.

11

Data Processing and Analysis

Once after completing the data collection, it is not an end for research process, as the researcher has to compile, analyze, interpret and present the findings. The earlier discussion reveals that, by the time the researcher get to the analysis of collected data, most of the really difficult work has been done. This is so because, identification and formulation of research problem, collection of review of literature, formulation of hypothesis, preparation of research design, selection of representative sample, selection of data collection methods etc., are much more difficult and if these works have been attended well, data analysis is usually a fairly straight forward affair. In fact, the analysis and interpretation of collected data will start during the process of data collection itself. The approach to data analysis depends on what methods of data collection were used (*e.g.*, schedules, interviews) and the type of data collected (*e.g.*, quantitative or qualitative). Data analysis can be defined as the process of systematically applying statistical and/or logical techniques to describe and illustrate, condense and recap, and evaluate data. According to Shamoo and Resnik (2003), *'various analytic procedures provide a way of drawing inductive inferences from data and distinguishing the signal (the phenomenon of interest) from the noise (statistical fluctuations) present in the data'*.

11.1. Processing of Data

Data processing may be defined as the process of checking, editing, classification and tabulation of collected data so that, it becomes amenable to analysis. Thus, 'processing' transforms survey responses obtained during collection into a form that is suitable for tabulation and data analysis. It includes all data handling activities after collection and prior to analyzing it for interpretation. In order to streamline processing of data there are five essential elements *viz.*, Checking, Editing, Coding, Classification and Tabulation. Thus, data processing facilitates for statistical analysis of the data and interpretation of the research findings. After collecting the data, now the researcher is concerned about the following aspects:

- ☆ He checks, whether the schedules are properly filled are not
- ☆ Reduce the data or information into a manageable form
- ☆ Code the answers into numerical form
- ☆ Classify the data into different groups and sub-groups
- ☆ Tabulate the data
- ☆ Analyze the data with the suitable statistical techniques
- ☆ Interpretation of the analytical findings

Thus, in data processing, the researcher converts the raw data or information into a meaningful form. This facilitates to present the data in the form of graphical representation so that, it can be easily understood by the readers. The processing of data can be done either by manual or electronic. In the former, the researcher employs the master sheet and tabulates the data in accordance with the application of statistical techniques for analyzing them. However, this method is more laborious and time consuming. However, data processing in electronic form is more preferred, as it facilitates the researcher to use the data as he requires and thus, enable him to apply various statistical techniques.

a. Checking of Data

This implies, verification of data collected by the researcher to know, whether they are relevant and appropriate for the question posed on the respondents. For example, if the researcher posed the question *'what is the cost of cultivation of paddy in your farm'?* The answer in the schedule is written as Rs. 1400/acre. But, can a farmer cultivate the paddy with a meager amount of only Rs. 1400/acre and this amount is not even sufficient to purchase chemical fertilizers applied on the paddy crop. So, the correct answer could be Rs.14000/acre. Similarly, if the answer to the question, '*what is the market price of the paddy in September?*' is Rs. 120/quintal. This answer is wrong, as the MSP of paddy for the year 2011-12 is Rs.1080/quintal. So, the correct answer could be Rs.1200/quintal. So, such type of answers needs editing.

b. Editing the Data

This refers to the process of scrutiny of raw data collected to detect errors and omissions, wherever possible. Data editing is done at two stages *viz.*, at the time of recording the data and at the time of compiling and analysis of the data. It involves the following steps:

- ☆ ***Deciphering:*** In some cases, the handwriting of the data collectors is difficult to read and in such cases, deciphering is done by the researcher to make the data readable.
- ☆ ***Checking entries of data into the schedule:*** The researcher must also scrutinize the data collected in the schedule such that, whether all the questions are answered by the respondent and if any data or information is missing.
- ☆ ***Approximation:*** In this step, the researcher should edit the data by rounding off into a meaningful form and to convenient form so that, they are suitable for statistical analysis.

The big figures given by the respondent or collected through secondary sources are rounded off. For example, if the cultivated area of paddy is 245989 ha in a Country 'X', the researcher can conveniently approximate this datum as 2.46 lakh ha. Note that, many times, it is not necessary to give actual numbers and approximate figures serve the purpose for analyzing the data and for drawing the conclusions. This approximation of data facilitates the researcher to have a clear and easy grasp of figures. The extent to which approximation should be done depends upon the degree of accuracy desired in the data.

Editing ensures that the data is complete, error free and are ready to tabulate either manually or electronic means. The process of editing starts in the field itself (Field editing), where the researcher informally interacts with the respondents and he frequently checks the completed schedules in the very evening on the same day the data are collected to know, whether all the questions are answered and they are error free. In majority cases, the researcher goes for editing the data while compiling the data in his office and this refers to Central editing. The importance of editing can be understood by the following example:

Question: Whether you are aware about the importance of bio-fertilizers in cost-effective production of paddy?

(*a*) Yes

(*b*) No

(*c*) No response

In the above question, if the researcher edits the answer 'No' as 'No response', it leads to wrong conclusions. The answer 'No' means the respondent is not aware of the importance of bio-fertilizers nutrition in paddy. The answer 'No response' means, the respondent is totally unaware or not familiar with this type of nutrition on paddy crop.

Besides above, the researcher should also consider the following aspects regarding the data editing:

- ☆ *Completeness*: The researcher should ensure that, all the questions framed in the schedule must be completely filled. Note that, even if a single question is left unfilled, it will adversely influence the quality of the report at the end.
- ☆ *Accuracy:* This is very important aspect of data processing. The researcher should note that, while filling the schedule, the data recorded must be accurate (say, with reference to variables like size of land holding, fertilizers cost, pesticides cost etc.) and if this is not the case, the processing of the data leads to invalid findings.
- ☆ *Uniformity:* While editing the schedules, the researcher should be vigilant, whether the data collected for a question are uniform across all the schedules. For example if the researcher wishes to collect data about 'income' of the farmer, he should carefully check the data, whether the data is related to agricultural income alone in that particular year or total income in that particular year and the data pertaining to the selected variable (say, agricultural income only) should be collected from all the sample respondents. Such uniformity of data across the sample respondent is important to get reliable findings.

c. Coding of Data

Coding refers to the process of assigning a numerical values to the responses given by the respondents, so as to facilitate data analysis. Codes are fixed for the responses at the time of preparation of schedules or after completing the data collection. This process of assigning a codes to given responses and organizing them into a limited number of categories or classes is termed as Codification. Of course, the categories so developed from the codes are relevant to the research objectives and variables under consideration. The classes or categories created by codification process must have following properties:

- ☆ ***Exhaustiveness***: Implies there must be a class for every item of data.
- ☆ ***Mutually Exclusiveness:*** Implies one response can be put in one and only one cell of a category set.
- ☆ ***Unidimensionality***: Implies each class is defined in terms of only a single concept.
- ☆ ***Data analysis:*** Permit efficient analysis of data.
- ☆ ***Manageability:*** Facilitate reduction of voluminous data into manageable form.

Types of Coding

There are two types of coding *viz.*, Alphabetic coding and Numeric coding.

Alphabetic Coding

In this coding, alphabets are used as codes. For example, if the variable is 'organization', then the coding is as follows:

- ☆ **Private:** 'Pr'
- ☆ **Public :** 'Pu'
- ☆ **Cooperative:** 'Co'

Numeric Coding

In this coding, numbers are used as codes. For example, if the variable is 'Age of the farmers', then the coding is as follows:

Upto 20 years: '1'

21-40 years: '2'

41-60 years : '3'

Note that, when giving Zero ('0') code, the respondent should be very careful. This zero ('0') is given by the researcher under two situations *viz.*, when the respondent has not answered a particular question or when the respondent answered 'No' for a question. So, carefully, the researcher has to allocate two different codes if faces such type of responses from the respondents.

Coding in Case of Closed Questions

In the case of closed questions, the classes or categories are determined before collection, with the numerical code usually appearing on the schedule beside each response category. That is, the response categories for closed questions can be pre-coded on the schedule itself. For example:

- ☆ Question related to Sex: 1 for Male, 2 for Female
- ☆ Question related to Religion: 1 for Hindu, 2 for Muslim, 3 for Christian —— and so on
- ☆ Question related to Marital Status: 1 for Married, 2 for Unmarried, 3 for Widowed, 4 for Divorced —— and so on

Code boxes can be placed either next to the item to be coded or down the margin of the schedule. With computer-assisted collection methods, the codes are automatically captured when the respondent selects an answer. The benefits of closed questions include: they are less burdensome to respondents, capture and analysis are cheaper, faster and easier than for open questions. But, one disadvantage of closed questions is that, the respondent's natural wording is unknown.

Coding in Case of Open Questions

For open questions, coding is done after data collection. The coding may be either manual or automated. Manual coding commands interpretation and judgement skills on the part of the coder. That is, when the open questions have to coded manually, the coder after completing the data collection must thoroughly go through or read the data, interpret and manually convert a written response to an open question into a numeric code. This numeric code is then either recorded on the schedule or entered into a computer. The numeric code is assigned based on considering the presence of a key word in the answer for the given question. Sometimes, coding is based on the answer to only one question and sometimes it is based on the answer to several related questions. In the latter case, care should be taken to note the numeric codes, where such key word appears across the answers of different open ended questions. This all depends upon the clarity and completeness of the written response and how well the coder has interpreted the answers *i.e.*, skill of the coder. Traditionally, coding open questions has been a manual operation. However, recently, due to rapid advances in computer technology, automated coding gained momentum, considering the cost-effectiveness and timeliness in completing the coding operation. Here, two files are input to an automated coding system: One file contains the survey responses that are to be coded (called as Write-in file) and the other file is called the Reference file, which contain typical written responses (or phrases) and their corresponding numeric codes. Most often, the first step of automated coding is Parsing. Parsing is the process of standardizing a phrase in order to allow the computer to recognize equivalent phrases across the responses. Parsing usually consists of the removal of extraneous characters, such as punctuation, and the removal of double words, trivial words, certain suffixes and prefixes, etc. Both the survey write-in files and the reference file are parsed before continuing. The next step involves, for every write-in in the survey file, a search of the reference file for an exact match. If found, the code in the reference file is copied to the survey file and the record is considered coded. However, if an exact match cannot be found, an attempt can be made to find the closest match possible among the reference file records. Every record on the reference file is assigned a score, which is a measure of how similar the reference file phrase is to the survey response. Scores are evaluated based on predetermined parameters (which are specified to reduce the risk of making an error) and if a close enough match is found, the code is transferred to the survey response and the record is considered coded. Sometimes several reference file records are found with similar scores and sometimes no reference file record is found to be close to the survey response. In these situations, the records are generally sent to a small manual coding operation staffed with expert coders who are responsible for coding any records left uncoded at the end of the automated phase and for quality control of the output of the automated system (Survey Methods and Practice, Catalogue no. 12-587-X, Statistics, Canada).

Whatever may be the type of coding *i.e.*, either for closed questions or open questions, it involves assigning numbers or other symbols to answers for the responses given by the respondents so that, they can be grouped into limited number of classes or categories. However, this classifying of data into limited categories sacrifices some data details. But, still coding is essential for efficient analysis.

d. Classification of Data

After checking and editing the data for its completeness, accuracy and uniformity, the researcher has to classify the data based on certain criteria. Classification is the process of arranging data in groups or classes according to their resemblances and similarities. Thus, a large heterogeneous data are grouped into different homogenous groups such that, the data units having similar characteristics are placed in one group. So, in this manner, the entire data is divided into a number of classes. Two criteria are used *viz.*, qualitative and quantitative criteria for classifying the data and they are exemplified here under:

- ***Classification of data based on qualitative criteria:*** For example, based on sex, the farmers are categorized into male farmers and female farmers.
- ***Classification of data based on quantitative criteria:*** For example, based on size of land holding, the farmers are categorized into Small, Medium and Large farmers (Table 11.1).

Advantages of Classification of Data

The classification of data ensures the following advantages:

- Helps to separate homogeneous data from a large heterogeneous data (Table 11.1).
- It facilitates the researcher to easily handle the data and to draw comparative picture across different classes or groups.
- The researcher can make use of classified data in various ways and thus, facilitate him to get a range of possible solutions to address the research problem.
- Facilitate easy tabulation of data.

Table 11.1: Size and Class-wise Distribution of Land Holdings of Farmers in a Village.

Size and Class of Land Holding (ha)	*Number*	*Per cent to Total Number*	*Area (ha)*	*Per cent to Total Area*	*Per Capita (ha)*
Small (< 2.00 ha)	3022	75.81	1358.72	41.57	0.45
Medium (2.00 – 10.00 ha)	641	16.07	865.18	26.47	1.35
Large (≥10.01 ha)	323	8.12	1044.62	31.96	3.23
Total	3986	100.00	3268.52	100.00	0.82

Source : Gram Panchayat Office.

e. Tabulation of Data

After checking, editing, coding and classifying the data properly, now the researcher can tabulate the data. Tabulation is the final stage in data processing and it is the stepping-stone to the analysis and interpretation of data. The tabulation ensures pooling of data or information into a tabular formats specially designed for the purpose. This facilitates for easy descriptive interpretation, analysis and graphical representation of the pooled data. These tables can be prepared manually and/or by computers. For example, a researcher is conducting a survey on the research topic, *'Performance of Kurnool regulated agricultural market in Rayalaseema region of Andhra Pradesh'* and he included a question in the schedule posed on the selected commission agents in the selected regulated markets *viz., 'How much quantity of cotton (kapas) you handled in the year 2011-12?'* For this question, the researcher has

Table 11.2: Tabulation of Raw Data in Microsoft Excel Sheet Regarding Size-Distribution of Number of Commission Agents Handling Cotton (*Kapas*) in Kurnool Regulated Market (Univariate table).

	A	B	C	D	E	F	G	H	I	J	K	L
1	Class (Quantity in quintals)	Number of commission agents										
2	<150	3										
3	151-300	2										
4	301-450	1										
5	451-600	4										
6	601-750	2										
7	751-900	4										
		↓										
	Total	30										

Table 11.3: Tabulation of Raw Data in Microsoft Excel Sheet Regarding Size-Distribution of Percentage of Commission Agents Handling Cotton (*kapas*) in Kurnool Regulated Market.

	A	B	C	D	E	F	G	H	I	J	K	L
1	Class (Quantity in quintals)	Number of commission agents	% to total commission agents									
2	<150	3	10.00									
3	151-300	2	6.67									
4	301-450	1	3.33									
5	451-600	4	13.33									
6	601-750	2	6.67									
7	751-900	4	13.33									
		↓	↓									
	Total	30	100.00									

collected information from 30 selected commission agents. In the computer using MS-Excel sheet, the information is tabulated as given below (Tables 11.2 and 11.3).

From the above Table 11.2, the researcher can draw useful information as given through Table 11.3. That is, the data portrayed in Table 11.2 can be analyzed (by using percentages) and showed through Table 11.3 and this helps to draw useful inferences about the market structure of cotton (kapas). From the collected data regarding the quantity of cotton (kapas) handled by the commission agents, it is tabulated as shown through Table 11.4.

Table 11.4: Tabulation of Raw Data in Microsoft Excel Sheet Regarding Size-Distribution of Commission Agents and Quantity of Cotton (*Kapas*) Handled in Kurnool Regulated Market (Bivariate table).

	A	B	C	D	E	F	G	H	I
1	Class (Quantity in quintals)	Number of commission agents	% to total commission agents	% of total quantity handled					
2	<150	3	10.00	2.10					
3	151-300	2	6.67	3.89					
4	301-450	1	3.33	6.50					
5	451-600	4	13.33	8.46					
6	601-750	2	6.67	11.06					
7	751-900	4	13.33	14.39					
	Total	30	100.00	100.00					

Table 11.5: Average Share (per cent) of Maize Area, Production and Productivity at All-India Level in different Country Groups and World Before and After WTO.

Sl.No.	*Item*	*Period*	*South Asia*	*Asia**	*Developing nations*	*World*
1.	Area	1961-70	83.00(1)		7.83	4.69
		1971-80	83.20(1)		7.97	4.83
		1981-94	78.86(1)		7.12	4.51
		1995-05#	78.13(1)	15.16	7.07	4.67
		1961-05	80.57(1)		7.45	4.66
2.	Production	1961-70	77.80(1)		6.26	2.22
		1971-80	78.71(1)		4.96	1.75
		1981-94	78.27(1)		4.23	1.74
		1995-05	77.79(1)	7.30	4.33	1.95
		1961-05	78.15(1)		4.87	1.90
3.	Productivity comparison	1961-70	-6.27(5)		-20.02	-52.60
	(per cent)	1971-80	-5.40(5)		-37.87	-63.76
		1981-94	-0.74(3)		-40.67	-61.51
		1995-05	-0.42(2)	-52.00	-38.86	-58.39
		1961-05	-2.93(3)		-35.01	-59.27

Note : * - Data not available from 1961-94.

Figures in parentheses indicate India's rank among SAARC nations.

Raw Data Source: <www.fao.org>

Types of Tables

There are different types of tables depending upon the number of variables included in the tables. They include: Univariate table (if the data pertaining to only variable are shown), Bivariate table (if the data pertaining to two variables are shown) and Multivariate table (if the data pertaining to more than two variables are shown). Table 11.2 represents Univariate table, Bivariate and Multivariate tables are shown through Tables 11.4 and 11.5 respectively.

After tabulation of data, the researcher should be concerned with the following aspects:

- ***Specifying the title for the table:*** Note that, title should be as brief as possible, but at the same time should convey the full meaning. It should encompass the subject matter and time period (Table 11.6).
- ***Indicating captions and stubs:*** A Good table is expected to have the following aspects:
  - ***Title:*** Every table must have a suitable title and it should be self-explanatory. Note that, in presenting the table, the 'title of the table' should be at its top.
  - ***Table number:*** Every table must have table number as it helps in locating it. The table numbers must run in a consecutive serial order. In thesis, the tables are numbered chapter-wise *i.e.,* chapter number followed by serial number of the table. For example, in Chapter 3, the table numbers are given as 3.1, 3.2, 3.3 and so on. In Chapter 5, the table numbers are given as 5.1, 5.2, 5.3 and so on.
  - ***Captions:*** The headings of the columns are called 'Captions'. A table may have one or more captions. The captions should be brief and self explanatory (Table 11.5.1).

  Sometimes, the captions may indicate the units of numerical values given in the columns. Captions should also be numbered.

  - ***Stub:*** The headings of the rows are called Stubs. A table may have one or more stubs. The captions should be brief and self explanatory (Table 11.5.1).
  - ***Body:*** The body of the table comprises of data or numerical numbers. If the entries in a particular row or in column have the same units such as quintals, tonnes, lakhs or thousands, length in metres, percentages etc., the same can be indicated in respective stub or caption (Table 11.5.1).
  - ***Ruling and Spacing:*** For clarity purpose, ruling should be done to separate rows and columns. Further, all the columns and rows should be separated at equi-distances.
  - ***Foot notes:*** Any additional information about the numerical data or about captions or stubs can be given as footnotes below the table.
  - ***Source:*** The source of data should be clearly mentioned in next page figure.
- The data entered in the columns and rows must be clear and the last row or last column, which generally shows the respective totals must be larger than the others. Similarly, first column and first row of all columns must be larger than others, as they generally indicate titles of each column or each row along with units of measurement.
- As far as possible, below the table, source of data should be clearly mentioned (Table 11.6).
- As far as possible, much of the information should be presented on the left hand side of the table, as the readers' eyes move from left to right.
- Text references should identify tables by table numbers rather than by expressions like 'As shown through table given below' or 'As indicated through above table' etc. So, it is correct to mention. 'It is clear from the Table 11.5.1', 'As shown through Table 4.1' etc.

Table 11.5.1: Features of Good 'Table.

	Caption	Caption	Caption	Caption	Caption	Caption	Caption	Caption	Caption	Caption	Captio n
Stub		Body									
Stub									Body		
Stub											
Stub			Body								
Stub							Body				
Stub											
Stub											Body

Foot Notes:
Source:

Table 11.6: Public and Private Sectors Investments in Agriculture (1999-00 prices).

Years	*Investments in Agriculture (Rs. Crores)*			*Per cent Share of*		*Investment in Agriculture as per cent of GDP*
	Public	*Private*	*Total*	*Public Sector in Agril*	*Private Sector in Agril*	
1999-00	7716	35757	43473	17.7	82.3	1.6
2000-01	7155	31580	38735	18.5	81.5	1.6
2001-02	8746	38297	47043	18.6	81.4	1.6
2002-03	7962	38861	46823	17.0	83.0	1.5
2003-04	9376	35756	45132	20.8	79.2	1.4
2004-05	10267	38309	48576	21.1	78.9	1.3
2005-06*	13219	41320	54539	24.2	75.8	1.3

Note: * - Quick estimates.

Source: Economic Survey – Various issues.

As shown in the Tables 11.2 and 11.3, Univariate table mainly helps in describing the frequency of data of a particular variable, where as Bivariate and Multivariate tables helps in studying the relationship between the variables under consideration. The tabulation of data further helps in the following aspects:

- It presents an overall view of the research findings.
- It indicates the trends in the variables over a period of time.
- It help to draw a comparative picture across the selected variables.
- It provides a basis for employing various statistical techniques.

11.2. Data Analysis

Data analysis may be defined as the process of computation of certain parameters along with identification of relationship patterns that may exist among data groups. In the process of analysis, relationships may be discovered that may support or conflict the original hypothesis. This analysis leads to valid conclusions only if the relationship pattern stands the statistical test of significance. Note that, analysis of data involves both processing and interpretation of data. Processing of data involves transformation of data into convenient form that can be used for analysis purpose. Interpretation of data involves drawing inferences about the analytical findings based on the *a priori* knowledge the researcher possess. Thus, analysis of data serves in establishing a rapport between theory and empirical incidence. There are a number of issues that researchers should be cognizant of with respect to data analysis. These include:

a. Having the necessary skills to analyze.
b. Concurrently selecting data collection methods and appropriate analysis.
c. Drawing unbiased inferences.
d. Determining statistical significance.
e. Providing honest and accurate analysis.
f. Manner of presenting data.
g. Data recording method.
h. Training on statistical techniques.
i. Reliability and Validity.
j. Extent of analysis.

a. Having Necessary Skills to Analyze

The researcher should have sufficient skills to analyze the collected data with appropriate statistical techniques. For this, he should have undergone prior training for learning different statistical techniques and in interpreting the analytical findings. A common practice followed by the researchers is to defer the selection and execution of analytical procedures to a 'Statistician'. But, the researcher should have substantially more than a basic understanding of the rationale for selecting one method of analysis over another. This can allow him to discuss with the Statistician to employ relevant techniques for analyzing the collected data.

b. Concurrently Selecting Data Collection Methods and Appropriate Analysis

Note that, though data collection methods and analysis differ across the type of research problem, the optimal stage for determining them occurs early in the research process and should not be an afterthought. In the words of Smeeton and Goda (2003), *'statistical advice should be obtained at the stage of initial planning of an investigation so that, for example, the method of sampling and design of questionnaire are appropriate'*. Planning statistical analysis in advance will enable the researcher to fit right technique for the right type of data collected and thus, leads to framing of realistic and meaningful conclusions.

c. Drawing Unbiased Inference

It is essential to note that, any bias occurring in the collection of the data or selection of method of analysis, will increase the likelihood of drawing a biased inference. So, the researcher must be careful in overcoming this biased inference through selecting representative sample, collection of data without any bias, employing right techniques for analyzing the data etc.

d. Determining Statistical Significance

Mere findings of analytical data will not serve the purpose of drawing useful inferences about the problem and these findings should be significant statistically. Unless the findings are statistically significant, there lies no meaning in drawing the valid conclusions and in generalizing the same for the target population.

e. Providing Honest and Accurate Analysis

While fitting the statistical techniques, the researcher should take adequate care in reducing the likelihood of occurrence of statistical error. Common challenges include the exclusion of outliers, filling in missing data, transformation of data, data mining (Data mining is a powerful new technology (computer-aided process) with great potential to help companies focus on the most important information in the data they have collected about the behavior of their customers and potential customers. It discovers information within the data that queries and reports can't effectively reveal. Specific uses of data mining include: Market segmentation, Customer churn (predict which customers are likely to leave the company and go to a competitor), Fraud detection, Direct marketing, Interactive marketing, Market basket analysis (understand what products or services are commonly purchased together *e.g.*, beer and diapers), Trend analysis etc.) etc. (Shamoo and Resnik, 2003).

f. Manner of Presenting Data

Instead of presenting the data in raw form, the researcher may enhance the impression of significant findings by showing them in a format that is clearly understood by the readers. This presentation of data and analytical findings in a systematic format counts much in improving the quality of presentation of report.

g. Data Recording Method

Analyses could also be influenced by the method in which data was recorded. For example, researcher can collect the data or information in one or more of the following ways:

- ☆ Primary data collected through schedule (*i.e.*, through survey method).
- ☆ Secondary data collected from published reports.
- ☆ By framing closed ended questions or open ended questions.
- ☆ Collecting in depth data or information with reference to a particular case.

While each methodology employed has rationale and advantages, issues of objectivity and subjectivity may be raised when data is analyzed.

h. Training on Statistical Techniques

A major challenge to the researcher after collecting the requisite data or information is to employ a right statistical technique for the right type of data collected by him during the course of investigation. For deciding the right type of statistical technique to be employed, he should receive prior training on different statistical techniques by using statistical packages on the computer and to interpret the analytical findings.

i. Reliability and Validity

Researchers performing either quantitative or qualitative analyses should be aware of challenges to reliability and validity. In general, three factors influence the reliability of analyzed data *viz.*, stability, reproducibility and accuracy. When the researcher is certain about what he is measuring, it refers to reliability. Validity of analyzed data refers to the generalizability of results.

j. Extent of Analysis

The researcher should be aware, whether he has to employ statistical or non-statistical methods in analyzing the data. While statistical analysis is typically performed on quantitative data, there are numerous analytical procedures specifically designed for qualitative material. Regardless of whether one studies quantitative or qualitative phenomena, researchers use a variety of tools to analyze data in order to test hypotheses, discern patterns of behaviour and ultimately answer research questions.

11.3. Phases of Data Analysis

From the above discussion, it is quite easy now to understand the meaning of data analysis, which refers to a process of inspecting, cleaning, transforming, and modeling data with the goal of highlighting useful information, suggesting conclusions, and supporting decision making. Thus, data analysis is a 'process', within which several phases can be distinguished *viz.*, Data cleaning, Initial data analysis and Main data analysis (Figure 11.1).

a. Data Cleaning (Data Cleansing or Data Scrubbing)

It is the process of amending or removing data in a database that is incorrect, incomplete, improperly formatted, or duplicated. It also refers to a procedure, where the data collected by the researcher both from primary and secondary sources are thoroughly inspected. That is, if the collected data comprises erroneous information, they will be discarded and again re-collected and similarly, if any data are unwanted, they are discarded, while compiling the data. This data cleaning is done by the researcher at the time of data compilation and entry into the computer. So, the aim of data cleaning is to ensure a clean data from which as many errors as possible have been removed prior to starting the statistical analysis. Data cleaning involves monitoring the following:

- ☆ Presence of duplicate data (*i.e.*, the same respondent occurring more than once).
- ☆ Out-of-range values should be removed.
- ☆ Incomplete, incorrect, inaccurate, irrelevant data will be removed and replacing or modifying them with reliable data.
- ☆ Removing typographical errors in the data.
- ☆ Cross checking the collected data with a validated data set.
- ☆ May also involve harmonization of data. For example, harmonization of short codes ('mt', 'qtls' etc.) to actual words ('meters', 'quintals' etc.).

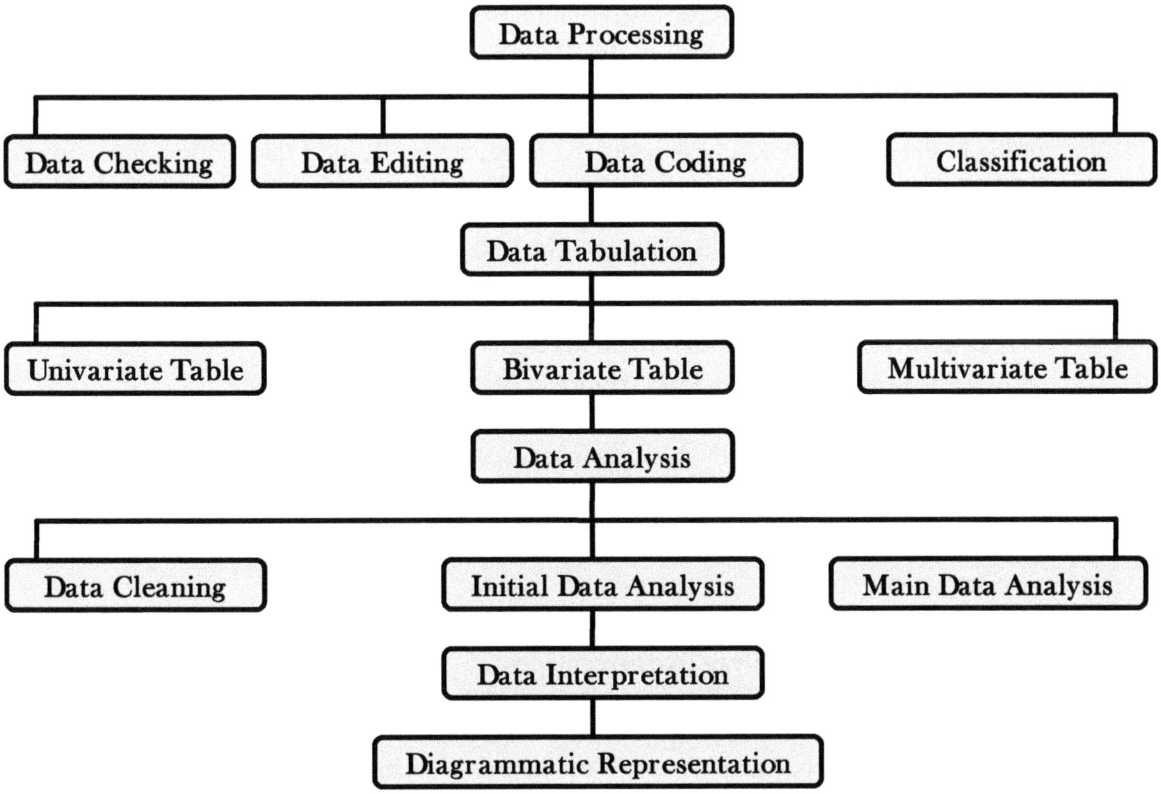

Figure 11.1: Steps in Data Analysis.

b. Initial Data Analysis

The most important distinction between the initial data analysis phase and the main analysis phase is that, in the former, the main objective is to search for errors and outliers in the collected data and in the latter, the objective is to go for in depth (statistical) analysis of the refined data to obtain valid findings. The initial data analysis phase includes the following four aspects:

- ***Quality of data:*** This should be given utmost importance, as the analytical findings are directly dependent on the quality of data used in the analysis. Data quality can be assessed in several ways, using different types of analyses: frequency counts, descriptive statistics (mean, standard deviation, median), normality (skewness, kurtosis, frequency histograms, normal probability plots), associations (correlations, scatter plots) etc. Further, the researcher should check about the missing data, if any. Note that, the choice of suitable statistical analysis depends upon the quality of data.
- ***Transformation of data:*** After assessing the quality of the data, the researcher should think about appropriate transformation of data to employ various statistical analyses. For example, if the researcher employs regression model to overcome autocorrelation problem, he will go for transformation of data by subtracting the times of estimated value of autocorrelation coefficient ($\hat{\rho}$) the value of variable in the previous period from the original observation. The other transformations include: Square root transformation (if the distribution differs

moderately from normal), Log-transformation (if the distribution differs substantially from normal), Inverse transformation (if the distribution differs severely from normal), make categorical (ordinal/dichotomous) (if the distribution differs severely from normal, and no transformations help).

In the Initial data analysis phase itself, the researcher will study the relationship between the variables, which form the basis for Main data analysis phase. The relationship between the variables can be assessed by looking at basic statistics of important variables, scatter plots, correlations, cross-tabulations etc. The findings of this initial data analysis are to be documented, and if necessary, corrective actions are also to be taken. Also, the original plan for the main data analysis can and should be specified in more detail and/or rewritten. A very brief list of the more popular methods is given here under:

- ***Measures of Central tendency:*** Arithmetic Mean, Median and Mode.
- ***Measures of dispersion:*** These are of two types *viz.*, Absolute Measures and Relative Measures. Most common absolute measures of variability include Range, Semi-Inter quartile Range or Quartile Deviation, Mean Deviation, Variance and Standard Deviation. Most common relative measures of variability include Coefficient of Variation, Coefficient of Mean Deviation, Coefficient of Quartile Deviation and Coefficient of Standard Deviation.
- Scatter diagram and Correlation coefficient.
- Other descriptive analysis like percentages.

c. Main Data Analysis Phase

In this phase, the researcher employs relevant statistical techniques to study the relationship between the variables, so as to validate the research findings. Many statistical techniques* have been used for statistical analyses. A very brief list of the more popular methods is given here under:

- Inferential tests for difference of means (Z-tests).
- Inferential parametric tests for significance (F-tests, Z-tests, t-tests).
- Inferential nonparametric tests for significance (chi-square, Mann-Whitney, Kruskal-Wallis).
- Multiple Linear Regression (MLR) Model.
- Logistic regression.
- Generalized linear models.
- General log-linear models.
- Analysis of Covariance (ANCOVA) Model.
- Time series analysis – Trend, Cyclical, Seasonal and Irregular variations.
- Auto Regressive Integrated Moving Average (ARIMA) Model.
- DEA technique.
- Linear programming.

---

* The students should refer the text books on 'Statistics' for studying different statistical techniques in-detail. The inclusion of these techniques is beyond the scope of this book.

Main data analysis mainly aims at two important aspects *viz.*, statistical estimation of population parameters from the results of sample data analysis and testing of H_0 to assess the probability of sample statistical findings under assumed population conditions i.e., the extent probability under which the sample statistical findings can be generalized back to the population. Presently, software packages are available like Q research software (quantitative data analysis software for market research), Quantum (part of the SPSS MR product line, mostly for data validation and tabulation in Marketing and Opinion Research), RATS (comprehensive econometric analysis package), SAS (comprehensive statistical package), SHAZAM (comprehensive econometrics and statistics package), Simul (econometric tool for multidimensional (multi-sectoral, multi-regional) modeling), SOCR (online tools for teaching statistics and probability theory), Speakeasy (numerical computational environment and programming language with many statistical and econometric analysis features), Stata (comprehensive statistics package), Statgraphics (general statistics package), STATISTICA (comprehensive statistics package), StatXact (package for exact nonparametric and parametric statistics), Systat (general statistics package), S-PLUS (general statistics package), Unistat (general statistics package that can also work as Excel add-in) etc., to analyze the compiled data for applications of various statistical techniques like correlation coefficients, MLR, time series analysis, ARIMA model, DEA approach etc. These software packages enable the researcher to employ various statistical techniques that run at ease, greater speed and with accuracy. For an Agricultural Economist (Scientist or Professor) or a research student in the field of Agricultural Economics need not have a thorough knowledge on software programme, but it is just sufficient to have knowledge in running the programme and interpretation of analytical findings.

11.4. Statistical Analysis to be Employed

Tables 11.7 and 11.8 shows general guidelines for choosing a statistical analysis. Note that, these are only general guidelines and should not be construed as hard and fast rules. As discussed earlier, the data could be analyzed in multiple ways, each of which could yield legitimate answers. They cover a number of common analyses and helps the researcher to choose among them based on the number of IVs and DVs, nature of IVs and DVs etc. The Table 11.9 shows one or more statistical tests commonly used given these types of variables (but not necessarily the only type of test that could be used).

11.5. Graphical Representation of the Analytical Findings

In social sciences research, graphical representations were given much importance to highlight the trends or relationship of the study variables. The importance of graphical or pictorial representation can be highlighted from the saying that, *'one picture is more than a thousand words'*. Various forms of graphical representations like bar diagram, histogram, line graphs, pre charts etc. (Figures 11.2 to 11.5) will be used and these help in interpreting the data quickly. Note that, in presenting the figure, the caption should be at its bottom.

11.6. Concept of Scaling

The word 'Scale' or 'Scaling' is generally used for measuring something. For example, we use Barometer to measure atmospheric pressure, we use Thermometer to measure temperature etc. Similarly, in Social Sciences research, we employ 'Scales' to quantify the qualitative data given by the respondents. However, measuring the physical phenomenon is relatively an easy task compared to measurement of responses in Social Sciences research. For example, if the research study is directed towards measuring the *'Effectiveness of Professors in rendering quality education in Agricultural Colleges of ANGRAU'*, then the

Table 11.7: Statistical Analysis to be Employed for different Types of Variables.

Number of Dependent Variables	*Nature of IVs*	*Nature of DV(s)*	*Test(s)*
1	0 IVs (1 population)	interval and normal	one-sample t-test
		ordinal or interval	one-sample median
		categorical (2 categories)	binomial test
		Categorical	Chi-square goodness-of-fit
	1 IV with 2 levels	interval and normal	2 independent sample t-test
	(independent groups)	ordinal or interval	Wilcoxon-Mann Whitney test
		Categorical	Chi-square test
			Fisher's exact test
	1 IV with 2 or more levels	interval and normal	one-way ANOVA
	(independent groups)	ordinal or interval	Kruskal Wallis
		Categorical	Chi-square test
	1 IV with 2 levels (dependent/	interval and normal	paired t-test
	matched groups)	ordinal or interval	Wilcoxon signed ranks test
		categorical	McNemar
	1 IV with 2 or more levels (dependent/matched groups)	interval and normal	one-way repeated measures ANOVA
		ordinal or interval	Friedman test
		Categorical	repeated measures logistic regression
	2 or more IVs (independent groups)	interval and normal	factorial ANOVA
		ordinal or interval	ordered logistic regression
		categorical	factorial logistic regression
	1 interval IV	interval and normal	correlation
		interval and normal	simple linear regression
		ordinal or interval	non-parametric correlation
		categorical	simple logistic regression
	1 or more interval IVs and/or 1 or	interval and normal	multiple regression
	more categorical IVs		analysis of covariance
		categorical	multiple logistic regression
			discriminant analysis
2+	1 IV with 2 or more levels (independent groups)	interval and normal	one-way MANOVA
	2+	interval and normal	multivariate, multiple linear regression
	0	interval and normal	factor analysis
2 sets of 2+	0	interval and normal	canonical correlation

Source: James D. Leeper, (Choosing the Correct Statistic').

Table 11.8: Choosing the Correct Statistical Test.

Numberof Dependent Variables*	*Number of Independent** Variables*	*Type of Dependent Variable(s)*	*Type of Independent Variable(s)*	*Measure*	*Test(s)*
1	0 (1 population)	Continuous normal	not applicable (none)	Mean	One-sample t-test
		Continuous non-normal		Median	One-sample median
		Categorical		Proportions	Chi-square goodness-of-fit, binomial test
	1 (2 independent populations)	Normal	2 categories	Mean	2 independent sample t-test
		Non-normal		Medians	Mann Whitney, Wilcoxon rank sum test
		Categorical		proportions	Chi-square test Fisher's Exact test
	0 (1 population measured twice) or	normal	Not applicable/ Categorical	Means	Paired t-test
	1 (2 matched populations)	Non-normal		Medians	Wilcoxon signed ranks test
		Categorical		Proportions	McNemar, Chi-square test
	1 (3 or more populations)	Normal	Categorical	Means	one-way ANOVA
		Non-normal		Medians	Kruskal Wallis
		Categorical		Proportions	Chi-square test
	2 or more (*e.g.*, 2-way ANOVA)	Normal	Categorical	means	Factorial ANOVA
		Non-normal		Medians	Friedman test
		Categorical		Proportions	Log-linear, logistic regression
	0 (1 population measured 3 or more times)	Normal	Not applicable	Means	Repeated measures ANOVA
	1	Normal	Continuous		Correlation, Simple linear regression
		Non-normal			Non-parametric correlation
		Categorical	Categorical or continuous		Logistic regression
			Continuous		Discriminant analysis
	2 or more	Normal	Continuous		Multiple linear regression
		Non-normal			
		Categorical			Logistic regression
		Normal	Mixed categorical and continuous		Analysis of Covariance General Linear Models (regression)
		Non-normal			
		Categorical			Logistic regression
2	2 or more	Normal	Categorical		MANOVA
2 or more	2 or more	Normal	Continuous		Multivariate multiple linear regression
2 sets of 2 or more	0	Normal	not applicable		Canonical correlation
2 or more	0	Normal	not applicable		Factor analysis

*: Outcome; **: Predictor

Source: James D. Leeper ('Choosing the Correct Statistic').

Table 11.9: Choosing the Correct Statistical Test.

		Dependent Variable	
		1. Attribute	2. Numeric
Independent Variable	1. Attribute	3. Chi-square Non-Parametric Techniques	4. ANOVA ANCOVA Repeated Measures ANOVA (Z-tests & t-tests)
	2. Numeric	5. Discriminant Analysis	6. Correlation Regression Multiple Regression

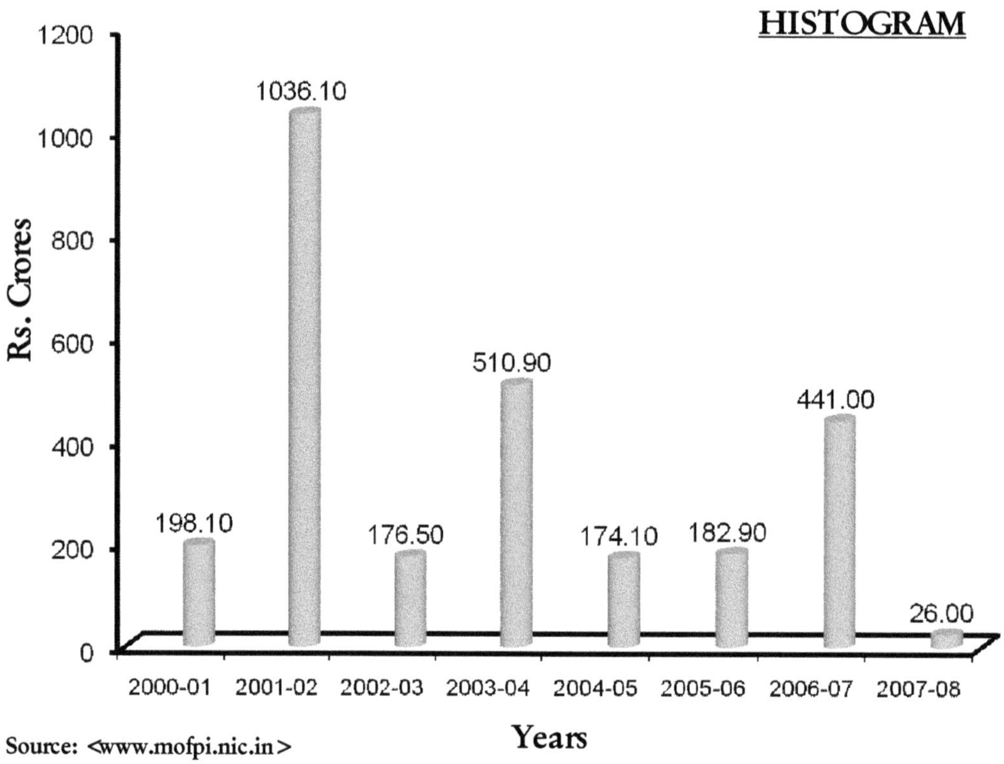

Source: <www.mofpi.nic.in>

Figure 11.2: FDI Flows into the Food-Processing Sector in India.

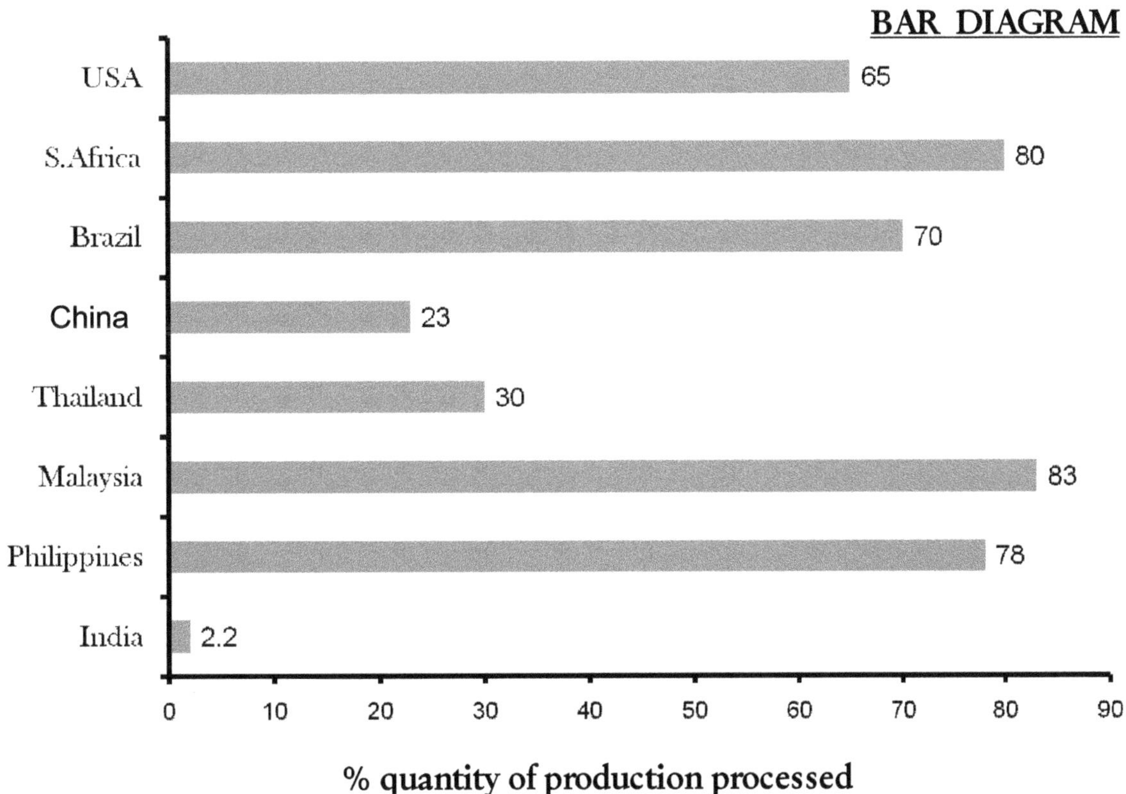

Source: <www.mofpi.nic.in>

Figure 11.3: Percentage of Quantity of Production of Fruits and Vegetables Processed in different Countries.

researcher has to conduct interviews on selected students in different Agricultural Colleges of ANGRAU. However, it is quite difficult to measure emotions, attitudes, faiths, values etc., through the devices that are used for measuring the physical data. In this context, Scaling of data comes to the hand of the researcher in quantification of the above qualitative responses. Even in Social Sciences research (*'Effectiveness of Professors in rendering quality education in Agricultural Colleges of ANGRAU'*), some data like payment made to the Professors, teaching experience, age, number of teaching hours etc., can be quantified, but some variables like emotions, social attitudes, commitment towards students etc., cannot be quantified. In this context, scaling or measurement is quite important in the field of Social Sciences research. The above explanation infers that, Scaling involves the conversion of the characteristics or qualitative data into quantitative data. After this conversion, the scaling is done. This has to be done because, qualitative data or measurement are mostly subjective and differ from researcher to researcher. Unless it is converted into quantitative data, the measurement would not be possible. The following are the other important merits of scaling:

- It makes the qualitative information quantitative and thus, makes the research study more scientific.
- It ensures objective measurement of information.

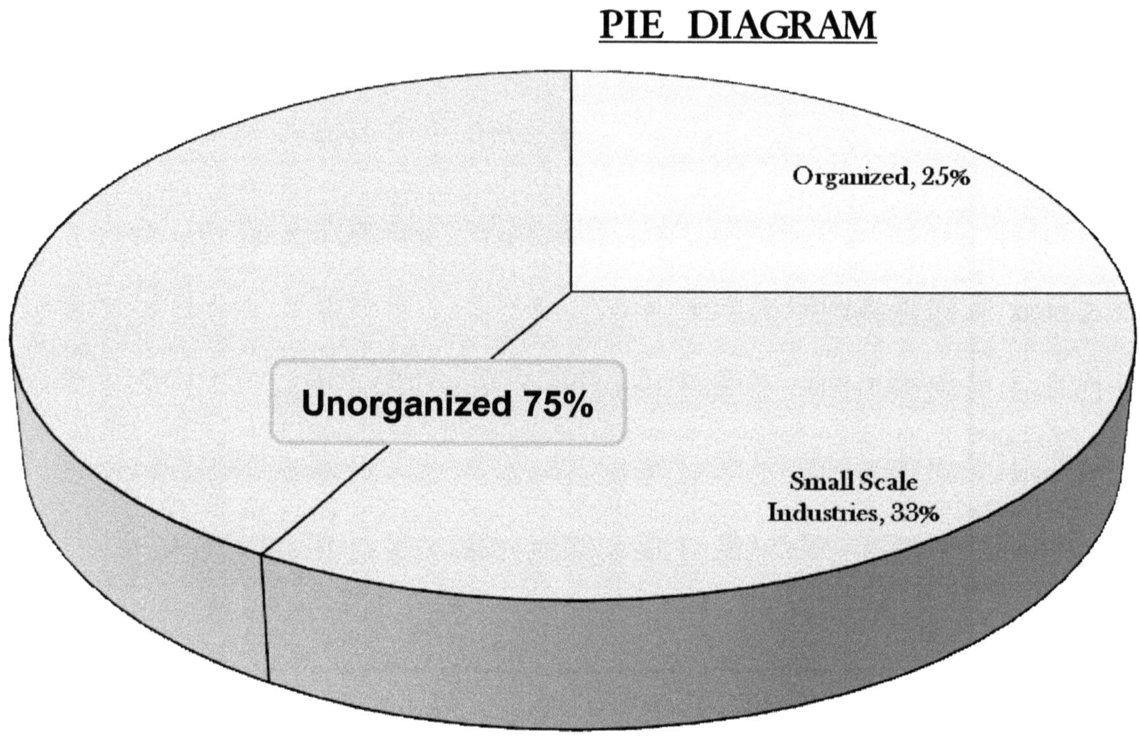

Figure 11.4: Structure of Indian Food Processing Industry.

Types of Scales

The scales may be broadly classified on one or more of the following basis:

1. Scaling based on Response Form

Based on this criterion, there are two types of scales *viz.*, Categorical scale and Comparative scale.

- *Categorical scale:* This is also called as Rating scale. In this scaling, the respondent is simply asked to rate the preference of a particular object without comparing with other related objects. For example, the researcher may ask the respondent to rate the SRI production technology of paddy as 'Very Good' or 'Good' or Not bad' or 'Bad' or 'Very Bad'.
- *Comparative scale:* In this scaling, the researcher ask the respondent to scale the object by comparing with other related objects. So, the respondent will rank different objects of comparison on priority basis. So, this scale is also known as Ranking scale. For example, the researcher may ask the respondent to rank different production technologies of paddy *viz.*, SRI, Transplanted technology, Semi-dry technology, Alternate wetting and Drying method etc., on preference basis for cultivating the crop.

2. Scaling Based on Subjectivity

Based on this criterion, there are two types of scaling the responses *viz.*, Subjective personal preferences and Non-preference judgments. For example, in the former type, the researcher may ask

LINE DIAGRAM

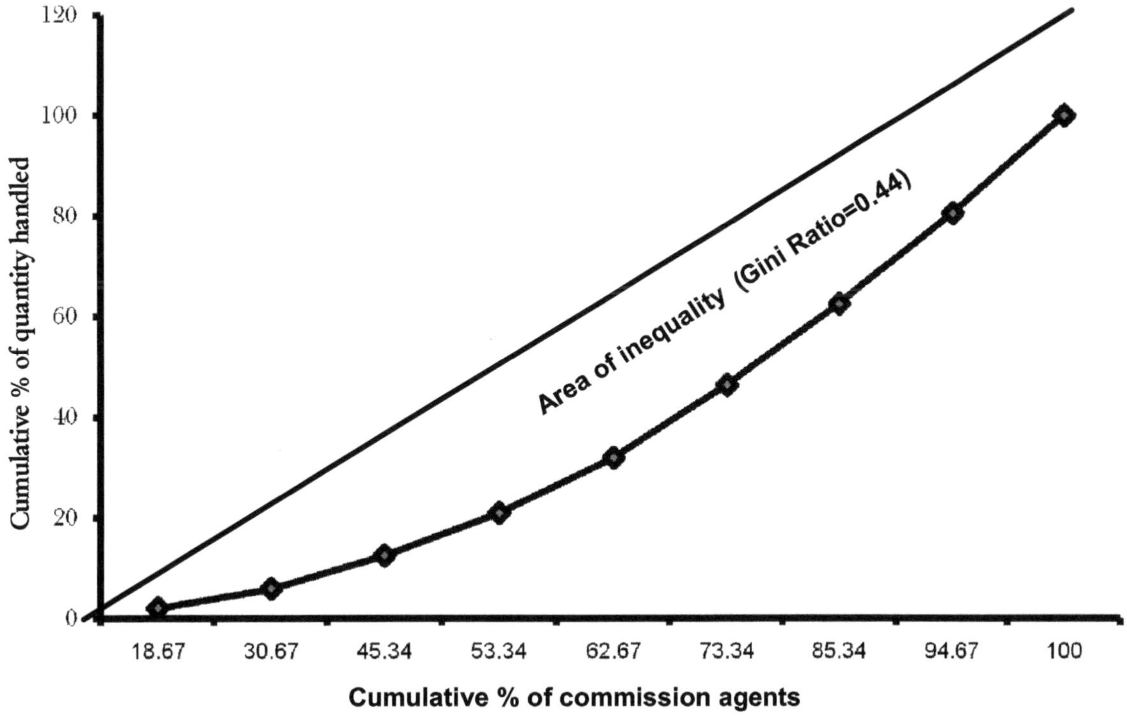

Figure 11.5: Lorenz Curve for Commission Agents Handling Paddy in Kurnool Regulated Market (2010-11).

the farmer, which form of pest control he prefers like application of chemical pesticides, biological control, neem based pesticides, IPM etc. In the latter type, the respondent may be asked to rank the above pest control measures based on the criterion of cost minimization or effectiveness in controlling the pests.

3. Scaling Based on Properties

Based on the property of mathematical comparisons such as Order (say, one number is greater than or less than or equal to another), Distance (*i.e.*, the difference between numbers are ordered. Say, the difference between any pairs of numbers is greater than or less than or equal to any other pair of numbers) and Origin (*i.e.*, the number series has a unique origin indicated by number zero), there are four types of scales *viz.*, Nominal, Ordinal, Interval and Ratio scales.

- ☆ ***Nominal Scale:*** This is the lowest level of measurement. In this scaling, numerical values are assigned to objects in order to label them. For example, cricket players are numbered on their jersey, which helps commentators and spectators to identify them. Similarly, the respondents can categorized into male and females, urban and rural, Hindus, Muslims and Christians etc. So, this scale helps to keep a track of objects (or events and people). Note that, assigning numbers or classifying into different categories have no meaning, as these numbers cannot be ranked or added. In this context, some experts consider this is not a measurement. But, since the members can be counted by assigning the numbers, it can be considered as a scale. The demerit of using this scale is that, not much statistical analysis can be done. Mode (a measure of central tendency) and Chi-square test are commonly employed to analyze the data (Table 11.10). Compared to other scaling techniques, this is more simple, as it helps to categorize the objects into different categories. However, this method of scaling is least powerful, as no statistical tests can be employed and hence, it has only limited use in research. This scale does not suggest any order or distance or origin relationship between the categories.

- ✰ ***Ordinal Scale:*** Ordinal scale includes the characteristics of the nominal scale plus an indicator of order. For example, If 'A' is greater than 'B' and 'B' is greater than 'C', then 'A' is greater than 'C'. Similarly, the researcher will place some criteria *viz.*, selection of crop enterprise, production technology, soil test based fertilizers application, IPM, adoption of recommended package of practices etc., to study the managerial ability of farmers in taking up farm business. The researcher will assign different scores to different criteria. By summing up these scores across different criteria, a final score will be measured to assess the managerial ability of the farmer. This type of scale is also known as Summated Rating Scale. Even ranks are also assigned for the objects (persons) in ascending order (from lowest to highest *i.e.*, 1, 2, 3 and so on) or descending order (from highest to lowest *i.e.*, 10, 9, 8 and so on). Note that, these numbers only indicate the rank order and they do not indicate the intervals between them. That is, the real difference between ranks 1 and 2 may be more than or less than 2 and 3. So, the ordinal scale indicates a statement of 'greater than' or 'less than' without stating how much greater or less. Other descriptors can be: 'superior to', 'happier than', 'richer than', 'more difficult', 'above' etc. For example, based on the income earned, the respondents can be classified as High, Middle and Low income groups. Thus, ordinal scale attempts to place objects or events in a particular order. That is, it permits the ranking of objects from lowest to highest or highest to lowest. However, in this scale, there is no attempt to have equal intervals of the scale. For example, the ordinal scale asks the respondent to rate the use of a product (say, new brand of pesticide) as 'Excellent', 'Very Good', 'Good', 'Not Bad', 'Bad' etc. But, this rating does not indicate, how many times 'Excellent' is superior to 'Good' or over 'Not Bad' etc. Thus, the simplest ordinal scale is a ranking. When a market researcher asks the respondent to rank 5 types of beer from most flavourful to least flavourful, he/she is asking the respondent to create an ordinal scale of preference. There is no objective distance between any two points on the subjective scale. For the respondent, the top beer may be far superior to the second preferred beer but, to another respondent with the same top and second beer, the distance may be subjectively small. So, ordinary scales are sometimes referred as Ranking scales or Rank Order Scales. Thus, ordinary scale indicates only ranks and ranks order and nothing more and even the absolute difference between adjacent ranks may not be equal. Regarding the use of statistical measures, 'Median' (a measure of central tendency) is commonly used for ordinary scale. Other techniques like measures of dispersion (percentile or quartile method), rank order coefficients are also employed (Table 11.10).
- ✰ ***Interval Scale:*** Interval scales have the power of nominal and ordinal scales plus one additional strength *i.e.*, they incorporate the concept of equality of interval. For example, the distance between 1 and 2 equals the distance between 2 and 3, and the elapsed time between 3 and 6 A. M. equals the time gap between 4 and 7 A. M. One cannot say, however, 6 A.M. is twice as late as 3 A.M. because 'zero time' is an arbitrary origin. It can be said that, an increase in temperature from 80^0C to 90^0C is same increase in temperature as an increase from 30^0C to 40^0C. However, it cannot be said that, the temperature of 80^0C is twice as hot as the temperature of 40^0C because, both numbers are dependent on fact that the zero on the

scale is arbitrary and fixed as the temperature of the freezing point of water. When the respondent is asked to rate his satisfaction in consuming a processed food product on a 7 point scale, from Dissatisfied to Satisfied, he was using an interval scale. So, in an interval scale, it is assumed to have equidistant points between each of the scale elements. This means that, we can interpret differences in the distance along the scale. We contrast this to an ordinal scale, where we can only talk about differences in order, not differences in the degree of order. The concept of interval scale can be further understood from the following simple example: If the IQ of two farmers is 100 and 150, in nominal scale, it implies, the two farmers have different IQs, in Ordinal scale, the first farmer has lower IQ than second farmer and in Interval scale, the IQ of first farmer is 50 per cent lower than second farmer or the IQ of second farmers is 50 per cent higher than first farmer. (In Ratio scale, discussed below), it implies, the ratio of IQs two selected farmers is 1:1.5. Altitude based on sea level and calendar years are other examples for interval measurement. Interval scales provide more powerful measurement than ordinal scale because, in the former, the concept of equality of interval is inherent. Arithmetic Mean is the appropriate measure of central tendency, while standard deviation is the most widely used measure of dispersion. Tests for statistical significance used in case of interval scales are the 't' test and 'F' test (Table 11.10).

☆ ***Ratio Scale:*** Also referred as Absolute scale. Ratio scales incorporate all three powers of the above previous scales *i.e.*, 'Naming' under Nominal scale, 'Naming' and 'Ranking' under Ordinal scale and 'Naming', 'Ranking' and 'Equal Intervals' under Interval scale plus the provision for absolute zero or origin. The absolute zero represents a point on the scale,

Table 11.10: Comparative Picture across Different Scales of Measurement.

Criteria	*Nominal*	*Ordinal*	*Interval*	*Ratio*
Properties of measurement	Naming	Naming and ranking	Naming, ranking and equal intervals	Naming, ranking, equal intervals and zero point
Characteristics	No order, no distance and no origin	Will have order, no distance and no origin	Will have order and distance, but no origin	Will have all the three characteristics *viz.*, order, distance and origin.
Nature of measurement	Categorical	Ranking	Scoring	Scoring
Nature of underlying construct	Discrete	Discrete or Continuous	Continuous	Continuous
Mathematical functions	None	None	Addition and Subtraction	Addition and Subtraction
Central tendency	Mode	Median	Mean	Mean
Popular statistical tests used	χ^2 test, Coefficient of contingency test	'U' test, Rank order coefficients, Percentile or quartile measure	Standard deviation, Pearson's 'r', 't'- test, 'F' test	All statistical techniques used at nominal, ordinal and interval scales can also be used in ratio scale.
Examples	*Gender:* Male and female *Religion:* Hindus, Muslims and Christians	*Income:* High, Middle and low	*IQ:* IQ of farmer A is 50 per cent lower than farmer B. *Age:* Age of farmer A is twice as old as farmer B.	*IQ:* IQ of farmer A to B is 1:2. *Age:* Age of farmer A to B is 2:1.

where there is an absence of the given attribute. Since the scale assumes absolute zero origin, it directly indicates the proportion of one value to another. For example, as discussed above, if the IQ of two farmers is 100 and 150, in Ratio scale, it implies, the ratio of IQs two selected farmers is 1:1.5. Similarly, if the ratio between the salaries of Assistant Professor and Associate Professor is 1:2, and if an Assistant Professor gets Rs.30000/month, then Associate Professor gets Rs.60000/month. So, the ratio scale facilitates a comparison between the objects, which was not possible with Interval scales. The merit of this ratio scale is that, all mathematical operations are possible. So, this is the highest and most ideal scale of measurement. Variables measured at highest level can also be measured at lower levels and not *vice versa.* That is, variables that can be measured at ratio level, can also be measured at interval, ordinal and nominal scales. Measures of central tendency can be computed by calculating geometric and harmonic means. Further, all statistical techniques used at nominal, ordinal and interval scales can also be used in ratio scale (Table 11.10).

It can be inferred from the above discussion that, starting from normal scale and progressing to ratio scale, the quantum of information increases progressively.

4. Scaling Based on Judging the Attitude

Based on this criterion, there are two types of scales *viz.*, Rating scale and Attitude scale.

☆ ***Rating Scale:*** In rating scale, the rater makes a judgment about the characteristics of the subject or attitude *'not on the basis of the opinions of the subjects, but on the basis of the opinions and judgment of the investigator himself'*. Thus, in rating scales discretion of the investigator comes into the picture of the measurement of characteristics of the subject. In rating scale, the rater collects the data through various means like non-verbal behavior, verbal behavavior, personal documents etc. Note that, compared to speechless behaviour, the easier method of understanding the attitude is that of conversation. In addition to conversation, various facial expressions and fluctuations in the volume of sound are also good indicators of attitude. In the words of Kietch and Cruchfield, *'it is a common belief that, what a man says may be less revealing than how he says it.'* Thus, the manner of expressing a response may become more informative than the verbal expression itself.

Rating scale is a very useful device in assessing quality, especially when quality is difficult to measure objectively. For example, *'How good was the modern technology of paddy production adopted by the farmer?* This question can hardly be answered objectively. So, rating scale record judgments or opinions and indicates the degree or amount of different degrees of opinions, which are arranged along a line on the scale. Considering the above, the following is the rating scale:

Excellent	*Very Good*	*Good*	*Average*	*Poor*	*Very Poor*

Rating scales have been successfully utilized for measuring the following:

❖ Farmer's performance.

- ❖ Effectiveness of a technology.
- ❖ Rating the stress, emotions, intelligence etc., of a person.
- ❖ Appraisal about an activity, courses etc., in the college.

However, rating scale suffers from the following limitations:

- ❖ In general, raters would not like to run down their own people or activities by giving low ratings. So, they give higher ratings to almost all cases. This false rating (*i.e.*, rating on a higher side) refers to 'generosity error of rating'.
- ❖ Some raters play the safe game and hence, they rate near the midpoint of the scale.
- ❖ Some raters are very strict and cautious and hence, they are hesitant in rating on average and higher sides. That is, they show the tendency to rate low in all cases. This is called Stringency error, which is opposite to generosity of error.
- ❖ Sometimes, rater rates one aspect that is influenced by rating another aspect. This is called Hallo error.
- ❖ Sometimes, rater rates the aspects without properly understanding the terms used in the question. Such error is called as the Logical error.

☆ ***Attitude Scale:*** In the words of Barr David Johnson, an attitude may be defined as *'a learned emotional response set for or against something'*. It is a tendency of an individual to read in a certain way towards a phenomenon. It is what a person feels or believes in. It is the inner feeling of an individual. It may be positive, negative or neutral. In general, the two terms 'Opinion' and 'Attitude' are used interchangeably, but these two are significantly different. An opinion may not lead to any kind of activity in a particular direction. But, an attitude compels one to act either favourably or unfavourably according to what they perceive to be correct. The respondents' attitude can be evaluated with the help of schedule . But, it is ill adapted for scaling accurately the intensity of an attitude. So, 'Attitude scale' is developed to minimize the difficulties of schedule in defining the attitude in terms of a single attitude object. All items, therefore, may be constructed with graduations of favour or disfavour.

Characteristics of Attitude Scale

The following are the essential characteristics of an attitude scale:

- ❖ It provides for quantitative measure on a uni-dimensional scale of continuum.
- ❖ It uses statements from the extreme positive to extreme negative position.
- ❖ It generally uses a five point scale.
- ❖ It could be standardized.

Types of Attitude Scale

The following are the popular and commonly used methods of measuring attitudes indirectly for research purposes:

(*a*) ***Thurstone Scale (Differential scale)***: This scale is used when attitude is accepted as a uni-dimensional linear continuum. This scale generates a group of indicators about a particular variable (say, attitude of farmers towards using bio-fertilizers) that can be empirically verified. This scale involves equi-interval scaling of statements of respondents, which facilitate to compare easily and draw the differences in their attitudes. The following is the general procedure of measuring attitude through this scale:

- ❖ Collection of attitude statements from the available literature (say, responses of farmers about their attitude towards using bio-fertilizers), professional extension workers, agricultural scientists etc. Attitude statement is the opinion made on a particular item.
- ❖ Careful scrutiny and editing of the attitude statements. These statements should be clear, short and contain only one idea. Double negatives and ambiguous statements must be avoided. Out of the large number of statements collected from the literature, finalize the generalized statements about the item under consideration.
- ❖ Mail these finalized statements to selected judges to rate these statements on 11 point equal appearing interval scale. The judges were requested to sort out the finalized statements about bio-fertilizers use on the 11-point scale. The number of judges used by Thurstone was 300. Each judge ranks each statement on the 11 point scale. Thus, one judge might rank a statement in 2nd category and the same response might be ranked by another judge on 7th category and so on. So, finally, all these rankings are combined into one master sheet of 11 categories. Note that, the judges are asked not express their opinion, but to sort them at their face value. The items, which bring out a marked disagreement between the judges are discarded.

1	2	3	4	5	6	7	8	9	10	11
Most Favourable				Neutral						Most Unfavourable

Compute the Scale Value of each Statement

The median of the distribution of ranking for each statement was considered as a Scale value of each statement. It is computed as below:

Attitude Statement 1	*Sorting Categories*										
	1	*2*	*3*	*4*	*5*	*6*	*7*	*8*	*9*	*10*	*11*
Frequency (f)								45	57	86	112
Proportions (p)								0.15	0.19	0.29	0.37
Cumulative proportions (cp)								0.15	0.34	0.63	1.00

Note: The sorting categories 1, 2, 3 ——, 11 refer to mid-points of class intervals 0-1.5, 1.5–3.0, 3.0-4.5 etc.

- ❖ In the above table, in the first row, frequency of the occurrence of Statement 1 in each category will be indicated.
- ❖ Proportion is computed through dividing the frequency of Statement 1 in each category by total number of judges (300).
- ❖ Cumulative proportions will be worked out by adding the proportion in a given category with the sum of all proportions below that category.

- The following is the formula given by Edwards (1969) to compute scale value of each statement:

$$S = L + \frac{(0.50 - \Sigma pb)}{pw} \times I$$

where, S = Median or Scale value of the statement, L = Lower limit of the interval in which median falls, Σpb=the sum proportions below the interval in which the median falls, pw=the proportion within the interval in which the median falls, I = width of the class interval.

- Since the statements are sent to 300 judges, the extent of disagreement among these judges with reference to each attitude statement has to be computed through Inter-quartile range (Q). This 'Q' is an index of dispersion of attitude statements on the scale (Edwards, 1969). The greater the Q value, the more is the dispersion for such statement among the judges and hence, that statement must be excluded from the scale. The 'Q' consists of middle 50 per cent of the judgements. So, it is important to find out two other point measures: the 75th Centile (C_{75}) and 25th Centile (C_{25}). The following are the formulae employed to compute 75th Centile (C_{75}) and 25th Centile (C_{25}):

$$C_{75} = L + \frac{(0.75 - \Sigma pb)}{Pw} \times I$$

$$C_{25} = L + \frac{(0.25 - \Sigma pb)}{Pw} \times I$$

where, L = Lower limit of the interval in which centile concerned falls, Spb=the sum proportions below the interval in which the centile concerned falls, pw=the proportion within the interval in which the centile concerned falls, I = width of the class interval.

- From both the centiles calculated above, Inter-quartile range (Q) can be calculated as: Q $= C_{75} - C_{25}$.
- Note that, the above procedure is to be repeated for all the attitude statements collected.
- The following are the criteria for selecting final attitude statements (Thurstone, 1946):
  - ❑ Statements with high 'Q' must be excluded.
  - ❑ Scale values should have equal appearing intervals.
- After excluding some attitude statements based on above criteria, arrange remaining statements in ascending order based on Scale values (S) and their corresponding 'Q' values are to be presented in the next column to draw the conclusions.
- The selected statements were later randomly arranged in the final format of the questionnaire for eliciting information from the farmers. This random arrangement is essential, so as to overcome response bias, which otherwise might lead to low reliability of the scale.
- Each of the statement must be provided with four-point response categories *viz.*, Strongly Agree, Agree, Disagree and Strongly Disagree. After recording the responses from the respondents (farmers), the scoring was done as shown though following table:

Attitude Statements	*Scores*			
	Strongly Agree	*Agree*	*Disagree*	*Strongly Disagree*
Positive (+) Statement 1 (Scale value (10)) Example: Increased application of bio-fertilizers improves soil health	(4)	(3)	(2)	(1)
Negative (-) Statement 10 (Scale value (5)) Example: The use of bio-fertilizers is less profitable compared to the cost incurred	(1)	(2)	(3)	(4)

- In the above table, Statement 1 is a positive statement. So, for the positive response highest score is given (4) and hence, the Score value = Scale value x Score given for the category of response. So, Score value = 10 x 4 = 40, and for negative response, the score value will be less *i.e.*, 10 x 1 = 10. Statement 10 is a negative statement. So, for the positive response lowest score is given (1) and hence, the Score value = Scale value x Score given for the category of response. So, Score value = 5 x 1 = 5 and for negative response, the Score value will be more *i.e.*, 5 x 4 = 20.
- Likewise, score values for all the statements will be computed. The summation of the score values under each response category indicates the respondent's attitude for the use of bio-fertilizers on the crops.
- Considering the time, cost and energy constraints on the part of the researcher in identifying the judges, that too, judges with required professionalized skills, now a days, the Thurstone scale is not often employed in the questionnaire.

(*b*) ***Likert Scale:*** For some questions framed in the schedule, the intensity of answering them by the respondents may vary more than mere *'agree/disagree'*. That is, the respondents may even answer some questions as 'Strongly agree' or 'Strongly disagree' (say, for example: *'do you believe news in news papers is politically biased?'*) and therefore Likert (1932) has introduced these variations in answering the questions in the scale. So, in Likert's scale, the coding of answers varies from 'Strongly agree' to 'Strongly disagree' on the scale instead of mere 'Agree' to 'Disagree'.

On Likert's scale, each question may have some positive statements and some negative statements. The respondents were asked to give the responses for different statements in a question in terms of several degrees, usually five degrees (occasionally, three or seven degrees are also used) *viz.*, Strongly Agree, Agree, Undecided, Disagree and Strongly Disagree. That is, at one extreme, there is strong agreement with the attitude statement and at the other extreme there is strong disagreement and in between them lie intermediate points. This is shown here under.

Statement	Strongly Agree	Agree	Undecided	Disagree	Strongly Disagree
Positive Statement	5	4	3	2	1
Negative Statement	1	2	3	4	5

Thus, Likert's scaling technique assigns a scale value to each of the five responses *i.e.*, Strongly Agree, Agree, Undecided, Disagree and Strongly Disagree for each statement. All positive statements are scored from maximum to minimum i. e., from a score of 5 to a score of 1 or 5 for strongly agree and so on 1 for strongly disagree. The negative statement or statement opposing the proposition would be scored in the opposite order *i.e.*, from a score of 1 to a score of 5 or 1 for strongly agree and so on 5 for strongly disagree. Thus, the scoring pattern is such that, agreement with favourable statements and disagreement with unfavourable statements are created as equivalent. After assigning the scores of each statement in a question, now the researcher has to compute total of these scores to measure the respondent's favour or disfavour towards the subject in the question. Assume the question consists of 30 statements then,

- The total score possible, if all the statements were rated positively is: 30 x 5 = 150.
- The total score possible, if all the statements were rated negatively is: 30 x 1 = 30.
- The total score possible, if all the statements were considered neutral is: 30 x 3 = 90.

Thus, Likert's scale involves summation of ratings of respondents for different statements in a given question. Hence, Likert's scale is also known as Summated scale. In the above example, the summated score for the question given by any individual range fall between 30 and 150. If the score given by the respondent is above 90, it implies positive response for the question, if the score is below 90, it implies negative response for the question and if the score is equal to 90, it implies, neutral position of the respondent with regard to the question. Thus, the total of the scores of different statements in a question indicates the measurement of attitude of the respondent towards the question. So, Likert's scale clearly indicates the strength of the respondent's approval or disapproval of the statement. This method removes the necessity of submitting items to the judges for working out scaled values for each item. Still this scale yields the same results as obtained from Thurstone's scale. Hence, it is considered superior over the Thurstone's scale.

Procedure for Constructing Likert's Scale

It is essential to note the following aspects before framing Likert's scale:

- The researcher should collect a number of statements considered relevant to the attitude being studied by the researcher.
- Statements may or may not be correct, but they must be representative of opinion held by a substantial number of people.
- The statements must express definite favourableness or unfavourableness to a particular point of view.
- Majority of the researchers are of the opinion that, neutral statements that are likely to be answered as 'Undecided' should not be included in the question, as neutral items do not work well in Likert's scale.
- The number of positive statements and the number of negative statements should be equal or approximately equal in the scale.

- ❖ In general, score values for different responses should not be indicated in the questionnaire, as it leads to response bias.
- ❖ A trial test (pilot study) may be administered on a random sample of respondents to check the responses on the selected statements in a question. Only those statements that correlate with the total test should be retained.
- ❖ The scores to different statements in a question are given in such a way that, for the positive statement, highest score is given for the most favourable response and *vice versa*.
- ❖ Compute the total score of all the statements in a question by adding the scores across different statements in a question.
- ❖ Finally, the response of the respondent for the given question are analyzed based on the total score value attained.

The following table exemplifies the format of a Likert's scale on attitude of farmer about SRI production technology of paddy:

SRI technology is superior over Transplanted technology in terms of yield	**SA**	**A**	**U**	**D**	**SD**
SRI technology demands more human labour for weeding	**SA**	**A**	**U**	**D**	**SD**
Water requirement under SRI technology is higher compared to Transplanted technology	**SA**	**A**	**U**	**D**	**SD**
Seed rate under SRI technology is lower compared to Transplanted technology	**SA**	**A**	**U**	**D**	**SD**
For taking up SRI technology, training is not required on the part of the farmer, as it is easier to adopt compared to Transplanted technology.	**SA**	**A**	**U**	**D**	**SD**
SRI technology involves application of more of manures compared to Transplanted technology.	**SA**	**A**	**U**	**D**	**SD**

Advantages of Likert's Scale

The Likert's scale have following advantages:

- ❖ It is easy and simple to construct compared to Thurstone' scale.
- ❖ It is more reliable and provides more information compared to Thurstone' scale.
- ❖ It can easily be used in respondent centred and stimulus centred studies. That is, this scale enable the respondent know, how well the responses differ across the respondents and across stimuli.
- ❖ It takes less time in construction.
- ❖ Consumes less time, as this method removes the necessity of submitting items to the judges for working out scaled values for each item.
- ❖ Since both Thurstone's scale and Likert's scale yield similar results, and the construction of the latter is simpler over the former, Likert's scale is mostly used by the students in social science research.
- ❖ Unlike Thurstone's scale, each statement given the question can be given an empirical test in Likert's scale.

- ❖ Likert's scale is more reliable compared to Thurstone's scale, as the former permits the expression of the respondent for the statements over several degrees ranging from Strongly Agree to Strongly Disagree. But, Thurstone's scale simply allows only two alternative responses *viz.*, agree or disagree.

Limitations

However, Likert's scale suffers from the following limitations:

- ❖ The intensity comparison of responses given by the respondents is not possible.
- ❖ The scale employed here is an ordinal scale and not the intervals scale.
- ❖ It is unlikely that the statements given in a question are of equal value in 'forness' or 'againstness'. Further, total score of the respondent for a question is not direct, as it is based on the score values of various statements in that question.
- ❖ The gap between the responses like Strongly agree and Agree is not the same as Agree and Undecided or Undecided and Disagree. So, there is no basis for believing that, the five categories of responses indicated in the Likert's scale are equally spaced.
- ❖ The respondent may not be a good judge of himself and may not be clearly aware of his real attitude. So, they tend to respond according to 'what they should feel' rather than 'what they really feel'.
- ❖ It takes longer to complete than other scales because, respondents have to read each statement.

Considering the above demerits, Lotus Guttman has proposed Cumulative scale, popularly called as Guttman's Scalogram analysis or Guttman's Cumulative scale.

(*c*) ***Guttman's Scalogram Analysis or Guttman's Cumulative Scale***: As discussed above, Likert's scale is the most popular response scale used in survey research, where survey respondents are asked to provide their opinions about various topics, usually in the form of agree or disagree response to a series of statements. This Likert scale is used to measure agreement with a statement and the strength of opinions in a single scale. On the other hand, as opposed to measuring intensity, Guttman scale measure extremity *i.e.*, how does each respondent rank on a scale of combined items? This scale comprises of a series of statements that are created along a *continuum*. The continuum is uni-dimensional, and the series of statements all measure a single concept. If a respondent agrees with a given statement, by definition that respondent should agree with each statement 'below' that level. And Guttman scales are cumulative, meaning that, the researcher is concerned with respondents' total score, not just their responses to single statements. This scale concept can be better understood from the following example:

1. 2 + 1
2. 3 + (3+4)
3. 5 [3(3-2)+2(7-5)]
4. 7[{(3-2)+(6+7) – (4-2)}{35/11}]

In the above example, if the respondent is capable of answering the fourth problem, he is capable of doing the above three problems as well. Similarly, if the respondent is capable of answering the third problem, he is capable of doing the above two problems as well. The same concept is followed in Guttman scale. That is, in a given set of items or statements, the respondent who agrees with any specific question in the list will also agree with all previous questions. Put more formally, the researcher can able to predict item responses perfectly knowing only the total score for the respondent. By using plus (+) and minus (-) signs, this can be shown by a diagram called 'Scalogram'. For example, imagine a ten-item cumulative scale. If the respondent scores a four, it should mean that, he/she agreed with the first four statements. If the respondent scores an eight, it should mean they agreed with the first eight. The object is to find a set of items that perfectly matches this pattern. In practice, we would seldom expect to find this cumulative pattern perfectly. So, the objective of scalogram analysis is to examine how closely a set of items corresponds with this idea of cumulativeness. The following are the steps involved in the construction of Guttman scale.

Define the Focus

As discussed in the above scaling methods, in Guttman also, the focus of the scale should be clearly defined. Assume the researcher wishes to develop a Guttman (Cumulative) scale that measures farmers' attitude towards SRI production technology of paddy.

Develop the Items

Now, the researcher has to develop a large set of statements or items that reflect the concept. This can be done by the researcher on his own or by taking the guidance of experts in the relevant field of focus. Let us assume, the researcher has developed following statements regarding farmers' attitude towards SRI production technology of paddy:

1. I adopt SRI production technology of paddy because, it yields more output.
2. I adopt SRI production technology of paddy because, it demands less irrigation water.
3. I adopt SRI production technology of paddy because, it requires less seed rate.
4. I adopt SRI production technology of paddy because, more manures application enhances soil fertility.
5. I adopt SRI production technology of paddy because, local Agricultural Officer suggested me.
6. I adopt SRI production technology of paddy because, technology is easy to understand and adopt.

Note that, the researcher can frame any number of statements even about 80-100.

Rate the Items

Next, the above statements are to be placed before a group of judges to rate the statements or items in terms of how favourable they are to the adoption of SRI production technology of paddy. They would give a 'Yes' if the item was favorable toward adoption of SRI production technology of paddy and a 'No' if it is not. Notice that, the researcher is not asking the judges, whether they personally agree with the statement. Instead, he asks them to make a judgment about how the statement is related to the construct of interest.

Develop the Cumulative Scale

In developing the cumulative scale, a matrix or table is to be constructed, which shows the responses of all the respondents on all of the listed items. After listing the responses, the matrix is so arranged in such a way that, the respondents who agree with more statements are listed at the top and those agreeing with fewer are at the bottom. The responses are arranged for the statements from left to right *i.e.*, from those that most agreed to those that fewest agreed to. The matrix so arranged is shown here under.

Respondents	Item 1	Item 6	Item 4	Item 2	Item 5	Item 3
5	+	+	+	+	+	+
13	+	+	+	+	+	-
3	+	+	+	+	-	-
25	+	+	+	-	-	-
17	+	+	-	-	-	-
4	+	-	-	-	-	-
11	-	-	-	-	-	-

Note that, the scale is very nearly cumulative on going from left to right across the columns (items). Specifically, if someone agreed with Item 6, they always agreed with Item 1. And, if someone agreed with Item 4, they always agreed with Items 1 and 6. In the above table, there are fewer items or statements to examine. In case, if there are lots of items, a data analysis called 'Scalogram analysis is used to determine the cumulative property.

Administering the Scale

Once after completing the final scale items, it is relatively simple to administer the scale. From the above table, the items might be listed in cumulative order as:

1. I adopt SRI production technology of paddy because, it yields more output.
2. I adopt SRI production technology of paddy because, technology is easy to understand and adopt.
3. I adopt SRI production technology of paddy because, more manures application enhances soil fertility.
4. I adopt SRI production technology of paddy because, it demands less irrigation water.
5. I adopt SRI production technology of paddy because, local Agricultural Officer suggested me.
6. I adopt SRI production technology of paddy because, it requires less seed rate.

Of course, when the questionnaire is given to the respondent, the items will be mixed up in their order. So, the final scale might look like:

Instructions

Place a check next to each statement you agree with.

_____ I adopt SRI production technology of paddy because, it yields more output.

_____ I adopt SRI production technology of paddy because, it demands less irrigation water.

_____ I adopt SRI production technology of paddy because, it requires less seed rate.

_____ I adopt SRI production technology of paddy because, more manures application enhances soil fertility.

_____ I adopt SRI production technology of paddy because, local Agricultural Officer suggested me.

____ I adopt SRI production technology of paddy because, technology iseasy to understand and adopt.

Each scale item has a scale value associated with it (obtained from the Scalogram analysis). To compute a respondent's scale score, simply sum the scale values of every item they agree with. In the above example, their final value should be an indication of their attitude towards adoption of SRI production technology of paddy.

Advantages of Guttman's Scale

They include:

- ✰ Measures only a singly dimension of attitude.
- ✰ Researcher's subjective judgment is not allowed to creep in the development of scale since the scale is determined by the replies of respondents.
- ✰ Small number of items are required to make the scale.

Limitations of Guttman's Scale

This Scale suffers from the following limitations:

- ✰ The procedure is tedious and complex.
- ✰ It is not very much reliable.
- ✰ It is not very objective scale.
- ✰ It is difficult to compare to other scaling methods.

Guttman constructed a number of attitude scales during World War II. Currently, however, as a method of constructing attitude scales, Guttman scaling is almost never used. Investigators prefer to use simpler methods of scale construction such as Likert's scaling.

(*d*) ***Bogardus Social Distance Scale***: The Bogardus social distance scale is a psychological testing scale created by Emory Bogardus. This scale is meant for measuring the willingness of people to participate in social relations with other kinds of people. Let us assume that, the researcher is interested to study the extent to which the vegetarian (Brahmins) are willing to associate with non-vegetarians. Then he formulates the following questions:

- ❖ Are you willing to live in the same colony as non-vegetarians?
- ❖ Are you willing to live in the same community as non-vegetarians?
- ❖ Are you willing to live in the same neighbourhood as non-vegetarians?
- ❖ Are you willing to live next door to a non-vegetarian?
- ❖ Are you willing to let your child marry a non-vegetarian?

The clear differences in intensity suggest a structure among the items. Presumably, if a respondent is willing to accept a certain association, he or she is willing to accept all those that precede it on the list (those with lesser intensities). In the above example, if the respondent agrees for the last option *i.e..*, 'he is willing to marry his child to a non-vegetarian', then he is willing to accept the earlier options. This implies that, if the respondent agrees with 'hard' items, he will also agree 'easy' items and if the respondent disagrees with 'easy' items', he will also disagree with 'hard' items. Let us look at another example: if the respondent disagrees 'dalit' as a colleague in his office, he will also disagree to invite him to his house.

On the contrary, if the respondent agrees to invite 'dalit' into his house, he will also agree him as a colleague in his office. Note that, Bogardus social distance scale is a cumulative scale (like a Guttman scale), because agreement with any item implies agreement with all preceding items. The scale has been criticized as too simple because, the social interactions and attitudes in close familial or friendship-type relationships may be qualitatively different from social interactions with and attitudes toward relationships with far-away contacts such as citizens or visitors in one's country.

Procedure for Constructing Bogardus Social Distance Scale

In making the social distance scale, the researcher has to prepare a list of single sentence descriptions (can frame around 50 to 60 sentences), nearly all of which were heard in ordinary conversations where a person was expressing himself about other persons. These statements represent several different types of social relationships; that is, they relate to contacts within the family, within social or fraternal groups, within neighborhoods, within churches, within schools, within play groups, within transportation groups, within occupational and business groups, within political or national groups. For example:

- Would have as my guests at public dinners.
- Would work beside in an office.
- Would entertain overnight in my home.
- Would decline to invite to my home.
- Would have as regular friends.
- As only visitors in my country.
- Would allow a few of their children to attend school with my children.
- Would have live outside my neighborhood.
- Would have none of their children attend school with my children.
- Would have live outside my country.
- Would dance with in public regularly.
- Would dance with in private regularly.
- Would have merely as speaking acquaintances.
- Would play bridge or golf with regularly.
- Would play bridge or golf with occasionally.
- Would decline to play bridge or golf with.
- Would forbid my children from playing with their children.
- Would have several families in my neighborhood.
- As citizen in my country'.

- ☆ A group of persons (say, 100 judges) were selected to rate each of the above statements according to the amount of social distance.
- ☆ Each of the 100 persons was asked to judge the amount of social distance which he thought existed between the person making, for example, statement No. 1 and the person concerning whom it was made, from the standpoint of the first two, persons involved. In the same fashion each statement was judged.

- Each of the statements was typed on a slip of paper. Each judge was given different slips of paper (number of slips equal to number of statements prepared) and asked to distribute them in seven boxes or piles representing seven different degrees of social distance.
- Each judge was asked to study carefully the slips in each box or pile and to reclassify any that might be rated more accurately. No rule was made looking toward an even distribution. It was requested, however, that at the close of the exercise, each of the seven boxes must contain at least one slip. In case more than 25 per cent of the total slips appeared in any box, it was concluded that sufficient discrimination had not been used and the work of this judge was discarded.
- Social distance was defined in this instance for each judge as the degree of sympathetic understanding that exists between two persons or between a person and a group (personal distance, or personal-group distance). The judge was urged to view the social distance situation described on each slip as objectively as possible. Each judge of course worked independently of the others.
- The judgments, ranging from 1 to 7 for each of the above statements by the 100 judges were added and the arithmetic mean is taken. Consider (select) the statements with the mean varying from 1.00 through a series of equal distances like (nearest to) 2.00, 3.00, 4.00, 5.00, and 6.00 for the scale. (Note that, a score of 1.00 for a group is taken to indicate no social distance).
    - Would entertain overnight in my home (1.00)
    - Would have as regular friends (2.00)
    - Would work beside in an office (3.00)
    - Would have several families in my neighbourhood (4.00)
    - Would play bridge or golf with regularly (5.00)
    - Would allow a few of their children to attend school with my children (6.00)

  Thus, Bogardus scale attempts to measure respondents' degree of warmth, intimacy, indifference, or hostility to particular social relationships, by having them indicate agreement or disagreement with a series of statements about particular (say) religious groups. This scale is usually applied to the study of ethnic relations, social classes and social values generally.

(e) Semantic Differential Scale (Factor Scale)

The semantic differential (SD) is a technique developed during the 1940s and 1950s by Charles E. Osgood to measure the meaning of language quantitatively. The SD measures people's reactions to stimulus words and concepts in terms of ratings on bipolar scales defined with contrasting adjectives at each end. An example of an SD scale is given below:

Good ___ ___ ___ ___ ___ ___ ___ **Bad**

3 2 1 0 1 2 3

Usually, the position marked 0 is labeled 'Neutral', the 1 positions are labeled 'Slightly', the 2 positions 'quite', and the 3 positions 'extremely'. Thus, this scale measures both directionality of a reaction (e.g., good versus bad) and its intensity (slight through extreme). The other bipolar objectives include: Fast vs Slow, Improved vs Traditional, New vs Old etc. The other example of judgement pertaining to a new brand of pesticide is shown below.

	Very Much	Somewhat	Neither	Somewhat	Very Much	
Improved		X				Not improved
Effective					X	Not Effective
Relatable			X			Unrelatable

Thus, SD scale asks respondents of a questionnaire to choose between two opposite positions using qualifiers to bridge the gap between them. This scale consists of a series of descriptive adjectives or phrases, which are polar opposite and these can be conveniently categorized into the following three dimensions:

- ✰ Evaluative (E): Good vs Bad, Clean vs Dirty, Fair vs Unfair.
- ✰ Potency (P): Long vs Short, Strong vs Weak, Thick vs Thin, Powerful- Powerless etc.
- ✰ Activity (A): Fast vs Slow, Active vs Passive etc.

A number of basic considerations are involved in SD methodology:

- ✰ A SD scale is a list of opposite adjectives. This method is invented by C.E. Osgood in order to measure the connotative meaning of cultural objects.
- ✰ Bipolar adjective scales are a simple, economical means for obtaining data on people's reactions. With adaptations, such scales can be used with adults or children, persons from all walks of life, and persons from any culture.
- ✰ Ratings on bipolar adjective scales tend to be correlated with reference to the above the three dimensions *viz..*, Evaluative, Potency, and Activity (EPA).
- ✰ Note that, unlike the rating scale, the SD scale does not have a neutral or middle selection. A person must choose, to a certain extent, one or the other adjective.

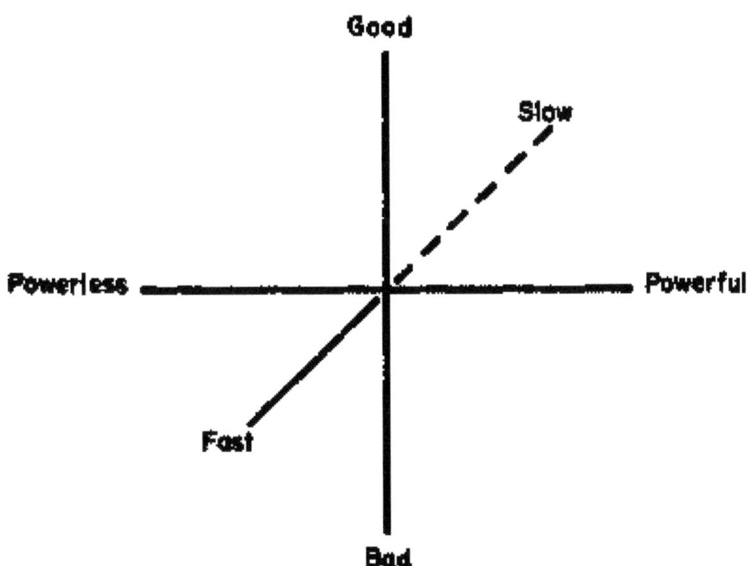

Figure 11.6: EPA Dimensions in SD Space.

- SD space—Sometimes it is convenient to think of the EPA dimensions as forming a three-dimensional space. The SD, or affective, space is illustrated in Figure 11.6; the origin or center of this space represents neutrality on all three dimensions. Treating EPA measurements of a stimulus as coordinates allows the stimulus to be positioned as a point in the space, and this point graphically represents the affective response to the stimulus.

Semantic differential scales are used in a variety of social science research but it also is used in marketing and practical, user experience research and therapy. Sometimes SDs are also known as Polarities.

Choosing the Scaling Technique

The following are the factors that influence the selection of scaling technique in measuring the qualitative information:

- The level of information (nominal, ordinal, interval, or ratio) desired.
- Knowledge level of the respondents.
- Method employed for data collection.

12

Testing of Hypothesis and Interpretation of the Findings

As discussed under Chapter 7, there are two important statistical hypotheses *viz.*, H_O and H_A. We already discussed that, the hypothesis formulated in the study must be proved statistically *i.e.*, through statistical inference the researcher has to accept or reject the hypothesis. We also know that, hypothesis is an assumption about a population parameter. This assumption may or may not be true. The best way to determine, whether to accept or reject the formulated statistical hypothesis is by examining the entire population. Since conducting census study (dealing with entire population units) is often impractical, researcher select a representative sample from the population. At the time of identifying the research problem, the researcher is in a confusion state and the formulation of hypothesis guide the researcher towards the formulation of the problem. The data collected pertaining to the variables considered under the problem will be subjected to statistical analysis and the findings must be tested for their statistical significance. If the findings were proved statistically it implies, the hypothesis is tested for validity. The logic of hypothesis testing is similar to the *'presumed innocent until proven guilty'*. So, in hypothesis testing, we assume that, the H_O is a possible truth, until the sample data analysis concludes that it is otherwise. So, in hypothesis testing, the researcher uses sample data to evaluate a hypothesis about a population.

12.1. Steps in Testing of Hypothesis

The hypothesis testing involves the following sequential steps:

A. Specify the statistical hypothesis *viz.*, H_O and H_A
B. Select the significance Level (denoted by the Greek letter alpha'α')
C. Find the critical value at the required level of significance and degrees of freedom.
D. Formulation of One Sided or Two Sided Hypothesis
E. Calculate the Test Statistic and Corresponding Probability (P) Value
F. Confidence Intervals for Parameters ($\hat{a}$ and $\hat{b}$)
G. Drawing a Conclusion

A. Specify the Statistical Hypothesis *viz.*, H_O and H_A

This step consists in making a formal statement of the H_o and also of the H_A. This means that, hypothesis should be clearly stated, considering the nature of the research problem. As discussed earlier, in H_O, the term 'null' often refers to the common view about the variables or 'no difference' or 'no correlation' or 'no association' of the variables, while the H_A is what the researcher really thinks is the cause of a phenomenon. That is, H_A of indicates 'there is a difference' or 'there is a correlation', or 'there is an association' between the variables. If the H_O is rejected by the statistical test, then only the H_A remains valid (because, H_O and H_A are mutually exclusive and exhaustive). The H_O is formulated to be tested for possible rejection. If statistical analysis leads the researcher to reject the H_O, then H_A could be supported. That is, if the H_O is not accepted, then the H_A is used. Thus, two conclusions generally arise for the researcher in the formulation of H_O *viz.*, accept the H_O and reject the H_O in favour of H_A. It is to be noted that, the researcher should carefully use the 'wording' of the null and ensure that, it is as specific as possible. Say, for example, a researcher has formulated the following H_O and H_A:

H_O: There is no difference between the efficacies of two pesticides, on average

H_A: The new pesticide is effective than the current pesticide, on average

In the above hypotheses, there is a major error. Say, if the farmers actually receive low crop yields due to new pesticide than applying current pesticide, here both H_O and H_A formulated are not valid. That is, if the H_O is rejected, even the H_A is also not accepted or it is also invalid. The researcher rejects the H_O, when the statistical tests generate the results such that, they do not fit the H_O at 1 per cent or 5 per cent level of significance. In such case, he has to accept the H_A. But, when the statistical tests generate the results such that, they fit the formulated H_O at 1 per cent or 5 per cent level of significance, the researcher has to accept the H_O. Note that, not rejecting H_O (or accepting H_O) does not really mean that, formulated H_O is true. This is because, there might not be enough evidence against H_O to accept H_A. Though the H_O is accepted, it is still a hypothesis, and must conform to the principle of falsifiability, in the same way that, rejecting the H_O does not prove the H_A. Further, even if the H_O is not rejected (*i.e.*, accepted), then the researcher must be careful to interpret the findings. The thinking on this is similar to a legal verdict. Just because a person has been declared 'not guilty' in a case, it does not mean that, he is innocent and he is judged 'not guilty' because of lack of evidence. In the same way, just because a H_O is not rejected, it does not mean that, the statement is true.

Even H_O can be formulated in the form of a negative statement. Say, for example, the H_O: *'There is no relationship between farmer's income and expenditure on luxuries'*. This H_O can be tested for statistical significance and the statistical tests would aim at 'nullifying' this hypothesis and hence, it is called as 'Null hypothesis'. The formulation of H_O will have two advantages. First, they have to be formulated in a more exact way. Second, it is quite easier to 'reject' the hypothesis than to confirm it with complete certainty. Hence, the concept of H_O is found to be very useful. In general, when the researchers accept the H_O, they presume that, the conducted survey or experiment is a failure. But, note that, when a researcher accepts the H_O, it simply implies, the initial assumption is only incorrect and not the entire survey or experiment. Hence, accepting or rejecting the H_O should be seen as a positive result.

Since, H_O is formulated with the objective of its rejection after conducting the statistical significance tests, the H_A indicates the relationship or difference that exists between two variables or groups and goes further to state the nature or direction of that relationship or differences. Note that, H_A can be stated in directional form (Directional H_A) or non-directional form (Non-directional H_A) and this can be easily understood from the following example:

Assume the researcher formulated the following H_O and H_A in conducting a research problem to study the differences in managerial abilities of men and women farmers in practicing paddy cultivation:

H_O: There is no difference in the managerial abilities of men (M_M) and women (W_M) farmers in practicing paddy cultivation *i.e.*, H_O: $M_M=W_M$ or H_O: $M_M-M_W=0$

H_A *(Non-directional):* There is difference in the managerial abilities of men and women farmers in practicing paddy cultivation *i.e.*, H_A: $M_M \neq W_M$.

H_A (Directional)

- The managerial ability of men farmers in practicing paddy cultivation is higher compared to the managerial ability of women farmers *i.e.*, H_A: $M_M>W_M$ (or)
- The managerial ability of men farmers in practicing paddy cultivation is lower compared to the managerial ability of women farmers *i.e.*, H_A: $M_M<W_M$.

Thus, a non-directional H_A (also the two-tailed hypothesis), is a hypothesis that does not specify the nature and direction of the relationship between the independent and dependent variable. A directional H_A refers to a hypothesis that predicts a direction, which will be taken by the difference. It has only one critical value that is being used, and it is commonly used by researchers. So, H_A is a statement that suggests a potential outcome that the researcher may expect after rejecting H_O. The sources of H_A comes from earlier research studies. It is established only, when a H_O is rejected. This H_A serves as the desired conclusion of the researcher. This formulation of H_A is an important step, which must be accomplished with due care in accordance with the object and nature of the problem under consideration. It also indicates, whether the researcher should use one tailed test or a two tailed test. If H_A is of directional type *i.e.*, 'greater than' or 'less than', we use one-tailed test, but when H_A is of non-directional type *i.e.*, 'not equal to' or whether greater than or less than', then we use a two-tailed test. The students must note the following aspects regarding directional and non-directional H_A:

- ***One-tailed directional***: A one-tailed directional H_A is concerned with the region of rejection for only one tail of the sampling distribution.
- ***Two-tailed directional:*** A two-tailed directional H_A is concerned with both regions of rejection of the sampling distribution.

- ☆ ***Non-directional:*** A non-directional H_A is not concerned with either region of rejection, but, rather, it is only concerned that null hypothesis is not true.

B. Select the Significance Level (Denoted by the Greek Letter Alpha 'α')

The hypothesis is tested on a pre-determined level of significance and the same should be specified for the statistical test conducted. For example, in the given regression equation, 't' test will be employed to test the significance of individual regression coefficients ('t'cal = Regression coefficient/Standard error). In the equation given below, variable (X_1) is statistically significant at one per cent level and variable (X_2) is statistically significant at five per cent level.

$$\begin{aligned} Y = \; & 7.18333 \; + \; 1.56718X_1^{**} + 1.09597X_2^{*} \\ & (1.01578) \quad (0.50918) \quad (0.50607) \end{aligned}$$

That is, generally, in practice, either 1 per cent (0.01-indicated by **) or 5 per cent (0.05-indicated by *) level is adopted in social sciences research. In case of 1 per cent level of significance, it means that, there is only one per cent chance that the researcher will accept the H_O (or 99% chance of accepting H_A), when the H_O framed is actually not true (or rejected). Similarly, in case of 5 per cent level of significance, there is only a 5 per cent chance that the researcher will accept the H_O (or 95% chance of accepting H_A), when the H_O framed is actually not true (or rejected). In other words, in case of 1 per cent level of significance, it means that, there is only one per cent chance that the researcher will accept the H_A (or 99% chance of accepting H_O), when the H_O framed is actually true (or accepted). Similarly, in case of 5 per cent level of significance, there is a 5 per cent chance that the researcher will accept the H_A (or 95% chance of accepting H_O), when the H_O framed is actually true (or accepted). A significance level of 1 per cent means that, in the long run, the risk or the probability of making the wrong decision (say, Type I error *i.e.*, rejecting the H_O when it is true or Type II error i.e., accepting the H_O when it is false) is only 1 per cent. That is, the researcher is likely to make the possible wrong decisions in only one out of 100 occasions. Similarly, a significance level of 5 per cent means that, in the long run, the risk or the probability of making the wrong decision (say, Type I error or Type II error) is only 5 per cent. That is, the researcher is likely to make the possible wrong decisions in only five out of 100 occasions. Thus, 1 per cent level provides more confidence to the researcher's decision than a 5 per cent level. So, the smaller the significance level chosen, the greater the burden of proof needed to reject the H_O or to support the H_A. The students should also note that, we can perform the test at any level (usually 1% or 5%). For example, performing the test ('t' test or 'Z' test) at a 5 per cent level means that, there is a 5 per cent chance of wrongly rejecting H_O (*i.e.*, committing Type I error). If we perform the test at the 5 per cent level and decide to reject the H_O, we say *'there is significant evidence at the 5 per cent level to suggest the H_O is false'*. Thus, the confidence at which a H_O is accepted or rejected (or H_A is rejected or accepted) is called Level of significance. The factors that affect the level of significance are:

- ☆ The size of the sample;
- ☆ The variability of measurements within samples;
- ☆ Whether the hypothesis is directional or non – directional.

In brief, the level of significance must be adequate in the context of the purpose and nature of inquiry.

C. Find the Critical Value (Table Value) at the Required Level of Significance and Degrees of Freedom

Critical value is the value that divides the region under normal distribution curve* into two regions *viz.*, Acceptance region and Rejection region. If the computed test statistic falls in the rejection region, we reject the H_O, otherwise, we accept the H_O. The critical value (table value) depends upon the level of significance. That is, if the calculated test statistic value (say, $'t'_{cal}$ value) is less than the critical value, we accept the H_O, otherwise, we reject the H_O.

D. Formulation One of Sided or Two Sided Hypothesis

The H_A is one sided, if the parameter is larger or smaller than the H_O value. It is two sided, when the parameter is different from the H_O value. So, the H_A can take one of three following forms:

- ✰ $H_O: b_1 = 0$
- ✰ $H_A: b_1 > 0$, is one-sided H_A (Directional hypothesis).
- ✰ $H_A: b_1 < 0$, also a one-sided H_A (Directional hypothesis).
- ✰ $H_A: b_1 \neq 0$, is two-sided H_A. That is, the true value is either greater or less than 0 (Non-Directional hypothesis).

E. Calculate the Test Statistic and Corresponding Probability (P) Value

Hypothesis testing generally uses a test statistic. If the calculated statistic (say, 't' calculated value) is equal to or greater than 't' tab value at 1 or 5 per cent level of significance, then reject the H_O (or accept the H_A), but if the calculated statistic (say, 't' calculated value) is less than 't' tab value at 1 or 5 per cent level of significance, then accept the H_O. This can also be explained as follows: *'if the calculated probability 'P' of a variable is equal to or smaller than 'α' value (level of significance i.e., 1 per cent or 5 per cent) in case of one tailed test (and' $\alpha/2$' in case of two-tailed test), then reject the H_O (i.e., accept the H_A), but if the calculated probability 'P' is greater than 'α' value in case of one tailed test (and' $\alpha/2$' in case of two-tailed test), then accept the H_O.'* So, the larger the 'P' value, the smaller will be the evidence against the H_O *i.e.*, we accept the H_O. This infers that, the probability that the statistic takes a value as extreme or more than extreme assuming that the H_O is true is called P- value. So, P-value is the probability of observing a sample statistic as extreme as the test statistic, assuming the H_O is true. This can be understood from the following example: The given regression equation is:

$$\begin{array}{llll} Y = 7.18333 & + \; 1.56718X_1^{**} & + \; 1.09597X_2^{*} & + \; 2.08736 \\ \quad (1.01578) & \quad (0.50918) & \quad (0.50607) & \quad (3.87530) \end{array}$$

In this case, variable X_1 is statistically significant, as the $'t'_{cal}$ value is greater than $'t'_{tab}$ value at 1 per cent level of significance. Similarly, variable X_2 is statistically significant, as the $'t'_{cal}$ value is greater than $'t'_{tab}$ value at 5 per cent level of significance. The same can be interpreted from the given Table 12.1.

The probability value of variable X_1 is 0.0022 and is less than 0.01 (*i.e.*, 1 per cent level of significance) and hence, we reject the H_O (b_1=0) or accept H_A at 1 per cent level of significance. The probability value

---

* The Normal Distribution has:

- ✰ Mean = Median = Mode
- ✰ Symmetry about the center
- ✰ 50 per cent of values less than the mean and 50 per cent greater than the mean.

of variable X_2 is 0.0305 and is less than 0.05 (*i.e.*, 5 per cent level of significance), but greater than 0.01 (*i.e.*, 1 per cent level of significance) and hence, we reject the H_O (b_2=0) or accept H_A at 5 per cent level of significance. However, for variable X_3, the probability value is 0.7139 and is greater than 0.05 (*i.e.*, more than 5 per cent level of significance) and hence, we accept the H_O (b_3=0).

Table 12.1: An Hypothetical Multiple Linear Regression Model.

Item	*Coefficient*	*Standard error*	*t-statistic('t'cal)*	*Probability (P)*
Intercept	7.18333	1.01578	7.07169	0.0000
X_1	1.56718	0.50918	3.07414	0.0022
X_2	1.09597	0.50607	2.16563	0.0305
X_3	2.08736	3.87530	0.53863	0.7139

Note: 't'tab at 1 per cent level of significance: 2.764 and 't'tab at 5 per cent level of significance: 1.812 (at 10 degrees of freedom, one tailed test)

Note that, sample size directly impacts the P-value. Large sample sizes produce small P-values and *vice versa*. Hence, it is essential to verify the practical relevance of research results. Go through the following examples:

- Average plant heights were significantly different between the treatment and control plots. (4.2m vs 4.7m; p = 0.005; n=1,000). If we observe the probability value, it implies there is significant difference between average plant heights in treatment and control plots. But, if we observe the average plant height difference, this is not the case. So, a larger sample size (n=1000) has resulted in a small P-value.
- Average plant heights were not significantly different between the treatment and control plots. (4.2m vs 8.9m; p = 0.17; n=20). If we observe the probability value, it implies there is no significant difference between average plant heights in treatment and control plots. But, if we observe the average plant heights in two plots, there could be difference, but, it can't be determined significantly, whether it is a true difference or just happened due to the natural variability in age within these two plots. So, a smaller sample size has resulted in a high P-value.

Thus, a larger sample size may mislead the researcher in getting lower probability value than 'α' value, though there is no significant difference between the two groups and a smaller sample size may mislead the researcher through getting higher probability values than 'α' value, though there is significant difference between the two groups. Hence, the researcher should plan for only adequate and representative sample size, so as to derive a meaningful relationship or difference if one exists.

One Tailed Test and Two Tailed Test

As discussed above, there are actually two different types of H_A *viz.*, Directional H_A and Non-directional H_A.

- ***Directional H_A (also termed 'One-tailed' test for testing direction of difference)***: In this hypothesis, the direction of deviation from the null value is clearly specified. That is, a specific predicted outcome is stated. A directional hypothesis consider only one tail (the other tail is ignored as irrelevant to H_A) thus, all of 'α' (level of significance) can be placed in that one tail. One-tailed tests will be generally used in the light of strong previous research, theoretical, or logical considerations. Hence a 'claim', 'belief', or 'hypothesis', or 'it was predicted that' is not sufficient to justify the use of a one-tailed test. A one-tailed test of a hypothesis would reject the H_O, if the sample statistic is either significantly higher (right tailed or upper tailed test) or significantly lower (left tailed or lower tailed test) than the hypothesized population parameter. Hence, in a one- tailed test, there is only one rejection region (Figure 12.1). So, one-tailed test may be right tailed (upper tailed) test or left tailed (lower tailed) test. In right tailed test, the hypotheses include: H_O: Sample mean ; $\overline{X}$ = Population mean (μ) and H_A: > μ. In left tailed test, the hypotheses include: H_O: $\overline{X} = \mu$ and H_A: $\overline{X} < \mu$.

☆ ***Right Tailed Test:*** The following example enables the readers to clearly understand about the one-tailed (right-tailed) test. Assume that, a random sample of n=49 was selected from a certain population with a population mean of 20. The sample mean and sample standard deviation (s) are 25 and 5 respectively. Using this information, assuming 5 per cent level of significance, compute whether sample mean differ significantly from population mean or not? In this case, given

Population mean = 20

Sample mean = 25

Sample standard deviation = 5

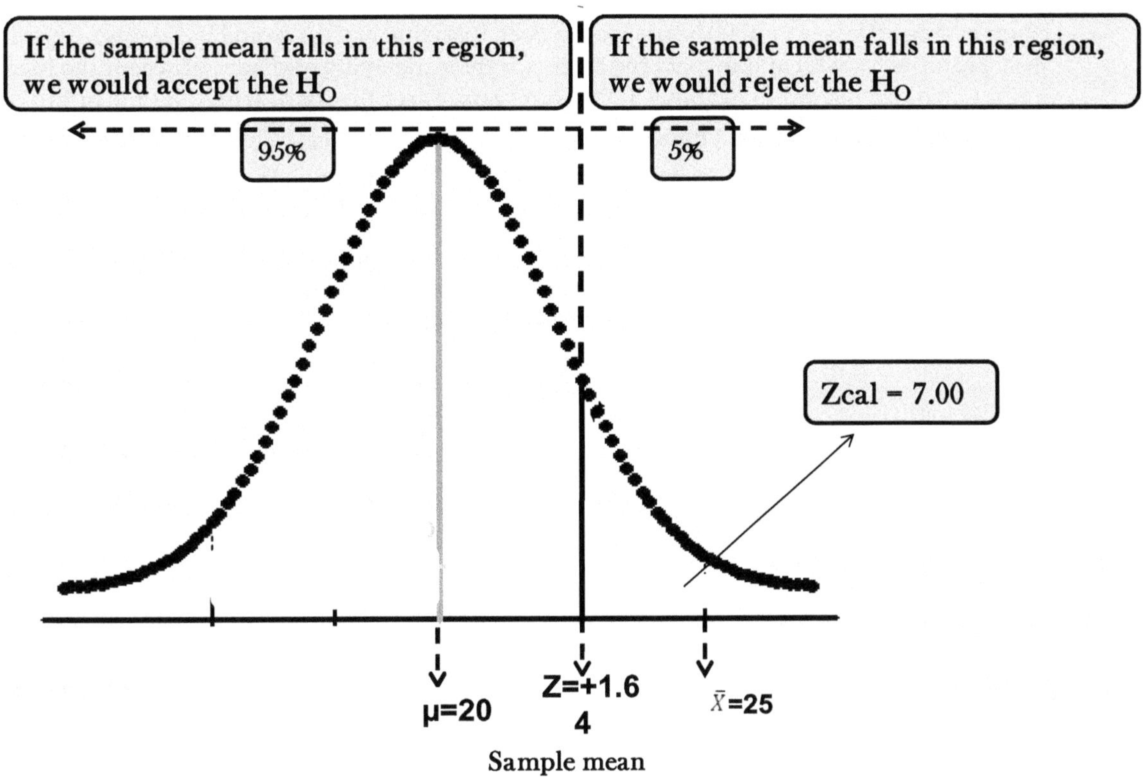

Figure 12.1: One-Tailed Test (Upper tailed or Right tailed test) – Acceptance and Rejection Regions.

Considering above, the following hypotheses are formulated:

- H_O: Sample mean $\overline{X}$ is statistically equal to Population mean (μ).
- H_A: Sample mean $\overline{X}$ is statistically higher than Population mean (μ) (One-tailed (right tailed) hypothesis).

In the above case, since the sample size n=49, the researcher has to employ Z-statistic* to study the statistical difference between sample mean and population mean. The following is the formula employed to compute Z-statistic:

$Zcal = (\overline{X} - \mu)/(s/\sqrt{n}) = (25-20)/(5/\sqrt{49}) = 7.00$

The above value Z=7.00 is called Z-calculated value. Since, the assumed level of significance is 0.05 and it is a one-tailed test because, the researcher framed the H_A that the sample mean will be statistically higher than population mean, the Z tabulated value is +1.64. This tabulated value is also called as Critical value. Since, the Zcal value (7.00) is greater than 'Z' tab value (+1.64), we reject the H_O** and accept the H_A. The same concept is depicted through Figure 12.1.

- ☆ ***Left Tailed Test:*** One-tailed test may also be left tailed (lower tailed) test. In left tailed test, the hypotheses include: H_O: $\overline{X} = \mu$ and H_A: $\overline{X} < \mu$. The following example enables the readers to clearly understand about the one-tailed (left-tailed) test: Assume that, a wholesaler in Nandyal purchase electric bulbs from Hyderabad and he will accept the bulbs only when their average life is only 1000 hours. So, on each load that arrives from Hyderabad, he will test a sample (n=400) and inspect the bulbs for their mean life and he will accept the load only when the mean life is above 1000 hours and reject the load, when the mean life of sample bulbs is below 1000 hours. So, the H_O: $\overline{X} = \mu$; and H_A: $\overline{X} < \mu$. The rest of the procedure to compute Zcal and its comparison with Ztab value will be the same as discussed above. In this case, since H_A: $\overline{X} < \mu$, we see towards the left side of the normal distribution curve. The Figure 12.2 illustrates this example.
- ☆ ***Non-directional H_A (or 'Two-tailed' tests – for testing difference only)***: These are stated in '≠' terms. That is, the direction of deviation of the alternative case in H_A is not specified. A two-tailed test requires us to consider both sides of the region under normal distribution

---

* Z–Test: Conditions under which it is applied: Sample size should be large (> 30), Population standard deviation must be known.

't'-Test: is given by W.S.Gosset (pen name 'Student'). Also called Student's 't' distribution. Conditions under which it is applied: Sample size should be small (≤ 30), Population standard deviation must be unknown, Normal or approximately normal population.

** If the calculated statistic is higher than the critical value from the table:

☆ The finding is significant.

☆ We reject the H_O.

☆ The probability is small that the difference happened by chance, and 'P' is less than the critical alpha level (P < alpha).

If the statistic is lower than the critical value from the table:

☆ The finding is not significant.

☆ We accept the H_O.

☆ The probability is high that the difference happened by chance, and P is greater than the critical alpha level (P > alpha).

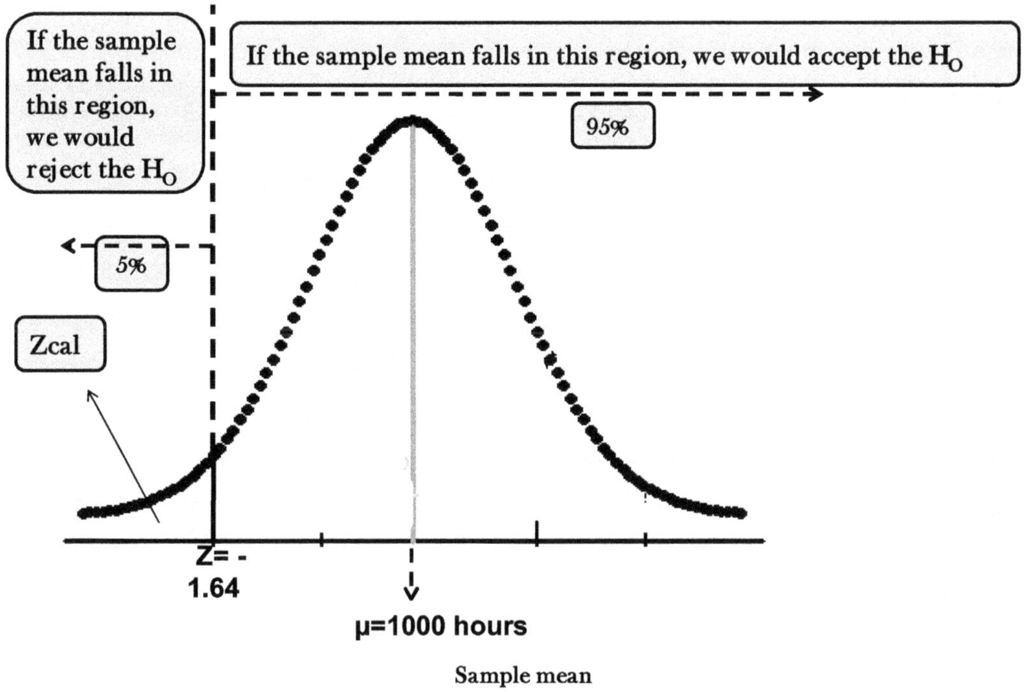

Figure 12.2: One-Tailed Test (Lower Tailed or Left Tailed Test) – Acceptance and Rejection Regions.

curve. So we split 'α' and place half in each tail. A two-tailed test of a hypothesis would reject the H_O, if the sample statistic significantly differs from the hypothesized population parameter. Hence, in a two- tailed test, there are two rejection regions. In two-tailed test, the hypotheses include: H_O: Sample mean $\overline{X}$ = Population mean (μ) and H_A: $\overline{X} \neq \mu$. The following example enable the readers to clearly understand about the two-tailed test: Assume that, a random sample of n=50 was selected from a certain population with a population mean of 45. The $\overline{X}$ and sample standard deviation (s) are 45.2 and 3.02 respectively. Using this information, assuming 5 per cent level of significance, compute whether sample mean differ significantly from population mean or not? In this case, given

- *Population mean = 45*
- *Sample mean = 45.2*
- *Sample standard deviation = 3.02*
- *Considering above, the following hypotheses are formulated:*
  - *H_O: Sample mean $\overline{X}$ is statistically equal to Population mean (μ) or $\overline{X}$ = 45.*
  - *H_A: Sample mean $\overline{X}$ is statistically ≠ Population mean (μ) or $\overline{X} \neq 45$.*

In the above case, since the sample size n=50, the researcher has to employ Z-statistic to study the statistical difference between sample mean and population mean. The following is the formula employed to compute Z-statistic:

$$Zcal = (\overline{X} - \mu)/(s/\sqrt{n}) = (45.2-45)/(3.02/\sqrt{50}) = 0.468$$

The above value Z=0.468 is called Z-calculated value. Since, the assumed level of significance is 0.05 and it is a two-tailed test because, the researcher framed the two critical regions both on left side and right side under the normal distribution curve, an area equal to $\alpha/2$ lies on both the tails, for a test of significance level α. Since, the Z_{cal} value (0.468) is less than Z_{tab} value (+1.96 at 5 per cent level of significance) on the right side of the tail of the normal distribution curve, the H_O is accepted and concluded that, $\overline{X}$ is equal to μ. The same concept is depicted through Figure 12.3.

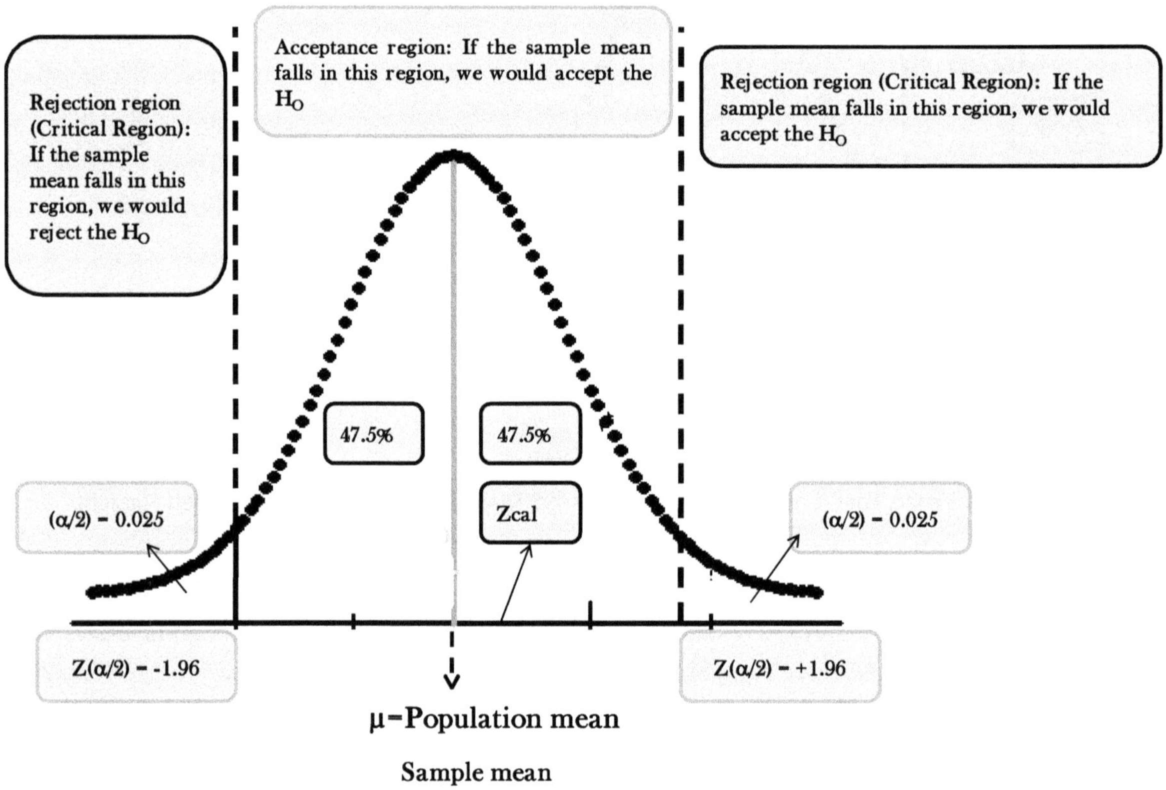

Figure 12.3: Two-Tailed Test – Acceptance and Rejection Regions.

It is essential to note that, in case of large samples (n>30) and in case of small samples ($n \leq 30$), the critical values for rejection of H_O are shown through Table 12.2. The calculated values of 'Z' in case of large samples and 't' in case of small samples must exceed these critical or table values to reject the H_O.

F. Confidence Intervals for *Parameters* $\hat{a}$ and $\hat{b}$

From the t-test so calculated earlier (Table 12.1), we reject the hypothesis if (t cal > t tab) or $\hat{b}$ value is twice more than standard error of $\hat{b}$ *i.e.* ($\sigma\hat{b}$) or ($\sigma\hat{b}$) is at least twice lower than $\hat{b}$ value and by rejecting the hypothesis Ho: True b=o, we conclude that $\hat{b} \neq 0$. So, the t-test indicate that, for the sample drawn from the population, true b is not equal to zero. But, from 't' statistic we cannot say that, whether the estimated $\hat{b}$ value will approach true b value. That is from 't' statistic, we cannot explain

Table 12.2: Critical Values of 'Z' and 't' Tests at 1 and 5 percents Levels of Significance.

Test	*1 per cent level of Significance*	*5 per cent level of Significance*
Large samples (Z test)		
One tailed test**	2.33 (α=0.01)	1.64 (α=0.05)
Two tailed test**	-2.58 to +2.58(α=0.005)	-1.96 to +1.96 (α=0.025)
Small samples (t test) (at 10 degrees of freedom)		
One tailed test**	2.764 (α=0.01)	1.812 (α=0.05)
Two tailed test*	-3.5814 to +3.5814 (α=0.005)	-2.6337 to +2.6337 (α=0.025)

Note: *: Apply the formula: TINV (probability, degrees of freedom).

**: See Appendix given at the end of this Chapter.

about the un-biasedness of the estimate. Since $\hat{b} \neq 0$, as we proved from 't' cal value, this value should lie in and around the true b value. So we have to work out the limiting (value) levels for $\hat{b}$ with in which the true b will lie. Such, limiting values or levels around the estimated $\hat{b}$ are called Confidence Intervals or Prediction Intervals or Confidence Limits of the estimate. So, these confidence intervals around $\hat{b}$ indicate, within these intervals the true b will lie. These confidence intervals are worked out around $\hat{b}$ by assuming a certain probability distribution around $\hat{b}$ with in which the true b will lie and this probability level is called Confidence level or Confidence co-efficient.

In general applied econometrics, we assume 95 per cent confidence level or probability around the $\hat{b}$ where the true b will lie. This 95 per cent confidence level implies that, out of 100 cases the researcher goes for repeated sampling process, the true b will lie with in the confidence intervals of $\hat{b}$ in 95 cases and only in five cases the true b falls outside the confidence intervals.

Confidence Intervals from 'Z'-test or from Standard Normal Distribution

Step 1

Z statistic for $\hat{b}$: $Z = \dfrac{\hat{b} - b}{\sigma(\hat{b})}$

Step 2

We assume 95 per cent confidence level, which implies 5 per cent level of significance. Since Z tab value at 5 per cent level of significance is 1.96, it will have two values in two tail Z test i.e, - 1.96 and + 1.96. We can write the confidence interval as follows

$$P\{-1.96 < Z < +1.96\} = 95 \text{ per cent}$$

Step 3

Let us substitute Step 1 for Z statistic in Step 2. We can write as

$$P\{-1.96 < \frac{\hat{b}-b}{\sigma(\hat{b})} < +1.96\} = 95 \text{ per cent.}$$

Step 4

Re-arranging Step 3, we will get

$P\{\hat{b} - 1.96(\sigma\hat{b}) < b < \hat{b} + 1.96(\sigma\hat{b})\} = 95$ per cent.

Step 5

The two confidence intervals for $\hat{b}$ at 95 per cent probability is :

$\hat{b}$ -1.96 $(\sigma\hat{b}) < b < \hat{b}$ +1.96 $(\sigma\hat{b})$.

This can also be expressed as: $b = \hat{b} \pm 1.96(\sigma\hat{b})$

$$b = \hat{b} \pm Z\,tab\,(\sigma\hat{b})$$

This can also be expressed as:

Confidence Interval 1: $b = \hat{b} + 1.96(\sigma\hat{b}) = \hat{b} + Z\,tab\,(\sigma\hat{b})$

Confidence Interval 2: $b = \hat{b} - 1.96(\sigma\hat{b}) = \hat{b} - Z\,tab\,(\sigma\hat{b})$.

Example

If $\hat{b} = 8.4$ and $\sigma\hat{b} = 2.2$ and assume confidence level as 95 per cent (*i.e.*, the level of significance as 5 per cent), then the confidence intervals for $\hat{b}$ is computed as follows:

We know, $b = \hat{b} \pm 1.96(\sigma\hat{b})$

1. $b = \hat{b} + 1.96(\sigma\hat{b}) = 8.4 + 1.96(2.2)$

   $b = 12.712$

2. $b = \hat{b} - 1.96(\sigma\hat{b}) = 8.4 - 1.96(2.2)$

   $b = 4.1$.

Confidence Intervals from Student t-test Distribution

The estimation of confidence intervals from Student 't'- distribution is similar to 'Z' distribution, but the main difference is that, in 't' distribution, we have to consider degrees of freedom because, 't'-test deals with small samples ($n \leq 30$).

Step 1

't' statistic for $\hat{b}$, $t = \frac{\hat{b}-b}{\sigma(\hat{b})}$ at (n-k) degrees of freedom

Where, k = number of parameters estimated or number of explanatory variables + intercept

Step 2

We assume 95 per cent confidence level, which implies 5 per cent level of significance. So, we have to verify 't' table value at 5 per cent level of significance at (n-k) degrees of freedom. The confidence intervals can be written as:

$P\{-t_{0.025} < t < +t_{0.025}\} = 95$ per cent at (n-k) degrees of freedom.

Step 3

Substitute Step 1 for 't' statistic in Step 2. We can write as:

$P\{-t_{0.025} < \frac{\hat{b}-b}{\sigma(\hat{b})} < +t_{0.025}\} = 95$ per cent at (n-k) degrees of freedom.

Step 4

Re-arranging Step 3, we will get:

$P\{\hat{b} - t_{0.025}(\sigma\hat{b}) < b < \hat{b} + t_{0.025}(\sigma\hat{b}) = 95$ per cent at (n-k) degrees of freedom.

Step 5

The two confidence intervals for $\hat{b}$ is given by:

$\hat{b} - t_{0.025}(\sigma\hat{b}) < b < \hat{b} + t_{0.025}(\sigma\hat{b})$ at (n-k) degrees of freedom. This can also be written as:

$b = \hat{b} \pm t_{0.025}(\sigma\hat{b})$ at (n-k) degrees of freedom

Confidence Interval 1:

$b = \hat{b} + t_{0.025}(\sigma\hat{b})$ at (n-k) degrees of freedom

Confidence Interval 2:

$b = \hat{b} - t_{0.025}(\sigma\hat{b})$ at (n-k) degrees of freedom.

Example

Suppose, we have estimated a regression line from a sample of 20 observations.

$$\hat{Y} = \underset{(38.2)}{128.5} + \underset{(0.85)}{2.88X}$$

Work out confidence intervals for both for $\hat{a}$ and $\hat{b}$.

Solution

For $\hat{a}$: ('t'$_{tab}$ at (n-k = 18) df = 2.101)

Confidence Interval 1 = $\hat{a} + t_{0.025}(\sigma\hat{a})$

$= 128.5 + 2.101(38.2)$

$= 208.7$

Confidence Interval 2 = $\hat{a} - t_{0.025}(\sigma \hat{a})$

$= 128.5 - 2.101(38.2)$

$= 48.2$

For $\hat{b}$:

Confidence Interval 1 = $\hat{b} + t_{0.025}(\sigma \hat{b})$

$= 2.88 + 2.101(0.85)$

$= 4.67$

Confidence Interval 2 = $\hat{b} - t_{0.025}(\sigma \hat{b})$

$= 2.88 - 2.101(0.85)$

$= 1.09.$

G. Drawing a Conclusion

If Probability (P) value is less than or equal to significance level (α), then reject the H_O in favor of H_A. So, the result is statistically significant. On the other hand, if P-value > Significance level (α), then accept the H_O. So, the result is not statistically significant. Assume the researcher wishes to analyze the improvement in cotton yield among the IPM technology adopted farmers compared to the farmers, who have not practiced that technology.

- H_0: There is no difference in yields among farmers with and without adoption of IPM technology.
- H_A: There is a difference in yields among the farmers due to the adoption of IPM technology on cotton.
- $\alpha = 0.05$; 20 per cent increase in yield of cotton due to adoption of IPM technology, P-value = 0.001.

Since, the P-value (0.001) < Significance level (α=0.05), we reject the H_O in favor of H_A. So, the result is statistically significant. That is, the difference in cotton yields between the IPM adopted farmers and those farmers, who have not adopted this technology, was statistically significant. We can conclude that, there was a significant (20 per cent) increase in cotton yields among IPM adopted farmers over other farmers, who have not adopted this technology.

General Procedure for Statistical Testing of Hypothesis

The above discussion infers the following the procedure involved in statistical testing procedure:

- State the H_O and H_A relevant to the framed objectives and identified variables for the study (say, the researcher wishes to study the degree of relationship between the variables *viz.*, prices of the commodity and its quantity demanded).
- Select the appropriate statistical test to be employed in this study (say, correlation coefficient is to be computed to analyze the degree of relationship between the variables *viz.*, price of the commodity and its quantity demanded).
- Select the desired level of significance. In most cases, the common level of statistical significance is 5 per cent or 0.05. Even 1 per cent or 0.01 level is also used. It all depends

upon how much a risk of Type I error the researcher is willing to accept and the effect that this choice has on the risk of committing Type II error.

- ☆ For the compiled and tabulated data of the selected variables, appropriate statistical test is to be employed to compute the statistic. (Correlation coefficient is computed for the selected variables *viz.*, price of the commodity and its quantity demanded and say, r = -0.94).
- ☆ Select the correct type of probability distribution. In general, there are two basic types of distributions *viz.*, normal distribution and 't' distribution. Among these two, normal distribution is an ideal model (Figures 12.1 to 12.3). The 't' distribution is much flatter than normal distribution and 't' distribution varies with every possible sample size. This 't' distribution is followed when the sample size is less than 30 and as the sample size increases, 't' distribution loses its flatness and gradually equals to normal distribution. Thus, normal distribution is the appropriate probability distribution, when the sample size exceeds 30.
- ☆ After calculating the statistic (say, correlation coefficient), now conduct the significance test to estimate the significant test value (say, 't' test or calculated value).
- ☆ Obtain the table or critical value (say, 't' table value) in the statistical tables at a given degrees of freedom (say, at n-2 degrees of freedom for two variables *viz.*, price of the commodity and its quantity demanded).
- ☆ Compare 't' calculated value and 't' table or critical value to make the decision. If the 't' calculated value is more than 't' table value, the researcher has to reject the H_O and accept H_A. On the other hand, if the 't' calculated value is less than 't' table value, the researcher has to accept the H_O and reject H_A.
- ☆ In order to generalize the sample statistic back to the population parameter, compute the confidence intervals.

12.2. Errors in Drawing Conclusions in Research

In testing the H_O, we come across the following types of alternatives:

i. H_O is correct, test accepts H_O
ii. H_O is correct, but test rejects H_O
iii. H_O is not correct, but test accepts H_O
iv. H_O is not correct, test rejects H_O

In the above alternatives, committing alternative (ii) is said to be committing Type I error and committing alternative (iii) is said to be the committing Type II error. The other two alternatives *i.e.* (i) and (iv) are correct decisions. The same concept is simplified in following Table 12.3.

Table 12.3: Committing Type I and Type II Errors in Hypothesis Testing.

Decision Taken by the Researcher	*H_O*	
	Correct (True)	*Incorrect (False)*
Accept H_O	Correct decision	Type II error
Reject H_O	Type I error	Correct decision

Note that, a reduction in the probability of committing Type I error increases the risk of committing Type II error and *vice versa*. That is, when the probability of rejecting the H_0 when it is correct (true) decreases (Type I error), the probability of accepting the H_0 when it is false (Type II error) increases and *vice versa*. Further, if we choose the significance level of 5 per cent compared to 1 per cent level in committing Type I error, then it would increase the probability or risk of committing the Type II error. Note that, if the researcher increases the sample size, it will decrease the risk of committing both Type I and Type II errors. We already know, the researcher makes the decision to reject or not reject (accept) the H_0 with a given probability of being correct. This probability of being correct is referred to as the Level of significance or Probability level of the test of significance, 0.05 (5 per cent) or 0.01 (5per cent). If 0.05 is set, the researcher will have a 5 per cent probability of making a Type I error, whereas, if 0.01 is selected, the researcher will have only 1 per cent of probability of committing Type I error. In other words, working at 0.05, the researcher has 95 per cent chance of making correct decision; working with 0.01 the researcher will have 99 per cent of chance of making correct decision. Let us discuss these two errors here under.

Type I Error

When we reject the H_0, when it is supposed to be correct or true, it implies Type I error. That is, the probability of rejecting the H_0 when it is true, is called Type I error. Probability of Type I error is denoted by 'α'. So, technically, 'α' is known as Type I error. Hence, Type I error is also called Alpha (α) error or Error of First Kind.

Type II Error

When we accept the H_0, even if it is incorrect or false, it implies Type II error. That is, the probability of accepting the H_0 when it is false or not correct is called Type II error. So, when a false hypothesis is erroneously accepted, it is known as Type II error. Probability of Type II error is denoted by β. So, Type II error is also called Beta (β) error or Error of Second Kind.

Example 1

The following is an example that simplifies your understanding about this concept. Assume that, a pesticide firm is introducing a new brand of pesticide into the market. Now the firm faces two alternatives, whether the 'pesticide succeeds in the market' or 'pesticide fails in the market'. Accordingly, the pesticide firm takes one of the following two decisions *viz.*, 'introduce new brand of pesticide into the market' or 'does not introduce new brand of pesticide into the market'. The same is explained through the Table 12.4.

Table 12.4: Type I and Type II Errors Regarding the Decision of Introducing a New Brand of Pesticide into the Market.

Decision taken by the Pesticide Firm	*H_0 is Correct (True)* *Pesticide Fails in the Market*	*H_0 is Incorrect (False)* *Pesticide Succeeds in the Market*
Introduce the new brand of pesticide into market	(Wrong decision) Type I error Reject H_0	Correct decision
Does not introduce the new brand of pesticide into market	Correct decision	(Wrong decision) Type II error Accept H_0

The Table 12.4 reveals that, there are chances of committing errors (or) taking wrong decisions in two ways:

- ☆ Introduce the new brand of pesticide, when it fails in the market and this is called as Type I error (as the researcher rejected the H_O when it is true).
- ☆ Does not introduce new brand of pesticide, when it succeeds in the market and this is called as Type II error (as the researcher accepted the H_O when it is false).

Example 2

Suppose, the researcher (student) wishes to study the effect of a seed treatment chemical (against fungus) on groundnut. If the disease is present, the test yields 'Positive' result and if the disease is not present, the test yields 'Negative' result. So, with reference to 'disease', the sub-categories include: 'Disease Present' and 'Disease Not Present'. Regarding the conclusion to be derived, there are again two sub-categories *viz.*, 'Test Positive' and 'Test Negative'. So, we have four possible types of sub-categories that may occur with the outcome of this experiment. Across the top of the Table 12.5, we have the reality (Company's conclusion). The reality is that, there either is an effect or there is no effect regarding the efficacy of the seed treatment chemical (as opined by the chemical manufacturing firm). Down the side of the Table (*i.e.*, rows), we have the researcher's conclusions. The study may conclude there is an effect or that there is no effect. So we have four possible outcomes. Two of these events are errors. Two of them are correct conclusions.

Table 12.5: Type I Error and Type II Error.

	Reality (say, Company's conclusion)	
Researcher's Conclusion	**True Effect**	**No Effect**
Disease is controlled (True Effect) (Reject H_O) or (Accept H_A)	**Truth**	**Type I or Alpha Error**
Disease is not controlled No Effect (Don't reject H_O) or (Accept H_O)	**Type II or Beta Error**	**Truth**

Let's look first at the upper left hand corner of the table. Let's suppose, the researcher sampled 50 seeds of groundnut and from the study, he concluded that, the chemical had a true effect. That is, it appeared to reduce the fungal infection in the seed. So, this result is the truth and we are in the upper left corner of the table. On the other hand, if the researcher concluded that, there is no effect of this chemical on the sample seeds *i.e.*, the chemical has not controlled the fungal infection, and it implies the researcher has made another correct decision. So, we have the truth and hence we are in the lower right hand corner of the table. So, if the test showed no effect and there really is no effect (company's conclusion), then we are in the lower right corner and if there is an effect and the researcher concludes there is an effect, he made the correct decision in the upper left corner. Thus, in both these cases, the researcher made the correct decision.

Looking at the table, it should be apparent that, the researcher can also make incorrect decisions or errors. If there is no effect (company's conclusion), but if the researcher concluded that, there is an effect, then he has made an error (indicated in the upper right corner) and this is called as Type I Error or alpha error (*i.e.*, rejecting the H_O when it is ought to be true). Alpha is the probability of a Type I Error. If there is an effect (company's conclusion), but if the researcher concluded that, there is no effect, then he has made an error (indicated in the lower left corner) and this is called as Type II Error or Beta error (*i.e.*, accepting the H_O when it is ought to be false). Beta is the probability of a Type II Error.

The above discussion infers that, in Type I error or Alpha error, the researcher concludes 'there is an effect' (*i.e.* reject H_O or accept H_A), but in reality, there isn't. So, Type I Error is a false positive. In Type II error or Beta error, the researcher concludes, 'there is no effect' (*i.e.*, H_O is true), but in reality, there is. So, Type II Error is a false negative. So, the researcher must be careful in taking each decision, as he run the risk of making an error. Unfortunately, when the researcher decide to reject the H_O, he always run the risk of making a Type I Error. That is, Type I error is possible, when the researcher's decision is to reject the H_O when in fact, the H_O is true. Using a more stringent alpha level (say, 0.01) decreases the likelihood to make a Type I error, but it also increases the likelihood to make a Type II Error. A Type II error occurs, when the researcher decide to accept the H_O when in fact, the H_O is false or alternative *i.e.*, H_A is true. So, Type I Error and Type II Error are inversely related. The students can still understand better by going through the following Table 12.6:

Table 12.6: Type I Error and Type II Error.

		True Situation in Population	
		H_0 True	*H_0 False*
Decision made based	**H_0 False (reject)**	Type I Error	Correct Decision
upon sample data	**H_0 True (accept)**	Correct Decision	Type II Error

Example 3

Let us look at another example. Assume that, the researcher wishes to test a new seed treatment chemical on groundnut and to find it out, whether this new chemical is effective or not effective over the earlier seed treatment chemical. So,

H_O: There is no difference in effects between the two seed treatment chemicals on average.

The following Table 12.7 gives a summary of possible results of any hypothesis test:

Table 12.7: Type I Error and Type II Error in Testing H_O.

		Decision	
		Reject H_O	*Don't reject H_O (Accept H_O)*
Truth	H_O	Type I Error	Correct decision
Truth	H_A	Correct decision	Type II Error

We already know that, in a hypothesis test, a Type I error occurs, when the H_O is rejected when it is in fact true; that is, H_O is wrongly rejected. So, a Type I error would occur, if the researcher concluded that, the two chemicals produced different effects when in fact there was no difference between them.

In a hypothesis test, a Type II error occurs when the H_O is not rejected (*i.e.*, accepted) when it is in fact false. So, a Type II error would occur, if the researcher concluded that, the two chemicals produced the same effect, *i.e.*, there is no difference in the effects of the two chemicals on average, when in fact they produced different ones.

Note that, a Type I error is often considered to be more serious, and therefore, more important to avoid, than a Type II error. The hypothesis test procedure is, therefore, adjusted so that, there is a guaranteed 'low' probability of rejecting the H_O wrongly; this probability is never zero. This probability of a Type I error can be precisely computed as

P (Type I error) = Significance level = α

If we do not reject the H_O, it may still be false (a Type II error) as the sample may not be big enough to identify the falseness of the H_O (especially if the truth is very close to hypothesis). For any given set of data, Type I and Type II errors are inversely related; the smaller the risk of one, the higher the risk of the other. Type II error is frequently due to sample sizes being too small. The exact probability of a Type II error is generally unknown, but is symbolized by β and written as:

P (Type II error) = β

12.3. Size of Test vs Power of a Test

a. Size of a Test

The size of a test refers to the probability of rejecting the H_O, when it is true. That is, it refers to the probability of committing Type I error. This is denoted by 'α'. So, the level of significance and size of test are synonym. Therefore,

Size of a test = P (reject H_O) = α

b. Power of a Test

The power of a test refers to the probability of rejecting the H_O, when it is actually false ie, when H_A is true. That is, the probability of *not* committing a Type II error is called the Power of a hypothesis test. Therefore,

$$\begin{aligned} \text{Power} &= P\ (\text{reject } H_O/H_A \text{ is true}) \\ &= P\ (\text{accept } H_O/H_A \text{ is false}) \\ &= 1 - (\text{Probability of committing Type II error}) \\ &= 1 - \beta \end{aligned}$$

The maximum power a test can have is one, the minimum is zero. Ideally, we want a test to have high power, close to 1. For a fixed level of significance, if we increase the sample size, the probability of Type II error decreases, which in turn increases the power. So, to increase the power, the best method is to increase the sample size.

Factors that Affect Power

The power of a hypothesis test is affected by three factors.

- ***Sample size (n):*** Other things being equal, the greater the sample size, the greater the power of the test.
- ***Significance level*** **(α):** The higher the significance level, the higher the power of the test. If you increase the significance level, you reduce the region of acceptance. As a result, you are more likely to reject the H_O. This means, you are less likely to accept the H_O when it is false; *i.e.*, less likely to make a Type II error. Hence, the power of the test is increased.
- ***The 'true' value of the parameter being tested:*** The greater the difference between the "true" value of a parameter and the value specified in the H_O, the greater the power of the test.

12.4. Benefits of Hypothesis Testing

A hypothesis testing gives the following benefits:

- It establishes the focus and guides the researcher in the direction for a research effort.
- It helps the researcher to clearly state the purpose of the research activity.
- It helps to identify, what variables are to be considered in the study.
- It helps to assess the impact of a comparative study.
- It makes the analysis simple.

12.5. Interpretation of Analytical Findings

After analyzing the data, the researcher should interpret the findings and this facilitates him to convert the data into information. This, in turn, leads to drawing conclusions about what and why of the problem and thus, helps him to formulate policy implications. But, the interpretation of analytical findings demands a greater skill on the part of the researcher. For proper interpretation of data, it requires the execution of the following steps systematically and in the following sequence:

- Select representative sample from the target population.
- Selecting proper data collection methods.
- Select correct variables for the study.
- Select suitable statistical techniques for analyzing the data.

Above all, the researcher should have adequate experience in interpreting the analytical findings. The following are the important precautionary measures that are to be keep in mind by the researcher before interpreting the analytical findings:

- The data must be collected from the reliable sources.
- Appropriate statistical techniques should be employed.
- The researcher must have an *a priori* knowledge about the problem and this guide him in the process of interpretation.

- ☆ The researcher must note that, throughout the course of investigation, there should be constant interaction between initial hypothesis, empirical observation and theoretical conceptions.
- ☆ The researcher should limit the interpretation of findings keeping in view of the objectives and variables considered under the study.
- ☆ It is essential for the researcher to interact with the experts in that field of research study and this enhance the correctness of interpretation and utility of the findings.
- ☆ The researcher must be in no hurry in interpreting the analytical findings. This is because, in some cases, the conclusions appear to be right at the beginning, but may change after analyzing the data. This facilitates to overcome false generalization and interpretation of data.

Appendix

Commonly used 'Z' values

Probability	*'Z' Value*	
0.20	0.85	* this is often used in in power calculations power=0.8, â=0.2
0.10	1.29	
0.05	1.65	* this is often used for testing statistical significane p<0.05
0.025	1.96	* this is often used to define the 95% confidence interval (2.5% each side)
0.01	2.33	
0.005	2.58	
0.0025	2.81	
0.001	3.10	

http://www.statstodo.com/ZTest_Tab.php

Critical '*t*' Values Complete Table : df = 1 to 200

The columns are probabilities, and the rows degrees of freedoms. Each cell contains the critical '*t*' value for that particular degrees of freedom and probability of Type I error (alpha α). α is the probability of Type I error. The t values are for the two tail test. The same t value is used for half the alpha value in a one tail test. For example, at 20 degrees of freedom, the t value of 2.086 is critical for α=0.05 (2 tail) and α=0.025 (1 tail). With the same degrees of freedom, 1.725 is critical for α=0.1 (2 tail) and α=0.05 (1 tail).

α (1 tail)	*0.05*	*0.025*	*0.01*	*0.005*	*0.0025*	*0.001*	*0.0005*
α (2 tail)	*0.1*	*0.05*	*0.02*	*0.01*	*0.005*	*0.002*	*0.001*
df							
1	6.3138	12.7065	31.8193	63.6551	127.3447	318.4930	636.0450
2	2.9200	4.3026	6.9646	9.9247	14.0887	22.3276	31.5989
3	2.3534	3.1824	4.5407	5.8408	7.4534	10.2145	12.9242
4	2.1319	2.7764	3.7470	4.6041	5.5976	7.1732	8.6103
5	2.0150	2.5706	3.3650	4.0322	4.7734	5.8934	6.8688
6	1.9432	2.4469	3.1426	3.7074	4.3168	5.2076	5.9589
7	1.8946	2.3646	2.9980	3.4995	4.0294	4.7852	5.4079
8	1.8595	2.3060	2.8965	3.3554	3.8325	4.5008	5.0414
9	1.8331	2.2621	2.8214	3.2498	3.6896	4.2969	4.7809
10	1.8124	2.2282	2.7638	3.1693	3.5814	4.1437	4.5869
11	1.7959	2.2010	2.7181	3.1058	3.4966	4.0247	4.4369
12	1.7823	2.1788	2.6810	3.0545	3.4284	3.9296	4.3178
13	1.7709	2.1604	2.6503	3.0123	3.3725	3.8520	4.2208
14	1.7613	2.1448	2.6245	2.9768	3.3257	3.7874	4.1404

α (1 tail)	0.05	0.025	0.01	0.005	0.0025	0.001	0.0005
α (2 tail)	0.1	0.05	0.02	0.01	0.005	0.002	0.001
df							
15	1.7530	2.1314	2.6025	2.9467	3.2860	3.7328	4.0728
16	1.7459	2.1199	2.5835	2.9208	3.2520	3.6861	4.0150
17	1.7396	2.1098	2.5669	2.8983	3.2224	3.6458	3.9651
18	1.7341	2.1009	2.5524	2.8784	3.1966	3.6105	3.9216
19	1.7291	2.0930	2.5395	2.8609	3.1737	3.5794	3.8834
20	1.7247	2.0860	2.5280	2.8454	3.1534	3.5518	3.8495
21	1.7207	2.0796	2.5176	2.8314	3.1352	3.5272	3.8193
22	1.7172	2.0739	2.5083	2.8188	3.1188	3.5050	3.7921
23	1.7139	2.0686	2.4998	2.8073	3.1040	3.4850	3.7676
24	1.7109	2.0639	2.4922	2.7970	3.0905	3.4668	3.7454
25	1.7081	2.0596	2.4851	2.7874	3.0782	3.4502	3.7251
26	1.7056	2.0555	2.4786	2.7787	3.0669	3.4350	3.7067
27	1.7033	2.0518	2.4727	2.7707	3.0565	3.4211	3.6896
28	1.7011	2.0484	2.4671	2.7633	3.0469	3.4082	3.6739
29	1.6991	2.0452	2.4620	2.7564	3.0380	3.3962	3.6594
30	1.6973	2.0423	2.4572	2.7500	3.0298	3.3852	3.6459
31	1.6955	2.0395	2.4528	2.7440	3.0221	3.3749	3.6334
32	1.6939	2.0369	2.4487	2.7385	3.0150	3.3653	3.6218
33	1.6924	2.0345	2.4448	2.7333	3.0082	3.3563	3.6109
34	1.6909	2.0322	2.4411	2.7284	3.0019	3.3479	3.6008
35	1.6896	2.0301	2.4377	2.7238	2.9961	3.3400	3.5912
36	1.6883	2.0281	2.4345	2.7195	2.9905	3.3326	3.5822
37	1.6871	2.0262	2.4315	2.7154	2.9853	3.3256	3.5737
38	1.6859	2.0244	2.4286	2.7115	2.9803	3.3190	3.5657
39	1.6849	2.0227	2.4258	2.7079	2.9756	3.3128	3.5581
40	1.6839	2.0211	2.4233	2.7045	2.9712	3.3069	3.5510

13

Writing Research Report

Research reports mainly take the form of a thesis, dissertation, journal article and a paper to be prescribed at a workshop or conference etc. These reports vary in their format and style. For example, there are differences found in a research report prepared as a thesis or dissertation* and a research report prepared as a manuscript for research article or a technical bulletin. The dissertation and thesis are more elaborate and comprehensive, while research papers prepared for journal articles and

---

* Though Thesis and Dissertation words are used interchangeably, there are few differences between them.

- ☆ Both are research oriented activities. In Indian, British and South African Universities, the researcher will have to write a dissertation for his post-graduate programme. He will write a thesis for the attainment of a doctorate or Ph.D. degree. (However, in American Universities, reverse was the case *i.e.*, a dissertation leading to a Ph.D. degree and a thesis is produced while enrolled in a Master's programme).
- ☆ Thesis implies that original research is involved. Dissertation implies that the researcher have looked into something and are setting down what he found and perhaps including his own thoughts on it. Thus a Thesis is what you will write to obtain a higher degree, but a dissertation is more likely to be a small part of a first degree.
- ☆ For a thesis, you have conduct original research, while for a dissertation you have to synthesize already existing literature.
- ☆ Thesis analysis is added to the already existing literature, while dissertation is an analysis of the existing literature.

Contd...

presented in conferences are more precise and concise. Though there are differences in formats or styles of presentations across research reports, their functions remains same like, serves as a basic reference material for future use in developing research, provides factual base in formulating policies, generates systematic knowledge about a research problem etc. The 'format' refers to the general pattern followed for organizing different chapters in the report. That is, a format refers to an outline of the report that includes several chapters; sections and sub-sections; headings and sub-headings followed under each chapter etc., in presenting the detailed information of the research study. In general, all research reports follow the sequence of chapters that are in parallel with the steps involved in the research process. In this Chapter, let us focus our attention regarding the submission of a research report (Thesis) submitted by a student pursuing M.Sc. (Ag) or Ph.D. programme in the Department of Agricultural Economics in SAUs. The following is the layout of the Thesis:

13.1. Layout of the Thesis

There is a scientific method for the layout of the thesis submitted by the student pursuing M.Sc (Ag) or Ph.D programme in the Department of Agricultural Economics in SAUs. Thesis or a dissertation is a statement of investigation or research presenting the author's findings and any conclusions reached, submitted by the author in support of his/her candidature for a higher degree, professional qualification or other award. A Ph.D. thesis must demonstrate the originality and ability of the student for an independent investigation and the results of the research must constitute a contribution to knowledge. The thesis must exhibit the students' mastery over the literature on the subject and familiarity with its sources. A M.Sc. (Ag) thesis must demonstrate the student's familiarity with the tools of research and ability to present the results of their investigation effectively. The layout of the thesis means as to what the research report should contain. The contents of the research report (thesis) contains the following three major sections:

- ☆ Thesis is lengthier thus, takes more time to be completed, while dissertation is short or concise; therefore, it does not consume too much time to be completed.
- ☆ In thesis, researcher have to include a hypothesis based on the research work. In contrast to thesis, researcher should have a decent knowledge of the new discoveries in order to infer conclusion effectively in dissertation.
- ☆ Thesis is written as an academic research paper, while dissertation is more like an academic book.
- ☆ Data collected in thesis is based upon the hypothetical analysis of a researchable problem, whereas dissertation is an academic work and is mainly comprised of theory and argumentation based on original research.
- ☆ Thesis is basically written by students to earn scholarships. In contrast to thesis, dissertations are not usually written to get scholarships, rather students write them for academic purposes.
- ☆ In dissertation, students have to list down the major findings concerning with their topic, whereas thesis demand students to conduct a hypothesis based on their research work.
- ☆ Arguments play the major role in a thesis. In contrast to thesis, students need to focus on their background research for dissertation.
- ☆ Students conclude their viewpoint on the basis of research work in thesis, while major findings aid students to infer their conclusion in dissertation.

The points listed above are the various ways of comparing thesis and dissertation, where thesis is submitted at the Ph.D level and dissertation is submitted at the post-graduate level. However, the above points listed for difference between thesis and dissertation will reverse in the countries (USA), where thesis is the basic requirement to earn the Master's degree and dissertation to earn a doctorate.

A. Preliminary Pages or Preliminaries

- ☆ Title page
- ☆ Declaration
- ☆ Certificates
- ☆ Acknowledgements
- ☆ List of contents
- ☆ List of tables
- ☆ List of figures
- ☆ List of symbols and abbreviations
- ☆ List of appendices
- ☆ Abstract

B. Main Text

- ☆ Introduction
- ☆ Review of literature
- ☆ Materials and Methods
- ☆ Agro-Economic Features
- ☆ Results and Discussion
- ☆ Summary and Conclusions

C. End Matter

- ☆ Literature cited
- ☆ Appendices

The names of these sections are only to facilitate the arrangement of various chapters and they are not to be indicated as headings. Chapter dividers shall be avoided.

A. Preliminary Pages

The following are the elements or sub-sections (in order) that comprise the preliminary material:

- ☆ ***Title page:*** This page comprises of
  - ❖ ***Title:*** Times New Roman font, bold with 20 point
  - ❖ ***Author Name:*** Times New Roman font, bold with 16 point
  - ❖ ***Author's qualification:*** Times New Roman Font, bold with 12 point
  - ❖ ***Submission statement:*** Times New Roman font, bold with 14 point
  - ❖ ***Degree:*** Times New Roman font, bold with 18 point
  - ❖ ***Discipline in brackets:*** Times New Roman font, bold with 14 point
  - ❖ ***Name of the chairperson:*** Times New Roman font, bold with 14 point
  - ❖ ***Emblem:*** 2.5 cm diameter
  - ❖ ***Imprint:*** Times New Roman font, bold with 14 point
  - ❖ Name of the Department

MARKETING EFFICIENCY IN TRANSACTING PADDY IN REGULATED AGRICULTURAL MARKETS OF ANDHRA PRADESH

BY
NIRMAL RAVI KUMAR KOTAMRAJU
B.Sc. (Ag.)

THESIS SUBMITTED TO THE
ACHARYA N.G. RANGA AGRICULTURAL UNIVERSITY
IN PARTIAL FULFILMENT OF THE REQUIREMENTS
FOR THE AWARD OF THE DEGREE OF
MASTER OF SCIENCE IN AGRICULTURE
(AGRICULTURAL ECONOMICS)

CHAIRPERSON: Dr. V. T. RAJU

DEPARTMENT OF AGRICULTURAL ECONOMICS
AGRICULTURAL COLLEGE, MAHANANDI
ACHARYA N.G. RANGA AGRICULTURAL UNIVERSITY
RAJENDRANAGAR, HYDERABAD - 500 030
2012

- Name of the College
- Name of the University
- Year of submission of thesis

The title of the thesis should be self explanatory and give an idea about the content of the thesis. It shall be placed near the top of the page in capital letters in bold. The sub-title, if any, shall follow the title after a colon in the same line. If the title exceeds one line, an inverted pyramid format shall be followed as far as possible without splitting the words in each line. When two or more lines are needed, should be double spaced. It is advisable to give the thesis a title in which the first noun expresses the most important topic that was studied: for this reason, it is not wise to start the title with the word 'Research'. Title should not contain redundancies such as 'A study of....or 'An investigation on....... Abbreviations should not appear in the title. If scientific names are to be used, they should be in italics. The title should contain 12 to 15 words. Avoid use of words like 'By.... 'From.... Preferred order of words in the researcher's name: Start with first, middle and followed by last name. Full name

DECLARATION

I, **Mr. NIRMAL RAVI KUMAR KOTAMRAJU**, hereby declare that the thesis entitled **"MARKETING EFFICIENCY IN TRANSACTING PADDY IN REGULATED AGRICULTURAL MARKETS OF ANDHRA PRADESH"** submitted to the **Acharya N.G. Ranga Agricultural University** for the degree of **Master of Science in Agriculture** in the major field of **Agricultural Economics** is the result of the original research work done by me. I also declare that no material contained in the thesis has been published earlier in any manner.

Date: 5-8-2012 **(NIRMAL RAVI KUMAR KOTAMRAJU)**
MA-10-001

should be used and initials should be avoided. Titles like Dr. or Mr. or Ms. or Mrs. etc., should not appear in the names. The year should follow at the bottom of the page. Paginate using roman numbers (small case) starting with this title page.

- ✰ ***Declaration:*** The researcher must give a formal confirmation that the work is solely that of the author.
- ✰ ***Certificates:*** This is given by both Major Advisor and also by the Advisory Committee of the student.
- ✰ ***Acknowledgements:*** This page permits the researcher to express appreciation of persons, who have contributed significantly to the completion of the research study. So, it is customary to thank the Advisor(s), who helped the student during the course of research investigation at each and every stage. The researcher should also thank the Sponsors of the research (say, Stipend offered by the SAU or Indian Council of Agricultural Research (ICAR) or International Crops Research Institute for Semi-Arid Tropics (ICRISAT) etc.), who provided financial assistance for undertaking research. Acknowledgements shall be a brief note for technical and financial (institutional only) assistance received by the candidate during his/her research work and preparation of thesis. It shall not exceed two pages. The student

CERTIFICATE

Mr. NIRMAL RAVI KUMAR KOTAMRAJU has satisfactorily prosecuted the course of research and that thesis entitled **"MARKETING EFFICIENCY IN TRANSACTING PADDY IN REGULATED AGRICULTURAL MARKETS OF ANDHRA PRADESH"** submitted is the result of original research work and is of sufficiently high standard to warrant its presentation to the examination. I also certify that neither the thesis nor its part thereof has been previously submitted by him for a degree of any university.

Date:

Place: Mahanandi

(V.T. Raju)
Chairperson
Professor & University Head
Department of Agricultural Economics
Agricultural College, Mahanandi-518502

should also thank his fellow students, college administration, College Librarian etc., in extending their support in taking up and conducting the research study. Note that, the student should acknowledge only those persons, who helped significantly for completion of research work. He should also acknowledge the sample farmers and other stakeholders, who helped him in providing the requisite data and information that are essential for research investigation. It shall not contain nick names and statements of dedication. Acknowledgements for paid services shall be avoided.

- ✰ ***List of contents:*** The List of contents is an outline of the thesis, which indicates page number of each major chapter. Entries listed in the list of contents should be identical to headings in the main report. For example, if a header reads Materials and Methods in the body of the thesis (Chapter III), it should read the same in the List of Contents, and not Material and Research Methods. This list of contents is essential, as it provides the reader a quick overview of the major chapters of the thesis, with page references, so that (s)he can go through the report in a definite order or skip certain sections.

CERTIFICATE

This is to certify that the thesis entitled, **"MARKETING EFFICIENCY IN TRANSACTING PADDY IN REGULATED AGRICULTURAL MARKETS OF ANDHRA PRADESH"** submitted in partial fulfilment of the requirements for the degree of **'MASTER OF SCIENCE IN AGRICULTURE'** in the major field of Agricultural Economics of the Acharya N. G. Ranga Agricultural University, Hyderabad, is a record of the bonafide original research work carried out by **Mr. NIRMAL RAVI KUMAR KOTAMRAJU** under our guidance and supervision.

No part of the thesis has been submitted by the student for any other degree or diploma. The published part and all assistance received during the course of investigations have been duly acknowledged by the author of the thesis.

(V.T. Raju)
Chairperson of the Advisory Committee

Thesis approved by the Student Advisory Committee

Chairperson:	**Dr. V.T. Raju** Professor & University Head Department of Agricultural Economics Agricultural College, Mahanandi-518502	--------------------------
Member :	--------------------	-------------------------
Member :	--------------------	-------------------------

Date of Final Viva-Voce: ------------

☆ ***List of tables:*** Note that, when Table 10 and above are inserted in the list, the decimal points following the 10 align with the decimal points for numbers 1 to 9. The rule of decimal point alignment also applies to any lists or tables using decimal points. The following examples indicate CORRECT and INCORRECT expressions of numbers above 10.

CORRECT

 9. Summary of.
10. Summary of.

INCORRECT

9. Summary of.
10. Summary of.

INCORRECT

9. Summary of.
 10. Summary of.

LIST OF CONTENTS

Chapter	Title	Page No.
I	INTRODUCTION	1-7
II	REVIEW OF LITERATURE	8-24
III	MATERIALS AND METHODS	25-35
IV	AGRO-ECONOMIC FEATURES	36-58
V	RESULTS AND DISCUSSION	59-115
VI	SUMMARY AND CONCLUSIONS	116-123
	LITERATURE CITED	124-130
	APPENDICES	131 - 139

- In general, table number starts with the Chapter number. Say, in Chapter IV, table numbers start with 4 *i.e.*, as 4.1, 4.2, 4.3 and so on. Similarly, in Chapter V, table numbers start with 5 *i.e.*, as 5.1, 5.2, 5.3 and so on.
- The table numbers and titles of the tables given in the list of tables must match with the table numbers and titles of the tables given in the main report.
- The page numbers given for each table in the list of tables must match with the page on which it can be found in the main report.

☆ ***List of Figures:*** As mentioned in case of List of tables, the numbers of figures also start with the Chapter number. Say, in Chapter III, figure numbers start with 3 *i.e.*, as 3.1, 3.2, 3.3 and so on. In Chapter IV, figure numbers start with 4 *i.e.*, as 4.1, 4.2, 4.3 and so on. Similarly, in Chapter V, figure numbers start with 5 *i.e.*, as 5.1, 5.2, 5.3 and so on.

- The figure numbers and titles of the figures given in the list of figures must match with the figure numbers and titles of the figures given in the main report.
- The page numbers given for each figure in the list of figures must match with the page on which it can be found in the main report.

LIST OF TABLES

Table No.	Title	Page No.
4.1	Administrative divisions in Srikakulam district	37
4.2	Rainfall distribution in Srikakulam district	39
4.3	Maximum and minimum temperatures in Srikakulam district	40
4.4	Demographic particulars of Srikakulam district	41
4.5	Literacy status in Srikakulam district	42
4.6	Occupational distribution of population in Srikakulam district	42
4.7	Land utilization pattern in Srikakulam district	44

- ☆ ***List of Symbols and Abbreviations (optional):*** If abbreviations or acronyms are used in the report, these should be stated in full in the text the first time they are mentioned followed by the abbreviation in brackets. If there are many, they should be listed in an alphabetical order and shown under List of Symbols and Abbreviations. For the titles of the periodicals, only full title should be given and no abbreviation shall be used.

  Note that, the List of contents, List of tables and List of figures and List of Symbols and Abbreviations should be prepared at the last, as they can only be finalized after preparing the complete report. Further, the researcher has to finalize the page numbers for locating tables and figures.

- ☆ ***List of appendices:*** Any supporting material not included, but referred to in the main text should be given as an appendix. Like List of Tables and List of Figures, the appendix number and title of the appendix given in the list of appendices must match with the appendix number and title of the appendix given in the end matter of the report. If there is more than one, each appendix shall be given alphabetic designation such as Appendix A, B, etc. and be titled.

LIST OF FIGURES

Figure No.	Title	Page No.
3.1	Sampling design regarding the selection of respondents	26
3.2	Researcher collecting primary data from the sample farmer	28
4.1	Andhra Pradesh map showing Srikakulam district	36
4.2	Map of Srikakulam district showing selected mandals	38
4.3	Map of Ponduru mandal showing selected villages	52
4.4	Map of Gara mandal showing selected villages	54
5.1	Component-wise shares of total family expenditure of marginal farmers	66

Note that, the exact format of the List of Contents and Lists of Tables, Figures and Appendices differs across the Universities.

☆ ***Abstract:*** It shall be written as Abstract in capital letters (in Times New Roman font with 18 point and bold). This page indicates the summary of the research report. It should be brief and informative. A reader, who has been attracted by the title of the thesis will usually look at the Abstract to decide whether the report is worth reading. The Abstract should be written only *after* the first or even the second draft of the report has been completed. Abstract should deal with the following components:

- 'WHY' this study was needed
- 'WHAT' has been studied (objectives)
- 'WHERE" the study was conducted (place of study)
- 'HOW' the study was conducted (the type of study and methods employed)
- Major findings and conclusions, followed by
- Major (or all) recommendations.

LIST OF SYMBOLS AND ABBREVIATIONS

%	:	Per cent
<	:	Less than
>	:	Greater than
^{0}C	:	Degree Celsius
Crs	:	Crores
et al.	:	and others people
etc.	:	and so on; and other people/things
ha	:	Hectare
i.e.,	:	That is
MLR	:	Multiple Linear Regression
mm	:	Millimeter
MSP	:	Minimum Support Price
No.	:	Number
Rs	:	Rupees
Rs/ha	:	Rupees per hectare

Note that, 'Abstract' is the first page that was studied by the readers and hence, it demands skillful writing on the part of the researcher. Note that, the abstract is normally limited to 500 words. It is not desirable to exceed this limit. However, do not make it too much shorter. The word 'ABSTRACT' has to be centered and the matter should be given in single line space. The abstract shall be preceded by the name of the author, title, degree to which it is submitted and the particulars of faculty, department, Chairperson, University and year of submission. The abstract shall be condensed but informative covering the essential points of the investigation. It shall include the nature and scope of the research undertaken, brief methodology of investigation, conclusions and contributions made to the knowledge of the subject treated. The presentation of 'Abstract' ensures the following advantages:

- It gives an outline about the results obtained in the study so that, the reader understands the subject matter without having to read the full thesis.
- The purpose of the abstract is to make subsequent search for information easier.
- It helps subsequent researchers to decide, whether it is worth to go through the entire report.
- It allows the other students for reference purpose.

LIST OF APPENDICES

Appendix No.	Title	Page No.
A.	List of farmers in the selected village and their size of land holdings	
B.	Map showing the irrigation potential of KWD	
C.	Photographs showing the interactions of the researcher with the Agricultural Produce Market Committee Officials of Kurnool district	
D.	Schedule employed for collecting data and information from the respondents	

ABSTRACT

Name of the Author	:	**NIRMAL RAVI KUMAR KOTAMRAJU**
Title of the thesis	:	**MARKETING EFFICIENCY IN TRANSACTING PADDY IN REGULATED AGRICULTURAL MARKETS OF ANDHRA PRADESH**
Submitted for the award of	:	**Master of Science**
Faculty	:	**Agriculture**
Discipline	:	**Agricultural Economics**
Major Advisor	:	**Dr. V. T. RAJU**
University	:	**Acharya N.G. Ranga Agricultural University**
Year of Submission	:	**2012**

In the era of globalization and economic liberalization, the farmer's focus shifted towards market led production and this called for revamping the agricultural production sector towards strengthening the requisite farm infrastructure, extensive use of machinery,

Majority of the students think that, in a thesis 'Introduction' is the first chapter that is read by the readers. Some others think that, conclusions of major findings are usually read first. But, note that, 'Abstract' is the page that the Committee and everybody else read first.

- ☆ ***Foreword*** (say, in case of research report (Technical bulletin))
- ☆ ***Preface*** (say, in case of research report (Technical bulletin))

Other Important General Instructions followed in Preparing the Preliminary Pages

The preliminary pages shall be numbered with lower case Roman numerals *i.e.*, i, ii, iii, iv, v, vi, etc., at the bottom centre of the page. The Title Page counts as page 'i', but the number is not printed on the page. The Copyright Page (if included) counts as 'page ii', but is also not numbered. The remaining preliminary pages are numbered consecutively.

B. Main Text

This section provides the complete outline of the thesis along with all details. This section is divided into different Chapters as given below:

I. Introduction
II. Review of literature
III. Materials and Methods
IV. Agro-Economic features
V. Results and Discussion
VI. Summary and Conclusions
Literature cited

Introduction ☆ Objectives	Why am I doing this? ☆ What do/did I hope to find out?
Review of Literature	What is known? What is unknown?
Materials and Methods	How is/was it found out?
Agro-Economic Features	What is the socio-agro-economic background of the sample village
Results	What did I find?
Discussion	What does it mean?
Conclusion	What is the significance? What applications? Where to next?

Chapter I–Introduction

Its purpose is to introduce the research topic to readers. This chapter highlights the importance and need for taking up the research study. It begins with background information regarding the problem under investigation. In general, the 'Introduction' begins with a broader perspective of the problem and becomes narrower as it proceeds. Talk about the global perspective followed by the local scenario of the selected problem. Generally, this chapter runs about three to six pages, but may vary considerably depending on the nature of the study. It shall be written as Chapter I (in Times New Roman font with 14 point and bold) followed by Introduction in capital letters (in Times New Roman font with 18 point and bold). The introduction should briefly state the lacunae in the subject and the gaps the thesis attempts to fill up. The background and purpose of the investigation should be indicated. The objectives of the investigation shall be clearly mentioned classifying them as general and specific objectives. This chapter is framed in such a way that:

- ☆ It must create interest to the reader in going through the topic.
- ☆ It must establish the problem that leads to the study.
- ☆ It must place the study within the larger context of the literature.
- ☆ It must reach out to a specific audience.
- ☆ It must provide readers with a brief summary of literature and research related to the problem being investigated, and should lead up to the statement of the problem.

This chapter highlights the background of the problem under consideration, gradually introduces the problem under consideration and also justifies its selection for the study. So, the research topic is gradually narrowed down in dealing with its focus of the study and provides a brief rationale for why that particular study is worth pursuing. This chapter is dealt through several sub-heads like:

- ☆ ***Problem statement:*** This is the important part of introductory chapter. As the heading implies, the purpose of the study is stated in this section. This sub-head forms the most critical part of the thesis, as it provides focus and direction for the remainder of the study and in writing subsequent components of the thesis. Note that, the problem statement must be written in a non-technical and simple language so that, even an ordinary reader (say, even to a literate farmer) may follow it. Generally, there is no one 'correct' or 'best way' to write the problem statement. However, the written problem statement must ensure the following:
  - ❖ It should facilitate the readers to get clarified about the problem and its severity.
  - ❖ It helps the researcher in identifying the variables to be investigated in the study.
  - ❖ It highlights main theme of the study.
  - ❖ Must indicate exactly what the problem is.
  - ❖ Indicate why and how it is a problem.
  - ❖ Give information to support the problem say, by use of statistics or evidence. This should be derived from background information to illustrate connectivity.
  - ❖ Length: Maximum one page.
- ☆ ***Objectives:*** The problem statement is further explicated in this sub-head. The problem statement formulated earlier enable the researcher to frame the objectives of the study. One general objective, which should be in line with the title and 4-6 specific objectives in relevance to the selected problem will be listed in this sub-head. This help him in identifying the variables to be considered in the study and in planning the methodology for the study The specific objectives should be related to the general objective. Note that, the objectives should not be in the form of questions. These specific objectives have to be numbered (1, 2, 3....).
- ☆ ***Hypothesis:*** For the specific objectives framed for the study, the researcher has to formulate hypotheses and these will provide direction for the researcher to proceed.
- ☆ ***Scope of the study:*** This is a kind of a disclaimer. Should cite the focus of the study either geographical area or target group/population.
- ☆ ***Significance of the Study (Problem):*** This section addresses the 'so what' of the study and report. It describes or explains the potential value of the study and findings to the social

sciences research in agriculture. This section, therefore, should identify the audience for the study and how the results will be beneficial to them. Note that, research is conducted to add to the existing knowledge base and/or solve a problem – how the present research study will do this should be articulated in this sub-head.

☆ ***Limitations (Optional):*** These are the factors, which are usually beyond the researcher's control, that may affect the results of the study or how the results are interpreted. Stating limitations of the study may be very useful for readers because, they provide a method to acknowledge possible errors or difficulties in interpreting results of the study. Limitations that are not readily apparent at the start of the research project may develop or become apparent as the study progresses. In any case, limitations should not be considered as excuses; they are simply factors or conditions that help the reader get a truer sense of what the study results mean and how widely they can be generalized. While all studies have some inherent limitations, the researcher should address only those that may have a significant effect on the selected study. Stating limitations of the study assists the readers in understanding some of the inherent problems encountered by the researcher. Research design that controls these limitations should be selected for the study such that, it reduces them to a reasonable number and scope. Note that, while this sub-head is optional, almost all theses include it. Not including this section implies that, the present study has accounted for all (or nearly all) variables, is generalizable to all populations and could be replicated accurately under all conditions. When discussing the limitations, the researcher should be sure to:

- ❖ Describe each limitation in detailed, but concise terms.
- ❖ Explain, why each limitation exists.
- ❖ Provide the reasons, why each limitation could not be overcome using the method(s) chosen to gather the data. If possible, cite to other studies that had similar problems when information is available.
- ❖ Assess the impact of each limitation in relation to the overall findings and conclusions of the study.

☆ ***Delimitations (Optional):*** As discussed earlier, technically, delimitations (factors which the researcher controls) are distinct from limitations (over which the researcher has little or no control). However, in some dissertations, the researcher includes delimitations within the sub-head of limitations. Since they both are technically different, they should be dealt in separate sections of Chapter I. These delimitations describe the scope of the study or establish parameters or limits for the study. For example, setting limits on the sample size, extent of the geographic region from which data are collected, response formats included in the schedules prepared for data collection, the time frame for the study etc., are the important delimitations that the researcher controls in the study.

☆ ***Assumptions (Optional):*** Assumptions stated by the researcher usually address limitations that the researcher is aware of that may affect the study, but he will not attempt to control. Generally, conditions that have already been stated as limitations or delimitations should not be addressed in this section. In social sciences research in agriculture, it is generally assumed that, all respondents will answer all survey questions honestly and to the best of their abilities. Note that, if respondents' honesty in providing the correct data has been listed as a possible limitation in the thesis, the same should not be stated as an assumption. The other example of assumption is that, it is assumed that, in this study, it is assumed that, participants' gender will not significantly affect their perceptions.

- ☆ ***Organization of the Study:*** This was discussed as last sub-head in Chapter I. This sub-head implies the plan of presentation of the thesis. That is, it summarizes the contents of each of the chapters that will comprise the study. This permits readers to know, what information will be found in each chapter and facilitates finding specific information without searching through the thesis page by page to do so. The following points presents an example of this sub-head:
  - ❖ The thesis is presented in six chapters.
  - ❖ The importance, the objectives and the limitations of the study are presented in this Chapter I.
  - ❖ Critical view of past work done is presented in the Chapter II.
  - ❖ The Chapter III deals with the Materials and Methods used in the study including the sampling procedure, method of collection of data, estimation methods and various concepts and terms used in the study.
  - ❖ The Chapter IV deals with Agro-Economic Features of the selected district and of the selected mandals.
  - ❖ The Chapter V encompasses a critical analysis of the results and discussion.
  - ❖ Summary and Conclusions emerged from the study and policy implications are presented in Chapter VI.
  - ❖ Literature cited is presented at the end of the thesis.

  Note that, as the List of contents have already been presented at the beginning of Chapter I, it is equally acceptable to label this section 'Organization of the Remainder of the Study', in which case the description will be limited to Chapters II through Literature cited.

Other Guidelines for Writing Chapter I

While writing Chapter I, the researcher must answer himself the following four questions:

- ☆ What was I studying?
- ☆ Why was this topic important to investigate?
- ☆ What did I know about this topic before I did this study?
- ☆ How will this study advance the knowledge?

The researcher must note that, a well-written introduction is important because, it creates first impression on the part of the readers about the logic of selected research problem, writing style, the overall quality of research and ultimately the validity of research findings and conclusions. Even though 'Introduction' is the first Chapter written in the main section of the thesis, it is often useful to finish this chapter very late in the writing process because, by that time, the analysis of Results and Discussion and the Conclusions will have been completed and it ensures that 'Introduction' matches the structure of the thesis. Note that, one sentence paragraphs are unacceptable. A paragraph should have a minimum of five sentences. Tables and figures should be labeled as per the chapter number. That is, in Chapter I, the first table should be labeled as Table 1.1 and first figure should be labeled as Figure 1.1.

Chapter II–Review of Literature

This Chapter is about what happened and when in the area of the selected research earlier. That is, it includes the relevant theories and models that have been used for similar cases or studies in the past. The literature review illustrates, summarizes, appraises and clarifies the background for the proposed research investigation. It will give a hypothetical foundation for the research and helps the researcher to establish the nature and direction of proposed research activity. When the researcher reviews the earlier research articles or books pertaining to the selected research problem, it facilitates him to collect relevant information, data and to identify variables for the selected research topic. This facilitates the researcher to point out the earlier authors, who worked in similar studies and mentioning their findings. This encourages the researcher to put his/her input and opinions and justifications for the study. This Chapter will serve as the reference for everything in the thesis. It is the database/ information-base/knowledge-base/event-base/fact-based/evidences-based that the researcher will rely on in framing research design (Chapter III) and discussing the analytical findings (Chapter V). The literature review make sense of a body of research and present an analysis of the available literature so that, the readers does not have to access each individual research report included in the review. This is important because, there is an increasing amount of literature available to the readers, who cannot be expected to read and assimilate all the information on the proposed topic. Everyone who works within the scope of the proposed research topic has a professional duty to be up-to-date with recent developments and ideas so that, they execute the proposed research topic on the systematic lines. The thorough analysis of the research topic with reference to a particular subject area and sometimes in a particular subject area within a certain time period helps the researcher to analyze, how each work is similar to and how it varies from the others thereby, guides the researcher way forward for further research. Further, it helps to gain knowledge and understand about the proposed research topic.

Note that, in writing the review of literature for the selected research topic, the researcher conveniently divides it into different sub-heads (say, in accordance with the variables he identified or in accordance with the objectives framed for the study) and under each sub-head he will collect the relevant review. For example, assume that, the researcher framed the following objectives for the selected research topic on *'An Economic Analysis of Farmers' Indebtedness in Srikakulam District of Andhra Pradesh:*

- ✰ To study the agro–economic scenario of sample farmers to assess the equity and non-equity capital components employed by the sample farmers.
- ✰ To study the source-wise and purpose-wise indebtedness of sample farmers in the cultivation of major crops.
- ✰ To analyze and prioritize the factors responsible for indebtedness of sample farmers.
- ✰ To study the impact of Agricultural Debt Waiver and Debt Relief Scheme implemented by the financial institutions on the indebtedness of farmers.

For the above framed specific objectives, the researcher collect the literature under the following sub-heads:

- ✰ Lending pattern to farmers.
- ✰ Extent of indebtedness.
- ✰ Factors responsible for indebtedness.
- ✰ Implementation of Agricultural Debt Waiver and Debt Relief Scheme.

The division of research review into different sub-heads facilitate the researcher to divide the lengthy research topic into different sections and this facilitate him to conveniently frame the past research findings in a proper sequence. The review is presented in such a way that, it includes the last name of the researcher with year of study in parenthesis followed by a brief description of his study. The logical sequence of presentation is that, the review should proceed from the past year to the current year. That is, there should be an historical discussion. The review (under each sub-head) will start with the first major work in the field (or from the year chosen by the researcher, if the history is long) and proceed to show how subsequent research and commentary has built on, modified, or improved up to the latest year possible, leading into where the present study fits. In case of more than two authors of a research article, use *'et al'* (italicize) at the end of the first author's last name followed by the year in the parenthesis. Note that, Chapter II is 'NOT' the place for the researcher to offer any personal ideas or theories. Further, it is difficult to estimate how long Chapter II should be. Generally, it may run between 15 to 50 pages depending upon the nature of the research topic and the number of framed objectives under consideration. At the end of each sub-head, the researcher should give a short summary of that sub-head in the context, how that particular topic has progressed over a period of time, what gaps are still existing in that area, how the present research will fill this gap etc. This is so because, mere description of past studies and findings has no meaning, unless those descriptions and findings are analyzed critically to find out research gaps to be bridged by the present study. Therefore, it is required that, after description of the previous research work under each sub-head, the researcher should give his/her critical appraisal to find out significance of the present study. For writing a good review of literature, the researcher should keep in mind the following 'five C's':

Review of Literature

Rao (1989) opined that there were some practical difficulties to the farmers' in obtaining no dues certificates, continuing security better, equitable deed, non-encumbrance certificate etc., various forms were required to be filled by the borrowers and considerable time was taken between applying for a loan and disbursing the loan.

Vaidya *et al. (1989) revealed that the cost incurred by the borrower in process of* availing and repaying the loan, besides the interest charges while calculating the cost of credit. They found that borrowers incurred 4.26 per cent as additional cost other than interest charges i.e. (1.14 per cent and 3.12 per cent) in process of availing loan and repaying the loan) which accounted to 23.54 per cent of average amount of loan. The cost in addition to interest charges was found to be higher for small farmers than medium farmers.

Ramakrishna (1990) reported that the major portion of borrowing from agriculturist moneylenders, the main source of non-institutional credit is used for consumption by both the types of borrowers. The repayment of institutional loans is relatively lower as compared to non-institutional loans.

Mohapatra and Mishra (1995) revealed that majority of respondents (9.11%) found difficult to pay back even the installments of credit amount taken. 72.2 per cent of respondents visualized that credit programme was complex and 35.5 per cent of respondents opined that wrong and biased identification of beneficiaries was a major constraint in credit programmes.

NABARD (2001) reported that there are some disquieting features of lending to small borrowers. The number of small borrower accounts in case of commercial banks has come down over time indicating shifting of their focus to large borrowers. The rate of growth in agricultural advances to small and marginal farmers of less than 5 acres farm size by scheduled commercial banks has come down as compared to other farm size categories.

- ✰ ***Cite:*** Keep the primary focus on the literature pertinent to the selected research problem.
- ✰ ***Compare:*** The earlier studies of different authors pertaining to a particular topic (sub-head) should be compared in terms of methodologies employed and findings expressed in the

literature. The researcher should also carefully note, what do the authors agree on? who applied similar approaches to analyzing the research problem?

- ***Contrast:*** The researcher should examine the past studies carefully to locate, what are the major areas of disagreement, controversy, or debate?
- ***Critique:*** The researcher should examine, which methodologies and findings seem most reliable, valid, or appropriate, and why?
- ***Connect:*** The researcher should draw the information, how well the past researches connect to the present selected research problem, how it differs from the past studies etc.

Other General Guidelines in Writing the Review of Literature

They include:

- The researcher should take the guidance of Major Advisor and other Professors for advice when selecting the books and research articles.
- This Chapter can be divided into five sub-heads: *viz.*,
  - 2.1 Introduction
  - 2.2 Theoretical review/Conceptual Framework: In this, the researcher will review the empirical and theoretical literature relevant to the problem being investigated. He will indicate, what has been done by other researchers including the methodologies used and identify gaps. The hypothesized variables should be subheadings of the literature review to form a framework that would help in analysis. Conceptual framework should demonstrate an understanding of what variable influences what. Consistency is important in citation. Each key variable should be 4-5 pages long.
  - 2.3 Critique of the existing literature relevant to the study.
  - 2.4 Summary.
  - 2.5 Identification of research gaps.
  - 2.6 How the present study fill the identified research gaps.
- Note that, a literature review is never perfect. Even when a researcher reviews his own article published earlier, he may notice at many places, where he could change to improve the presentation of the findings. This implies that, the researcher must revise the literature review multiple times. He should continuously go through the past articles to ascertain whether the review tells a story and generate an idea, whether the review is sufficient enough to identify key variables for the present study?
- It is better for the researcher to consider peer reviewed articles for writing the review.
- After collecting the literature, the researcher should ascertain:
  - Have sufficient numbers of research studies been critically reviewed?
  - Does the literature review end with a good summary?
  - Does the literature review make a convincing argument for the purpose of the study?
  - Does any information need to be added to the literature review to help the reader understand the key variables?
  - Does any information need to be removed because, it is irrelevant to the study?
  - The researcher should stick to the point that is actually needed to review in the thesis. This is because, Professors are busy people and do not want to read more than is directly relevant to the study.

- To avoid plagiarism*, it is important to 'Paraphrase' and 'Summarize' the articles that are found. Paraphrase involves rephrasing a passage from a source into own words. Paraphrases are usually shorter by slightly condensing the information. *'Paraphrasing involves putting a passage from source material into your own words. A paraphrase must also be attributed to the original source. Paraphrased material is usually shorter than the original passage, taking a somewhat broader segment of the source and condensing it slightly'.* Summarizing involves writing only the main ideas in own words. Summaries are considerably shorter with only a broad overview of the source. Summaries are often used when the researcher want to draw one major conclusion from an entire research article.
- The title shall be written as Chapter II (in Times New Roman font with 14 point and bold) followed by Review of Literature in capital letters (in Times New Roman font with 18 point and bold). Review of literature shall be up to date and include only those works that have been published, scheduled for publication or deposited in libraries as theses/ dissertations.

Chapter III–Materials and Methods

It shall be written as Chapter III (in Times New Roman font with 14 point and bold) followed by Materials and Methods in capital letters (in Times New Roman font with 18 point and bold). This chapter forms the background regarding the methodological aspects of addressing the research problem under consideration with reference to sampling procedure say, for selection of district, mandals, villages and sample farmers, tools of data collection and statistical techniques employed for the analysis. The basic terms and concepts involved in the study are also dealt here, as it facilitates to understand the concepts in a scientific manner. Methodological limitations or procedural weaknesses, if any, shall be given. If any improvement or modification of the method over others has been done, it shall be mentioned. The names of the insecticides, fungicides, herbicides, chemicals etc., shall begin with capital letter in the beginning of the sentence and thereafter with lower cap only. Trade names however, are to be indicated with capital letter (first letter) either at the beginning of the sentence or anywhere in the text. Certain crop names shall be indicated uniformly as Pigeonpea (Redgram), Chickpea (Bengal Gram), Cowpea, Green Gram, Black Gram, Soybean, Groundnut, Pearl millet, Finger millet etc. This chapter is dealt under the following sub-heads:

- ***Sampling procedure:*** This sub-head describes the population used in the study and the process followed in selecting a representative sample (Figure 13.1). As discussed earlier, unless the population is extremely small, a sample usually will be drawn from the population. In general, the sample will accurately represent the population from which it is drawn if (a) sample selection carefully follows an appropriate sampling design, (b) the sample is randomly selected from the population, (c) a large enough sample is selected in relation to the total population (d) the sample size is adequate for the data-collecting instrument being used etc.
- ***Data collection:*** This sub-head describes the type of instrument followed to gather data from the selected sample respondents. We know, in agro-economic research, schedule is the data

* Plagiarism is the presentation of the work, idea or creation of another person as if it is your own. It is considered cheating and is a very serious academic offence that may lead up to expulsion from the academic or research programme. Plagiarised material can be drawn from, and presented in, written, graphic and visual form, including electronic data, and oral presentations. Plagiarism occurs when the origin of the material used is not appropriately cited.

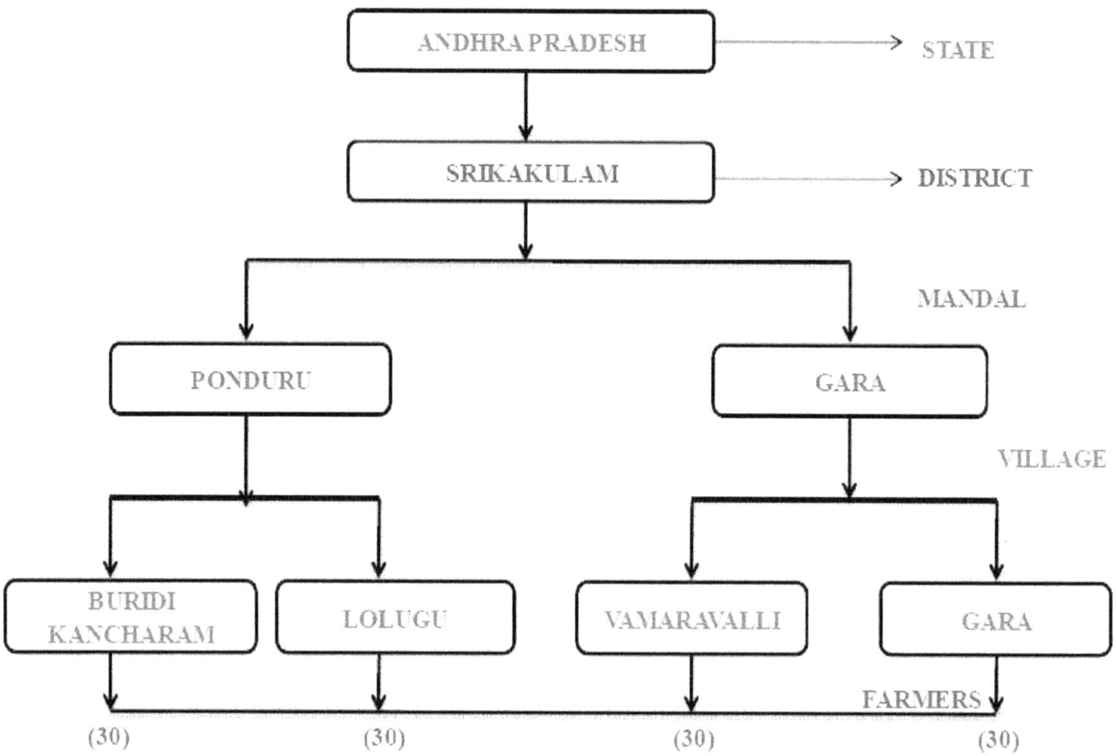

Figure 13.1: Sampling Design Regarding the Selection of Farmer-Borrowers.

collection instrument employed on the sample respondents selected for the study. The researcher should include the items or questions in the schedule that are derived from the information gleaned from the literature review or may be an adaptation of a previous study. Note that, the schedule so designed by the researcher should always be pre (pilot) tested or field tested to ensure its validity and clarity of instructions and items. In general, subjects similar to those who will be in the study sample (but not included in the actual sample) may serve as subjects for pilot testing. Results of pilot testing and accompanying comments should be used, if necessary, to revise the schedule before discussing with the actual sample. The schedule should also be juried or critiqued by having several 'experts' examine it and make recommendations prior to, or in lieu of pilot testing. This pilot testing though consume some time, but it provides the researcher much greater information that leads to a more reliable and valid data collection instrument. Further, the researcher should also deal with the type of data collected for the variables under consideration *i.e.,* primary data or secondary data or both.

- ✰ ***Statistical tools and techniques:*** This deals with the type of statistical tools and techniques employed in the study to analyze the data of the selected variables. Of course, the nature of data collected by the researcher influences the type of statistical techniques to be employed in the study. Note that, if any complex statistical procedures are being used, they should be briefly described and its source must be referenced. Tests of significance should be accompanied by a statement of the level of significance that will be used (for example, all statistical analyses will use up to 5 per cent level of significance). The statistical software package being used for data analysis should also be stated. The most commonly used descriptive statistics include mean, standard deviation, frequency counts, and percentages. Among the most commonly used inferential statistics are Chi-square (χ^2), '*t*' test, ANOVA and correlation coefficients. More complex statistical procedures include ANCOVA, MANOVA, Canonical Correlation, MLR analysis, Discriminant analysis, DEA etc.

- ✰ ***Terms and concepts used in the study:*** In this sub-head, the researcher briefly explains important terms and concepts used in the study that are unusual or not widely understood. In addition, common terms that have special meaning in the study should also be defined in this sub-head. Acronyms (except those in common usage) frequently require definition at this point. A brief introductory statement usually precedes the actual list of definitions of the terms used in the study: *'The following definitions are provided to ensure uniformity and understanding of these terms throughout the study'*. This sub-head enables the readers to know the meanings of different terms and concepts so that, they can better understand the information and analytical findings. Note that, the researcher should follow an alphabetical order in defining terms and concepts used in the study and when defining terms, it is important to cite appropriate references, if all or some of the definitions are taken from other sources.

Other General Guidelines in Writing Materials and Methods

They include:

- ✰ The researcher must note that, this chapter must be well-written and logically organized because, he/she are not actually doing the research with reference to the information provided in this chapter. As a consequence, the reader will never have a study outcome from which to evaluate whether the researcher's methodological choices were the correct ones. The objective here is to ensure that the reader is convinced that the selected research design and tools of analysis will correctly address the research problem. The selected research design and tools of analysis should be absolutely and unmistakably tied to the specific aims of the study.
- ✰ The researcher must be specific about the methodological approaches that he/she plan to undertake to collect information, about the techniques to analyze it, and about tests of external validity (*i.e.*, the trustworthiness by which the present research findings can be generalized to other people, places or times).
- ✰ As far as possible, quote the references of the review of the literature from where the techniques of analysis are referred.
- ✰ Indicate the data sources clearly. The limitations if any, in collecting both primary and secondary data must be clearly mentioned.
- ✰ Above all, the researcher must note that,
  - ❖ Is the sample representative of the target population?
  - ❖ Is each measure in the study valid to quantify the key variables as identified in the research hypotheses, questions, and/or objectives?
  - ❖ Is each measure appropriate for the sample?

- ❖ Were the research procedures appropriate for the research hypotheses, questions, and/or objectives?
- ❖ Were the research procedures clearly stated, so they can be replicated?

At the end of this chapter, a brief summary of the methodology should be given highlighting the main points and this is to be done only when the methodology section is very long or complex. This summary section should provide a smooth transition to Chapter IV.

Chapter IV–Agro-Economic Features

This chapter provides a brief information of the important agro-economic features of the study area (say, about selected district, mandals and villages) with the objective of providing necessary background of the sample respondents selected for the study. As discussed under Chapter 2 (Research), before planning any explanatory or analytical research, a reconnaissance survey of the study area is essential to highlight the seriousness of the study problem, amenities and facilities available in the study area, the extent to which the present research findings can be generalized etc., and this indicates the readers about the seriousness of the research problem in the study area.

Chapter V–Results and Discussion

It shall be written as Chapter V (in Times New Roman font with 14 point and bold) followed by Results and Discussion in capital letters (in Times New Roman font with 18 point and bold). The results should be based on a careful consideration of collected data. This is the main body of research report. It contains statistical summaries and analysis of data. There should be logical sequence in the analysis of data. It deals with the application of the statistical techniques to the data collected and interpret the analytical findings at preselected alpha levels (levels of significance). So, this chapter makes the researcher to confirm, whether the formulated hypotheses earlier are accepted or rejected. Thus, this section is about hypotheses testing. All probability values have to be mentioned and reported for each test. The researcher must comment and justify, why he rejected or fail-to-reject a particular hypothesis. If in the thesis, the researcher accepted a lot of the H_O, the researcher need not be dejected and may consider one or more of the following options:

- ✰ The schedule questions were not well designed, formulated or understood by the respondents, or
- ✰ The respondents were careless/not-serious in responding to the researcher, or
- ✰ The sample size was not enough, or
- ✰ It is the reality.

Note that, this chapter should be structured around the objectives and hypotheses developed in Chapter I. The researcher should start writing this chapter by reviewing the purpose of the study. The purpose of the study was described clear back in the Chapter I (Introduction). The researcher has to link or compile all results in accordance with the framed objectives for the study. It is essential that, the researcher should clearly indicate, whether the analytical findings of the present study will tally or deviate from the past studies as briefed under the Chapter II (Review of literature). That is, the researcher has to refer to the literature in Chapter II to support his/her findings. Further, he should not forget that, this Chapter V is researcher's own ground, *i.e.*, he should use all his own language and opinion in the interpretation and discussion of the analytical findings, of course with the support of the literature collected. So, this chapter is all about benchmarking the present research results with respect to the results and findings from the past literature. But, this chapter is not the proper place for drawing conclusions of the findings and this should be passed to Chapter VI (Summary and Conclusions).

Other General Guidelines in Writing Results and Discussion

They include:

- ✰ As a rule of thumb, confine the information to what the researcher have found out (placed under Chapter II - Review of Literature) for discussing the analytical findings. 'Discussion' of analytical findings is considered as the heart of the thesis. This is where, the researcher tie together the research questions or hypotheses, the data analyzed and the previous research findings. So, the researcher should carefully link the present findings with the past reviews in the sense to explain, to what extent the present findings tally with the past findings? or to what extent the present findings contradict with the past findings?
- ✰ The analytical findings should be discussed objective-wise. While discussing the findings, care should be taken not to introduce new findings *i.e.,* findings not mentioned in the tables and graphs. The discussion may include findings from other related studies that support or contradict the present findings. Limitation of the study and generalizability of the findings should also be mentioned.
- ✰ While discussing the analytical findings, the researcher must relate these to both the purpose of the study and to published results from other studies examined in the literature review. This section may be used to forward theories and/or models, or raise questions regarding previously developed theories.
- ✰ It is important to note that, this chapter provides the researcher to explore ideas and possibilities. Unlike most other chapters in the thesis whose content and form are dictated by fairly rigid standards, this chapter may be open-ended and take the form that researcher desires.
- ✰ Analysis of data is made under sub-headings pertaining to each hypothesis. Once the statistics have been calculated, the next step is to organize the findings into tables and figures so that, the reader can easily read and interpret them. The key to making both tables and figures is to make them clear, both in appearance and in interpretation. When the findings are presented through a table or a figure, the researcher should critically look over it to ascertain, whether the reader be able to understand what everything in the table/figure means? A good practice is to ask a peer to review the table or figure and give their comments about how it can be more clear.
- ✰ Table titles should be at the top of the tables.
- ✰ Tables collected from elsewhere from authentic publications should have source(s) below them.
- ✰ Any table generated by the researcher should not have the source quote.
- ✰ The titles of the table should be in brief and clear indicating the nature of information presented.
- ✰ Column headings and row headings of a table should be clearly labeled.
- ✰ If no data is available for a particular cell in a table, indicate the lack by a dash (-), rather than a zero (0).
- ✰ The tables should be well constructed, and without anomalies such as percentages which do not add up to 100 per cent.
- ✰ Avoid too many decimal places for the data given in the tables.

- ✰ Every table must be discussed in the text. This means that, the researcher must explain the key elements of a table in the body of the research report. When discussing tables, refer to them by the table numbers only, not where they appear in reference to the text. This is acceptable: '*As can be seen in Table 3.*' This is not acceptable: '*The table above shows.*'. The rationale for this rule is that, the placement of the table may change when the final document is produced.
- ✰ Readers should be able to interpret a table just by looking at the table, without reading the body of the research report itself. Therefore, each table should have a clear title that focuses on the key statistics within that table and all acronyms and abbreviations should be explained in the table notes.
- ✰ The researcher has to look at the following guidelines when creating a table:
  - ❖ Table titles should be brief, but clearly explain the table.
  - ❖ All similar entries in the table should carry the same number of decimal points. In other words, every entry within the same column should have the same number of decimal places. In general, whole numbers such as degrees of freedom should have '0' decimal points, p-values and correlations should be rounded to three decimal places (thousandths place value), and all other numbers (means, standard deviations, t-values, F-values) should carry two decimal places (hundredths place value).
  - ❖ Tables typically have a horizontal rule at the top, bottom, and after the table labels. Vertical lines are rarely used.
  - ❖ Once the tables have been assembled and placed in the text, number the tables starting from 1 in the order that they appear in the body. In the thesis, before the table number, chapter number precedes. For example, table numbers 1, 2, and 3 in Chapter V are numbered as 5.1, 5.2 and 5.3 respectively.
  - ❖ The decimal points must be lined up in each of the columns. This makes the table easier to read.
  - ❖ At the time of finalizing the thesis document before its submission to the Advisory Committee, the researcher must pose himself the following questions:
    - ❑ Is the table necessary?
    - ❑ Is the entire table, including the title, headings, and notes double spaced?
    - ❑ Are all comparable tables in the manuscript consistent in presentation?
    - ❑ Is the title brief but explanatory?
    - ❑ Does every column have a column heading?
    - ❑ Are all abbreviations, special use of italics, parentheses, and dashes, and special symbols explained?
    - ❑ Are all probability levels correctly identified, and are asterisks attached to the appropriate table entries? Is a probability level assigned the same number of asterisks in all tables in the manuscript?
    - ❑ Are all vertical rules eliminated?
    - ❑ Will the table fit across the width of the page?

- ❑ If all or part of a copyrighted table is reproduced, do the table notes give full credit to the copyright owner?
- ❑ Is the table referred to in the text?

☆ Indicate the data sources clearly below the table. If the table comprises the results of analyzed data collected from a secondary source say, www.fao.org, then indicate below the table as 'Raw Data Source: www.fao.org'.

☆ Figures usually contain graphics such as graphs, diagrams or photographs. Figures can have different shadings.

☆ Just like tables, figures can also be an excellent way for readers to quickly understand and easily interpret the statistical findings. Figures given in the body should also be fully understandable without reading the text, but also be referenced in the text. Figures should also be numbered consecutively like tables. Legends and notes should be included, so the reader can easily interpret the figure. One difference is that, figures do not have titles. Instead, a caption at the bottom of the figure functions both as a title and an explanation of the figure.

☆ Graphs should clarify and not complicate, and care should be taken that they do not mislead.

☆ The researcher must use his/her expertise and judge to determine whether to use a line graph, pie graph, bar graph, etc., in presentation of graph. Again, it is helpful to have a peer review the figures to determine, whether there might be a better way to assemble the graph.

☆ At the time of finalizing the thesis document before its submission to the Advisory Committee, the researcher must pose himself the following questions regarding the presentation of figures:

- ❖ Is the figure necessary?
- ❖ Is the figure simple, clean, and free of extra detail?
- ❖ Are the data plotted accurately?
- ❖ Is the grid scale correctly proportioned?
- ❖ Is the lettering large and dark enough to read? Is the lettering compatible in size with the rest of the figure?
- ❖ Are terms spelled correctly?
- ❖ Are all abbreviations and symbols explained in the figure legend or figure caption? Are the symbols, abbreviations, and terminology in the figure consistent with those in the figure caption? In other figures? In the text?
- ❖ Are all figures mentioned in the text?
- ❖ Are figures that are being reproduced or adapted from another source given proper credit in the figure caption?

☆ Tables and figures enable the reader to comprehend and interpret data quickly. It is advisable to use several tables rather than to use one table that is crowded. Good tables and figures are self–explanatory. Each table should be presented on the single (same) page. The word 'Table' or 'Figure' is followed by the corresponding table or figure number in Arabic numerals.

☆ Tables and figures are numbered continuously, but separately for each chapter *i.e.*, chapter number should precede the table number.

☆ It is recommended that, extremely long tables or figures or very detailed information not be included in this Chapter V. Due to space requirements (and questionable interest to most readers), it is better to place this information in an Appendix and not in Chapter V. For example, the size of land holdings of list of farmers surveyed in each village would be more appropriately placed in an Appendix, while a summary table showing the number of farmers across different sizes of land holding in a village (please see Figure 11.1) should be included in Chapter V.

- Avoid breaking tables and figures onto two or more pages whenever possible. This is because, it is difficult for the reader to have to flip back and forth to cover all of the information contained in the table or figure.
- If appropriate statistical tests are used, P-values alone are not very helpful. Confidence intervals and the type of tests used should be indicated.
- When reporting statistical results of data analyses (particularly inferential statistics), it is appropriate to include sufficient information in the table and accompanying text to permit the reader to corroborate the results of the analyses. Therefore, appropriate statistical symbols should be utilized to report these results. These statistical symbols should be italicized. Words, rather than symbols, should be used in the narrative, while symbols may be used in tables and inside of parentheses within the narrative. For example, *'the mean score of 89.25 marks for men farmers was higher than the mean score of 87.50 marks for women farmers in the sample with reference to their knowledge towards IPM technology of paddy'*. The same inference can be expressed otherwise. *'The men farmers in the sample scored higher (M = 89.25) than the women farmers (M = 87.50) regarding their knowledge on IPM technology'*. The commonly used statistical symbols include:
  - M = mean
  - df = degrees of freedom
  - SD = standard deviation
  - t = t statistic (t test)
  - f = frequency
  - F = Fisher's statistic (ANOVA)
  - p = probability
  - r = correlation coefficient (Pearson)
  - $\chi 2$ = Chi-square statistic
- One of the most common errors the researcher performs in this chapter is a lack of clear reporting. Numbers are reported in a table, but it is unclear how the numbers were calculated based on the items from the schedule. The reader should know, how each statistic was calculated based on the schedule items. A second error is, the statistics employed by the researcher are not thoroughly explained. For example, was the finding statistically significant? If not, what does that mean? If it was significant, where were the significant differences?
- The findings are also variously referred to as generalizations, implications, inferences, interpretations and recommendations for further research, if any.
- Foot notes is a citation used when an exact quotation or when a passage of another writer is paraphrased in the thesis. This establishes the existence of evidence for the information generated in the thesis. This footnote citation is indicated by placing an index number as a

superscript at the point of reference. Even symbols like asterisk (*) may be used. The footnote comprises of the following:

- Author's name followed by comma
- Title of the work followed by comma
- Place of publication followed by comma
- Publisher's name followed by comma
- Year of publication followed by comma
- Page numbers. Page number to be preceeded by 'p' for a single page and by 'pp' for two or more pages.

In order to shortened repeated references even more, there are a few abbreviations in use. They include *'Ibid'*, *'Idem'*, *'op. cit'* and *'loc.cit'*.

'Ibid'

If the same reference is used consecutively, but information is drawn from a separate page(s), then in the foot note following the first reference, the second reference is indicated by *'Ibid'*., followed by page number. The following is the example:

'Kumar[1] opined that, the Public Distribution System in the country facilitates the supply of food grains to the poor at a subsidized price. Further, a minimal amount of social security must be provided to those who are old, sick or disabled and cannot take on work even if it is available. Special schemes must ensure that they do not go hungry. Therefore, the PDS needs to be restructured[2] and there is a need to explore the possibility of introducing innovative ideas such as smart cards, food credit/debit cards, food stamps and decentralized procurement, to eliminate hunger and make food available to the poor wherever they may be in cost-effective manner.

---

1. Kumar K.N.R., *Food Security in India: Reality or Myth*, Daya Books Limited, 2009, p.176
2. *Ibid.*, p. 181

Thus, Ibid is used when citing a new reference that is from the same text as the preceeding reference. In the first reference to the text, give the full details of the source and use *Ibid* for the second. If the citation is from the same page as the previous reference, simply use *Ibid*. If it is from a different page, use ibid + page number (*e.g.*, ibid p. 56). From Latin, *'Ibid' means 'in the same place'*. This repeats the previous author and title and whatever else is identical. It is still frequently used.

Without 'Ibid'	*With 'Ibid'*
Kumar K.N. Ravi, Research Methodology in Agricultural Economics, Daya Publishing House, New Delhi, p. 42	Kumar K.N. Ravi, Research Methodology in Agricultural Economics, Daya Publishing House, New Delhi, p. 42
Kumar K.N. Ravi, Research Methodology in Agricultural Economics, Daya Publishing House, New Delhi, p. 42	*Ibid.*
Kumar K.N. Ravi, Research Methodology in Agricultural Economics, Daya Publishing House, New Delhi, p. 87	*Ibid.*, p. 87

Without 'Ibid'	*With 'Ibid'*
Meenakshi K, Research Methodology for Agricultural Extension, Daya Publishing House, New Delhi, p. 117	Meenakshi K, Research Methodology for Agricultural Extension, Daya Publishing House, New Delhi, p. 117
Kumar K.N. Ravi, Research Methodology in Agricultural Economics, Daya Publishing House, New Delhi, p. 87	Kumar K.N. Ravi, Research Methodology in Agricultural Economics, Daya Publishing House, New Delhi, p. 87
Kumar K.N. Ravi, Research Methodology in Agricultural Economics, Daya Publishing House, New Delhi, p. 42	*Ibid.*, p. 42

'Idem'

From Latin *idem* means '*the same*'. This repeats the previous author. It is rarely used today.

'Op.cit'

Derived from Latin '*opere citato*', *which means* 'in the work cited'. This repeats the previous title for a given author. When a second but non-consecutive reference is made as the earlier source but from different page number(s) (*i.e.,* when a second reference is made from an earlier source but from different page(s) after one or more citations from other references), then write the author's name, followed by '*op. cit.,*' and page number(s). That is, '*op. cit*' is used to indicate that an author has already been cited, but it is not the preceding reference. It is not often used today.

'Loc.cit'

Derived from Latin '*loco citato*', which means 'in the place cited'. This repeats the title and page number for a given author. When a second but non-consecutive reference is made as the earlier source but from same page number (*i.e.,* when a second reference is made from an earlier source in the same page(s) after one or more citations from other references), then write the author's (second) name, followed by '*loc. it.*'. Here, page number(s) is not necessary. It is not often used today.

Without loc.cit.	*With loc.cit.*
Kumar K.N. Ravi, *Research Methodology in Agricultural Economics*, Daya Publishing House, New Delhi, p. 42	Kumar K.N. Ravi, Research Methodology in Agricultural Economics, Daya Publishing House, New Delhi, p. 42
Meenakshi K, Research Methodology for Agricultural Extension, Daya Publishing House, New Delhi, p. 117	Meenakshi K, Research Methodology for Agricultural Extension, Daya Publishing House, New Delhi, p. 117
Ravi, *Research Methodology in Agricultural Economics,* p. 87	Ravi, *op.cit.*, p. 87
Ravi, *Research Methodology in Agricultural Economics,* p. 87	*Ibid.*
Meenakshi K, Research Methodology for Agricultural Extension, p. 117	Meenakshi, *loc.cit.*
Ravi, *Research Methodology in Agricultural Economics,* p. 87	Ravi, *loc.cit.*
Ravi, *Research Methodology in Agricultural Economics,* p. 42	*Ibid.*, p. 42

Example

1. Kumar K.N. Ravi. (2008)., *Research Methodology for Agricultural Economics, Daya Publishing House, New Delhi, p 141.*
2. *Ibid.*, pp. 191-192.
3. Sharma, Ranjit. (2005)., *Research Methodology for Agricultural Extension*, Newbury Park, CA: Sage Publication.
4. *op.cit.*, p. 197.
5. Tagore Varma G. (2005). Essentials of Research Design and Methodology, New York City, NY : Wiley, p. 49.
6. Lehmann, E.L. (2008). Testing Statistical Hypothesis. New York, NY : Springer Publishing Company. P. 67.
7. Varma, *loc. cit.*

Thus, *'Ibid'* refers to the immediately preceding reference; *'op. cit'* refers to the prior reference by the same author (not immediately preceding reference) but of different page number, *'loc. cit'* refers to prior reference by the same author (not immediately preceding reference) and of the same page number.

Abbreviation	*Use*	*Example*
ibid. (in the same place)	To indicate the same source as in the footnote immediately above. If referring to a different page number write Ibid p.44.	1. Kumar K.N. Ravi. (2008). *Research Methodology for Agricultural Economics, Daya Publishing House, New Delhi, p . 141.*
		2. *Ibid.*
		3. *Ibid.*, pp. 191-192.
op.cit. (in the work cited)	To indicate a different page number of a source already cited, but one not immediately above.	1. Kumar K.N. Ravi. (2008). *Research Methodology for Agricultural Economics, Daya Publishing House, New Delhi, p . 141.*
		2. *Ibid.*, pp. 191-192.
		3. Sharma Ranjit (2005). *Research Methodology for Agricultural extension*, Newbury Park, CA: Sage Publication.
		4. Ravi, *op.cit.*, p. 197.
loc.cit. (in the place already quoted)	To indicate a source and page number already cited, but one not immediately above.	1. Tagore Varma G. (2005). Essentials of Research Design and Methodology, New York City, NY : Wiley, p. 49.
		2. Lehmann, E.L. (2008). Testing Statistical Hypothesis. New York, NY : Springer Publishing Company. p. 67.
		3. Varma, *loc. cit.*

Footnotes may be numbered 1,2,3 .... in each chapter. These footnotes are provided at the bottom of the page by drawing a dividing line (as shown in the above example) of one and a half inches long. The first footnote should be provided at double space below the dividing line.

☆ Avoid needless words that cause verbosity. For example:

Instead of	*Use*
A large number of	Many
In large number of cases	Often
In small number of cases	A few
It has been reported that	As reported
An adequate amount of	Enough

☆ Avoid superfluous phrases in sentences such as, *'As agreed by the authorities'*, *'It may be pointed out that'*, *'As agreed by many'* etc.

☆ Use concrete words instead of abstract words. For example:

Instead of	*Use*
The position regarding the supply of edible oils in the market may be serious.	There may be a serious edible oils shortage in the market.
The situation with regard to shortage of supply of skilled labour is not likely to improve even in the near future.	The shortage of skilled labour supply is likely to continue even in the near future.
Conditions for employment in agricultural sector have improved compared to the past period.	Employment prospects in agricultural sector are better than many years past.

☆ Avoid *tautology* or repetition in the sentences. For example, 'returned *back*', 'declined to *accept*' etc. In these, the words in italics are redundant and should be omitted.

Chapter VI–Summary and Conclusions

It shall be written as Chapter-V (in Times New Roman font with 14 point and bold) followed by Summary and Conclusions in capital letters (in Times New Roman font with 18 point and bold). This shall be the final chapter of the thesis. Summary serves as a synopsis of study. In this, the researcher once again reminds the reader with the introduction, problem statement, objectives, hypotheses, literature review, methodology (sampling and analytical models), brief discussion on the analytical findings and drawing useful conclusions emerged from the study. So, the reader reading this section would have a good overview of why the study was done, the specific purpose of the study and hypotheses/research questions, what the literature relates about the problem under investigation (very briefly), the methods used to gather data for the study, and findings emerging from analysis of the data. Note that, the findings are part of the Summary and are not included in a separate section. While not mandatory, this section is generally sub-divided into subsections (with corresponding subheads) related to purpose, literature review, methodology (procedures), and findings. Although there is no 'best' length for this chapter, usually it runs for about three to six pages provides an adequate overview of the study. If the section gets much longer than this, it fails to remain a 'Summary'. Note that, the researcher must summarize the findings that are based on data analysis.

The conclusions drawn from the findings must be such, which may lead to generalization and its applicability in similar circumstances. The conditions of research work limiting its scope for generalization must be made clear by the researcher. Frequently, conclusions provide answers to hypotheses or research questions posed in Chapter 1. While conclusions may be written in narrative form or listed, listing them is generally easier for readers to follow and helps maintain clarity of focus for each conclusion. It is important to note that, *'conclusions are not the same as findings and should not simply be restatements of findings from Chapter V'*. A conclusion should be broader and more encompassing than a specific finding, and several findings may be incorporated into one conclusion. While several findings may be used to support one conclusion, it is also possible that one finding might give rise to several conclusions (although this is somewhat less common). Generally, while specific findings are stated in the past tense (*e.g.*, farmers expressed greatest satisfaction with SRI production technology of paddy), generalization or conclusions are stated in the present tense (*e.g.*, farmers are most satisfied with SRI production technology of paddy). The following example illustrates the relationship between summarized finding and conclusion drawn in the study: A study of farmers participating in a training programme on irrigation water management in Mutluru village of KWD in Andhra Pradesh revealed the following findings:

- ✰ Only five per cent of marginal farmers attended the training programme.
- ✰ Only seven per cent of small farmers attended the training programme.
- ✰ Nearly 80 per cent of large farmers attended the training programme.

From the above findings, the following conclusion was drawn: *'Unlike large farmers, marginal and small farmers continue to remain underrepresented in the training programme on irrigation water management'*. This example clearly illustrates that, the conclusion has combined all the above three findings into a single broad statement that appears well supported by the study findings.

At the end of this Chapter VI, the researcher offers recommendations that emerge from the study. Generally, recommendations are of two distinct types: Recommendations for action or practice (based on the study's findings and conclusions, and sometimes headed Recommendations from the Study or Recommendations for Practice) and Recommendations for Further Study. Separate headings for each type can be used to present recommendations. Recommendations for practice are generally prescriptive in nature and address what could or should be done by the sample members and related members to address the problem. Recommendations for further study contain suggestions regarding follow-up studies or replication studies. These recommendations usually acknowledge limitations or delimitations that the study included and which further studies could help explain or clarify. However, the researcher must note that, *the recommendations 'have' to come from the research findings only*. Further, they should be related to the study area and for the sample (or target population) and should not be presented considering the macro level perspective. The researcher must be realistic in his recommendations *i.e.*, they must be implemented to overcome the problem and note that, they should not be mere vague statements. For example, 'Farmers should organize themselves', 'Government should regulate practices' etc., are considered vague in their expressions. Instead, the researcher should rather indicate 'What', 'How' and other aspects of the action or recommendation suggested.

Other General Guidelines for Presenting Summary and Conclusions

They include:

- ✰ The Summary section of Chapter VI should provide a brief recap of the entire study.
- ✰ The conclusions and recommendations should follow logically from the discussion of the findings.
- ✰ Conclusions can be short, as they have already been elaborately discussed in Chapter V.

- ☆ As the discussion will follow the sequence in which the findings have been presented (which in turn depends on objectives), the conclusions should logically follow the same order.
- ☆ It makes easy reading for a reader, if the recommendations are again placed in roughly the same sequence as the conclusions. However, the recommendations may at the same time be summarized according to the groups towards which they are directed. For example:
  - ❖ Policy-makers
  - ❖ Farmers
  - ❖ Private input agencies
  - ❖ NGOs
  - ❖ Officials of Irrigation department
  - ❖ Officials of Electricity department
- ☆ The conclusions and recommendations that are possible in a study are those that are based out of the data that were collected and analyzed and they should be meaningful and unique for the study. Remember, the valid recommendations given must be generated from the statistical findings of the data. After the recommendations have listed out, re-read each recommendation to ascertain what statistical results are used to support that recommendation? If there is no statistical result to support it, then cancel that recommendation.
- ☆ At the end, the researcher may add a paragraph highlighting how the present research study will opens the door for five to ten additional questions. Based on the Limitations of the study discussed in Chapter I, Results and Discussion of Chapter V and Summary and Conclusions of Chapter VI, it makes the researcher to suggest, what additional research should be conducted in the future? What questions arose because of the major findings of the present study? How can other research studies improve over the limitations that were described in the Limitations section of the study?

III. End Matter

This section includes Literature Cited and Appendices.

Literature Cited

It shall be written as Literature Cited in capital letters (in Times New Roman font with 18 point and bold). Literature cited shall contain complete reference list, citing all the literature and other sources referred to in the thesis and appendices including websites arranged by the name of the author in alphabetic order. Individual reference entries shall not be split over two pages. Format and placement of reference citations shall be consistent throughout the thesis. Similar system shall be followed in the text of synopsis also. It is very important that a researcher should carefully prepare the References section. The researcher should provide a complete reference list of all materials used in the study. Every work cited in the body of the thesis should be included in this sub-head. This is important for two major reasons: to aid the reader in finding additional resources about the topic of work, and also to demonstrate that the researcher did the research carefully and honestly. If a number of works are not listed in the references section, then the reader becomes suspicious that the researcher either did not does the research project carefully or that he plagiarized the work. Format for that list should follow a certain guidelines as established by the University. The references are to be presented in the sequence of Alphabetical order of the principal author's last name. The same format should be used consistently throughout the report. Careful attention should be paid to the reference section in terms of differences in dates or the spelling of authors' names between the citation and the reference listing, journals names, volume number, issue number, page numbers etc. Note that, there should be no references listed that did not appear as citations within the thesis. 'Bibliography' and 'Literature cited' are the two words that are used interchangeably, but they actually differ. The researcher might have consulted a lot of books, essays and websites for writing the thesis. Though you might have referred to these, while preparing a write up, the content of these might not have been included in the actual text. This is what refers to Bibliography. Literature cited are those that are directly included in the thesis. Similarly, literature cited also differs from footnotes. In the former, the list of materials from which the information was referred by the researcher and included in the thesis are included. Footnotes is provided at the bottom of the respective pages (say, in Chapters like Introduction, Materials and Methods, Results and Discussion etc), that specifically indicate the source and pages from which the quotes or paraphrases are found. Footnotes are used to cite the sources from which the statements or paraphrases are quoted in the text, while the literature cited lists the list of references used by the researcher in an alphabetic order. The format of presentation of footnotes is similar to the literature cited except one difference *i.e.*, in case of books and reports listed in literature cited, page numbers are not indicated, whereas the page numbers are indicated in footnotes. Lastly, in footnotes, author's name begins with the first name (Ravi K.N. Kumar), while in literature cited author's name begins with second word or surname (Kumar K.N. Ravi). The following information will be useful in citation (Literature cited):

- ✰ In case of an institutional publication, the name of the institution or organization shall be treated as the author. No entry will appear as an anonymous publication.
- ✰ The main parts of a complete entry for a book are (i) Name(s) of author(s), (ii) Year of publication, (iii) Title of book in italics, (iv) Name and city of publishers, (v) Pages referred to in the book. An example is:
  - ❖ Kumar K.N. Ravi. 2012. Microeconomic analysis in Agriculture. Daya Publishing House, New Delhi, 110002.
- ✰ Sometimes a book does not have a stated author. In such a case, use the title to start the entry. An example is: Macroeconomics, 2nd ed. 2007. New York.
- ✰ Sometimes books contain essays by a group of people, each author contributing one chapter or essay, (i) Name(s) of author(s), (ii) Year of publication, (iii) Title of the chapter (iv) Author(s) of the book, (v) Title of book in italics, (vi) Name and city of publishers, (vii) Pages referred to in the book.
- ✰ The main parts of entry for an article in a journal/periodical are (i) Name(s) of author(s) (ii) Year of publication, (iii) Title of article, (iv) Name of Journal in italics with volume number and pages of the article.
  - ❖ Sriram K.S and Meenakshi, K. 1995. Performance of regulated agricultural markets in Andhra Pradesh. *Indian Journal of Agricultural Marketing*. 2(1): 45-49.
- ✰ The main parts of entry of a website are (i) Name of the institute, (ii) Title of the article/information/web page in italics and the URL. Note that the title of the web-page should be italicized.
- ✰ The first line of the references starts from the margin and the subsequent lines continued from the fifth letter (space) of the above line. Each reference is single spaced: a double space should be provided between individual references.

- If there are several references with different titles by the same author, the name(s) of the author(s) are to be typed again.
- More than one publication of the same author in the same year, a small alphabet shall be mentioned at the year as mentioned below.
  - Sriram, K and Meenakshi, K. 2011a. Marketing efficiency in transacting paddy in Kurnool regulated market of Andhra Pradesh. *Agricultural Marketing. 3:23-28*
  - Sriram, K and Meenakshi, K. 2011b. Structure and Performance of groundnut markets in Rayalassema region of Andhra Pradesh. *Agricultural Situation in India.* 10:20-26.
- The following format is followed for citing the theses references.
  - ***For Unpublished thesis:*** Author, A. A. "Title of Thesis: Subtitle." Unpublished thesis type, University, Year.
  - ***For Published thesis:*** Author, A. A. *Title of Thesis: Subtitle.* Place of publication: Publisher; Year of publication.
  - ***From a full text database:*** Author, A. A. "Title of Thesis: Subtitle." Thesis type., University, Year. Database Name (Identification number if given).

The following format is employed to present the list of references in the thesis as guided by the Acharya N.G. Ranga Agricultural University.

Literature Cited

Mohapatro, B.P and Mishra, R.K. 1995. Evaluation study on acceptance of farm credit by small farmers in rural setting of Orissa. *Financing Agriculture. 27 (4):7-10.*

NABARD, 2001. *Report of the Expert Committee on Rural Credit, Mumbai.*

Ramakrishna, P. 1990. Agricultural households and institutional finance. *Indian Journal of Agricultural Economics. 45 (2): 187-188.*

Rao, V.K.B. 1989. Problems and Prospects of banks in financing Agriculture, *Agricultural Banker. 12 (4):15.*

Vaidya, C.S., Sikka, B.K and Swarup, R. 1989. Cost of credit in hill agriculture-A spatial analysis. *Indian Journal of Agricultural Economics. 43 (3):439-440.*

There is another important term that is relevant to discuss at this juncture *i.e.,* Annotated Bibliography. In this bibliography, each entry is followed by a one-sentence or one paragraph description about the content of the work cited. The following are the examples:

Gulati, Ashok, and Sudha Narayanan. 2003. *The Subsidy Syndrome in Indian Agriculture.* New Delhi: Oxford University Press.

As a result of a combination of domestic and trade policies, Indian farm prices have been lower than international prices.

Mullen, Kathleen, David Orden, and Ashok Gulati. 2005. *Agricultural Policies in India: Producer Support Estimates 1985–2002*. MTID Discussion Paper 82. Washington, D.C.: International Food Policy Research Institute.

Despite these input subsidies, Indian agriculture remains 'disprotected' overall, meaning that it suffers from Government policies more than it benefits from them.

In general, the following is the order to be followed in presenting the Bibliography (say, in a text book). Note that, all entries are alphabetically ordered by Surname of Author. The sequence is as follows:

- ☆ Global references texts such as dictionaries, encyclopedias, translation manuals etc.
- ☆ Books
- ☆ Refereed Journals
- ☆ Journals
- ☆ Refereed (Peer reviewed) reports
- ☆ Reports
- ☆ Papers
- ☆ Letters and diaries
- ☆ Newspapers
- ☆ Radio and TV productions
- ☆ Websites, Blogs etc.
- ☆ Other

Appendices

The Appendices harbour lengthy tables (say, list of farmers, big charts, schedule employed for data collection, copies of any letters of permission required for the study, tables that are of only minor importance to the study, glossary of terms (optional) etc. Thus, the final attachments to the thesis are shown in the Appendices. They contain pertinent (and often supplementary) materials that are not important enough, do not fit appropriately in any specific section of the body or are too long to include in the main body of the thesis, but which may be of interest to some readers. That is, the information that would be useful to special categories of readers, but is not of interest to the average reader can be included in appendices. Each appendix should be numbered and should also have a title. The *Appendices* come after the Literature cited. Some research reports harbour 'Biography' (optional as the last page in the report). Here, the researcher writes a brief biography about himself, which includes his birth date, his education path and his career path to date. This takes around one paragraph (10-12 lines).

Other Important Guidelines the Researcher Should Follow in Presenting the Style of Thesis

Before planning the thesis writing, it is essential for the researcher to divide into different manageable sections. Later, each section can be further divided under different headings or chapters. As discussed above, in general, the thesis has three main sections *viz.*, Preliminary pages, Main text and End matter. Section 2 (Main text), for example, is sub divided into different chapters like Introduction, Review of Literature, Materials and Methods etc. After dividing each section into different headings or chapters, now the researcher will plan different sub-headings under each chapter and this guide him to frame the relevant content (*i.e.*, literature reviewed, data collected and analytical findings) in a logical order. Now, after planning the thesis out line and its contents, it is high time for the researcher to discuss with his Major Advisor. The guidance offered by the Major Advisor regarding the logical arrangement of material should be carefully noted and the same should be incorporated in the rough copy of the thesis to be prepared. The researcher at this juncture should not forget the time duration that he has to spare to complete each section and chapter and he has to finalize this time frame in consultation with the Major Advisor. While entering the thesis in word-file, open two files for each chapter, one file for writing the matter for each chapter and other file to place or copy the relevant back ground literature and citing references. So, for each chapter, a separate literature cited file will be created. Thus, by the end of completing the thesis, the researcher simultaneously complete the list of references (literature cited) and later these references will be arranged in a proper sequence, as discussed earlier. Please note to make a back up these files at least every day. In addition to the above planning, the researcher should also follow the following important guidelines in writing the thesis:

- ***Text:*** The text pages shall be numbered with Indo Arabic numerals consecutively throughout the thesis including the photographs, diagrams, etc. at the bottom right corner of the page at final submission stage.
- ***Position of Page Numbers:*** Page numbers shall be located at the lower corner of the page approximately 2 cm from the right edge. Page numbers shall be given without brackets. For pages with landscape setup, the page numbers shall be indicated in the same position as all other page numbers in the main body of the thesis.
- ***Chapter Heading (First level heading):*** Every chapter in body of the manuscript shall begin with a new page. Chapter should be numbered with Roman numerals and the chapter heading shall be placed in capitals with 18 point in Times New Roman font. It should be centered on five 1.5 line spaces below the top edge of the page.
- ***Paragraphs:*** The text should begin with two 1.5 space lines below the chapter title. A paragraph indention of eight spaces (*Indent paragraph means* to *indent* it using the "Tab" key) shall be used. Subsequent pages of the text should begin 2 cm below from the top of the page. No page should end with the beginning of a paragraph i.e., less than one full line in length.
- ***Sub-heading within chapters:*** Theoretically any number of sub-heading levels are possible, but in practice more than three levels are confusing. All sub-headings and numbers shall have left alignment.
- ***Centre Heading (Second level heading):*** This heading shall be in bold capitals and placed at 1.5 space below the last line of the previous section with 18 point in Times New Roman font. The text following this heading should appear in regular paragraph form. The heading preceded by the number shall be typed flush with the left margin.
- ***Side Heading (Third level heading):*** This heading in bold shall be placed at 1.5 space below the last line of the previous section with 14 point in Times New Roman font. Initial letters of words except articles and prepositions (other than at the beginning) should be in capital. The text following this heading should appear in regular paragraph form. The heading preceded by the number shall be typed flush with the left margin.
- ***Run-on-Heading (Fourth level heading or Paragraph heading):*** The initial letters of the words except for articles, conjunctions and prepositions should be in capital. The heading should be preceded by the number in bold with 12 point in Times New Roman font. It must be followed by a colon and the text should follow immediately on the same line.

- ☆ ***Numbering of Divisions:*** The structural elements (Headings) shall be numbered as 1, 2, 3 and subdivided as 1.1, 1.2, 2.1, 2.2, etc. Further sub-divisions may be as 1.1.1, 1.2.1, etc. Only Arabic numerals shall be used. A full stop shall be placed between numbers designating subdivisions of different levels. At the end no full stop shall be placed.
- ☆ ***Mechanical Conventions:*** The conventions listed in the CBE (Commander of the Order of the British Empire) Style Manual (1972) should be followed for punctuations: full stop, ellipsis, comma semicolon, question mark, numbering, illustrations, diagrams etc.
- ☆ ***Decoration:*** No ornamentation or bordering of the sides shall be permitted.
- ☆ ***Paper:*** White bond paper of A_4 size (210 x 297 mm) should be used.
- ☆ ***Typing:*** Type characters shall not be less than 12 point with Times New Roman font. However, subscript, superscript and if essential, tables shall not be less than 8 point type in Times New Roman font. Typing should be of even quality with clear black characters. Copies produced by Xerox or comparably permanent process are acceptable. Only one side of the paper shall be used.
- ☆ ***Margins:*** The left side margin shall be 4 cm wide and the top, bottom and right side margins shall be 2 cm wide.
- ☆ ***Spacing:*** A spacing of 1.5 lines shall be used in type script except for quotations of footnotes where single line spacing can be used. Spacing in abstract shall be in single line space.
- ☆ ***Style of Writing:*** The thesis should be written in the past tense, passive voice and in the third person. The guidelines given by the CBE(Commander of the Order of the British Empire) Style Manual should be followed. For referral purpose the latest Webster's New International Dictionary should be followed. While writing 'Introduction' and 'Policy Implications' in the thesis, present tense may be used.
- ☆ ***Transliteration:*** When quotations, authors names, and titles of works have originally appeared in a non-Roman alphabet, the information should be transliterated into Roman alphabet for use within the text.
- ☆ ***Illustrations:*** Illustrations should immediately follow the textual reference on the following page. The caption shall be placed at the bottom of the figures with the initial letter in capital and shall maintain a single space separating the figure and the caption. The caption should be single spaced closed by a full stop. Photo print or Art paper should be used for printing of photographs. Photographs shall not be pasted directly in the thesis.
- ☆ ***Tables:*** Tables longer than half a page shall be placed on a separate page. Shorter tables should be placed on the page with the text above or below. A table running longer than one page may be continued on two or more pages by indicating the continuation *e.g.* Table 3.2 (cont:). The table number shall end with a full stop without any other punctuations both in the text and at table title. The title shall be placed above the table following single space separating the table and the title. The first letter of the title is capitalized and single spaced in sentence form. A table is closed with a horizontal line and footnote (if any) pertaining to the table shall be placed below in single line space. In table, the units of captions or stubs should be indicated with units say, yield should be indicated in kg ha^{-1} without any decimals. Note that, for the data given in the table, the decimals shall be uniform either one or two or three for all the values.

- ***Citations:*** In citing works in the text, the following system (author, year) shall be followed. Similar system shall be followed in the text of synopsis also. Ex., Reddy (1985), Sharma (1999), Singh and Singh (1999), Ravi *et al.* (2010), Sriram *et al.* (2000), Meena *et al.* (1996) etc. For the citation of an institutional publication such as Annual Report, Committee Report, Statistical Bulletin etc., the organization shall be treated as the author (say, WTO, 1995; FAO, 2011).
- Use of quotes should be minimal, if at all; learn to paraphrase.
- Use transitional words to enable the reader to anticipate the next idea and connect that idea with prior ideas.
- Thesis report should be presented in a creative, clear, concise and comprehensive style. The sentences should be grammatically correct. Colloquial expressions such as 'write up' for report and 'put in' for insert should be avoided.
- Literary style of writing is to be replaced by scientific and scholarly style reflecting precise thinking.
- Descriptions should be free from bias, ambiguity and vagueness.
- Ideas need to be presented logically and sequentially so that, the reader finds no difficulty in reading.
- Note that, thesis is a scientific document, but not a novel or treatise. It should not contain any subjective and emotional statements. Instead, it should contain factual and objective statements.
- Personal pronouns such as 'I', 'Me', 'We', 'Our', 'Your' etc., should be avoided. For example, instead of writing 'I randomly selected 60 farmers', it is advisable to write 'sixty farmers are selected randomly by the researcher'. Passive voice is preferred over active voice. That is, always use the 'third person'. Another good example is, write, 'It could be considered that', and not, 'I think that'.
- Sexist language should be replaced by non-sexist language while writing thesis report. Male or female nouns and pronouns (he and she) should be avoided by using plurals. For example, write children and their parents have been interviewed rather than child and his parents were interviewed.
- Instead of using titles and first names of the cited authors, last name is needed. For example, instead of writing Professor Mark Williams, write 'Williams'.
- Constructed forms of modal auxiliaries and abbreviations should be avoided. For example, shouldn't, can't, couldn't etc., should not be used. However, abbreviations can be used to avoid repetition, if the same has been spelled out with the abbreviation in parentheses. For example, researcher can write MLR, if he/she has used MLR in parenthesis in his/her first sentence like Multiple Linear Regression (MLR).
- Economy of expression is important for writing a research report. Long sentences and long paragraphs should be avoided. Short, simple words are better than long words. Try to have an average sentence length of 15-20 words, as long sentences are hard to follow. Check spelling and grammar; if this is a weak area then improve by asking for feedback from your tutor, major advisor; reading texts on grammar, punctuation and spelling, etc.
- Fractions and numbers, which are less than ten should be expressed in words. For example, six markets were selected or 50 per cent of total number of commission agents are selected into the sample.

- ☆ The derivations of statistical formula should not be given in the report.
- ☆ Research report should not be written hurriedly. It should be revised many times before submission for evaluation. Even typed manuscripts require to be thoroughly proofread before final typing.
- ☆ The researcher must note the target audience for whom the thesis or report is intended to say, Scientific community or Sponsor of the research or farming community. Accordingly, he must present the form and style of presentation. In general, for student's thesis, the target audience will be the scientific community. For the scientific community working in SAUs or ICAR or ICRISAT, the target audience includes fellow scientific personnel or Sponsor of the research or farming community.
- ☆ There are certain conventions that determine when to use numbers and when to use words.
  - ❖ Use words in the following cases:
    - ❑ When the numbers are smaller than 10. For example: It is clear from the data that, three farmers are receiving marketing information from co-farmers.
    - ❑ When to use round numbers.
    - ❑ When to express number in terms of fraction. For example: *One-fifth* of the farmers are receiving marketing information from television.
    - ❑ Numbers at the beginning of the sentence. It is to be noted that, no sentence should begin with a number. For example: *Thirty five* per cent of the farmers opined that,
  - ❖ Use the word *'per cent'* as one word with a number and the word *'percentage'* without reference to a number. For example: 'About 35 *per cent* of the farmers are of the opinion that ————'. A significant *percentage* of the farmers opined that, training had major influence on ————'.
  - ❖ Use per cent symbol (per cent) only in tables and figures.
  - ❖ Use numerals to express formulae of chemicals, dosages of fertilizers and pesticides applied etc.
  - ❖ Express in numerals even though the numbers are less than 10 in case of table numbers and figure numbers.
- ☆ Certain English and Latin abbreviations are commonly used in Literature cited/Bibliography and footnotes to eliminate tedious repetition. The researcher should be well-versed with these abbreviations, so as to recognize them as well as to use them in his/her report. The following is the list of common abbreviations used:

anon.,	:	anonymous
ante.,	:	before
art.,	:	article
aug.,	:	augmented
bk.,	:	book
bull.,	:	bulletin
cf.,	:	compare
ch.,	:	chapter
col.,	:	column

diss.,	:	dissertation
ed.,	:	editor, edition, edited
ed. cit.,	:	edition cited
e.g.,	:	exempli gratia: for example
eng.,	:	enlarged
et al.,	:	and others
ex.,	:	example
f.,ff.,	:	and the following
fig(s).,	:	figures(s)
fn.,	:	footnote
ibid.,ibidem.,	:	in the same place (when two or more successive footnotes refer to the same work, it is not necessary to repeat complete reference for the second footnote. '*Ibid*'. may be used. If different pages are referred to, pagination must be shown
id.,idem.,	:	the same
ill.,illus.,or illust(s).,	:	illustrated,illustration(s)
intro.,intro.,	:	introduction
l or ll.,	:	line(s)
loc.cit.,	:	in the place cited;
no pub.,	:	no publisher
no(s).,	:	number(s)
o.p.,	:	out of print
op.cit: opera citat	:	in the work cited (if reference has been made to work and new reference is to be made to different works, op. cit. must be used. the name of the author must precede
p or pp.,	:	page(s)
post.,	:	after
rev.,	:	revised
tr.,trans.,	:	translator,translated,translation
vid or vide.,	:	see, refer to
viz.,	:	namely
vol. or vol(s).,	:	volume(s)
vs., versus.,	:	against

☆ The researcher must carefully distinguish between singular and plural words to be used in the thesis or report. For example:

Singular	:	Plural
Datum	:	Data

Stratum	:	Strata
Thesis	:	Theses
Hypothesis	:	Hypotheses
Appendix	:	Appendices
Basis	:	Bases
Analysis	:	Analyses
Index	:	Indices

The researcher must follow all the above guidelines meticulously for preparing the report. Before submitting the draft to the Advisory Committee, he should go through with a fresher outlook. He should read the report word by word, so as to check any mistakes to ensure absolute accuracy. The sequence of page numbers, table numbers, figures numbers, footnote numbers etc., should be checked. The footnotes and literature cited must be checked carefully against the original. He should take considerable time for checking the report till it was finally ready to submit to the Advisory Committee. Later, the suggestions and corrections offered by the Committee must be incorporated in the report for its final evaluation by the external examiner(s).

References

Aday, L.A. 1996. Designing and conducting health surveys. 2nd ed. San Francisco (CA): Jossey-Bass.

Anonymous, Thesis Guidelines, ANGRAU.

Aurangzeb Z. Khan, Research Methodology, (Business Research Methods), MBA III (Research Methodology).

Aveyard, H. 2007. Doing a literature review in health and social care: A practical guide. Maidenhead. NY: Open University Press.

Badke, W. B. 2008. Research strategies: Finding your way through the information fog (3rd ed.). Lincoln, NE: iUniverse.

Barzun, J and Graff, H.F. 1977. The modern researcher (3rd edn). New York: Harcourt, Brace, Jovanovich.

Benevieve, D., Villanueva and Joannaopena, Problem Formulation and Hypothesis Testing A Module for AERS 282 -Social Research Design.

Beyea, S and Nicholl, L. 1998. Writing an integrative review. *AORN Journal* 67(4):877–80.

Bogardus, Emory S., Social Distance in the City.*Proceedings and Publications of the American Sociological Society*.20, 1926, 40–46.

Borg, W.R and Gall, M.D. 1979. Educational research: An introduction (3rd edn.). New York: Longman.

Borg, W.R and Gall, M.D. 1989. Educational research: An Introduction. 5th edn. White Plains, NY: Longman.

Brink H, Van der Walt Ch, Van Rensburg G.Fundamentals of Research.Methodology for Health-care Professionals, Second Edition 2006, ISBN 0 7021 6680 4, Juta and Co. (Pty) Ltd. books.google.sk/ books?isbn=0702166804.

Burns, N and Grove, S.K. 2005. The practice of nursing research, Elsevier Saunders, Missouri.

Burns, N., Grove, S.K. 2007. Understanding Nursing Research – Building an Evidence- Based Practice. 4th edn. Saunders Elsevier, St. Louis.

Catherinedawson, How To Books Ltd, 3 Newtec Place, Magdalen Road, Oxford OX4 1RE. United Kingdom.

Charles, E. 1948. Notar, *International Journal of Education* ISSN 1948-5476 2010, Vol. 2, No. 2: E2.

Choi, B.C.K and Pak, A.W.P. 2005.A catalog of biases in questionnaires. Prev Chronic Dis [serial online] 2005 Jan.

Collis, J.and Hussey, R.(2003) Business Research: a practical guide for undergraduate and postgraduate students, second edition.Basingstoke: Palgrave Macmillan.

Cooper., Donald, R., Schindler and Pamela, S. 2006. Business Research Methods, Tata McGraw Hill Companies; India.

Cooper, H.M. 1998. Synthesizing research: A guide for literature reviews. Thousand Oaks, CA: Sage.

Coughlan, M., Cronin, P and Ryan, F. 2007. Step-by-step guide to critiquing research. Part 1: quantitative research. *Br J Nurs* 16(11): 658–63.

Craswell, G. 2005. Writing for academic success: A postgraduate guide. Thousand Oaks, CA: Sage.

David R.Heise, The Semantic Differential and Attitude Research, Chapter 14 in *Attitude Measurement*. Edited by Gene F.Summers.Chicago: Rand McNally, 1970, pp.235-253.

Dawson and Catherine. 2002. Practical Research Methods, New Delhi, UBS Publishers' Distributors,

Day, R.A. 1998. How to write and publish a scientific paper. 5th edn. Phoenix (AZ): Oryx Press; 1998.

Dillman, D.A. 1978. Mail and telephone surveys: the total design method. New York: John Wiley & Sons, Inc.

Diolina Z.Mercado, A Manual on Processing and Reporting of Participatory Rural Appraisal (PRA) Data for Natural Resource Management, Pacific-German Regional Forestry Project, German Technical Cooperation.

Edwards, A.L. 1969. Techniques of attitude scale construction. Vakils, Feffer and Simons Inc, New York.

Eisenhardt and Kathleen, M. 1989. Building Theories From Case Study Research, Academy of Management. The Academy of Management Review; Oct 1989; 14, 4; ABI/INFORM Global pg. 53.

Fink, A. 1998. Conducting research literature reviews: From paper to the Internet. Thousand Oaks, CA: Sage.

Foddy, W. 1993. Constructing questions for interviews and questionnaires: Theory and Practice in Social Research. Cambridge (United Kingdom): Cambridge University Press.

Fraenkel, J. R and Wallen, N. E. 2006. How to design and evaluate research in education (6th ed.). Boston, MA: McGraw Hill.

Frerichs, R.R. 2008. Rapid Surveys (unpublished), Simple Random Sampling.

Galvan, J.L. 2004. Writing literature reviews. Glendale, CA: Pyrczak.

Garrard, J. 2007. Health sciences literature review made easy: The matrix method. Sudbury, MA: Jones and Bartlett.

Getudegu and Tegbaryigzaw. Research Methodology, University of Gondar, In collaboration with the Ethiopia Public Health Training Initiative, The Carter Center, the Ethiopia Ministry of Health, and the Ethiopia Ministry of Education.

Good, Carter V.(1963).Introduction to Educational Research, New York, Appleton Century-Crofth, 542 pp.

Gowers, EA. 1954. The complete plain words. London: Penguin (Pelican) Books.

Hart, C. 1998. Doing a literature review: Releasing the social science research imagination. London, UK: Sage Publications.

Hart, C. 2001. Doing a literature search: A comprehensive guide for the social sciences. Thousand Oaks. CA: Sage.

Holbrook, A.P., Bourke, S.F., Fairbairn, H., & Lovat, T.2007.Examiner comment on the literature review in Ph.D.theses.Studies in Higher Education, 32, 337-356.

Holbrook, A., Bourke, S., Lovat, T. and Dally, K. 2004. Qualities and characteristics in the written reports of doctoral thesis examiners. Australian Journal of Educational and Developmental Psychology, 4, 126-145.

Horowitz, L. 1984. Knowing where to look: The ultimate guide to research. Cincinnati, OH: Writer's Digest Books.

http://www.cdc.gov/pcd/issues/2005/jan/pdf/04_0050.pdf

http://explorable.com/non-probability-sampling

http://hgso.curtin.edu.au/pdf/Creative%20Production%20as%20Research.pdf

http://ipsos-mori.com/Assets/Docs/Techniques/rmc-sampling-techniques-surveys-and-samples.pdf

http://libguides.uwi.edu/content.php?pid=398376&sid=3272491.

http://owl.english.purdue.edu/owl/resource/560/20/.

http://people.du.ac.in/~pkdas/RM/rm-3.pdf

https://statistics.laerd.com/statistical-guides/types-of-variable.php

http://web.pdx.edu/~bertini/pdf/literature_review.pdf.

http://www.ncbi.nlm.nih.gov/pmc/articles/PMC2996198/

http://www.nongnu.org/bibulus/bibcit.html

http://www.sass.uottawa.ca/writing/kit/grad-literature-review.pdf

Hunt, D.M. Magruder, S and Bolon, D.S.1995. Questionnaire format bias: when are juxtaposed scales appropriate: a call for further research. Psychol Reports 1995;77:931-41.

James, D. and Leeper. 'Choosing the Correct Statistic', <http://bama.ua.edu/~jleeper>, University of Alabama.

Jenniferrowley.2002. Using Case Studies in Research, Management Research News, Volume 25 Number 1 2002.

Jenniferrowley, Conducting a Literature Review, School for Business and Regional Development, University of Wales, Bangor.

Johngill and Philjohnson 2010. ''The role of theory in management research', Research Methods for Managers, Chapter 3, Research Methods, University of London.

Karen, J., Jansen., Kevin, G., Corley and Bernard, J. J. E-Survey Methodology.

Kennedy, M. 2005. What makes a research question answerable? Retrieved from Michigan State University Doctoral Program in Teacher Education Web site: http://ed.

Kothari, C.R.1985. Research Methodology-Methods and Techniques, New Delhi, Wiley Eastern Limited.

Krishnaswami, O.R. and Ranganatham, M. 2005. 'Methodology for Research in Social Sciences', Himalaya Publishing House. Mumbai.

Kumar., Ranjit. 2005. Research Methodology-A Step-by-Step Guide for Beginners, (2nd.ed), Singapore, Pearson Education.

Leedy, P and Ormrod, J. 2001. Practical research: Planning and design (7th ed.). Upper Saddle River, NJ: Merrill Prentice Hall. Thousand Oaks: SAGE Publications.

Leedy, P.D and Ormrod, J. E. 2010. Practical Research: Planning and Design, Ninth Edition. NYC: Merril.

Lijphart, A.1971. "Comparative Politics and the Comparative Method", American Political Science Review, Vol. 65, no. 3, pp.

Lunenburg, F. C and Irby, B. J. 2008. Writing a successful thesis or dissertation: Tips and strategies in the social and behavioral sciences. Thousand Oaks, CA: Corwin.

Mariano M Ariola, 2007, Principles and Methods of Research, Rex Book Store

Marilynsimon.2011 Assumptions, Limitations and Delimitations, Excerpted from Simon, M. K. (2011). Dissertation and scholarly research: Recipes for success (2011 Ed.). Seattle, WA, Dissertation Success, LLC. <www.dissertationrecipes.com>

Mark, A. B. Guidelines for Writing Research Proposals and Dissertations, Division of Educational Administration, University of South Dakota.

McMillan, J. H and Schumacher, S. 2010. Research in Education: Evidence-Based Inquiry. (7th ed.). Boston, MA: Pearson.

Merriam, S. B. 1988. Case study research in education: A qualitative approach. San Francisco: Jossey Bass.

Mertler, C. A and Charles, C. M. 2005. Introduction to Educational Research. (5th ed.). Boston, MA: Pearson Education, Inc.

Miles., Matthew, B., Huberman, A. and Michael. 1994. Qualitative Data Analysis. Thousand Oaks: Sage.

National Open University of Nigeria, Course Code: AEM 501, Course Title : Statistics And Research Methods in Extension.

Neill, J. 2003. Tips for Writing an Outline (or Annotated Table of Contents).

Olufemi . A. Oladunni, Agricultural and Rural Management Training Institute (ARMTI), Ilorin.

Pan, M.L. 2004. Preparing literature reviews. Glendale, California.

Patricia Cronin, Frances Ryan, Michael Coughlan, Literature Review Organizer, British Journal of Nursing, 2008, Vol 17, No 1.

Paul, D. L and Jeanne, E. O. Practical Research: Planning and Design, Ninth Edition,. Published by Merrill. Pearson Education, Inc.

Polit, D and Beck, C.T. 2004 Nursing Research: Principles and Methods. 7th edn. Lippincott, Williams & Wilkins, Philadelphia.

Polit, D and Beck, C. T. 2006. Essentials of Nursing Research: Methods, Appraisal and Utilization. 6th edn. Lippincott Williams and Wilkins, Philadelphia.

Polit, D and Hungler, B.P. 1999. Nursing Research: Principles and Methods (6th Ed.) Philadelphia, Lippincott.

Rajasekar S., P.Philominathan and V.Chinnathambi, Research Methodology <arxiv.org/pdf/physics/0601009ý>.

Ramahuja. 2003. Research Methods, Rawat Publications, Jaipur.

Rosenfeld, P., Boothkewley, S., Edwards, J. E., and Thomas, M. D. 1996. Responses on computer surveys: Impression management, social desirability, and the big brother syndrome. Computers in Human Behavior, 12(2), 263-274.

Rudestam, K and Newton, R. 1992. Surviving your dissertation. London:Sage.

Sarantakos, S.1998. Social Research (2nd edn.), Macmillan Press, London.

Shamoo, A.E., Resnik, B.R. 2003. Responsible Conduct of Research. Oxford University Press.

Shawnee Vickery, Research Issues, Electronic Surveys: Advantages and Disadvantages Over Traditional Print Surveys, Eli Broad Graduate School of Management, Michigan State University, Decision Line, July 2001.

Simon, M. K. 2011. Dissertation and scholarly research: Recipes for success (2011 Ed.). Seattle, WA, Dissertation Success, LLC.

Smeeton, N., Goda, D. 2003. Conducting and presenting social work research: some basic statistical considerations. Br J Soc Work, 33: 567-573.

Stake, R. 1995. The art of case research. Thousand Oaks, CA: Sage Publications.

Sudman, S and Bradburn, N.M.1982. Asking questions: a practical guide to questionnaire design. San Francisco (CA): Jossey-Bass Publishers.

Sungachalwe. 2011. Factors Influencing Bean Producers' Choice of Marketing Channels in Zambia, University of Zambia.

Survey Methods and Practice, Catalogue no. 12-587-X, Statistics, Canada.

Synodinos, N. E., Papacostas, C. S., and Okimoto, G. M. 1994. Computer-administered vs. paperand-pencil surveys and the effect of sample selection. Behavior Research Methods, Instruments, and Computers, 26(4), 395-401.

Verma., Gajendra, K and Kankamallick.1999. Researching Education: Perspectives and Techniques. London: Falmer Press.

Wellington, J., Bathmaker, A., Hunt, C., McCulloch, G. and Sikes, P. 2005. Succeeding with your doctorate. London:Sage.

Witte, R. S. and Witte, J. S. 2009. Statistics. John Wiley & Sons.

www.amarrelli@beaconassociates.net

www.articlesbase.com/writing-articles/inductive-and-deductive-methods-of-teaching-1059831.html

www.authorstream.com/Presentation/vpantula-1392339-praý.

www.courses.ttu.edu/hdfs3390-reifman/samp.htm

www.cybernos.com/Walden/Simon-problem.ppt

www.deakin.edu.au/~agoodman/sci101/chap7.html

www.iriweb.org/

www.ispi.org/pdf/suggestedReading/ArticleFive_LiteratureReviews.pdf

www.jfmed.uniba.sk/./4Phases_of_research_process_-_Hanacek.docý.

www.kit.nl/net/KIT_Publicaties_output/showfile.aspx?e=738ý.

www.korbedpsych.com/R00Steps.html

www.libguides.usc.edu/content.php?pid=83009&sid=23

www.libguides.usc.edu/content.php?pid=83009&sid=616083

www.library.stkate.edu

www.people.du.ac.in/~pkdas/RM/rm-3.pdf.

www.ph.ucla.edu/epi/rapidsurveys/RScourse/RSbook_ch3.pdf

www.pitt.edu/~super7/43011-44001/43911.ppt

www.regent.edu/acad/schedu/pdfs/residency/.../dissertation_guidelines

www.sagepub.com/upm-data/40803_5.pdf

www.slideshare.net/./distinguish-between-parametric-vs-nonparametric, Sai Prakash.

www.socialresearchmethods.net

www.surveysystem.com/sdesign.htm

www.transcriptionwave.com/dissertation-thesis.html

www.uis.edu/ctl

www.utoronto.ca/writing/litrev.html ,The Literature Review: A Few Tips on Conducting It: http://www.uvic.ca/hsd/nursing/assets/docs/current/./lit_review_basics.swfý.

www.zeepedia.com

Yairlevy and Timothy, J. E. A Systems Approach to Conduct an Effective Literature Review in Support of Information Systems Research Informing Science Journal Volume 9, 2006.

Yamane, T. 1967. Statistics, An Introductory Analysis, 2nd Ed., New York: Harper and Row.

Yin, R., 1984. Case Study Research. Sage, Beverly Hills, California.

Yin, R. 1989. Case study research: Design and methods (Rev. ed.). Newbury Park, CA: Sage Publishing.

Yin, R. 1994. Case study research: Design and methods (2nd ed.). Beverly Hills, CA: Sage Publishing.

Young, Michael (1965), Innovation and Research in Education, London, Routledge and Kegan Paul, 183 pp.

Zaidahzainal. 2007. Case study as a research method, Journal Kemanusiaan bil.9, Jun 2007 Faculty of Management and Human Resource Development, Universiti Teknologi Malaysia.

www.ingramcontent.com/pod-product-compliance
Ingram Content Group UK Ltd.
Pitfield, Milton Keynes, MK11 3LW, UK
UKHW050952290926
15470UKWH00005B/1785

9 789351 305606